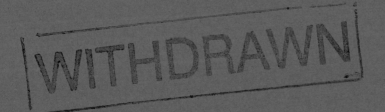

THE FORESTS HANDBOOK
Volume 1

The Forests Handbook

VOLUME 1
AN OVERVIEW OF
FOREST SCIENCE

EDITED BY
JULIAN EVANS

OBE BSc, PhD, DSc, FICFor

T.H. Huxley School of Environment, Earth Sciences and Engineering,
Imperial College of Science, Technology and Medicine, The University of London

**Blackwell
Science**

© 2001 by
Blackwell Science Ltd
Editorial Offices:
Osney Mead, Oxford OX2 0EL
25 John Street, London WC1N 2BS
23 Ainslie Place, Edinburgh EH3 6AJ
350 Main Street, Malden
  MA 02148 5018, USA
54 University Street, Carlton
  Victoria 3053, Australia
10, rue Casimir Delavigne
  75006 Paris, France

Other Editorial Offices:
Blackwell Wissenschafts-Verlag
  GmbH
Kurfürstendamm 57
10707 Berlin, Germany

Blackwell Science KK
MG Kodenmacho Building
7–10 Kodenmacho Nihombashi
Chuo-ku, Tokyo 104, Japan

First published 2001

Set by Excel Typesetters Co.,
Hong Kong
Printed and bound in Great Britain at
MPG Books Ltd, Bodmin, Cornwall

The Blackwell Science logo is a
trade mark of Blackwell Science Ltd,
registered at the United Kingdom
Trade Marks Registry

A catalogue record for this title
is available from the British Library

ISBN 0-632-04821-2 (vol. 1)
     0-632-04823-9 (vol. 2)
     0-632-04818-2 (set)

Library of Congress
Cataloging-in-Publication Data

The forests handbook/edited by
Julian Evans.
      p. cm.
    Includes bibliographical
references.
    Contents: v. 1. An overview of
forest science.
    ISBN 0-632-04821-2
    1. Forests and forestry.
I. Julian Evans.
SD373 .F65   2000
333.75 — dc21   00-021516

DISTRIBUTORS
Marston Book Services Ltd
PO Box 269
Abingdon, Oxon OX14 4YN
(Orders: Tel: 01235 465500
              Fax: 01235 465555)

USA
Blackwell Science, Inc.
Commerce Place
350 Main Street
Malden, MA 02148 5018
(Orders: Tel: 800 759 6102
              781 388 8250
          Fax: 781 388 8255)

Canada
Login Brothers Book Company
324 Saulteaux Crescent
Winnipeg, Manitoba R3J 3T2
(Orders: Tel: 204 837-2987)

Australia
Blackwell Science Pty Ltd
54 University Street
Carlton, Victoria 3053
(Orders: Tel: 3 9347 0300
          Fax: 3 9347 5001)

For further information on
Blackwell Science, visit our website:
www.blackwell-science.com

Contents

Colour plate section appears between pp. 306 and 307

List of Contributors
(Volume 1)

PETER M. ATTIWILL *School of Botany, University of Melbourne, Parkville, Victoria 3052, Australia*
p.attiwill@botany.unimelb.edu.au

JIM B. BALL *Coordinator, Forest Programmes Coordination and Information Unit, Forestry Department, UN Food and Agriculture Organization, Rome, Italy*
James.Ball@fao.org

STEPHEN BASS *International Institute for Environment and Development, 3 Endsleigh Street, London WC1H 0DD, UK*
steve.bass@iied.org

L.A. (SAMPURNO) BRUIJNZEEL *Faculty of Earth Sciences, Vrije Universiteit, De Boelelaan 1085, 1081 HV Amsterdam, The Netherlands*
brul@geo.vu.nl

WILLIAM CAVENDISH *Head of Policy, The Labour Party, Millbank Tower, Millbank, London SW1P 4GT, UK*
William_Cavendish@new.labour.org.uk

HUGH F. EVANS *Forest Research, Alice Holt Lodge, Wrecclesham, Farnham, Surrey GU10 4LH, UK*
H.Evans@forestry.gov.uk

DAVID G. FOWLER *Centre for Ecology and Hydrology, Edinburgh Research Station, Bush Estate, Penicuik, EH26 0QB, UK*
d.fowler@ceh.nerc.ac.uk

PETER H. FREER-SMITH *Forest Research, Alice Holt Lodge, Wrecclesham, Farnham, Surrey GU10 4LH, UK*
P.freer-smith@forestry.gov.uk

JABOURY GHAZOUL *T.H. Huxley School of Environment, Earth Sciences and Engineering, Imperial College, Silwood Park, Ascot, Berkshire SL5 7PY, UK*
j.ghazoul@ic.ac.uk

RONALD L. HENDRICK *Daniel B. Warnell School of Forest Resources, University of Georgia, Athens, Georgia 30602 USA*
rhendric@arches.uga.edu

PAUL G. JARVIS *Institute of Ecology and Resource Management, Darwin Building, University of Edinburgh, Mayfield Road, Edinburgh EH9 3JU, UK*
P.Jarvis@ed.ac.uk

MATHEW P. KOSHY *Department of Forest Sciences, Faculty of Forestry, 3041-2424 Main Mall, University of British Columbia, Vancouver, British Columbia, Canada V6T 1Z4*

ANTHONY R. LUDLOW *7 Kings Road, Alton, Hamshire GU34 1PZ, UK*
Tony.Ludlow@modelresearch.cix.co.uk

GENE NAMKOONG *Department of Forest Sciences, Faculty of Forestry, 3041-2424 Main Mall, University of British Columbia, Vancouver, British Columbia, Canada V6T 1Z4*
gene@unixg.ubc.ca

KJELL NILSSON *Danish Forest and Landscape Research Institute, Horsholm Kongevej 11, 2970 Horsolm, Denmark* kjn@fsl.dk

GEORGE F. PETERKEN *Beechwood House, St Briavels Common, Lydney, Gloucestershire GL15 6SL, UK* George.Peterken@care4free.net

HANS PRETZSCH *Chair of Forest Yield Science, University of Munich, Am Hochanper 13, 85354 Friesing, Germany* H.Pretzsch@lrz.tu-muenchen.de

THOMAS B. RANDRUP *Danish Forest and Landscape Research Institute, Horsholm Kongevej 11, 2970 Horsholm, Denmark* tr@fsl.dk

EUNICE A. SIMMONS *Environment Department, Wye College, University of London, Wye, Kent, TN25 5AH, UK* e.a.simmons@wye.ac.uk

BARBARA M. WANDALL *Danish Forest and Landscape Research Institute, Horsholm Kongevej 11, 2970 Horsholm, Denmark* blmw@fsl.dk

RICHARD H. WARING *College of Forestry, Oregon State University, Corvallis, Oregon, OR97331 USA* Richard.Waring@orst.edu

CHRISTOPHER J. WESTON *School of Forestry, The University of Melbourne, Creswick, Victoria 3363, Australia* c.weston@landfood.unimelb.edu.au

Preface

The Forests Handbook joins an eminent series of Blackwell's 'Handbooks'. However, the connotation of 'Handbook', for those brought up in the English tradition, is that of a practical manual telling the reader how to do things. This 'handbook' is not like that at all. What we have done in these two volumes is to assemble a unique compilation, and I employ that overused adjective advisedly. It is unique in the sense of bringing together eminent foresters, biologists, ecologists, scientists, academics and managers from many countries to tell a story that embraces much of the span of what we call 'forest science' though not all of operational forestry.

Volume 1 seeks to present an overview of the world's forests from what we know about where they occur and what they are like, about the way they function as complex ecosystems, as 'organisms' interacting with their environment, and about their interface with people—at least in part. We have tried to present the state-of-the-art science, and to present it in a way accessible to the general reader with an interest in trees, forests and forestry. It is, after all, sound science that underpins sustainable forest management.

Volume 2 seeks to apply this science to good practice. It is focused on operations and their impacts, on principles governing how to protect forests and on how to harness in sustainable ways the enormous benefits forests confer in addition to supply of wood. Volume 2 also contains valuable and highly informative case studies drawn from several countries to illustrate key points and show good management in practice.

In only two volumes one cannot cover everything, nor everything in detail and in depth. What is presented is a series of overviews of interrelated topics written by highly competent authorities. To achieve this authors were given considerable latitude to interpret their topic. In effect, although chapters are linked logically they do not necessarily trace a coherent flowing story, but are more a suite of essays that move through their subjects as the authors have chosen to address them. The value of this approach is to allow development of topics and themes by those best placed to write about them in ways that are largely unfettered.

While it is impossible to claim worldwide coverage, there is a deliberate attempt to make the text relevant across the globe and to draw on examples, cite practices and relate experience from a great many countries. Omissions are largely what might be termed 'forestry' such as forest economics, forest logging and harvesting operations, the whole matter of utilization and trade in forest products, and there is only limited coverage of forest fires and the pressing problem of tropical deforestation *per se*. There is insufficient treatment of social and policy-related factors in forest development. And, of course, not all the riches in the world's forest types can be mentioned. Indeed, apart from Chapter 2 in Volume 1, authors do not treat their subjects by taxonomic groupings but take a more functional approach. In contrast to perceived omissions, several chapters deal with aspects of ecophysiology and forest–environment interaction and some three chapters address the important subject of forest soils and/or their management. Foresters and forest scientists are notorious for neglecting this aspect of the ecosystem and the emphasis on soil is deliberate and welcome even if in a few places there is some repetition as the different authors provide their own perspectives. We recognize all these imbalances and ask for readers' forbearance. While coverage may not be comprehensive, we hope anyone interested in the world's forests will find much here to welcome. We hope, too, that it helps forward the goal of sustainable forest management

But why have a Forests Handbook? We believe that nowhere else will such a broad sweep of modern forest science and its application be found in two books, and in two books that bring together 45 authors from 12 countries to present their own perspectives. It was a monumental task, but worth it.

Julian Evans
April 2000

Acknowledgements

I am greatly indebted to Douglas Malcolm and Larry Morris for taking on the role of editorial advisers. Not only did they guide development of the book itself but in many instances undertook the arduous task of refereeing chapters. Thanks expressed here cannot do justice to the significance of their contribution to *The Forests Handbook*. That said, I must accept responsibilities for any mistakes, inaccuracies or omissions in the text.

A book of this type is primarily judged by the quality and diligence of its many authors, not of the editor. I am full of admiration for what the authors collectively have achieved and I record my thanks without reservation for all their hard work and dedication. We even kept close to the set deadlines!

I must also thank Claire Holmes of Forest Research (UK) and Delia Sandford of Blackwell Science who successively fulfilled that crucial role of editorial assistant. Their persistence in chasing up authors and keeping a check on this sizable undertaking was enormously appreciated. Added to this, Blackwell Science's editors were supportive throughout: Susan Sternberg and subsequently Delia herself.

I would also like thank Gus Hellier, who helped with chapter formatting and undertook the tedious task of checking references for me.

Finally, I am grateful to Imperial College of Science, Technology and Medicine for enabling me to continue this project during the two years since joining their staff.

Part 1
Forest Resources and Types

The world is blessed with both abundance and variety of trees and forests. But great changes are taking place in our time and it is essential to begin *The Forests Handbook* by describing the resource itself and the changes now impacting on our forests. This is a role that the UN Food and Agriculture Organization (FAO) is charged with undertaking, and we greatly welcome the significant contribution of Chapter 1 by Jim Ball, head of FAO's Forest Programmes Coordination and Information Unit. His analysis draws on FAO's impressive database of forest statistics.

Chapter 2, by Ron Hendrick, is a masterly overview of the main types of forest and woodland found across the globe. It complements the analysis of the physical extent of forest resources and describes the immense variety to help set the scene for understanding the processes that govern their growth and development.

1: Global Forest Resources: History and Dynamics

JIM B. BALL

1.1 HISTORICAL PERSPECTIVE

Forests covered about half of the earth's surface up to the development of early civilizations but today cover less than one-third of that area (FAO 1993). Massive forest clearances were a feature of many early cultures, including ancient Assyria, Babylon, China, Egypt, Greece and Rome. Much of this forest clearance was to provide land for agriculture, although these societies were also voracious consumers of wood for cooking, heating, copper smelting, pottery making, brick-firing, house construction and shipbuilding, and this led to deforestation wherever the forest did not have the opportunity to regenerate. For more than 10 000 years the forests of the Mediterranean region have been cleared; they now cover about one-sixth of the region and those that remain, often on land that cannot readily be cultivated, have been degraded by humans and their animals through unmanaged grazing. The export of high-quality cedar wood from Lebanon to Egypt began nearly 5000 years ago and led, eventually, to the virtually complete destruction of the cedar forests. The clearance of forests is thus not a new phenomenon in human history, except in its present scale and spread.

At times, the pressure on the forests eased. The decline and ultimate fall of the Roman Empire following the sack of Rome by the Goths in the fifth century AD, and the subsequent economic decline of much of Europe, led to the regrowth of forests in some areas. The same happened at the time of the Black Death in the mid-fourteenth century when the population of western Europe was reduced by over one-third, and in some places by half. In Germany, people fled rural life and the oppression of rural overlords into the depopulated cities, while many responded to the invitation of the Polish kings and moved to the eastern territories (the Baltic, East and West Prussia, Pomerania). As a result, the forests of central and western Europe colonized and regenerated over large areas, and it was not until the sixteenth century that population pressure and the demand for forest products again increased to a level comparable to that at the end of the thirteenth century.

Following the Thirty Years War (1618–48), famine and disease again reduced the population of rural Germany, on this occasion by 40%. This pattern was repeated during the Spanish Heritage War (1701–14) and the Seven Years War (1756–63). Following the order of the French King Louis XIV, nearly all towns and castles in south-western Germany were burnt. Their reconstruction consumed an enormous amount of timber so that by the middle of the eighteenth century much of the forest wealth of central Europe had been destroyed, a timber shortage ensued and there was a loss of soil fertility and soil erosion. Vast tracts in northern Germany (e.g. Lüneburger Heide) were degraded to deserts with shifting dunes.

In Britain, where population growth, land clearance and economic development were especially rapid, the natural forests had been heavily depleted by the late Middle Ages. Iron smelting, relying on charcoal, and other wood-based energy users encountered supply difficulties. By the sixteenth century most of the accessible wood resources had been exhausted in England and iron-makers turned to the woodlands of Scotland, Ireland and Wales. Timber began to be imported into England in large quantities for every purpose.

Destruction of the natural forests has thus taken place over most of Europe. It accelerated as the demand for fuelwood and timber grew rapidly in the early years of the Industrial Revolution, but with the continuation of economic growth and industrialization the pressures on the forests have gradually eased. By the early years of the present century, the total area of the European forests had more or less stabilized and in the past decade has even increased slightly. In particular, the pressure on the Mediterranean forests for the supply of fuelwood and land for agriculture and grazing has declined, although the risk of damage from fires continues.

A similar process of forest clearance occurred in the USA (Salwasser *et al.* 1994). At the time of European settlement in the early seventeenth century about 46% of the land was forested; by 1992 this area had declined to about one-third. Huge areas of forest were cleared in the eighteenth and nineteenth centuries as the population grew and industrialization progressed. Early clearing was for agriculture; in 1800, 95% of the people lived on the land, mostly practising subsistence farming. Most wood was used for firewood even in the late nineteenth century, which depleted forest resources especially around population centres. Fencing was the next most important use of wood, followed by early industrial use. The spread of the railroad then further increased demand for wood, which was used for locomotive fuel until the mid-eighteenth century but above all for sleepers or railroad ties. For example, in 1900 it was estimated that the replacement of ties on existing track required the output of 6–8 million ha of forest annually.

The importance of the sea for defence and trade led to laws being passed in several European countries in an effort to protect and expand wood resources for naval vessels and merchant fleets, especially oak for building the hulls and pine for masts. The laws were only partially successful, even for these limited objectives. The forest cover of France had shrunk from perhaps 35% of the country's area at the beginning of the sixteenth century to about 25% by the middle of the seventeenth century, and despite the reforms of the administration of the forests and of the forest laws by Louis XIV and his minister of finance, Colbert,

in 1661, the situation continued to deteriorate due to demands for shipbuilding and the smelting of iron ore. Nevertheless these reforms began a forestry tradition and laid the basis for a strong forest service.

A similar realization that the forest resource was not limitless led to the emergence of rules to control the use of forests in Germany, Denmark and The Netherlands in the eighteenth century, in India in the mid-nineteenth century, in Sweden and Finland in the late nineteenth century and in the USA in the early twentieth century. At that time colonial powers were reserving forests, introducing forest laws and establishing forest services in East Africa, Malaysia and many other now-independent countries. In the eighteenth century, efforts to restore forest cover and guarantee future timber supplies were made. Today Germany has 30% forest cover, which originated in the plantations of the seventeenth century; initially these stands were largely of single species (oak, beech, Scots pine and Norway spruce) but today are slowly being converted into mixed stands with uneven age structure.

The development of management techniques for natural forest began in Europe for a number of purposes. Sometimes this process was accelerated for specific objectives, for example to reduce the risk of avalanches, to control erosion, to stabilize sand dunes or to protect agricultural crops, as well as to provide firewood for towns. Management planning in the sense of regulation of yield had been introduced in France by Colbert, while in Denmark management plans with rotations of oak of 100 years were instituted in 1763.

Originally recruited from retired soldiers, the foresters of the specialized forest services created by the rulers of various German states (e.g. Brandenburg-Preussen, Hanover, Saxony) were to have a profound influence on forest science and practice even up to the present. The experience they gained, combined with traditional forestry knowledge, was published and widely disseminated. One of the best-known publications of that time was *Sylvicultura Oeconomica*, published in 1713 by Hans von Carlowitz (1645–1714). Von Carlowitz was a mining engineer in Saxony preoccupied with the insufficient timber supply for the

Saxon mines. His book dealt not only with timber harvesting, species selection, utilization of timber and sowing of conifer seeds to reforest clear-fellings, but also with promoting the introduction of foreign (exotic) species and he was the one who first coined the term *Nachhaltigkeit* (sustainable forest management). The handful of German foresters who became the founders of forest science included Heinrich Cotta (1763–1844), Wilhelm Pfeil (1783–1859), Johann Christian Hundshagen (1783–1834), Carl Heyer (1797–1856), Gottlob König (1779–1849) and George Ludwig Hartig (1764–1837), who as professor at the University of Berlin introduced forestry training into the Prussian forest service and had a strong impact on the forest management principles of his day, not only in Prussia but all over central Europe. Hartig's rule was that no more nor less timber should be extracted annually from state forests than could guarantee perpetual supply, which became the principle of central European forestry, i.e. sustainable forest management.

The reduction in forest area had been halted or even reversed in the industrialized countries by the 1920s. Since then the standing volume of forests has increased with better management and increased protection, and in Europe the yearly forest harvest of many countries is smaller than the annual increment. The reasons for this historical transition from rapidly diminishing to stable, or even increasing, forest areas in many countries of the Northern Hemisphere with industrial or service economies are complex. The substitution of coal for fuelwood for industrial and domestic use was one obvious factor. However, more important were the major demographic and economic changes that took place as industrialization and economic growth proceeded.

Rural populations fell drastically as people migrated to the cities and towns. By the turn of the present century, the average proportion of the population living in the rural areas of the industrial countries was around 40% compared with 80% or more at the beginning of the Industrial Revolution. Today, in Europe and the USA, the rural population is about 25% of the total. National population growth rates have dropped to around replacement level in most of the in-

dustrial countries. Equally important were the changes that took place in agricultural production. Subsistence farming gave way to an increasingly capital-intensive form of agriculture relying on mechanization and high inputs of fertilizers and pesticides and, in animal husbandry, the stall-feeding of animals. Rather than requiring ever-expanding amounts of land, agriculture became more productive, leading eventually to the present politically charged problems of crop surpluses and the need to 'set aside' or take increasing amounts of farmland out of food production. Additionally, increased public awareness of the value of forests and of environmental issues in urbanized societies of the postindustrial or service economies, combined with increasing demand for new services from forests such as recreation and amenity, have been important recent trends in stabilizing forest areas.

Although history does not repeat itself exactly, the experience of the industrial world helps shed light on what is happening in many developing countries in relation to the pressures on their forests. Expanding populations (especially those existing at subsistence level), accompanied by significant rural–urban migration, result in increased demand for agricultural production. Even highly punitive forest protection legislation has been largely ineffectual in the face of the desperation of people trying to grow enough food, or graze sufficient animals, to survive.

It is only with increased agricultural production supplying stabilized populations that the pressure to clear forest land diminishes. Indeed, sustainable forestry may be possible only where agriculture is, if not sustainable, at least stable. This transition can be seen taking place today in some of the rapidly industrializing countries of the developing world. The Republic of Korea, for example, where the rural population has fallen sharply while agricultural production has increased dramatically, has reached the turning point at which the area of forest can begin to expand again, aided by massive reforestation programmes. In others, such as the newly industrializing countries of the Pacific Rim, forest depletion continues, although the economic and demographic conditions and social demands for stabilizing the remaining areas of forest are beginning to emerge.

In many other developing countries, however, rural population growth continues virtually unabated, agricultural productivity is low and economic stagnation means that there are few employment opportunities for those wishing to leave the land. In such cases, increasing pressure to clear forest for agriculture is inevitable whether legalized by land reform or not.

The parallels with the experience of the industrial world are thus obvious but what makes today's position in the developing world different is the sheer impact of population numbers, the pace of change and the fragile nature of many tropical soils and their inability to be used for long-term agriculture. Regarding population numbers, for example, the population of the European continent grew from about 140 million in 1750 to 265 million by 1850, 392 million by 1950 and 499 million by 1990. The population of the USA grew from 23 million in 1850 to 151 million in 1950 and 248 million in 1990. Today, on the other hand, around 50 million people are being added to the rural populations of the developing world *every year*. Where there were about 12 ha of forest per head of the global population of about 500 million people in 1750 there is now about 0.75 ha per head, and the rate at which the forests are disappearing, especially tropical

forests, is far faster than at any other time in human history, despite net gains of forest cover in some developed countries and some evidence of a slightly slower rate of deforestation in developing countries in recent years.

1.2 RECENT ESTIMATES OF GLOBAL FOREST AREA

The following section is based on 1990 baseline figures, updated to 1995, prepared by the Food and Agriculture Organization (FAO) Forest Resource Assessment Programme, reported in and largely drawn from *State of the World's Forests 1997* (FAO 1997a). Box 1.1 describes two components of the assessment process.

In 1995 forests were estimated to cover 3454 million ha, or 26.6% of the total land area of the world (Greenland and Antarctic excepted). The distribution of forests (natural forest and plantations) by region in 1995 is shown in Fig. 1.1. Nearly 57% of the forests lay in developing countries, which for the most part are tropical and make up 58.9% of the total land area of the world.

Almost two-thirds of the world's forests were located in seven countries: Russia, Brazil, Canada, USA, China, Indonesia and Zaire; 29 countries

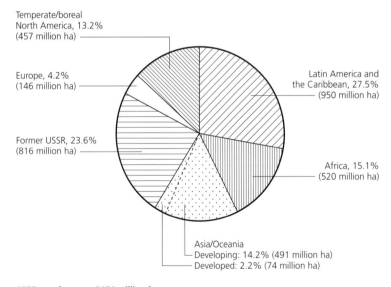

Temperate/boreal
North America, 13.2%
(457 million ha)

Europe, 4.2%
(146 million ha)

Former USSR, 23.6%
(816 million ha)

Latin America and
the Caribbean, 27.5%
(950 million ha)

Africa, 15.1%
(520 million ha)

Asia/Oceania
Developing: 14.2% (491 million ha)
Developed: 2.2% (74 million ha)

1995: total area = 3454 million ha

Developed countries: 1493 million ha
Developing countries: 1961 million ha

Fig. 1.1 Forest areas by main regions in 1995.

Box 1.1 Two essential components of global assessments of forest resources: national forest inventory systems and a common classification system for forests

National forest inventory systems
The quantity and quality of information provided by global and regional assessments depend to a large extent on the capacity of national forest inventory systems to collect and analyse data at national and subnational levels, and to adjust the information so that it is compatible with global and regional reporting parameters. Of the 143 developing countries covered by the Forest Resource Assessment 1990, all but seven had performed one reliable nationwide estimate of forest cover at some time between 1970 and 1990. However, only 25 had performed more than one national forest cover assessment, and very few of these had carried out more than one comprehensive national forest inventory. The database for developed countries is more complete; practically all developed countries were able to provide information not only on areas but also on biomass, volume and other forest parameters. However, work remains to be done to make the inventories carried out in many industrialized countries suitable for analysing changes over time. UNCED (United Nations Conference on Environment and Development) recognized that assessment and systematic observations were 'often neglected aspects of forest resources management, conservation and development . . .' and that 'in many developing countries there was a lack of structures and mechanisms to carry out these functions'. Consequently, it devoted one of the four programme areas of the forest chapter of Agenda 21 to the 'strengthening and establishment of systems for the assessment and systematic observations of forests and forest lands' and the 'provision of sound and adequate updated information on forest and forestland resources to economists, planners, decision-makers and local communities'. FAO, with the support of several donor countries, is pursuing an active programme of country capacity building in forest assessment, which ultimately will contribute to improving the quality of global forest resource assessments.

A common classification system for forests
A common set of concepts and classifications that can be applied to all wooded lands of the world is essential for securing consistency between national, regional and global forest resource assessments. The FAO and UN-ECE (United Nations Economic Commission for Europe) secretariats have tried over the years to build consensus on a minimum core of definitions and classifications within and between developed countries (all of which lie in the temperate and boreal zones) and developing countries (most of which fall within the tropical belt). In June 1996, FAO, in cooperation with UN-ECE and UNEP (United Nations Environment Programme) and with the support of the government of Finland, organized an expert meeting in the Finnish town of Kotka at which agreement was reached on a common core set of parameters to be assessed within the framework of the next global forest resource assessment in 2000 and on some of the concepts and classifications to be used to that end. (From FAO 1997a.)

---

had more than half of their land covered by forest, of which 21 were in the tropical belt. However 49 countries, in addition to the many non-forested small island states and territories, had less than 10% of their land covered by forests. Five entire subregions were in this category: North Africa (1.2% of the land area), Near East (1.9%), Temperate Oceania (6.2%), Non-tropical Southern Africa (6.8%) and West Sahelian Africa (7.5%).

The latest information on the distribution of forest and of forest cover change by ecological zones was provided in 1990; the results of the Forest Resources Assessment 2000 were released by FAO in 2000 (see endnote on p. 22). In 1990, temperate and boreal forests occupied 1.64 billion ha and tropical forests 1.76 billion ha. A breakdown by ecological zone was available only for tropical forests (Table 1.1). The figures on tropical forests show that most (88%) are in lowlands; of these, tropical rain forests accounted for 47% of all lowland tropical forests, followed by moist deciduous forest (38%) and dry and very dry formations (15%). The loss of forest cover through deforestation is discussed later.

Table 1.1 Forest cover area and rate of deforestation by main tropical ecological zone as estimated by the Forest Resources Assessment 1990. (From FAO 1997.)

Ecological zone	Land area (million ha)	Population density 1990 (km⁻²)	Annual population growth 1981–90* (% year⁻¹)	Forest cover 1990		Annual deforestation 1981–90*	
				Million ha	Percentage of land area	Million ha	% year⁻¹
Forest zone	4189.7	57	2.4	1748.2	42	15.3	0.8
Lowland formations	3476.6	57	2.3	1543.9	44	12.8	0.8
Rain forest	937.1	41	2.2	718.3	76	4.6	0.6
Moist deciduous	1298.6	55	2.4	587.3	46	6.1	1.0
Dry and very dry	1241.0	70	2.3	238.3	19	2.2	0.9
Upland formations (hill and mountain forest)	713.1	56	2.6	204.3	29	2.5	1.1
Non-forest zone (alpine areas, deserts)	588.6	15	3.1	8.1	1	0.1	1.0
Total tropics†	4778.3	52	2.4	1756.3	37	15.4	0.8

* Compound interest formulae were used for population growth and deforestation rate calculations.
† Totals may not tally due to rounding.

1.2.1 Natural forests

The forest area estimates described above include undisturbed forests, forests modified by humans through use and management (or 'seminatural' forests) and forests created artificially by humankind (i.e. forest plantations) by afforestation or reforestation. (Afforestation is defined as the establishment of a tree crop on an area from which it has always, or for a very long time, been absent. Reforestation is defined as the establishment of a tree crop on forest land.) In most industrialized countries, particularly in continental Europe, forests are being managed in such a way that at management-unit level a continuum exists from low-intensity management, involving natural regeneration, through more intensive methods involving some artificial planting to highly intensive methods with complete planting and cultivation; this makes it difficult to isolate figures for natural forest and plantations. The distinction between natural or seminatural forests and forest plantations can more easily be made for developing countries and some industrialized countries such as New Zealand in which forest plantations have been established using introduced species.

Interest in natural forests, particularly their role in the conservation of biological diversity, has led to efforts to compare forests today with what is thought to be their original character and to give complete protection to areas of forests which have had no, or minimal, human interference. Although there are difficulties in identifying the extent of natural forest, compounded by problems of definition (see Box 1.2), some information exists that can be used as an indication of broad patterns of natural forests in various regions.

Box 1.2 'Naturalness' of forests

A concept of 'naturalness' is being used as an indicator of biological diversity in some international programmes that are defining criteria and indicators for sustainable forest management. Naturalness is characterized by such elements as complex spatial structure, composition and distribution of species natural to the site, wide range of ages in tree species and presence of dead and decaying trees. Loosely defined terms such as 'virgin forest', 'primary forest', 'old growth forest', 'natural forest' and (ancient) 'seminatural forest' are also often used, making a consistent review of the state of 'naturalness' difficult.
(From FAO 1997a.)

An attempt has been made by the World Wide Fund for Nature (WWF) to quantify the area of forests in western Europe that has been relatively undisturbed by humans or which has retained much of its natural character. Their report (WWF 1994) distinguishes between 'virgin forest', defined as 'forest ecosystems whose characteristics are determined exclusively by natural location and environmental factors . . . without human influences present or visible any more', and 'natural and ancient seminatural forests', which 'have not been planted or sown by man for the past two centuries' and 'which continue to have a large number of the natural elements'. The study found that only a small proportion (probably <1%) of the total forest land in northern and western Europe could be considered as virgin forest, which has arisen since the last glaciation. Almost all was located in Sweden, Finland and Norway, with small areas in Greece, Austria and Switzerland and (according to another author) in France. In eastern Europe, Slovakia and Belarus have considerable areas of virgin forest while Poland and Croatia have small areas. In addition, in northern and western Europe, the WWF report identifies natural and ancient seminatural forests representing 2.1% of the total forest cover (1990) of the 16 countries concerned. There are a further 3 million ha of land in the region in national parks and other protected areas and another 50 000 ha in small forest reserves (mostly for nature conservation and scientific research), whose use is tightly restricted.

The situation in temperate and boreal North America is quite different from that of densely populated Europe and Japan, where use and management of forests for many centuries have left very little of the original forest area untouched. 'Old growth forests', as they are called in North America, still cover extensive areas. On the lands managed by the US National Park Service alone, old-growth forests covered 1.97 million ha in 1988. Although not a direct measure of the extent of old growth forest, it may be noted that the area of forest included in national parks and other protected areas in North America (USA and Canada) was reported to be nearly 49 million ha in 1990.

The FAO Forest Resource Assessment 1990 did not distinguish between undisturbed and disturbed 'natural' forests in developing countries. However, the Forest Resource Assessment 1980 made estimates of the areas of undisturbed closed forests (primary forests and old secondary forests where there had been no logging for the last 60–80 years) and of closed forests that were included in national parks and other protected areas (thus relatively undisturbed, at least in theory). At that time, these two categories together represented 60% of the total closed forest area in the tropics, a proportion varying from 39% in tropical Asia to 59% in tropical Africa and 69% in tropical America. These different proportions by region reflected a slower development of large-scale harvesting in tropical America compared with tropical Africa and Asia, and also the fact that, in tropical America, spontaneous colonization did not follow in the wake of logging as systematically as in the two other regions due to lower population pressure. Although the two categories do not match the concept of 'virgin forests' of Europe and of 'old growth forests' of North America discussed above, the sum of the two categories nevertheless gives an indication of the amount of forest disturbance (or management, depending on the point of view of the observer) that existed around 1980 in the humid tropics. Although corresponding estimates for 1990 and 1995 are not available, it is likely that the share of undisturbed forests remains higher in the three tropical regions than in Europe and probably in North America too.

1.2.2 Forest plantations

Recent data in this section are also available in *State of the World's Forests 1999* (FAO 1999), which also discusses some of the issues concerned with forest plantations in more depth.

Early development

By the seventeenth century the decline in the area of native forest in European countries led to the planting of trees, largely to provide alternative sources of timber supply. Sometimes plantations were established for the provision of services or other products than timber. They might be planted as shelterbelts, for dune stabilization, for amenity or for the supply of firewood.

Plantations of native species were established at first in areas where forests occurred naturally, for example France, Germany, England and Scotland, but later also on previously unforested land. In France, for example, planting of *Pinus pinaster* was started on the sand dunes of the Landes and with *Pinus sylvestris* and *Picea abies* on former agricultural land in the Vosges; in Germany likewise agricultural land was planted with Norway spruce (*Picea abies*) in Saxony. In tropical countries such as Myanmar (then Burma), teak (*Tectona grandis*) was planted as a native species in the *taungya* system.

Teak was one of the first exotic forest plantation species to be used, being planted in Sri Lanka and the island of Java (Indonesia) from early in the nineteenth century. Exotic species, such as Douglas fir (*Pseudotsuga menziesii*), Sitka spruce (*Picea sitchensis*), Japanese larch (*Larix kaempferi*), lodgepole pine (*Pinus contorta*) and poplar (*Populus deltoides*), were introduced to Europe during the nineteenth century, when they began to play an increasingly important role in forest plantation programmes. Other exotic species, particularly the eucalypts (*Eucalyptus* spp.) and wattle (*Acacia* spp.) from Australia, were introduced as exotics in tropical and subtropical countries from the middle of the nineteenth century, while *Pinus radiata* was introduced from California to New Zealand and other countries such as Chile from the early years of the twentieth century. Box 1.3 describes in detail the experience with forest plantations in Denmark.

The main reason for the establishment of plantations remained the decline in natural forest area and a scarcity of wood and, as discussed above, the conversion of woods and forests to agricultural and grazing use. In recent years increased areas of natural forest have been managed for nature conservation, recreation, wildlife parks, etc., and commercial wood production has been either reduced or eliminated. For example, over 25 000 ha of high-yielding hybrid poplar plantations have been established in the north-western USA between 1992 and 1997 in response to both increased demand for poplar wood for orientated-strand board and decreased supply from public forests (Anon. 1996). On the other hand, some new areas have become available in European countries as land is taken out of agriculture due to trade and market considerations.

Objectives

Forest plantations are tree crops that are in some, but not all, ways analogous to agricultural crops. They often have a simple structure, at least in youth, and are usually composed of one or a few species (but not varieties as in agriculture, except in a few cases such as the intensively bred poplars) chosen for their fast growth, yield of specified products and ease of management. When established for wood production they have a higher productivity of usable wood than natural forests, but due to the way they are managed they do not, indeed cannot, provide the full range of goods and services that natural, seminatural or even secondary forest can provide.

Although the objective of many plantations is the production of industrial roundwood and/or fuelwood, many are established for environmental protection or other services (e.g. soil and water conservation, enhancement of agricultural production in agroforestry systems, carbon sequestration) and some also provide non-wood forest products, such as fodder, various foodstuffs, medicines, etc.

Area

The area of forest plantations throughout the world started to increase in the 1970s as many governments became concerned about wood supplies for industry, and in developing countries fuelwood supplies, and has continued to increase since. However, there are no reliable global figures for plantation areas because forests of native species in several developed countries in the temperate and boreal regions, especially in continental Europe, are frequently regenerated naturally and it is not possible to distinguish those areas where supplementary artificial planting has been done. Furthermore many countries consider their plantations as 'seminatural' forests over a certain age, because as stands mature the clear initial row layout of the trees is lost and other species naturally regenerate under the

Box 1.3 Forest plantations and the experience of Denmark

The forest history of Denmark illustrates the strategies that have been undertaken since the eighteenth century to promote tree planting, both to meet wood shortages and to provide benefits other than wood, and demonstrates the way in which original objectives may change with time as technologies improve and new needs and priorities arise.

Denmark was one of the first countries to stabilize sand dunes by planting trees, starting in the 1720s on the northern coast of Zealand. By the 1850s, techniques had been developed and the species selected to initiate large plantation programmes to stabilize sand dunes in the western part of the country. The area covered by these plantations developed as follows: 1878, 800 ha; 1898, 20 000 ha; 1920, 30 000 ha; 1971, 47 000 ha. In some places a great variety of species were tried, e.g. hardwoods such as oak and beech, as well as conifers such as *Pinus sylvestris* and introduced firs such as *Abies nobilis*, *A. alba* and *A. nordmanniana*. In others, large areas of the introduced *Pinus mugo* were established. Later introductions included the North American lodgepole pine, *Pinus contorta*, and Sitka spruce, *Picea sitchensis*. In addition to their continuing function of sand-dune fixation, some of the older plantations now have considerable amenity value.

Denmark, like France and Britain, required large quantities of oak wood for maintaining its navy. The disastrous seizure of the entire fleet by the British in 1807 led to intensive replanting and even at one time the regulation in the dukedoms that an intending husband had to establish oak trees a certain number of years before he could marry, the so-called 'groom copses'. These oak trees matured after a rotation of 140 years, long after 'wooden walls' had given way to ironclads, and although the forest service made tongue-in-cheek efforts to get the modern Danish Navy to accept the oak that had been planted for them so long before, these areas of oak are now retained for their considerable amenity and recreational value.

In fact the timber supply situation had become serious long before 1807, when it was found that the forest cover of the moorland heath areas in Jutland was only 2% in 1780, and the state started a plantation programme with *Pinus mugo*. In the period following the loss of parts of southern Jutland to Germany after the war of 1864, and up to 1935, larger plantation programmes of mainly pine and spruce were carried out on these poor soils. The loss of this territory led to the adoption of the motto *Hvad udad tabes skal indad vindes*, which translates literally as 'external losses must be compensated internally', meaning that the loss of land had to be compensated for by intensified use of what remained. Plantations were established not only by the state but also by individuals and private organizations like Hedeselskabet. The conversion and diversification of extensive areas of *Pinus mugo* is now one of the current challenges in these areas.

The Danes were also among the pioneers of shelterbelts. Hedeselskabet, the Danish Land Development Service, established in 1866, is a private organization with the original objective of transforming the infertile heath areas of Jutland to productive agriculture and forest areas. This organization, which was subsidized by the state and which had massive support from the public, was responsible for the establishment of shelterbelts and the development of agricultural land in Jutland. The purpose of the belts was both to provide shelter and, in the early 1800s, to demarcate individual holdings. Later, from 1938 to 1963, 43 000 km of shelterbelts and live fences were established with the additional objective of providing employment. The main species planted in the shelterbelts was *Picea glauca*, at least in parts of Jutland. These now need replacement and are in many places being converted and given the additional role of wildlife corridors, through the planting of more species and the creation of greater variation in structure.

(From information provided by Mette Løyche-Wilkie and Soren Hald, forestry officers in FAO's Forestry Department.)

Table 1.2 Reported forest plantation areas (10^3 ha) in developing countries, 1995. (From FAO 1999.)

	Reported areas			Estimated net areas	Reported yearly rate
	Industrial	Non-industrial	Total		
Africa	3 787	3 025	6 812	5 861	288
Asia and Pacific	31 781	21 216	52 997	40 471	2 330
Latin America	7 826	2 134	9 960	8 898	401
Total	43 394	26 375	69 769	55 230	3019

1 The figures refer to forest plantations established for wood supply. They do not include, for example, over 11 million ha of plantations established for the supply of non-wood products in China. The figures incorporate new and more reliable inventories of forest plantations done in some countries such as Brazil and Indonesia.
2 Reported areas refer to gross areas derived from various published sources. The estimated net areas are derived from the reported areas through the application of reduction coefficients to allow for poor survival or other losses, based on inventory results where available and on expert opinion where not.

canopy. For example, Austria, Czech Republic and Finland, in responding to a questionnaire to collect data for the ECE/FAO Temperate and Boreal Forest Resource Assessment (TBFRA), stated that they had no plantations in their countries as defined by the TBFRA process.*

In 1995, an approximate estimate of the area of plantations in developed countries was 60 million ha, comprising 13.7 million ha in North America, 22.2 million ha in the Commonwealth of Independent States, 12.1 million ha in Europe and 13.2 million ha in Oceania (Australia, New Zealand, Japan). The most significant areas of plantations were in the Russian Federation (17.3 million ha, 2.1% of the country's total forest area), the USA (13.7 million ha, 6.3%), Japan (10.7 million ha, 44.4%), Ukraine (4.4 million ha, 46.8%) (FAO/ECE 1996), Spain (1.9 million ha, 14.9%), New Zealand (1.54 million ha, 19% of the total forest area in 1996, 91% of which was *Pinus radiata*) (Anon. 1997a) and Australia (1.04 million

ha in 1994, 85% of which was softwood species, mainly *P. radiata*) (Anon. 1997b).

Due to ecology and choice of species the distinction between natural forest and forest plantations is more clear-cut in developing countries and Table 1.2 shows the regional totals derived from a recent FAO review of published information for individual countries (FAO 1999).

The estimated 'net' plantation area of 55 million ha in developing countries in 1995 was about 2.8% of the total area of forests in developing countries. In 1980 the net plantation area was assessed at about 40 million ha. It has thus increased by about 15 million ha in 15 years, but even this figure may be liable to error. Many developing countries, especially those with large forest plantation programmes, provided updated information on their present plantation plans to FAO in 1996 and 1997 from which the reported annual rate of new plantations of 3 million ha was derived. Note, however, that this is the reported or planned rate, which may not necessarily have been achieved. Most of the countries with large plantation estates indicated that they intended to double their plantation areas between 1995 and 2010.

It was estimated from reported figures that 57% of the forest plantation area consisted of hardwood species and 63% was established for industrial purposes. Nearly three-quarters of these plantations were in the Asia–Pacific region, where China (21 million ha) and India (20 million

* Plantations are defined in TBFRA 2000 as 'forest stands established by planting or/and seeding in the process of afforestation or reforestation'. They are either 'of introduced species (all planted stands) or intensively managed stands of indigenous species which meet all of the following criteria: one or two species at planting, even age class, regular spacing'. However, plantations that have been 'without intensive management for a significant period of time' and may thus have developed extra species, lost the original row structure, etc. are considered as 'seminatural' forest stands.

ha) dominate, while about 15% were in Latin America and 10% in Africa.

One of the trends in the tropics has been that the proportion of industrial plantations established in large blocks fell from 40% of the total plantation area in 1980 to 35% in 1990. The proportion of smaller plantations established through farm forestry or agroforestry programmes grew in importance during the period 1980–95, particularly in the Asia–Oceania region. Unfortunately, the figures on small-scale private or community-owned forest plantations are even less reliable than for large-scale plantations. Some of these farm forestry or agroforestry plantations supply industrial wood markets for pulpwood (e.g. Brazil and Thailand) or peeler logs (e.g. India). In many countries, particularly those with limited forest area, planted trees grown outside the formal forest area often provide the bulk of fuelwood, poles, construction wood, utility wood, as well as fodder and other non-wood forest products for household use.

Of the area of hardwood plantations planted for industrial use, 30% or nearly 10 million ha consists of eucalyptus, followed by acacias (3.9 million ha or about 12% of the hardwood area) and teak (about 7%). Short rotation plantations of hardwood species such as the eucalypts, acacias and *Gmelina arborea* have been grown for many years by the private sector, but the establishment of plantations of teak or other valuable hardwood species has been carried out only by government forest services because of their slow growth and hence delayed returns. However, the likelihood of reduced supplies of high-quality hardwood logs derived from natural forests, combined with increasing purchasing power and expected higher prices for logs, is leading to increasing interest in investment by the private sector in valuable hardwood species, especially teak, in a number of countries, for example India, Malaysia, Costa Rica and Ghana.

Of the softwoods grown for industrial purposes, fast-growing pines such as *Pinus radiata*, *P. patula and P. caribaea* constitute about 25% of the area while other, often slow-growing pines (e.g. *P. kesiya*, *P. massoniana*, *P. merkusii*, *P. rox-burghii*, *P. halepensis*, *P. pinaster* and *P. wallichi-ana*) make up about 36%.

Responsibility for the monitoring and regulation of plantation crops grown for food and certain other purposes has long been the job of the agricultural sector. In recent years, however, several of these crops (the main ones being rubber, coconut and oil palm) have been providing 'forest' products that have been used for wood and fibre. For example, rubber wood is now used for the manufacture of about 80% of the furniture made in Malaysia, while coconut and oil palm trunks and the branches of rubber wood are used for various forms of reconstituted 'wood'. Rubber wood and coconut stems are derived from the conversion of old plantations formerly disposed of by burning, while oil palm fruit residues are used for medium-density fibre board. The development of these new markets has thus not only improved financial returns but also used the resources in an environmentally friendly manner.

The area of these species appears to have increased from the 14 million ha reported in 1990. The increase may be due to better coverage of the data, although it is known that oil palm areas are increasing rapidly, rubber tree areas are also increasing and coconut plantations are decreasing. Coconut plantations comprise the largest area (about 42% of the total), rubber 36% and oil palm 22%. Most of the coconut plantations are in Indonesia (33% of the area) and the Philippines (28%); most of the rubber plantations are in Indonesia (34%), Thailand (20%) and Malaysia (18%), while most of the oil palm plantations are in Malaysia (44%) and India (29%) (see Table 1.3). Not all of the areas mentioned above are suitable or available for substitute 'timber' production but they illustrate the potential.

Contributions of forest plantations to wood supply

The continuing increase in the area of forest plantations which have been established for industrial wood supply has been to meet the reduction of outturn foreseen from natural forests arising from deforestation and changes in land use (largely in the tropics and subtropics) or from natural forest being taken out of production and devoted to service functions such as conservation. It has been believed that the outputs from

Table 1.3 Reported plantation areas (10^3 ha) of 'non-forest' species in tropical and subtropical countries in 1995. (From FAO 1999.)

Region	Rubber	Coconut	Oil palm	Total
Latin America	238	269	265	772
Africa	529	461	922	1912
Asia and Pacific	8718	10546	4587	23851
Total	9485	11276	5774	26535

forest plantations can help to reduce the pressure on natural forests as sources of industrial wood supply; while the logging of tropical natural forests is not the prime cause of deforestation, logging roads often provide the means for farmers to gain access to forests. The reduction of logging, combined with effective protection, may thus help to reduce deforestation in certain locations until land use and ownership are clarified. However, none of these palliatives will remove the underlying causes of deforestation: high rates of population growth, poverty, hunger and a shortage of fertile land to cultivate.

The potential of forest plantations to meet demand for industrial roundwood is considerable; it has been estimated that the present global demand for paper pulp could be met from an area equivalent to only 1.5% of the world's closed forest area (IIED 1996). No global estimates of current output of timber from forest plantations are available, although FAO's global fibre supply model (GFSM) (FAO 1998) estimated that the potential annual growth of industrial wood from forest plantations in developing countries was about 5% of the increment of natural forests in 1995. In some countries, plantation production already makes a highly significant contribution to the industrial wood supply, for example in New Zealand 99% of industrial roundwood in 1997 was grown in plantations, while in Chile the equivalent figure was 95%, in Brazil and Argentina 60%, and in Zambia and Zimbabwe 50%.

Estimating the future contribution of forest plantations to wood supply is at present imprecise and is based on many more or less unreliable assumptions, particularly concerning the rate at which afforestation will continue. By the year 2010 the GFSM (op. cit.) estimated that the potential increment from forest plantations would be about 40% of that from natural forests in Asia, Oceania and Latin America and about 15% in Africa, under rates of deforestation and afforestation largely the same as today.

1.3 CHANGES IN FOREST COVER AND CONDITION

An analysis of changes in the cover and condition of the world's forests requires a differentiation between:
1 the increase in forest cover (by afforestation or natural colonization of trees on non-forest land) or decrease (by deforestation) of forest area; and
2 changes in forest condition, either positive (recovery of degraded stands, stand improvement treatments) or negative (decline, defoliation or dieback, effects of forest fires, degradation through unsustainable exploitation for wood, overgrazing, effects of pests and diseases).

1.3.1 Changes in forest cover

Between 1980 and 1995, the extent of the world's forests decreased by some 180 million ha, an area about the size of Indonesia or Mexico. This represents a global annual loss of 12 million ha, an area equivalent to the size of Greece or Bangladesh. During this 15-year period, developing countries lost nearly 200 million ha of natural forests, mostly through clearing for agriculture (shifting cultivation, other forms of subsistence agriculture, the establishment of cash crop plantations such as oil palm, and ranching). This was only very partially compensated for by the establish-

Table 1.4 Forest area change in developed countries (1990–95).

| | Forest area change | | |
| | | Annual | |
	1990–95 (million ha)	Million ha	Percentage
Europe	+1.94	+0.39	+0.27
Former USSR without Russia*	+2.78	+0.56	+1.12
Temperate/boreal North America	+3.82	+0.76	+0.17
Developed Asia–Oceania	+0.24	+0.05	+0.06
All developed countries (without Russia)	+8.78	+1.76	+0.12
All developed countries (without former USSR)	+6.00	+1.20	+0.18

* Russia is not included in this table since no reliable estimate for change in forest area between 1990 and 1995 could be made for this country.

ment of new forest plantations. Over the same period, forests in the developed world expanded slowly (by some 20 million ha) through afforestation and reforestation, including natural regrowth on land abandoned by agriculture.

Forests appear to have expanded in all regions/subregions of the developed world as shown in Table 1.4. Forest plantation and natural regeneration on abandoned agricultural land more than compensated for clearing of forests due to urbanization and infrastructure development in most industrialized countries, outside the former USSR. While the gain of forest cover in developed countries was 20 million ha in the period 1980–95, 9 million ha of this increase occurred in the 5 years from 1990 to 1995.

The situation is quite different in the developing world, where deforestation exceeded net afforestation/reforestation, particularly in the tropical zone. Table 1.5 shows the estimated balance for tropical and non-tropical regions of the developing world for the periods 1980–90 and 1990–95. As a whole, the annual rate of deforestation in the developing world between 1990 and 1995 was 0.7%, equivalent to 12.6 million ha, with the highest rate in tropical Asia–Oceania, closely correlated with population and income growth. The lowland forest formations were the most affected by deforestation (Table 1.1 shows that it occurred at a rate of 12.8 million ha annually between 1981 and 1990), although the proportional loss during this period was greater in upland

formations (1.1% compared with 0.8% in the lowland formations). There is some evidence that the rate of loss of forest cover in developing countries was slowing towards the end of the 15-year period 1980–95. The annual rate of forest loss in the period 1980–90 was 15.5 million ha compared with 13.7 million ha between 1990 and 1995.

1.3.2 Conversion of forests to other land cover

The Forest Resource Assessment 1990 carried out an assessment of the relative importance of the various factors involved in deforestation at regional and global levels between 1980 and 1990 in the whole tropical belt. Among the most significant outputs of the study were the 'area transition matrices' of the type shown in Table 1.6, which indicate transfers from one land cover class to another over the 3068 million ha of the tropical zone covered by the sample. For instance, the first row shows that 1275.9 million ha of the total area of 1368 million ha of closed forest in 1980 remained as closed forest up to 1990 and also shows the fate of the 92.1 million ha converted to other land cover classes (open forest; long fallow; fragmented forest; shrubs and short fallow; other, i.e. non-wooded, land cover; and forestry or woody forest and agricultural tree plantations). Figure 1.2 indicates the various types of change in forest cover and condition that occurred in each of the three tropical regions. All these transfers involved changes of woody biomass, which Fig. 1.3

Chapter 1

Table 1.5 Annual forest area changes in the developing regions estimated for 1990–95 and 1980–90. (From FAO 1999.)

| | 1990–95 | | | | 1980–90* | |
| | Natural forests | | Total forests | | Natural forests‡ (million ha) | Total forests† (million ha) |
	Million ha	Percentage of 1990 area	Million ha	Percentage of 1990 area		
Africa	−3.75	−0.71	−3.75	−0.71	−4.28	−4.12
Tropical	−3.70	−0.72	−3.70	−0.72	−4.19	−4.10
Non-tropical	−0.05	−0.41	−0.05	−0.37	−0.09	−0.02
Asia–Oceania (developing)	−4.17	−0.89	−3.47	−0.67	−4.41	−1.70
Tropical	−3.51	−1.14	−3.21	−0.98	−3.97	−2.49
Non-tropical	−0.66	−0.42	−0.26	−0.14	−0.45	0.79
Latin America and Caribbean	−5.81	−0.61	−5.81	−0.60	−6.77	−6.44
Tropical	−5.69	−0.62	−5.69	−0.62	−6.48	−6.21
Non-tropical	−0.12	−0.29	−0.12	−0.28	−0.29	−0.23
Developing world	−13.73	−0.70	−13.03	−0.65	−15.46	−12.26
Tropical	−12.91	−0.74	−12.59	−0.71	−14.63	−12.80
Non-tropical	−0.83	−0.39	−0.43	−0.18	−0.82	0.54

* Note that the estimates for 1980–90 are based on new information and differ slightly from those produced by the Forest Resources Assessment 1990.
† The difference between an increase in area due to plantation establishment and a decrease in area due to deforestation.
‡ The negative figures denote deforestation.

Table 1.6 Area transition matrix for the tropical zone for the period 1980–90. (From FAO 1996.)

| Land cover classes in 1980 | Land cover classes in 1990 (million ha) | | | | | | | |
	Closed forest	Open forest	Long fallow	Fragmented forest	Shrubs and short fallow	Other land cover*	Plantation (agricultural and forest)	Total 1980 (million ha)
Closed forest	1275.9	9.0	9.3	9.2	24.1	36.6	3.9	1368.0
Open forest	0.9	283.3	1.3	5.2	3.8	10.2	0.2	304.9
Long fallow	1.1	0.3	48.6	1.1	3.2	2.2	n.s.	56.5
Fragmented forest	0.6	0.6	0.6	159.3	1.9	11.7	0.4	175.1
Shrubs and short fallow	0.7	0.5	0.7	0.5	273.0	26.9	0.3	302.6
Other land cover*	0.8	0.7	0.3	1.4	4.0	837.3	0.5	845.0
Plantation	0.1	n.s.	0.0	0.0	n.s.	0.1	15.7	15.9
Total 1990 (million ha)	1280.1	294.4	60.8	176.73	310.0	925.0	21.0	3068.0

* Including water (relatively small areas).

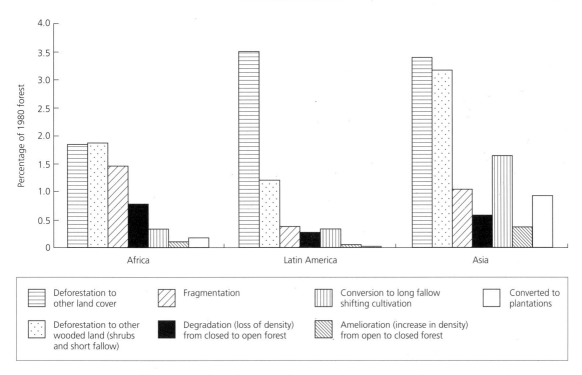

Fig. 1.2 Main categories of forest cover change by geographic region. (From FAO 1997.)

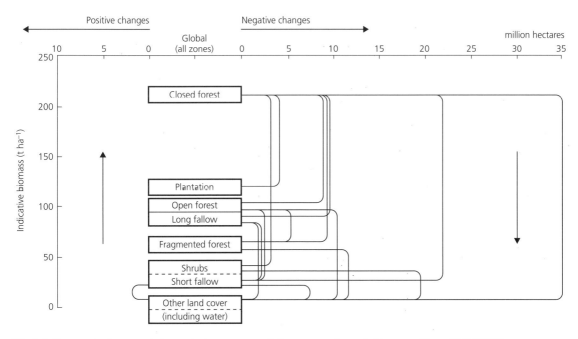

Fig. 1.3 Pan-tropical area and change in forest area and above-ground woody biomass. (From FAO 1997.)

attempts to capture by replacing each class along the *y*-axis (average biomass per hectare) and showing the importance of transfers along the *x*-axis.

1.3.3 Changes in forest condition

This section reviews the environmental threats to forests and the risk of attack from pests and disease; these are more fully discussed in Volume 1, Chapters 6 and 11, and in Volume 2, Chapters 8 and 9.

Factors affecting the health and vitality of all types of forests have attracted increasing attention in recent years. Outbreaks of fires have made headline news in developed and developing countries alike, while insect and disease attack and generalized declines in forest condition have caused more localized concern. Insects and disease and forest fires exist under 'natural' systems and conditions; what has caused concern has been their spread in artificial systems and in conditions considerably modified by humankind, resulting in direct economic loss.

Fire arising from natural causes (e.g. lightning) has been a major influence and is a driving evolutionary force in several forest ecosystems, such as the pine forests of Central America. Fires are important to the health and maintenance of these ecosystems and the total exclusion of fire from them can lead to the build-up of fuel and ultimately abnormally destructive fires that cause the loss of the ecosystem, as happened with the fire exclusion policy of the National Parks Service of the USA and the disastrous fires of 1988. However, many fires arising from human activities, whether planned or unplanned, have damaged and destroyed large areas of forests and woodlands. Resource managers face a demanding public often with conflicting needs and incomplete information, leading to the development of new policies that take into account public attitudes towards fire management.

At the global level, an agreement was signed at a meeting in 1990 to initiate the International Decade for Natural Disaster Reduction and the target date of 2000 was set for signatory countries to effectively monitor and manage wildfires in their respective countries. In view of the lack of information on causes, extent and the number of fires at present, and the difficulties in collecting data, this target date is unlikely to be met.

Developed countries

Total forest area is slowly increasing in the developed world, although the need for protection from fire, insects and disease has continued and some aspects of forest condition give cause for concern. Although the widespread decline and death of certain European forests due to air pollution predicted by many in the 1980s did not occur, deteriorating forest condition has remained a serious concern in Europe and North America. The main causes of the decline of forest condition in Europe have been low soil moisture availability due to drought, and high temperatures. These factors, combined with airborne pollution and the use of non-adapted seed sources in forest plantation development, have predisposed forests to attack by insects and disease and to decline. Forest damage due to air pollution is severe in some parts of central and eastern Europe in particular. Air pollution is also said to have a role in the decline observed in certain species in other industrialized countries, such as sugar maple (*Acer saccharum*) in eastern Canada, the high-elevation forests of red spruce (*Picea*) in eastern USA and *Cryptomeria japonica* stands in Japan.

The exclusion of fire may have caused the decline of some forest ecosystems, such as oak stands in central USA or natural eucalypt forests in Australia where fire is an integral part of the ecology of some plant communities and here it may be used to facilitate their natural regeneration.

Despite an overall increase in the number of fires (by some 40% in Europe, 8% in North America and 120% in the former USSR), wildfires over large areas of forest have become less frequent in the industrialized countries. Further, the average area burned by wildfire decreased in most regions between 1983–90 and 1991–94. There has been a slight reduction in total areas burned in the developed world overall as a result of improved prevention, detection and control systems, although there has been an increase of 15% in the former USSR. In 1990, the average area of forest and other wooded land affected annually by fires in Europe, North America and the former USSR together was about 4.26 million ha or 0.22% of their total area of forest and other wooded land.

Pests and diseases remain constant threats, particularly to the semi-natural forests and plantations in the industrialized world. Trees are more likely to be attacked when under stress, whether from abnormal weather conditions such as drought or high temperatures, airborne pollution, uncontrolled movement of germplasm or lack of management. Recent outbreaks of pests and diseases that may have been related to stress include the decline of various species of oaks in many parts of Europe and in central Russia, the recurrent infestation of forests in Poland by the nun moth, the attack on beech by the beech scale in western and central Europe, birch and ash diebacks in north-eastern USA, the littleleaf disease of shortleaf pine in southern USA, and the 'x-disease' of pines in southern California.

Developing countries

Less attention has been paid to forest condition in developing countries because reduction in forest area has attracted more attention. The area transition matrices for the tropical world (of the type shown in Table 1.6 and Figs 1.2 and 1.3) indicate that there has been a loss of density resulting from the transfer of forest areas from closed forest to open forest; this affected 6.6% of the total closed forest during the period 1980–90. This reduction in density has often been the result of over-exploitation for timber and fuelwood. Overgrazing and repeated bush fires are other significant causes of decrease in density, especially in the dry tropical and non-tropical zones.

Every year, very large areas of savannah woodland and mixed forest–grassland formations are affected by fires set by herders to provide an early flush of green grass when the rains start, particularly in the dry zones of Africa and South America, although no reliable data are available on their extent. Forests in the humid tropics have also, at times, been affected by large fires, the most serious in recent years having been those associated with the dry weather conditions arising from the El Niño Southern Oscillation

(ENSO). In 1997–98, these caused large-scale destruction of logged and secondary forest in Indonesia (particularly Kalimantan, Sumatra and Irian Jaya), Mexico and Central American countries, as well as leading to smoke pollution over wide areas beyond the borders of these countries. The cause of these fires has been burning for land clearance, but it should be noted that the fires are not necessarily in forest; although described as 'forest fires' they are frequently in grassland or on land that has been cleared of forest. It is *estimated* that of the 2 million ha burned in Indonesia in 1997, 150000–200000 ha may have been in forest. Other serious fires associated with ENSO effects have included that in East Kalimantan, which burned 3.6 million ha in 1983.

Coniferous forests in the humid tropics have often been affected by fires, although it must be noted that fire is frequently necessary to maintain and regenerate them. In the 1980s, the area of pine forest in Honduras and Nicaragua burned annually amounted to some 65000 ha (or about 3.5% of the total pine forest area of these two countries), and widespread fires in natural and artificial tropical pine forests occur in other countries such as Guatemala, Mexico and Indonesia (northern Sumatra). In the subtemperate and temperate zones of the developing world, fire is also a permanent threat to forests, particularly when they are no longer used by local people for grazing and other purposes. In the 1980s, the average area of forest and other wooded land burned annually was 140000 ha in the temperate/subtemperate zones of South America (including southern Brazil). From 1950 to 1990, fires in China are reported to have affected an average of 890000 ha annually, the most damaging one having been the 'May 6' fire, which burned some 1.85 million ha in the north-eastern province of Heilongjiang in 1987. In the absence of a global statistical fire database, it is difficult to provide an overall estimate of the annual extent of fires in forests and other wooded lands. A very crude estimate for the temperate/subtemperate and humid tropical zones of the developing world (leaving aside the significant dry tropical zone, for which little reliable information exists) would be of the order of 2 million ha of forest and other wooded land annually during the 1980s. Given the lack of sufficient capacity in fire prevention and control in most developing countries, no significant reduction in wooded areas burned is likely to occur in the near future.

Outbreaks of pests and diseases in developing countries are generally reported for those plantations and planted trees where the impact is most apparent. Introduced pests are usually extremely destructive and several of them have had very damaging effects in recent years in the developing world. Examples of introduced insects affecting plantations and planted trees include the *Leucaena* psyllid, *Heteropsylla cubana*, which has spread into Asia and the Pacific islands and is now extending across Africa; the cypress aphid, *Cinara cupressi*, which is established in eight eastern and southern African countries and is causing heavy mortality among a number of exotic and indigenous species but especially the important plantation species *Cupressus lusitanica*, threatening its future role in the plantation programmes of these countries; and the European woodwasp, *Sirex noctilio*, which has spread into Argentina, Uruguay and southern Brazil, affecting principally *Pinus taeda* but which may become a threat for the large Chilean plantation estate of *Pinus radiata*.

Strategies to control introduced insects from attacking forest trees include regulatory measures, eradication, integrated pest management, and monitoring of population levels and occurrence of potentially harmful species. Effective monitoring of insect populations will be especially important in view of the large plantation programmes being carried out to produce industrial roundwood, which often rely on one or a few exotic species. Such programmes may be able to afford the investment that is necessary for control measures. Smaller-scale programmes may have to insure against failure due to unexpected attack by insects or diseases by diversifying into several species adapted to the sites and end-uses.

There is also evidence of forest decline due to a combination of biotic and abiotic factors in the developing world, with air pollution likely to play an increasing role in some cases as industrial and transportation infrastructure develops. Examples of such decline include neem (*Azadirachta indica*) in the Sahel, framiré (*Terminalia ivorensis*) in Ivory Coast and Ghana, and *Eucalyptus globulus* plantations in Colombia and Peru, while airborne pollution is affecting forests near large cities and industrial areas in China (e.g. *Pinus massoniana* stands near Nanshan). Despite the overall significance of the corresponding damage and losses, surveys of forest decline and diebacks in developing countries remain all too rare.

1.4 CONCLUSIONS

The world's forests cover an area of 3454 million ha or approximately 26.6% of the land surface, with 56.8% of this area in developing countries, which are mostly tropical. Forest cover has largely stabilized in most industrialized countries but deforestation continues in many developing countries. Between 1990 and 1995, the area of natural forests in developing countries decreased by an estimated 13.7 million ha per year, although this rate of loss appears to be slightly less than in the period 1980–90. Furthermore, forest management is relatively little practised at present in natural tropical forests.

The extent and condition of the global forest resource are determined by many economic, social and political factors external to the forestry sector, including, in particular, continued population growth and higher rates of global economic growth. Population growth, which will occur mainly in tropical developing countries, will continue to be combined with urbanization, while global economic growth will continue to be combined with changing consumption patterns, especially in the regions with fastest economic growth.

In the coming decades, pressures for increased food production are expected to lead to continued conversion of forest land to agriculture in many developing countries. It is estimated that in developing countries 90 million ha of land, of which about half may be forest land, would need to be converted to arable crop production alone between 1990 and 2010 (FAO 1995). Agricultural land expansion is projected to be faster in sub-Saharan Africa than in the past; given the unsuitability of much of this zone for agriculture, there must be a continuing threat to natural forest cover from agricultural expansion as land continually goes out of agricultural production. Infrastructure development will also contribute significantly to the continued loss of forests. In addition to deforestation in developing countries, large areas of forest worldwide are being degraded by overharvesting, overgrazing, pests, disease, wildfires and airborne pollution.

At the same time as forest cover globally is decreasing and forests are being degraded, demands on forests to supply wood and non-wood products and social and environmental services are increasing. Global consumption of wood increased by 36% between 1970 and 1994, and is expected to increase by another 20% by 2010. Greater emphasis is being put on the services that forests and trees can provide, including soil and water conservation, sequestration of carbon for the mitigation of climate change, conservation of biological diversity, support in combating desertification, enhancement of agricultural production systems, improvement of living conditions in urban and peri-urban areas, and provision of educational and recreational opportunities. Forests will remain an essential source of the livelihood of the poorer sectors of the world's population and a home to indigenous peoples for some time to come.

The challenge of meeting the growing demand for forest products while safeguarding the ability of forests to provide a wide range of environmental services will increasingly be met through the planting of trees, either within the forest or outside. However, less natural or seminatural forest will be converted to plantations due to the emerging values of such ecosystems, and suitable land for plantation development will therefore be in short supply in many places. Governments will thus aim to encourage large- and small-scale landowners to plant trees outside forests, often integrated into agricultural systems, through policy measures including incentives. There are likely to be moves towards making blocks of trees more 'natural' with a diversity of species, in some instances in order to provide a wider range of goods and thus serve as an insurance against the possibility of a single species failure or to guard against possible loss of soil fertility or site degra-dation, as well as to provide for improved amenity and recreational potential. There should be less emphasis on the area or quantity of forest cover and more attention paid to forest health and condition.

Forest cover in industrialized countries will continue to expand, and several newly industrialized countries will see their forest cover stabilize and even increase. Industrialized and newly industrialized countries will place greater emphasis on the conservation of natural forest where it still exists and on the conservation of seminatural forest. Plantation areas will continue to expand, either as intensively managed systems akin to farming practices or management will tend to move towards a more 'ecosystem' approach.

REFERENCES

Anon. (1996) *Activities Related to Poplar and Willow Cultivation, Exploitation and Utilisation.* Report of the National Poplar Commission of the USA, 1992–96. Forest Service, US Department of Agriculture.

Anon. (1997a) *A National Exotic Forest Description.* Ministry of Forestry, New Zealand.

Anon. (1997b) *National Forest Inventory.* National Plantation Inventory of Australia, Bureau of Resource Sciences, Canberra.

Anon. (1998) *Status of Sustainable Forest Management in Europe.* Report to the Ministerial Conference on the Protection of Forests in Europe, Ministry of Agriculture, Rural Development and Fisheries, Lisbon, June 1998.

FAO (1993) *The Challenge of Sustainable Forest Management: What Future for the World's Forests.* FAO, Rome.

FAO (1995) *World Agriculture Towards 2010.* John Wiley & Sons.

FAO (1996) *Survey of Tropical Forest Cover and Study of Change Processes (FRA 1990).* Forestry Paper 130. FAO, Rome.

FAO (1997) *State of the World's Forests 1997.* FAO, Rome.

FAO (1998) *Global Fibre Supply Model.* FAO, Rome.

FAO (1999) *State of the World's Forests 1999,* FAO, Rome.

FAO/ECE (1996) *Long-term Historical Changes in the Forest Resource.* Geneva Timber and Forest Study Papers no. 10. UN-ECE/FAO, Geneva.

IIED (1996) *Towards a Sustainable Paper Cycle.* IIED (International Institute for Environment and Development), London.

Salwasser, H., MacCleery, D.W. & Snellgrove, T.A.
(1994) New perspectives on managing the United
States national forest system. In: *Readings in
Sustainable Forest Management*, pp. 235–66. FAO,
Rome.
WWF (1994) *The Status of Old Growth and Semi-
natural Forests in Western Europe*. Worldwide Fund
for Nature, Gland.

ENDNOTE

The final draft of this chapter was completed before publication of the FAO Global Forest Resources 2000. The latest information on the area and condition of the world's forests became available from FAO in August 2000 (Forest Resource Assessment Programme, Forest Department, FAO, Viale delle Terme di Caracalla, Roma 00100, Italy, fax 39 06 57052151 or e-mail forestry-information@fao.org. The Internet site, which includes the Global FRA2000 results, is http://www.fao.org/fo).

2: Forest Types and Classification

RONALD L. HENDRICK

2.1 INTRODUCTION

Closed canopy forests and savannahs occupy a tremendous amount of land, approximately 3540 million ha (Perry 1994) or about 26% of the earth's terrestrial surface (FAO 1997). Tropical forests comprise approximately 35% of the total, temperate forest 17%, boreal forest 16%, and the remainder is savannah and open woodlands. Among countries and continents, about 520 million ha of forest are in Africa (15.1%), 465 million ha in Asia and Oceania (16.4%), 146 million ha in Europe (4.2%), 950 million ha in South America, Central America and the Caribbean (27.5%), 457 million ha in North America (13.2%) and 816 million ha in the countries of the former USSR (23.6%). There is tremendous floristic diversity in these forests, with more than 10 000 known species of trees and shrubs, and tens of thousands of species of non-woody vascular and non-vascular understorey plants and epiphytes. The tropical forests are the most diverse, with over 2500 species of trees in both the Malay Peninsula and South America for example.

The tremendous expanse and diversity of the world's forests do not lend themselves well to detailed description in a review like this. Moreover, while there is a long history of vegetation inventory and classification in temperate and some boreal forests, much less has been done in the tropical regions, outside of Australia. Fortunately, there are a number of excellent and comprehensive descriptions and summaries of regional forests; I have made extensive use of many of them here, and most are readily accessible. For readers interested in more detailed descriptions of regional floras, Frodin's (1984) guide to published floras is an excellent, though somewhat dated, reference.

The approach I have taken in classifying and describing the world's forests is primarily climatic–geographical–physiognomic in nature, with the forests separated into broad types based primarily on prevailing climate (e.g. tropical, temperate, Mediterranean, etc.). Within each type, I describe the regional formations, noting major species, cover types and physiognomic forms. The discussion of boreal and temperate forests contains the most specific references to individual species and forest types, as many of the species native to these regions are abundantly represented and widely distributed, and the regional floras are especially well known. I focus primarily on the naturally occurring vegetation in each region, especially mature forests.

I describe the forests of Australasia, eastern Asia, Europe, Fennoscandinavia and North America in the most detail, as these regions have been subject to the most research. They are also the regions for which the literature is most abundant and accessible, at least for an English-speaker like myself. I preface discussion of the regional formations by general information on the extent, distribution, prevailing climate and soils of each forest type. Throughout, I cite references to which interested readers can refer for more information about a region or forest type of particular interest to them.

2.2 BOREAL FORESTS

Boreal forests are vast in size, widely distributed and remarkably variable despite a relatively low diversity of woody plants. Growing conditions range from extremely cold, low-moisture environments to perennially wet bogs and swamps of near-temperate climates. The trees are predominantly coniferous (especially *Abies*, *Picea* and *Larix* spp.), although a few deciduous genera (*Populus*, *Betula* and *Salix*) are nearly ubiquitous in their distribution and are often locally common. Vast areas of boreal forest remain relatively undisturbed by humans, although various disturbance agents, especially fire, have altered the composition and dynamics of these forests for millennia. Pressure from commercial harvesting is increasing, particularly in Siberian forests.

2.2.1 Distribution and extent

Boreal forests, or taiga, are distributed globally across the high latitudes of the Northern Hemisphere (Fig. 2.1). Estimates of their extent range from 9 to 17 million km² (Lieth 1975; Perry 1994). The precise area is difficult to estimate, due to incomplete inventories and varying definitions of 'boreal forest', but the upper values (15–17 million km²) are probably most accurate.

The northern limit of the North American boreal forest extends from about 68° N in Alaska to about 58° N at Hudson Bay and eastern Canada (Larsen 1980). It extends north to the Arctic Ocean in Scandinavia (above 70° N) and parts of western Russia and then follows a line ranging a few to several degrees above or below the Arctic Circle (66° 33′ N). Boreal forests occur discontinu-

ously southward to about 35° N at high elevations in the Appalachian Mountains of eastern North America and to similar latitudes in central Asia around the Indian subcontinent. Boreal forests grade into tundra at their northern edge, into montane coniferous, grasslands, parklands and mixed forests at their southern edges in North America and south-western Eurasia, and into woodland/steppe in the more central and eastern regions of Eurasia. The north–south width of the boreal forests ranges from about 500 to 2300 km in Eurasia and averages about 1000 km in North America (Larsen 1980; Archibold 1995).

2.2.2 Climate

The climate of the boreal forest is characterized by low temperatures and relatively low precipitation. The growing season is short (90 or fewer frost-free days in northern areas), and some forests are underlain by permafrost. However, the long days during the growing season partially compensate for the comparatively unfavourable growing conditions.

The northern edge of the boreal forest corresponds roughly to a July isotherm of 13°C in both Canada and Eurasia (Larsen 1980). The southern border generally follows an 18°C July isotherm, except where precipitation is too low, as in the northern prairie region of western Canada. Winter temperatures can be exceedingly low, less than –70°C in parts of Alaska and Siberia. At the Bonanza Creek Long-term Ecological Research (LTER) site near Fairbanks, Alaska, the mean

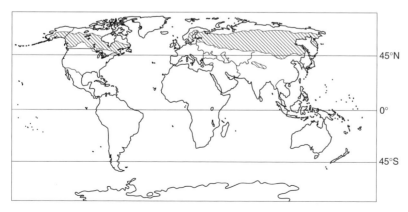

Fig. 2.1 Global extent and distribution of boreal forests.

daily temperature is –24.9°C, whereas the July mean is 16.4°C (Bonanza Creek LTER database). Mean annual temperature is 3.3°C, with an average of 233 days at below freezing temperatures annually. Summer temperatures in much of Canada and Eurasia are somewhat cooler (Larsen 1980).

Precipitation throughout the boreal region is typically less than in most temperate forests but is highly variable. In interior Alaska and Siberia, precipitation ranges from 100 to 270 mm (Rumney 1968; US National Oceanic and Atmospheric Administration records); in eastern Canada, however, it can be as great as 900 mm. Precipitation typically is greatest in summer, but where there is a maritime influence (e.g. eastern Norway and portions of Canada) it is distributed somewhat more evenly throughout the year and the autumnal drier periods characteristic of regions with a more continental climate are absent or less pronounced.

Common to all boreal forests are light regimes characterized by short winter and long summer days. Forests growing at or above the Arctic Circle experience at least 1 day each of both total light and total darkness. At the Bonanza Creek LTER site, average daily solar radiation ranges from a low of 750 kJ m^{-2} in January to 22 375 kJ m^{-2} in June (Slaughter & Viereck 1986). Sun angles are low year round (49.5° on the summer solstice and 2.6° on the winter solstice in Fairbanks, Alaska) and accentuate the effects of topography. The low sun angle also dramatically increases the amount of twilight, which comprises an average of 10.4% of every 24 h and about 17% of all light at 65°N compared with corresponding values of 3.1 and 6% at the equator.

2.2.3 Soils

Spodosols, or podzols, are the dominant soils in boreal forests. Spodosols form as iron, aluminium and organic matter move from the A horizon into the B (spodic) horizon, a process known as podzolization. The word 'podzol' is of Russian origin and refers to the ashy-coloured, highly leached E horizon that lies between the A and B horizon(s). Spodosols are typically rather acidic, in the 3.5–4.5 pH range. Organic matter deposited

around soil particles in the B horizon enhance cation exchange capacity, although Spodosols are typically somewhat nutrient-poor. Because of the recalcitrant coniferous litter and harsh climate, decomposition is comparatively slow, and a thick litter horizon overlays the mineral soil in many upland environments. Moss is often abundant in the understorey, decomposing as much as 10 times more slowly than other plant litter.

Although dominant in area, Spodosols are by no means the only soils associated with boreal forests. Organic soils (Histosols) form in low areas where decomposition is extremely slow and the soil is wet all or most of the year. In the broad floodplains that occur along river corridors, Entisols and Inceptisols are often found. In montane environments, Alfisols are common (Archibold 1995), which have higher pH and base saturation than Spodosols.

Found in no other forest soils, permafrost is a permanently frozen layer of soil that underlies boreal forests growing on low, wet areas or on north-facing upland slopes. It forms when the soil does not warm sufficiently to fully melt during the summer, and may range from a few meters in thickness under forests to over 240 m under treeless tundra. Permafrost forms gradually, and is associated with an accumulating insulating cover of litter and especially moss on the forest floor during forest development. The active layer (the upper soil that melts during the summer) may be less than 1 m thick. *Picea mariana* and *Larix* spp. are among the few trees that grow over permafrost, and even they are of low stature (<10–15 m) (Fig. 2.2). Where the active layer becomes too thin, the boreal forest or taiga gives way to tundra.

2.2.4 Dominant forest types

Upland forests of various types dominate much of the boreal region. Common upland genera include the conifers *Picea*, *Abies* and *Pinus*, and the hardwoods *Populus*, *Betula*, *Sorbus* and *Alnus*. Deciduous boreal forests often have an open understorey, owing to their characteristically early place in succession and the fact that many of them establish after catastrophic fires. Coniferous

Fig. 2.2 Low-stature (<6 m) *Picea mariana* forest growing on permafrost along the Tanana River in interior Alaska, USA. (Photo by the author.)

forests are often associated with a denser understorey and a well-developed moss layer.

Lowland communities include both forested and treeless communities. In addition to floodplain terraces, swamps and muskeg may also be forested. Muskeg is a nutrient-poor environment, and wet-mesic to wet; it is an open environment and species-poor. Swamp is more nutrient-rich and supports a more diverse flora, including some deciduous species.

2.2.5 Regional floras

Eurasian

Ahti *et al.* (1968) divided the boreal region of Fennoscandinavia into four zones, arranged from south to north: the hemiboreal, southern boreal, middle boreal and northern boreal zones. The hemiboreal zone is dominated by broadleaf species of more temperate affinity, such as *Fraxinus excelsior*, *Quercus robur*, *Tilia cordata* and *Ulmus glabra*. The southern boreal zone is transitional between the hemiboreal and middle/ northern zones, and contains a mix of broadleaf and coniferous species. The middle and northern zones are largely coniferous, with *Betula* spp. the primary deciduous trees. The northern zone is characterized by a distinctly boreal flora (including abundant *Salix* spp.) and the absence of the low-herb spruce forests of the middle zone (Eseen *et al.* 1997). The conifers *Pinus sylvestris* and *Picea abies* are the dominant regional species. Both grow on a wide variety of sites, from xeric to wet-mesic. *Pinus sylvestris* tends to be more abundant on drier sites, which are more frequently disturbed by fire and have a more continental climate. Conversely, *Picea abies* grows on moister sites where fires are less frequent and a maritime climate often prevails.

Relative to parts of the North American and eastern Eurasian boreal forests, shrubs are infrequent in Fennoscandinavian forests. The woody understorey is more apt to comprise small trees and saplings. *Salix starkeana*, *Salix xerophila* and *Juniperus communis* occur, especially after disturbance, and *Sorbus aucuparia* and *Salix caprea* occupy mesic sites. The dwarf shrubs *Calluna vulgaris* and *Empetrum hermaphroditum* are common on dry sites, and *Vaccinium* spp. and *Ledum palustre* are widespread and may be locally common. Bryophytes and lichens are typically more abundant than vascular plants, and bryophytes often dominate the forest floor on mesic sites (Eseen *et al.* 1997). Two species of moss also common to the western North American boreal forests, *Holocomium splendens* and *Pleurozium schreberi*, are abundant on low–medium productivity sites. On dry sites and to the north, lichens become more important, especially *Cladonia* spp. and *Stereocaulon* spp. Various species and growth forms of *Cladonia* characterize early, mid and late successional communities (Ahti 1977; Oksanen & Ahti 1982; Oksanen 1983).

Deciduous species are more abundant in early and mid successional communities, and apparently occupied a much greater area of Fennoscandinavia prior to active fire suppression (Zachrisson 1985). Mature deciduous forests are important centres of biodiversity for a variety of life forms, including epiphytic non-vascular plants growing on *Populus tremula* and *Salix caprea* and numerous insects and several birds, including the white-backed woodpecker (*Picoides leucotos*).

The forests of western Russia are similar to those of Fennoscandinavia, with abundant *Picea abies* and *Pinus sylvestris*, along with *Picea obovata* and *Larix siberica*. While *Picea abies* does not grow eastward of the Kama River, *Pinus sylvestris* occurs throughout Eurasia (although its northern range becomes more restricted in the central and eastern regions). In Siberia, it forms nearly monospecific stands on low-nutrient sandy, rocky and boggy soils. In the mountains, it grows predominantly on eastern and western slopes, and will grow on permafrost where the active layer exceeds 2 m (Nikolov & Hemisaari 1992).

Numerous other species occur throughout the central and eastern portions of the Eurasian boreal forest. *Abies siberica* has a near-continental distribution, primarily below the Arctic Circle, and grows on productive upland sites (Nikolov & Hemisaari 1992). Common overstorey associates include *Picea obovata* and *Pinus siberica*; *Abies siberica* also occupies the understorey of *Pinus sylvestris* and *Larix* spp. forests. *Abies siberica*, *Betula pendula*, *Betula pubescens* and *Pinus siberica* grow on fertile upland sites in central Siberia, and *Betula pendula* forms early successional communities with *Populus tremula* after fire or other disturbance. *Picea obovata* also has a continental distribution, with a wider north–south range, and forms both pure (river floodplains) and mixed (upland) stands. It, *Abies siberica* and *Pinus sylvestris* are the principal species of the closed Siberian taiga forest (Nikolov & Hemisaari 1992).

In addition to the continentally distributed species described above, the far eastern boreal forest contains *Pinus pumila*, a shrubby species that grows in rocky mountain soils and in peat and alluvial soils in the lowlands. Common mountain associates include *Betula* spp., *Rhododendron* spp. and *Alnus fruticosa*, among others. *Pinus pumila* most often grows in the understorey of *Larix gmelinii* (*Larix duhurica*) forests, which typically form a monospecific canopy. *Larix gmelinii* is more widely distributed, growing in mountains, river valleys, coastal environments, uplands, at the forest–tundra border and at the southern extent of the boreal forests in forest–steppe communities (Nikolov & Hemisaari 1992). *Picea ajanensis* grows in both pure and mixed stands in the southeast boreal forest, occurring with *Larix gmelinii*, *Picea obovata* and *Betula* spp. in the overstorey on upland sites at high latitudes and elevations, and with a *Ledum* spp. and *Sphagnum* spp. understorey in low areas in the north-west of its range. In the more southerly portion of its range, it grows in mixed forests that variously include other *Abies* spp., *Acer* spp., *Betula* spp., *Fraxinus mandshurica*, *Juglans mandshurica*, *Populus* spp., *Tilia* spp. and genera of more temperate climates.

North American

Larsen (1980) has extensively summarized the regional variation in the North American boreal forest, and I refer the reader to his text and the references contained therein for a detailed description. Viereck *et al.* (1992) have classified Alaska's forests in great detail, and their scheme is a useful reference for both Alaska and north-western Canada.

Although many species in the North American boreal forest are near continent-wide in their distribution, there is considerable regional variation in species associations and habitat preference (Table 2.1). In Alaska and Canada west of the Rocky Mountains, *Picea glauca* is the dominant late-successional species on upland sites devoid of permafrost and undisturbed by fire. *Populus tremuloides* and *Betula papyrifera* frequently colonize burned upland sites, and without further disturbance are succeeded by *Picea glauca* within about 100 years (Van Cleve *et al.* 1996). On floodplain terraces, *Populus balsamifera* instead of *Populus tremuloides* is present, and *Betula papyrifera* is often less abundant (see Fig. 2.2). Moreover, the formation of a dense, insulating moss layer (especially the feather mosses *Holocomium splendens* and *Pleurozium schreberi*) leads to the formation of permafrost and the

Table 2.1 Common trees of the North American boreal forest, their geographic distributions and typical site characteristics.

Dominant species	Geographic extent	Site characteristics
Picea glauca	Transcontinental	Uplands without permafrost or recent fires; elevated river floodplain terraces
Picea mariana	Transcontinental	West: on permafrost found on north-facing slopes and depressions of uplands and river floodplains. East: thin soils, wet sites
Betula papyrifera	Transcontinental	Successional post-fire uplands (with *P. tremuloides*) and disturbed floodplains (with *B. papyrifera*)
Populus balsamifera	Transcontinental	Successional floodplain terraces
Populus tremuloides	Transcontinental	Successional post-fire uplands
Pinus banksiana	Atlantic coast to 120°W	Various, especially sandy glacial and thin granitic soils
Pinus contorta var. *latifolia*	Pacific coast to 100°W, north to 64°N	Various, especially sandy glacial and thin granitic soils; mountains
Abies balsamea	Atlantic coast to 115°W	Upland mineral and low organic soils
Larix laricina	Transcontinental	Wet soils throughout

replacement of *Picea glauca* by *Picea mariana* on less frequently burned sites.

Picea mariana and occasionally *Larix laricina* dominate low wet areas, as well as north-facing slopes underlain by permafrost. These forests can degrade over time to open, wet communities dominated by low-stature ericaceous shrubs with few or no trees. However, fire can expose the mineral soil, resulting in permafrost melting and the re-establishment of forest vegetation. Further south, *Abies lasiocarpa*, *Picea engelmannii*, *Larix lyallii* and *Populus balsamifera* dominate the overstorey of high-elevation and subalpine forests.

In the boreal forests of the Mackenzie–Yukon region to the east, both *Picea glauca* and *Picea mariana* occupy the uplands, while the latter is more common along rivers and in lowlands. In west-central Canada, *Pinus banksiana* and *Pinus contorta* var. *latifolia* become common in the south, particularly on sandy soils of glacial origin and thin, coarse soils weathered from granitic outcrops. Both sandy and rocky soils share a number of species in common (Raup 1946; Thieret 1964). The ranges of *Pinus contorta* var. *latifolia* and *Pinus banksiana* overlap but the

former is more abundant to the west (see Table 2.1).

In the Canadian Shield region to the east, *Picea mariana* dominates the thin soils common throughout this area. *Pinus banksiana* occupies drier sites, and *Larix laricina* is an associate on wetter sites, although open bogs and fens are common. Further east and south around the Great Lakes sub-boreal species at the northern edge of their range, including *Pinus resinosa*, *Pinus strobus*, *Thuja occidentalis*, *Acer rubrum*, *Acer saccharum* and *Betula alleghaniensis*, intermix with the prevailing boreal species. The Hudson Bay lowlands are largely wet, and are covered by open bogs and fens and open *Picea mariana* forests.

Picea glauca and *Abies balsamifera* are the most dominant upland species conifers in the central and southern regions of eastern Canada. Common hardwood associates include *Populus tremuloides*, *Populus balsamifera* and *Betula papyrifera*. Further north, *Picea mariana* covers most of the uplands, although *Pinus banksiana* occupies sandy sites and early successional *Picea mariana* sites. Forests near the forest–tundra transition are predominantly *Picea mariana*–lichen or *Picea mariana–Sphagnum* communities, while central, southern and coastal *Picea mariana* forests more commonly have a low shrub understorey. *Populus balasamifera* is common in Quebec, the Canadian Maritime Provinces and the north-eastern USA, particularly Maine (Eyre 1980); it is replaced by *Picea mariana* at higher elevations, and further west in Ontario comprises only a limited portion of the forest landscape.

2.3 TROPICAL AND SUBTROPICAL FORESTS

The importance and fate of tropical and subtropical forests are topics of widespread and enduring interest. Long recognized as centres of high biotic diversity (perhaps containing more than half of the world's terrestrial plant and animal species), considerable effort has been expended on quantifying rates of deforestation, cataloguing species assemblages and identifying natural products of medicinal or other commercial importance. However, most of these efforts are of rather recent origin and tropical forests remain perhaps the least understood of all terrestrial ecosystems.

2.3.1 Distribution and extent

The tropics proper lie between the Tropic of Cancer (23.5°N) and the Tropic of Capricorn (23.5°S). About 23–35° north and south of these latitudes are regions similar in climate and vegetation type and form, commonly referred to

as subtropical (Fig. 2.3). There are perhaps 28–32 million km^2 of (sub)tropical forest, about 60% of which is dry or seasonal forest, the remainder being humid or wet.

Tropical forests are further defined by their location within one of three geographical formations: American (or Neotropical), African and Indo-Malesian. Within the American formation, forests are most abundant in the Amazon drainage basin. Brazil itself contains 352 million ha of intact moist tropical forest, 16 million ha of other closed canopy forests, 155 million ha of intact woodlands and about 35 million ha of degraded secondary forests of various types (Schroeder & Winjum 1995). Other areas with extensive forests are north-west South America, Central America and Mexico. Numerous islands in the Caribbean Sea have, or did have, (sub)tropical forests, and forests along the coast of the Gulf of Mexico in the USA are climatically and floristically subtropical, particularly in Florida and south Texas. In total, the American formation comprises about 50% of the world's tropical forests.

About 20% of tropical forests are, or were, located in Africa and Madagascar. Rain forests in the western and central portions of Africa range from Sierra Leone to Kenya and the high plateau of Uganda. The remaining rain forests in Africa are in the far eastern border, confined to wet areas along the coast and high elevations. They are much smaller in extent and typically drier than West African rain forests. Dry, closed canopy forests historically occurred in both western (e.g.

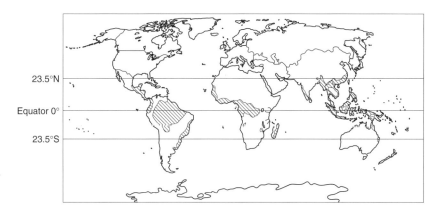

Fig. 2.3 Global extent and distribution of tropical forests.

Guinea) and eastern (Kenya to Mozambique) Africa, often bordering the rain forests. However, most have been lost to agricultural conversion or deliberate burning. Dry open forests are more abundant than closed forests, and occur, or occurred, widely in southern Africa from Angola to Mozambique and in portions of western central Africa. What remains of the forests of eastern Madagascar is rain forest, whereas the forests of western Madagascar are dry, closed canopy forests.

Rain forests in the Indo-Malesian formation occur in western India and Sri Lanka, in the mainland of South-East Asia across the Malay Peninsula and into southern China, the islands of the Malay archipelago, Philippines and Melanesia, and in Australia. Dry tropical forests occur further inland in South-East Asia, across eastern India and in portions of northern and eastern Australia.

2.3.2 Climate

The tropics generally correspond to frost-free areas, especially those where the mean annual temperature is 18°C or above (Walter 1985) or where the mean monthly temperature difference between the three warmest and three coldest months is 5°C or less. Together the tropical and subtropical regions comprise about 58% of the earth's surface area (Kellman & Tackaberry 1997).

Precipitation throughout the tropics is highly variable. Most rainfall is convective in origin, of short duration and high intensity, and patchily distributed (Kellman & Tackaberry 1997). At least 2000 mm of annual precipitation is generally regarded as necessary for the occurrence of rain forests (Lauer 1989). Of even greater importance, however, is that rainfall is distributed rather evenly throughout the year. Thus, rain forests occur in portions of Africa where annual precipitation is only 1600 mm but falls throughout the year. In areas where there are dry (<100 mm) periods of a few to several months in duration, and as precipitation declines to 1000 mm or less annually, rain forests are replaced by dry forest or savannah.

Topography also influences climate, because temperatures decrease as elevation increases whereas precipitation typically increases. Except in monsoon environments, maximum precipitation in tropical environments typically occurs at elevations of 800–1500 m, although smaller maxima may be found at elevations of 2700–3000 m (Lauer 1989). In South America, tropical forests are limited to elevations of less than 1700 m (Kricher 1997), because although moisture is abundant at higher elevations, temperatures are too low. Mountains also moderate the effect of air masses, as in South America, where the Andes Mountains prevent moist air masses from the Pacific reaching the interior, thereby casting a rain shadow on the leeward side. Similar effects occur in the mountainous regions of Central America, western India and western Burma, as well as in mountainous islands like Hawaii, Fiji and the West Indies (Kricher 1997).

2.3.3 Soils

Most soils in the tropics are old and highly weathered. Dominant lowland soils are Oxisols and Ultisols, which together comprise about 23% and 18% of all non-desert tropical soils respectively (Richter & Babbar 1991). Other common soils include Alfisols (10.5%), Vertisols (3.4%) and young soils of volcanic or alluvial origin (22.4%). Mountain soils of various orders comprise about 18% of tropical soils, while organic soils occur on less than 1% of the landscape.

Oxisols form in warm, wet environments and are commonly associated with tropical rain forests. There is typically little horizon differentiation. Iron and aluminium oxides are abundant, while potassium, magnesium and available phosphorus are low. Both pH (4.5–5.5) and cation exchange capacity (<16 mEq per 100 g of clay) are low. Large concentrations of Oxisols are found in the Amazon and Congo basins. Ultisols are typically associated with dry tropical forests and develop under seasonal precipitation regimes. Ultisols have a distinctive argillic horizon with a high concentration of iron sesquioxides. Organic matter content and cation exchange capacity are typically low, with a base saturation of less than 35% in the lower horizons. Ultisols are common in south-eastern North America and South-East Asia, although both Oxisols and Ultisols frequently occur together, along with one or more of the other dominant soil orders.

2.3.4 Dominant forest types

Although the term 'tropical forest' is often used as a surrogate for tropical rain forest, there are in fact a variety of regional forest types. Longman and Jenik (1987) identified five primary forest types: tropical evergreen, (sub)tropical seasonal, (sub)tropical semideciduous, sub(tropical) evergreen and mangrove.

Tropical evergreen forests (i.e. rain forests) include:

1 lowland subtypes, characterized by tall emergent trees (>30 m) projecting above a multilayered canopy, with sparse undergrowth and few epiphytes;
2 montane forests with lower stature trees and a rich woody understorey and herbaceous ground cover;
3 cloud forests with a closed canopy with gaps, low stature (<20 m) canopy trees, abundant lianas and climbers, and a rich herb, moss and fern cover;
4 alluvial multilayered forests (sometimes referred to as gallery forests) growing in flood-prone environments, with abundant herbaceous vegetation and epiphytes;
5 swamp forests growing in frequently or perennially flooded areas with abundant ferns and herbs;
6 peat forests with low-stature trees, poorly developed ground cover and occasionally abundant palms.

After disturbance, tropical evergreen forests frequently develop a thick undergrowth.

The (sub)tropical seasonal forests are predominantly evergreen, with some trees of deciduous habit. They range from lowland to montane environments, with climbers and evergreen shrubs present and tree ferns present at lower elevations. The (sub)tropical semideciduous forests are dominated by overstorey trees that lose their foliage during the dry season, although the understorey is often evergreen. Ground cover is abundant but

epiphytes are not. The (sub)tropical evergreen forests are structurally similar to tropical evergreen forests, but less productive due to periods of colder temperatures during the winter. Finally, mangrove forests are found in tidal areas, are evergreen in habit, are adapted to the salt-water environment, and have an understorey nearly devoid of vegetation but often containing non-vascular plants.

2.3.5 Regional formations

American

The forests of the Neotropics are exceptionally diverse, particularly the rain forests, and their phytogeography is complex. Archibold (1995) summarized Longman and Jenik's (1987) and Mabberly's (1992) data on the most significant woody plant families in the region, which include the Euphorbiaceae, Lecythidaceae, Leguminosae, Meliaceae, Moraceae, Myristicaceae and Sapotaceae. Prance (1989) identified 18 regions or floristic provinces within South American tropical forests, based upon species distribution and areas of high endemism; he defined four broad categories, into which the 18 provinces could be classified (Table 2.2). Of perhaps more general usefulness is a classification scheme based on climate, topography and vegetation type: (sub)tropical moist or broadleaf evergreen (i.e. rain forests), montane broadleaf, wetland, and seasonal/(semi)deciduous.

Moist (sub)tropical forests are diverse, multi-storeyed communities dominated by large-stature trees. Vines and other climbers are often abundant. These are low-elevation forests, growing in areas of high year-round rainfall. The largest concentration of these forests is in the Amazon basin (including Guyana, Surinam and French Guiana), although they occur in five other principal areas (Prance 1989): (i) Mexico and Central America (predominantly along the Gulf of Mexico and along the Caribbean), (ii) several Caribbean islands (Cuba, Puerto Rico, Hispaniola and Jamaica and some of the Lesser Antilles), (iii) the Pacific coast of South America (northern Colombia to northern Ecuador), (iv) northern Venezuela and (v) the Atlantic coast of Brazil. Northern

Table 2.2 Floristic provinces of Prance (1989).

Description	Province
Regional centres of high endemism	Panama–Chocó; Magdelena–Venezuela Gulf; northern and southern Andean submontane; French Guiana–eastern Amazonia; Amazonia; north-eastern Brazil; Planalto; Atlantic coast; southern Pacific; Chaco; southern Brazil
Archipelago-like centres of montane endemism	Northern and southern Andean montane; central Cordillera Venezuela; Guyana Highland
Regional transition zones between areas of high and low endemism	Amazon transition zone, bordering south Amazonia
Regional mosaics that are neither areas of especially high endemism or transition zones	Pantanal; Catatumbo–Llanos; Venezuelan Amazonas–savannah

Venezuela, Amazonia and associated areas and both coastal regions are characterized by high levels of endemism, while the other two regions have mixtures of both endemic and widely distributed species.

Montane forests occur in both humid and arid areas, and the associated forest types are consequently quite variable. Subtypes include lower montane and upper montane and subalpine forests (Prance 1989). Andean lower montane forests begin at 700–1200 m and range up to 1800–2400 m in altitude. Species in the genera *Licania* (Chrysobalanaceae) and *Eschweilera* (Lecythidaceae) occur here as well in the lowland forest. In northern Peru, podocarp (especially *Podocarpus oleifolius*) forests occur. Situated between the lower montane and subalpine forests (beginning at 3400 m or less), the Andean montane forests comprise more distinctive high-elevation species, including *Brinellia occidentalis*, *Symplocos pichindensis* and *Weinmannia balbisiana*. Subalpine forests may grow at elevations up to 3800 m and are dominated by low-stature species in families such as Ericaceae and Rhamnaceae. Cloud forests are characterized by particularly dense canopies and dark under-stories, abundant epiphytes (orchids, bromeliads, ferns, mosses and lycopods), and abundant shrubs but sparse vines. These forests are common along the eastern slope of the Andes, and are found in south-eastern Brazil, high elevations in the Greater and Lesser Antilles, and parts of Central America (chiefly Guatemala, Panama, Costa Rica

and Nicaragua) (Fig. 2.4). In Central America, the montane flora is more similar to that of southern North America, including species in the genera *Quercus* (Fagaceae) and *Liquidambar* (Hamamelidaceae) (see p. 49).

American tropical wetland forests include permanently flooded forests, periodically flooded forests and gallery forests. Permanently flooded forests are often dominated by palms (*Mauritia* and *Euterpe*). Periodically flooded forests include mangrove forests, which grow in marine tidal areas, and various freshwater and tidewater swamp forests. Mangrove forests occur from southern Florida southward along both the Atlantic and Pacific coasts of Central and South America. Eight species of mangrove occur in the American formation, among which *Rhizophora mangle*, *Avicennia germinans*, *Laguncularia racemose* and *Conocarpus erecta* are common (Kricher 1997). Mangrove forests are typically low in stature, although *Avicennia germinans* may reach heights of 20 m. Pacific coast mangrove forests are typically more diverse than their Atlantic counterparts. Detailed descriptions and comparisons of mangroves and mangrove forests can be found in Pool *et al.* (1977), Tomlinson (1986) and West (1977). Tropical tidal swamp forests are most abundant on the Amazon river delta, and are rather diverse. Palms (e.g. *Astrocaryum murumuru*, *Euterpe oleracae*, *Jessenia bataua*, *Manicaria saccifera*, *Mauritia flexuosa*, *Maximiliana regia*, *Oenocarpus* spp. and *Raphia taedigera*) are common (Prance 1989). Further

Fig. 2.4 Montane cloud forests in Costa Rica, with clearcut areas throughout (light-coloured patches). (Photo by Brian Palik.)

Fig. 2.5 Seasonal dry forest in Costa Rica. (Photo by Brian Palik.)

descriptions of these forests can be found in Anderson and Mori (1967).

Seasonal and semideciduous forests occur where there is a significant dry season and are found throughout the Neotropics (Fig. 2.5).

Significant areas occur from north-western Mexico to north-western Costa Rica (Kricher 1997). They also grow in Amazonia, southern Brazil, Venezuela, Paraguay and Bolivia. Rates of endemism are typically low, although there are significant geographical differences in species composition. Vast areas that formerly supported seasonal forests have been converted to the cultivation of coffee (*Coffea arabica*). The most abundantly woody plant families in Neotropical dry forests are generally Leguminosae and Bignoniaceae. Mytaceae is dominant in some forests of the Caribbean West Indies and often abundant elsewhere (Gentry 1995). Also common are representatives of Rubiaceae, Sapindaceae and Euphorbiaceae.

African

Aubréville (1956) recognized three primary regions of African forest: the Guinea Forest, the Nigerian Forest and the Equatorial Forest. The Guinea Forest extends about 1280 km from the Guinea coast and Sierra Leone eastward across Liberia, Ivory Coast and Ghana. Although the forests of Nigeria become more species-poor from east to west, the floristic distinction between the Nigerian and Equatorial forests is no longer recognized as significant enough to warrant separation between the two (White 1983). However, the savannahs (now largely agricultural) in Togo and Benin form a natural floristic boundary between Aubréville's Guinea Forest and the forests to the east. This area extends over the Nigerian lowland to Cameroon (Aubréville's Nigerian), and from Cameroon eastward over 2400 km into the African interior and the Ugandan plateau (Equatorial). Most of the forested area lies in the Congo Basin of northern Zaire (Democratic Republic of Congo) (Archibold 1995). Monod (1957, in Hamilton 1989) suggested that the rain forests of Africa comprise one large Guineo-Congolian Region. He further proposed that this region can be divided into two groups (the west-central Atlantic-Congo Domain and the eastern Oriental Domain), because the rain forests of western Africa are floristically distinct from the central and eastern forests.

African forests bear some taxonomic similarity to the American formation. For example, legumes (Leguminosae) are abundant, particularly in dry forests. Other families found in both formations include Euphorbiaceae, Meliaceae, Moraceae and Sapotaceae, all of which are believed to be old families that occurred on both continents prior to the break-up of Pangaea (Hamilton 1989). Generally, lowland tree species richness is much less in African forests than in either the American or Indo-Malesian formations, and single species often dominate the canopy. These include *Brachystegia laurentii*, *Gilbertiodendron dewevrei* and *Cynometra alexandri* (Hamilton 1989). However, there are generally recognized centres of floristic diversity (and, typically, endemism) within Africa proper. These occur in Cameroon and Gabon; eastern Zaire (Democratic Republic of Congo); eastern Ivory Coast and western Ghana; Sierra Leone and Liberia; and the vicinity of the East African coast (Hamilton 1989). The areas of high floristic diversity are associated with high rainfall, but even more importantly these regions probably had mesic climates during severe Quaternary climatic fluctuations that caused widespread extinctions in other areas during very arid periods. Patterns of species richness in montane forests are somewhat different, most notably in that montane forests in and around Cameroon lack much endemism, whereas there is a high degree of endemism in lowland forests in the area. Forests are floristically diverse in the Eastern Arc Mountains of Tanzania, and species change progressively with elevation (Lovett 1996). However, community diversity is not strongly correlated to elevation or latitude; instead, disturbance, low precipitation and a pronounced seasonal climate all reduce species richness.

The forests of southern Africa are variously referred to as either subtropical or temperate, but here I consider them subtropical due to their taxonomic similarity with Malesian forests, largely evergreen habit and low-latitude occurrence. The species assemblages at higher elevations and along the south coast resemble those of the montane forests of East Africa. *Podocarpus* spp. and *Olea laurifolia* are common, as is *Apodytes dimidiata*, *Curtesia dentata* and *Ocotea bullata* (Adamson 1956). Shade-tolerant shrubs and ferns occupy the understorey. Wetter montane forests contain *Cunonia capensis*, *Ilex mitis* and *Platylophus trifoliatus*. The endemic *Widdringtonia* (including *W. juniperoides* and *W. schwarzii*) forests grow in small areas north of Cape Town.

On the south-west coast are forests of *Podocarpus henkelii*, *Peroxylon utile* and *Xymalos monosperma* and the shrubby *Buxus macowani* (Adamson 1956). The east coast forests of Natal further to the north are more species-rich than those of the south coast. Podocarps are again common, and among the associated species are *Calodensrum capense*, *Celtis africana*, *Combretum krausii*, *Olea capensis*, *Ptaeroxylon obliquum* and *Zanthoxylu dayvi*.

Indo-Malesian

The Indo-Malesian tropical forest formation is the second largest of the three and has been extensivcly studied. Geographically, the formation ranges from India through Polynesia, and includes the east coast of Australia and the South-East Asian mainland. The regional forests are diverse in both type and composition, and include rain forests, monsoon forests and savannahs among others. The wide array of landforms, climates, soils and the degree and types of human activity within the region interact to add further layers of floristic and structural complexity. Regional floristic diversity is extremely high. About half of all plant families (220) and one-quarter of all genera (2400) are represented just within Malesia (the region from Sumatra and Malaya eastward to the Bismarck archipelago, north to the Malaysia–Thailand border and south to between New Guinea and Australia) and there are 8000–10000 species of small to large trees (Whitmore 1989). In Malaysia and Singapore, there are 82 families and 388 genera of trees; 27% of the tree species are endemic (Ng & Low 1982). Indo-Malesian tropical forests differ taxonomically from the American and African formations in that a number of conifers are well represented. Two families comprise most coniferous species and genera, the Araucariaceae and Podocarpaceae, although *Pinus* extends into western Malesian monsoon forests (Whitmore 1989).

The monsoons are the dominant climatic factor governing precipitation in the region, and hence determine the occurrence and distribution of evergreen and seasonal forests. Moist evergreen forests are centred in the Malay archipelago (Whitmore 1989). Whitmore (1975, 1984) identi-fied 13 types of rain forest (both evergreen and semi-evergreen) in the region, differing primarily in the type of sites they occupy (lowland, swamp, montane, etc.). Moist evergreen and semi-evergreen forests occur throughout Malaysia and Indonesia, although extensive logging and land-use changes have significantly reduced their extent. Dominant families include Anacardiaceae, Apocynaceae, Burseraceae, Bignoniaceae, Lecythidaceae, Leguminosae, Meliaceae and Rhamnaceae, among others (Sewandono 1956). Dipterocarpaceae are common in the western portion of Malesia on the landmasses of the Sunda Shelf (especially Sumatra, Malaya, Borneo, Palawan and Java), which projects out from the Asian mainland. Groups of dipterocarps may reach canopy heights of 60 m. Representatives of Fagaceae are common in montane forests.

Rain forests extend northward through the mainland and into China and a more subtropical climate, where common tree genera include *Quercus*, *Diospyros*, *Machilus*, *Castanopsis* and *Cinnamomum camphora* (see p. 40). The inland plains of Burma (Myanmar), Thailand, Cambodia and Laos lie in the rain shadow of the mountains of western Burma, and drier forests are more common. In the south, dipterocarps and legumes predominate at lower elevations, with dipterocarps and *Quercus* spp. at high elevations. Monsoon forests are scattered throughout, especially in south-eastern Indonesia and portions of New Guinea, but are greatly reduced in extent owing to human disturbance.

In western India, moist evergreen forests occupy the coastal region from about 21° S to near the southern tip of the peninsula. The Myrtaceae and Lauraceae are well represented, as are dipterocarps and species in the genus *Hopea*. Rain forests also occur in western Pakistan and eastern Burma. Inland of the west Indian coastal rain forest lie the monsoon forests, which are dominated by a deciduous overstorey and evergreen understorey. Important timber species in the region include teak (*Tectona grandis*) and rosewood (*Dalbergia latifolia*) (Chaturvedi 1956). Larger areas of Indian monsoon forest occur in the eastern central region and in the foothills of the Himalayas. Xerophytic forests dominate north-western India and the western peninsula (inland of the rain forests and monsoon forests). Various species of *Acacia* are common throughout. The remainder of the forests in and around India are predominantly dry deciduous forests. Teak is present in places, and in southern India sandalwood (*Santalum album*) is one of the most important timber species.

Australian tropical rain forests occur predominantly along the north-eastern (Queensland) coast and occupy less than 1% of the continent (Webb & Tracey 1994). Canopy trees include conifers in the Araucariaceae and Podocarpaceae and at least 33 dicotyledonous families. Among the better represented are Elaeocarpaceae, Lauraceae, Moraceae, Myrtaceae, Proteaceae, Sapotaceae and Sterculiaceae (Beadle 1981). Vines are characteristically abundant, while tree ferns and palm are occasionally present lower in the canopy. Monsoon, or semi-evergreen, forests are typically small in size (1–25 ha) and occur in the Arnhem peninsula, Melville Island and portions of Western Australia (Beadle 1981). Canopy species found in these forests vary geographically and by moisture regime, but include *Bombax ceiba, Buchaniana arborescens, Canarium australianum, Gmelia dalrympleana* and various species of *Ficus*. Common shrub and small tree

genera include *Canthium, Croton, Diospyros, Exocarpus, Litsea, Livistona, Myristica, Pongamia* and *Trema*. Many of the northern Australian dry forests contain genera characteristic of moist tropical forests and are believed to be derivatives of former rain forests that occupied the area during a period of more favourable climate.

Most of tropical Australia receives insufficient rain to support rain forest or monsoon forest and is instead dominated by *Eucalyptus* forest and woodland (Fig. 2.6). Along the eastern coast, intermixed with the moist tropical forests previously described, are tall *Eucalyptus* forests that grow on fertile soils. Canopy species are occasionally found in the moist tropical forests, and rain forest species are commonly found in the understorey of the tall *Eucalyptus* forests (Beadle 1981). Low-stature (sub)tropical *Eucalyptus* forests and woodlands occur on nutrient-poor soils of the north-eastern coast, the northern portion of the eastern inland lowlands, and the north central region. *Eucalyptus* forests of varying stature also occupy the subtropical south-western corner of Australia. Mallee, woodlands dominated by multistemmed shrubby *Eucalyptus*, occur primarily in the south-west and along the south-western coast. In drier interior regions,

Fig. 2.6 Australian *Eucalyptus* (*E. camaldulensis*) forest, a dry tropical community found along water courses. (Photo by Robert Teskey.)

Eucalyptus is replaced by woodlands typically dominated by various species of *Acacia* or *Casuarina*.

2.4 TEMPERATE BROADLEAF AND CONIFEROUS FORESTS

Of closed canopy forests, perhaps none has been as extensively utilized and altered by humans as the temperate forests. Significant portions of Europe and eastern Asia that once supported forests have long since been converted to pasture and agriculture, and little of the original vegetation remains. Similarly, much of the temperate forest in the USA has been cut, although a large percentage returned to forest after cutting or agricultural abandonment. None the less, forests still occupy moderate to significant portions of the temperate regions of North and South America, Europe, Asia and Australia, and are of significant economic and ecological importance.

2.4.1 Distribution and extent

There are about 1.1 million km² of temperate broadleaf deciduous forests in the world, with perhaps another 3–8 million km² of temperate coniferous and broadleaf evergreen forests and woodlands (Fig. 2.7). In North America, temperate broadleaf forests are confined primarily to the eastern USA and south-east Canada. West of the central prairie region are the interior and mountainous coniferous forests, and beyond them the coastal temperate coniferous rain forests of the Pacific North-west. Latitudinally, they occur between the subtropical zone (about 28–30°N) and the boreal zone (46–47°N), although the temperate coniferous rain forests of the Pacific coast of Canada and Alaska grow a little north of 56°N. Smaller areas of temperate forest occur in southern Mexico and into Belize and Guatemala, principally at high elevations and along river courses at 0–2500m (Röhrig 1991a). The South American temperate forests grow primarily along the far south-western coast and the southernmost tip of the continent. The deciduous forests east of the Andes grow south of 45°S, while the coastal rain forests grow between 41 and 56°S. Temperate broadleaf evergreen forests grow principally in Chile, with a minor extension into Argentina.

European temperate forests occur predominantly in western and central Europe, continuing in a narrow band through eastern Europe to the Ural Mountains in Russia. Temperate deciduous forests also occur in the Near East in a band around the southern Caspian Sea, broadening to the east into the Iranian highlands, Turkmenistan, Uzbekistan and Kazakhstan (Röhrig 1991b). Mountainous areas are forested primarily by conifers.

Asian temperate deciduous forests are found primarily in the central and south-eastern region, including mainland China, Taiwan, Korea and Japan. The region lies between about 26° and 50°N, with conifers increasing in importance with latitude and elevation. Temperate forests, primarily coniferous but often with abundant

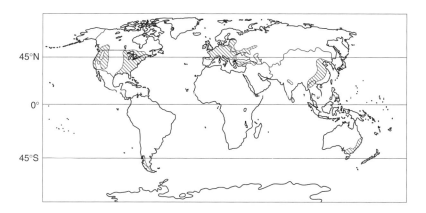

Fig. 2.7 Global distribution and extent of temperate broadleaved and coniferous forests.

deciduous canopy species, also occur in central Asia, principally at higher elevations in the Himalayan region. In Australia, moist temperate deciduous forests are found predominantly in northern and western Tasmania, while the mainland is occupied principally by the evergreen *Eucalyptus* forests of the south-east and south-west. Much of New Zealand is temperate broadleaf evergreen forest.

2.4.2 Climate

As expected for a forest type distributed so widely, prevailing climatic conditions vary substantially. Generally, though, temperate environments are characterized by warm, relatively humid summers and cool–cold winters. Where precipitation is insufficient to support closed canopy forests, temperate forests variously grade into grassland (steppe or prairie), savannah or open woodlands. To the north, where both temperature and precipitation decline, they merge with the boreal forests, and to the south with (sub)tropical vegetation.

At lower elevations, mean temperature during the coldest month typically ranges between –5 and 10°C, with mean temperatures between 10 and 18°C. In mountainous regions, the temperatures can be significantly lower, approaching those of the more northerly boreal zone. Mean annual temperatures typically range between 8 and 13°C but can be lower or higher, depending on geographical location and local elevation (Röhrig 1991c).

Precipitation is year-round, although there are significant geographical differences in seasonal distribution. Winters are wetter than summers in much of the temperate forest region, although precipitation is more evenly distributed in portions of eastern North America, western Europe, far southern South America and New Zealand. In the temperate coniferous rain forests of the north-western USA and south-eastern Canada, summers are dry and most precipitation falls during the late autumn and winter. Mean annual precipitation is less than 500 mm in some portions of South America and Russia. At the other extreme are the temperate coniferous rain forests of the north-west North American Pacific coast (cover photo), where

annual precipitation approaches 2500 mm. More typical values for the temperate forests of eastern North America, western Europe and South-East Asia are 550–1300 mm (Röhrig 1991c; Archibold 1995).

2.4.3 Soils

Soils are naturally variable, although the dominant soil orders found under broadleaf temperate forests are Alfisols, Inceptisols and Ultisols, with substantial areas of Entisols in floodplain environments (Archibold 1995). Spodosols and Ultisols are most common under temperate coniferous forest, but ash-derived Andosols are common in volcanic regions. Alfisols, Inceptisols and some Ultisols are relatively fertile, with high base saturation. Soil pH typically ranges from slightly acidic to slightly basic, although under conifers and some hardwoods, or in low-lying areas with impeded drainage and decomposition, soil pH may be 4 or less.

2.4.4 Regional formations

Although there are distinct regional compositional differences, there is remarkable taxonomic similarity among the Asian, European and North American formations. Numerous genera are common to all three, including *Acer*, *Betula*, *Fagus*, *Fraxinus*, *Populus*, *Quercus* and *Ulmus*. The flora of North America and eastern Asia are most similar to one another, and share several genera of forest trees not found in Europe, including *Catalpa*, *Diospyros*, *Liriodendron*, *Nyssa* and *Sassafras* (Röhrig 1991c). There are further such affinities in the herbaceous flora of the forest floor (Li 1972).

Although the modern floras evolved under similar climatic regimes, the taxonomic similarity is due primarily to their common geological past. Both the North American and Eurasian continents were connected until the mid-Tertiary, after the period (Cretaceous) during which most of the genera common to the temperate forests of the Northern Hemisphere differentiated. The South American and Australasian formations do not share a common geological past and differ taxonomically from those of the Northern Hemisphere. However, they do share some families (e.g. Fagaceae). Of the three Northern Hemisphere formations, the Asian is the richest, followed by the North American and the European distantly third. The comparatively depauperate European temperate flora is believed to have resulted largely from the east–west orientation of the region's mountain ranges, which inhibited migration and reduced available refugia during past glaciations.

Asian

On the Asian mainland, temperate deciduous forests occupy a coastal band along the China Sea from the Korean Peninsula and into China in the vicinity of Tianjin. From here, deciduous forests broaden throughout the Chinese lowlands to the west (to the Mongshan Mountains) and south to the Yangtze River. Intermixed along the Chinese coast and in the interior highlands are mixed broadleaf–coniferous forests. Further to the south, evergreen broadleaf trees become more abundant. Coniferous forests become more prominent in the northern parts of both China and Korea. Japan's forests consist primarily of temperate broadleaf deciduous communities in the central–north region, with conifers more abundant at higher elevations and latitudes. To the south, between about 26° 30′ and 35° 30′ N, lie the temperate broadleaf evergreen forests (Satoo 1983). Ching (1991) has described the composition of the region's forest in some detail and I only briefly summarize them here.

China. In China's north-eastern temperate forest region (eastern Heilongjiang and Liaoning provinces), conifers occur at the highest elevations and the northern latitudes, lending the forests a (sub)boreal character (Table 2.3). Below these forests (up to 1100 m elevation) are the *Pinus koraiensis*–mixed hardwood forests, some of which are rather species-rich (Table 2.3). Subtypes include the mixed hardwood–*Pinus koraiensis* forest, oak forests (*Quercus mongolica*) and *Betula* forests, the latter of which occur at higher latitudes and elevations (Ching 1991).

China's northern temperate forest is roughly triangular in shape, extending from 32° 30′ to 42° 30′ N and from 103° 30′ to 124° 10′ E (Ching 1991). The major types within the region include the northern oak forest and the southern oak forest. Common throughout the northern oak forest is *Quercus liaotungensis*, with increasing *Quercus mongolica* to the east. The Liaotung hills and plains comprise the central portion of the northern oak forest area, with forest composition varying by landform and geography (Table 2.3). In the mountainous regions of the Hebei and Laioning hills, broadleaf trees grow to elevations of about 1600 m, the forests being dominated by *Quercus* spp. along with *Betula* and *Populus*. The forest of the high loess plateaus of Shaanxi and Shanxi consist of *Pinus* and various hardwoods (Table 2.3). The *Picea meyeri* and *Picea wilsonii* characteristic of the original primary forest have been widely replaced on the plateau by *Betula* in disturbed areas (Ching 1991).

The southern oak forest of the northern region includes not only deciduous species but also broadleaf evergreens. It is distinguished from the northern oak forest by the scarcity of *Quercus liaotungensis* and the presence of the evergreen *Quercus variabilis*. The vegetation of eastern Shandong bears close similarity to many of Japan's forests and is floristically diverse (Table 2.3). Frost-tolerant *Tilia* spp. and *Corylus* spp. are found to the north, and *Machilus thunbergii* is present further south. The central and southern Shandong hills are forested with *Quercus*-dominated communities, as are the plains and hills of southern Shanxi and central Shaanxi (Table 2.3). *Betula* and *Salix* become progressively more important with elevation, as do the conifers *Pinus armandii*, *Pinus tabulaeformis* and *Abies fargesii* (Ching 1991).

The Yangtze River drainage in east central China is a transitional zone between the temperate forests to the north and the subtropical forests to the south. There is a high degree of endemism. The genus *Quercus* is well represented, and the tree flora is a mix of more northerly and southern species. Deciduous species and conifers are intermixed with the broadleaf evergreens, which include the genera *Actinodaphne*, *Castanopsis* and *Lindera*. Bamboo 'forests' grow here, including *Aruninaria* spp. and *Phyllostachus* spp. The upper river valley is particularly diverse, and includes 50 broadleaf and 12 coniferous genera.

Table 2.3 China's major temperate forest regions, with common overstorey tree genera and species.

Forest region	Geographic extent	Common tree species
North-eastern	Eastern Heilongjiang and Jilion provinces	High elevation (>1100 m): *Larix gemelinii, Picea jezoensis, Abies nephrolepsis*
		Low elevation (<1100 m): *Pinus koraiensis, Betula costata, Tilia amurensis, Acer, Alnus, Betula, Carpinus cordata, Fraxinus, Juglans mandshurica, Maackia amurensis, Tilia*
Northern	Triangular, extending between 32° 30′ to 42° 30′ N and 103° 30′ to 124° 10′ E	Northern oak forest: *Quercus liaotungensis, Q. mongolica, Acer mono, Betula duhurica, Phellodendron amurense, Q. acutissima, Q. aliena, Q. dentata* and *Q. variabilis* (uplands); *Juglans mandshurica, Alnus japonica, Populus simonii* (lowlands); *Picea meyeri, Picea wilsonii, Pinus tabulaeformis, Pinus bungeana, Platycladus orientalis,* various species of *Acer, Quercus, Tilia, Ulmus* (mountains and high loess plateaus)
		Southern oak forest: *Quercus* (including *Q. variabilis*), *Acer* spp., *Ailanthus altissima, Catalpei bungei, Machilus thunbergii, Pawlonia fortunei, Pistacia chinensis, Populus, Ulmus* and *Zelkova serrata* (eastern Shandong, southern Shandong hills); *Quercus variabilis, Q. acutissima, Q. aliena, Q. dentata, Q. liouana* (600–1600 m on plains and hills of southern Shanxi and central Shaanxi provinces); *Acer, Carpinus, Pinus armandii, Quercus aliena, Q. baronii, Tilia* (low elevations of southern Shanxi and central Shaanxi provinces); *Abies fargesii, Betula, Pinus armandii, Pinus tabuliformis, Salix* (higher elevations in region)
Yangtze Valley	Yangtze River drainage (east central China), bordered by subtropical forests to the south	*Quercus, Actinodaphne, Castanopsis, Lindera,* and over 50 broadleaf and 12 coniferous genera at lower elevations; *Abies, Picea* and *Tsuga* (high elevations)

Forest of *Abies, Picea* and *Tsuga* occupy the high elevations.

Korean Peninsula. There is disagreement over the classification and latitudinal zonation of Korea's forests (Hyun 1956; Yim 1977a,b; Ching 1991) but the following types are relatively distinctive: warm temperate broadleaf evergreen, southern deciduous, northern deciduous and cool temperate/boreal coniferous.

The warm temperate broadleaf evergreen forests occupy the southernmost coast of the Korean Peninsula, and include *Quercus* spp., *Castanopsis* spp., *Cinnamomum camphora* and *Machilus thunbergii* (Hyun 1956). Temperate deciduous forests lie north of approximately 35° N and account for about 85% of Korea's

forested land base (Hyun 1956). Species common to the southern deciduous forests are *Acer formosum, Carpinus laxiflora, Carpinus tschonoskii, Quercus mongolica* and *Quercus serrata.* The northern deciduous species include *Celtis koraiensis, Choseia bractesa, Hemiptera dividii, Larix olgensis, Magnolia parviflora* and *Populus maximowiczi.* Other species found variously in the temperate region include *Acer mono, Betula* spp., *Carpinus* spp., *Fraxinus* spp. and numerous *Quercus* spp. In the plateaus and mountains of northern Korea, coniferous species dominate the overstorey, including *Abies nephrolepis, Abies holophylla, Larix dahurica* and *Picea jezoensis* (Hyun 1956; Archibold 1995). *Pinus pumila,* a dwarf species, grows at the highest elevations (above 1200–1600 m). Broadleaf understorey species include *Acer* spp., *Betula* spp., *Tilia* spp. and *Ulmus* spp.

Japan. The forests of Japan include warm temperate broadleaf evergreen, temperate deciduous and cool/cold coniferous. The warm temperate broadleaf evergreen forests occur on the islands of Honshu, Shikoku and Kyushu, and extend from the lowlands to 600–850 m (Satoo 1983). Species characteristic of the region include *Castanopsis cuspidata*, *Cycobalanus* spp., *Cinnamomum* spp., *Machilus* spp. and *Quercus* spp. In more mature forests, *Quercus gilva* grows on deeper soils, *Quercus slicina* on steep slopes and *Quercus glauca* over limestone (Archibold 1995). Following disturbance, *Quercus acutissima*, *Quercus serrata* and *Quercus variabilis* establish in mid-succession. The conifers *Abies firma*, *Podocarpus macrophyllus*, *Podocarpus nagi*, *Torreya nucifera*, *Pinus thunbergii* and *Tsuga sieboldii* variously occur, depending on latitude and proximity to the coast.

Temperate deciduous forests on the four main Japanese islands (Hokkaido, Honshu, Shikoku and Kyushu) are diverse in species and habitat, too much so to adequately describe here. In general, though, *Fagus crenata* dominates many forests. Its associates vary, depending on topography, disturbance history, soils, etc. Sites facing the Sea of Japan are commonly cold, snowy and windy, and species like *Acer tschonoskii*, *Acer matsumarae*, *Alnus maximowiczii*, *Quercus mongolica* and *Sorbus commixta* are common (Ching 1991). Various species of *Alnus*, *Populus* and *Salix* grow in floodplain environments, and the conifers *Abies sachalinensis*, *Abies firma*, *Abies mariesii*, *Picea jezoensis*, *Pinus thunbergii* and *Tsuga sieboldii* variously grow with species of *Betula*, *Carpinus*, *Fraxinus*, *Magnolia*, *Quercus*, *Tilia* and *Ulmus* in montane regions. *Quercus cripsula* is also a common dominant, particularly on Hokkaido, and grows well on somewhat drier sites than *Fagus crenata*. Its associates include *Abies sachalinensis*, *Acer mono*, *Betula ermanii*, *Kalopanax pictus*, *Ostrya japonica*, *Picea jezoensis*, *Tilia japonica* and *Ulmus davidiana* var. *japonica* (Ching 1991). *Pinus densiflora* sometimes forms pure stands, with a broadleaf understorey. At higher latitudes and altitudes, the cool/cold coniferous forests are dominated by *Picea* spp. and *Abies* spp. Broadleaf associates include *Betula* spp., *Fraxinus mandshurica*, *Kalopanax septemlobus* and *Fagus crenata*.

Australasian

Australia and Tasmania. The majority of Australia's forests are considered (sub)tropical, some of which (the south-western and eastern *Eucalyptus* forests) have been described above (see p. 36). Webb and Tracey (1994) consider Australian forests south of about 38° S to be temperate, and I adopt that latitudinal boundary here. However, others consider the temperate forest zone to extend as far north as about 25° S (Bailey 1989), and in reality the transition from (sub)tropical to temperate is characteristically indistinct.

Regardless of the precise line of demarcation, the temperate forests of the Australian mainland are dominated primarily by *Eucalyptus*, and account for approximately 29 million ha of a total 42 million ha of closed forest, both temperate and tropical (Ovington & Pryor 1983). Groves (1994) has described the dominant vegetation in some wet sclerophyll Australian temperate *Eucalyptus* forests distributed among four main geographical regions (Table 2.4). Similar tabulations, including both wet and dry sclerophyllous *Eucalyptus* forests divided into major alliances, have been made by Ovington and Pryor (1983). Despite the diversity of *Eucalyptus* (>450 species), temperate forests are often dominated by only a few species, for example *E. obliqua–E. viminalis* forests in parts of Victoria and *E. marginata–E. calophylla* forests in Western Australia (Ovington & Pryor 1983).

Temperate rain forest occupies small areas of south-eastern Australia, principally in the territories of Victoria, the north-eastern corner of New South Wales and far south-eastern Queensland. The canopies of Victorian forests are dominated by *Nothofagus cunninghamii*, while *Athersperma moschatum*, *Acacia melanoxylon*, *Eucriphia lucida* and *Phyllocladus asplenfolius* are occasional or common associates (Ovington & Pryor 1983). *Nothofagus cunninghamii* does not grow in easternmost Victoria, and *Athersperma*

Table 2.4 Some *Eucalyptus* species common to Australian and Tasmanian temperate forests in different geographic regions. (From Groves 1994.)

Geographic region	Climate	Dominant species
Central and southern New South Wales	Warm temperate	*E. saligna, E. paniculata, E. pilularis, E. smithii*
Central and southern Victoria	Cool temperate	*E. regnans, E. viminalis, E. obliqua, E. cypellocarpa*
Southern Tasmania	Cool temperate	*E. regnans, E. obliqua, E. vaota, E. delegatensis*
South-western Western Australia	Warm temperate	*E. diversicolor, E. calophylla, E. jacksonii, E. guilfoylei*

moschatum, Eleaocarpus holopetalus and *Telopea oreades* are most common (Busby & Brown 1994). Eucalypts dominate the drier uplands in the region (Howard & Ashton 1973; Ashton & Attiwill 1994). In the forests of New South Wales, *Eucryphia moorei, Nothofagus moorei* and *Doryphora sassafras* are common, with the former two species occurring at somewhat higher elevations.

Western Tasmanian temperate rain forests are dominated by *Nothofagus cunninghamii*, which may contain any of about 70 associates, only a few of which typically occur together in a stand (Ovington & Pryor 1983). *Athrotaxis* spp. (especially *A. cupressoides*) occur at altitudes greater than 900 m. In drier environments, like eastern Tasmania, *Eucalyptus* spp. are dominant; common ones include *E. delegatensis, E. dalrympleana, E. gunnii, E. archeri, E. coccifera, E. urnigera* and *E. pauciflora*.

New Zealand. The climate of New Zealand ranges from subtropical to cool/cold temperate, but here I consider it mainly temperate. Generally, the native taxa are similar to those of the South American and Tasmanian forests. Most are evergreen in habit, including both conifers and broadleaf species. There are two broad categories of New Zealand forest: conifer–broadleaf, and beech (Wardle *et al.* 1983) (Fig. 2.8).

The conifer–broadleaf forests are predominantly lowland and are dominated by podocarps, including the genera *Dacrycarpus, Dacrydium, Phyllocladus* and, of course, *Podocarpus* (Wardle *et al.* 1983). Also present are two species of *Libo-*

cedrus and the kauri (*Agathis australis*). The warmer inland forests support primarily *Podocarpus spicatus* and *Podocarpus totara* as canopy dominants, along with a more minor component of *Podocarpus dacrydioides* and *Dacridium cupressinum* (Entrican & Holloway 1956). Hardwood associates of lower stature include the genera *Beilschmiedia, Knightia, Laurelia, Litsea* and *Nestegis* (Wardle *et al.* 1983). Forests range from relatively open to closed canopy, depending on species and habitat. For example, *Dacridium cupressinum, Podocarpus hallii* and *Podocarpus totara* are most common on drier sites and may form near-monospecific stands, while *Podocarpus dacrydioides* often dominates on wetter, swampy sites (Cockayne 1958). In the coastal hills and plains, *Dacridium cupressinum* is commonly dominant, although *Podocarpus hallii* and *Podocarpus totara* are likely to occupy more xeric sites.

Genera common and often dominant in coastal lowland forests include *Elaeocarpus, Metrosideros* and *Weinmannia* (Cockayne 1958; Wardle *et al.* 1983). The conifer–broadleaf shrub community in these forests is often well developed, with abundant species. The overstorey tree *Weinmannia racemosa* is abundant and widely distributed on the wetter, western side of both islands. Another important tree species is the giant kauri (*Agathis australis*), much reduced in extent and growing primarily north of 38° S, particularly on ridges (Wardle *et al.* 1983). It grows with podocarps and various hardwoods (Cockayne 1958), and occurs both singly and in groups or thickets on poorer soils.

Fig. 2.8 *Nothofagus* forest in New Zealand (South Island). (Photo by Robert Teskey.)

The beech, or *Nothofagus*, forests occupy cold, wet, montane environments and are characteristically lower in stature and less species-rich than the conifer–broadleaf forests (Fig. 2.8). Four species of *Nothofagus* grow in New Zealand: *N. solandri*, *N. menziesii*, *N. fusca* and *N. truncata*. *Nothfagus solandri* is the most widespread of the four species and there are two recognized varieties: var. *solandri* and var. *cliffortoides*. The former occupies warm, dry lowland environments primarily from East Cape on North Island to the centre of South Island (Cockayne 1958; Wardle *et al.* 1983), while var. *cliffortoides* grows at higher altitudes or on poor soils under high rainfall to the south. *Nothfagus solandri* forms nearly pure stands, with a sparse understorey composed of shrubs (especially *Coprosma* spp.), lichens and mosses (Wardle *et al.* 1983), although both varieties co-occur where their geographical ranges overlap. *Nothofagus fusca* is distributed from East Cape to the southern end of South Island, and grows best on deeper, well-drained soils. *Nothofagus menziesii* has similar site requirements and dominates the subalpine zone of western South Island and the eastern mountain ranges of North Island

(Wardle *et al.* 1983). *Nothofagus truncata* grows on drier, less fertile sites, primarily north of 42°S.

European

Temperate forests once covered much of Europe and the British Isles, but most of the original forest has long since been converted to agriculture. Moreover, many introduced species have been extensively planted, for example plantations of the western North American conifers *Picea sitchensis* and *Pseudotsuga menziesii* in the British Isles. None the less, temperate forests remain an important resource and there are numerous recognized community types. Jahn (1991) has extensively described the European deciduous forests, and I make considerable use of his excellent work in my more abbreviated summary here.

Two regions of temperate forest can be recognized: the Middle European, which occupies a band between about 43° and 55–60°N, and the Submediterranean, which generally follows the southern border of the Middle European south to a latitude of about 40°N.

Middle European. In the Middle European region are four somewhat distinctive provinces: the Atlantic, Subatlantic, Central European and Sarmatic. Throughout the region are genera common to most of the temperate forest of the Northern Hemisphere, including *Acer, Betula, Carpinus, Fraxinus, Pinus, Populus, Prunus, Quercus, Sorbus, Tilia* and *Ulmus* (Table 2.5). The genus *Fagus* is represented principally by *Fagus sylvatica*, which is distributed throughout most of the Middle European region and is, alongside *Quercus*, the most conspicuous and perhaps most ecologically important tree in Europe.

The Atlantic province includes the British Isles and the coastal regions of western Europe, and a number of characteristic native species are, or

were, widely distributed (see Table 2.5). In the British Isles, *Quercus* spp. still dominate many sites, although *Acer* spp., *Carpinus betulus, Fraxinus excelsior, Tilia* spp., and other species are abundant. *Fraxinus excelsior* is often the dominant species on high pH soils, with *Quercus robur* and *Quercus petraea* on sites of lower pH and fertility (Jahn 1991). *Fagus sylvatica* and *Fagus sylvatica–Quercus petraea* woodlands occur in south-eastern England. *Fagus sylvatica* is also common in north-western France, portions of Belgium and throughout the Middle European and portions of the Submediterranean regions. In the Atlantic province, *Fagus sylvatica* is typically absent only on sites that are wet, steep, calcareous or very sandy (Jahn 1991).

The Subatlantic province is orientated along

Table 2.5 Common trees of the four Middle European provinces of the European temperate forest.

Province	Geographic extent	Dominant species
Atlantic	British Isles, Ireland, coastal western Europe	*Fagus sylvatica, Quercus robur, Alnus glutinosa, Castanea sativa, Pinus sylvestris, Q. petraea, Tilia platyphyllos, Ulmus glabra*
Subatlantic	North-east to south-west band from southern Fennoscandinavia to southern France and northern Spain	Northern: *Quercus robur* (with *Carpinus betulus* on fertile wet sites and *Betula pendula, B. pubescens, Sorbus aucuparia* on less fertile); *Fagus sylvatica* (with *Abies alba* on better soils, *Q. petraea* on warmer slopes, and *Acer campestre, A. platanoides, A. pseudoplatanus, Fraxinus excelsior, Carpinus betulus, Prunus avium* on moist, fertile sites); *Acer pseudoplatanoides, Fraxinus excelsior* (rocky slopes); *Picea abies, Pinus mugo, Pinus sylvestris* (cold sites)
		Southern: *Quercus pubescens, Q. petraea, Q. robur* (low elevation); *Fagus sylvatica, A. alba* (high elevations)
Central European	Eastern Germany, Poland and Czech Republic	*Fagus sylvatica, Acer* spp., *Carpinus betulus, Fraxinus excelsior, Prunus avium, Quercus* spp., *Tilia cordata* (lower elevations); *Fagus sylvatica, Abies alba, Larix decidua* (montane); *Betula* spp., *Quercus* spp. (infertile, low pH soils); *Betula pubescens, Q. robur* (wet soils); *B. pendula, Q. petraea* (dry soils)
Sarmatic	Band east of Central European nearly 60° E, tapering from about 50–60° N in the west to about 53°–54° N at the eastern edge	Northern: *Pinus sylvestris, Picea abies, Abies alba* (western), *Abies sibirica, Carpinus, Populus tremula, Quercus robur, Tilia* spp.
		Southern: *Quercus robur, Tilia cordata, Carpinus betulus, Acer platanoides, Fraxinus excelsior, Ulmus glabra, U. laevis, U. minor*

a north-east to south-west axis, covering an east–west region between about 8 and 15°W in southern Fennoscandinavia and about 3°W and 3°E in southern France and northern Spain. *Quercus robur* is common in the north and co-occurs with a number of associates depending on soil conditions (see Table 2.5). *Fagus sylvatica* is prominent on upland sites in the north and in the hills and mountains to the south, and forms a complex regional mosaic of communities, determined by a combination of climatic, soil and physiographic factors (Jahn 1991) (see Table 2.5). In the southern portion of the Subatlantic province, *Quercus* spp. become increasingly prominent, particularly at lower elevations. In the eastern Pyrenees, *Quercus pubescens* is found at the lowest elevations, *Fagus sylvatica* and *Abies alba* at higher elevations, with *Abies alba* increasing in importance with altitude (Jahn 1991).

The Central European province covers the former East Germany, western Poland and western Czech Republic (Jahn 1991) and is characterized by a more continental climate than the Atlantic or Subatlantic provinces. *Fagus sylvatica* reaches its easternmost extent here, and occurs extensively north of the Baltic Sea. It has many associates, depending on elevation (see Table 2.5), and often has a more species-rich ground layer than comparable forests in the Subatlantic province. *Carpinus betulus–Quercus petraea* communities formerly occupied rich, mesic plains and uplands, but most have been converted to agriculture. *Betula–Quercus* forests grow on nutrient-poor, low pH soils, with *Betula pubescens* and *Quercus robur* common on wet sites and *Betula pendula* and *Quercus petraea* on drier sites. *Pinus sylvestris* and *Picea abies* are locally intermixed in the north and east, with *Larix decidua* common to the hills of southern Poland (Jahn 1991).

The Sarmatic province extends in a band eastward to nearly 60°E, its north–south boundaries tapering from about 50–60°N in the west to about 53–54°N at its eastern edge. To the south lies steppe and to the north boreal forests (see p. 26). Generally, the Sarmatic province is characterized by a progressive west-to-east loss of the deciduous species characteristic of the three provinces to the west and an increase in the importance of conifers. For example, there is a tendency for nutrient-poor sites to shift from domination by *Betula* and *Quercus robur* in the Subatlantic province to domination by *Pinus* in the Sarmatic. Similarly, on more fertile sites, *Fagus sylvatica* is no longer present, and has been replaced by *Tilia*, *Quercus robur*, *Carpinus* and *Pinus*. *Pinus sylvestris* is distributed throughout the Sarmatic province and is an associate or dominant on all but the most fertile, moist soils. On sandy soils it forms nearly pure stands, without deciduous associates. *Picea abies* is also common, occurring with *Pinus sylvestris*, and increasing in importance with soil fertility and moisture. *Abies alba* occurs in the eastern part of the province, with *Abies sibirica* in the far western portion. *Populus tremula* and *Quercus robur* are present along the southern boundary, although the latter is uncommon in the north. In contrast to the northern portions of the Sarmatic province, the south is largely deciduous. *Quercus robur* and *Tilia cordata* are dominant, but *Carpinus betulus* is an important (though often low stature) associate (see Table 2.5).

Intermixed or adjacent to these four provinces of the Middle European region are the forests of the Alps and the Carpathian Mountains (Table 2.6). The Western Alps (orientated north–south and running from the Mediterranean to Grenoble, France) are occupied by *Quercus pubescens* forests in the west-facing foothills, with *Fagus sylvatica*, *Quercus petraea* and *Pinus sylvestris* occurring at higher elevations (see Table 2.6). High-elevation interior forests contain a mix of deciduous and coniferous species. East-facing forests are predominantly deciduous hardwoods, with a large *Quercus* component.

In the Middle Alps, the north-facing foothills are forested with *Quercus petraea* and *Carpinus betulus*, which are replaced by *Fagus sylvatica* and *Acer pseudoplatanus* at higher elevations, *Abies alba* and *Picea abies* in the interior, *Fagus sylvatica* on upper south-facing slopes and *Castanea sativa* and *Quercus petraea* on the south-facing foothills (see Table 2.6). The Eastern Alps (Verona, Italy to the Dinara mountains of the former Yugoslavia) are occupied by a variety of hardwoods, with an increasing abundance of

Table 2.6 Tree species and genera common to the European Alps and Carpathian Mountains. (After Jahn 1991.)

Geographic region	Principal species or genera
Western Alps (and running north–south from the Mediterranean to Genoble, France)	*Quercus pubescens, Acer monspessulanum, A. opalus, Buxus sempervirens, Cotinus coggyrria, Sorbus aria* (west-facing foothills); *Fagus sylvatica, Q. petraea* and *Pinus sylvestris* (mid-elevations); *Abies alba, Pinus sylvestris, Picea abies, Fagus sylvatica* (upper elevations); *Larix decidua* (subalpine); *Acer pseudoplatanus, Fagus sylvatica* and *Abies alba, Castanea sativa, Quercus cerris, Q. ilex, Q. pubescens, Q. robur* (east-facing slopes)
Middle Alps	*Quercus petraea, Carpinus betulus* (north-facing foothills); *Fagus sylvatica, Acer pseudoplatanus* (higher elevations); *Abies alba* and *Picea abies* (interior); *Fagus sylvatica* (upper south-facing slopes); *Castanea sativa, Q. petraea* (south-facing foothills)
Eastern Alps (Verona, Italy to the Dinara mountains of the former Yugoslavia)	*Acer pseudoplatanus, Abies alba, Fagus sylvatica, Picea abies, Pinus sylvestris* (northern foothills); *Picea abies* (high elevations); *Larix decidua* (subalpine); above with *Fraxinus ornus, Ostrya carpinifolia, Quercus* (low foothills and foothill–plains transition)
Carpathian Mountains (an arc from Slovakia east through south-west Ukraine and into central Romania)	Western: *Carpinus betulus, Fraxinus excelsior, Prunus avium, Q. petraea, Tilia cordata* (minor); *Fagus sylvatica* (foothills and up to about 600 m)
	Eastern: *Acer tataricum, Q. robur* woodlands alternating with steppe (low elevations); *Carpinus betulus, Fagus sylvatica, Q. petraea, Tilia* (montane); *Picea abies* and *Abies alba* (high elevation); *Pinus mugo, Pinus cembra, Larix decidua* (subalpine at elevations exceeding about 1800 m)

conifers at higher elevations (see Table 2.6). Similar species occur on the south-facing slopes, with the addition of *Fraxinus ornus, Ostrya carpinifolia* and *Quercus* spp. in the low foothills and foothill–plains transition.

The Carpathian Mountains extend in an arc from Slovakia east through south-west Ukraine and into central Romania. The forests of the foothills and submontane zone (up to about 600 m) of the western Carpathians are predominantly mixed hardwood forests (see Table 2.6), with a minor component of *Fagus sylvatica*. On the eastern side, forest and steppe alternate at the lowest elevations, and are progressively replaced by mixed hardwoods, and then coniferous forests, as elevation increases (Jahn 1991).

Submediterranean. The Submediterranean region is characterized by a preponderance of deciduous oaks (*Quercus* spp.) and represents the southern distribution of many of the species common to the mixed forests of the Middle European region to the north. To the south, where summer droughts are more extended, grow the evergreen Mediterranean forests. The vegetation is quite variable and complex throughout and is sensitive to climatic and site factors in this region, which borders temperate and Mediterranean climatic zones. Deciduous species of the Submediterranean region include *Carpinus orientalis, Fagus sylvatica, Fagus moesica, Fagus orientalis, Fraxinus ornus, Ostrya carpinifolia, Quercus cerris, Quercus faginea, Quercus frainetto, Quercus pubescens* and *Quercus pyrenaica*. Various species of *Acer* and *Sorbus* are also present. *Fagus* spp. are characteristic of montane regions and co-occur with *Abies* spp. and *Juniperus* spp. (Jahn 1991). *Pinus* spp. are present, with *Pinus nigra* growing with *Quercus cerris, Quercus faginea, Quercus pubescens* and *Quercus pyrenaica* in submontane environments, *Pinus sylvestris* common in montane zones, and *Pinus peuce, Pinus heldreichii* and *Pinus mugo* growing at high elevations.

Near Eastern

The Near Eastern temperate forest lies between 35 and 45° N, from about 27° E in eastern Turkey to about 58° E in eastern Iran. The forests of this region are a mixture of both European and Colchian species (Röhrig 1991b), with elements of (sub)Mediterranean and steppe flora.

In western Turkey, oak forests are common, composed of various species of *Quercus* and other hardwoods (Table 2.7). *Fagus orientalis* grows at higher elevations, though not in abundance. In the western portion of the Anatolian plateau in Turkey, *Quercus*-dominated temperate broadleaf forests grow between 1000 and 2000 m (Röhrig 1991b) and contain a number of other genera as well (see Table 2.7). Extending into Iran further to the east are other forests dominated by *Quercus* (Röhrig 1991b). Various species of *Acer*, *Pyrus* and *Prunus* also occur, as well as a few species, like *Fraxinus excelsior*, more commonly found in the Middle European and Submediterranean regions. On alluvial soils at lower elevations (<400 m) grow more diverse forests, which in addition to *Quercus* spp. include *Alnus barbata* and *Carpinus* spp.

Moist forests between the Caspian Sea and Black Sea contain a number of species, including many missing from the European forests to the west (see Table 2.7) (Walter 1985). Further east, the species and assemblages change but a few genera predominate, with a number of low-stature woody species occupying the understorey (e.g. *Buxus*, *Crataegus*, *Corylus*, *Ilex*, *Ligustrum*, *Rhododendron* and *Sambucus*) (Röhrig 1991b).

At high elevations (600–1100 m) are the beech (*Fagus orientalis*) forests. Near pure *Fagus orientalis* forests grow between the Sakarya River in western Turkey and the Turkey–Georgia border. Between 600 and 1300 m on calcareous northern slopes on the mountains of the southern Crimean Peninsula, *Fagus orientalis* grows alone or with *Carpinus betulus*, *Fraxinus excelsior*, *Tilia cordata* and *Ulmus glabra* (Röhrig 1991b). On southern slopes, it is found with *Acer hycranum*, *Pinus nigra*, *Carpinus orientalis* and *Quercus petraea*. In the Caucasus Mountains, it grows between 1000 and 1500 m with *Acer* spp., *Fraxinus excelsior*, *Tilia platyphyllos* and *Ulmus*

Table 2.7 Forest regions and principal species and genera of the Near Eastern temperate forest.

Geographic region	Principal species and genera
Western Turkey	*Quercus cerris*, *Q. frainetto*, *Q. hungarica*, *Carpinus orientalis*, *Castanea sativa*, *Fagus orientalis* (higher elevations, not abundant), *Populus tremula* (dry forests)
Western Anatolian Plateau (Turkey)	*Quercus castaneifolia*, *Q. petraea*, *Q. pubescens*, *Q. cerris*, *Q. hartwissiana*, *Q. haas*, *Q. delachampii*, *Acer platanoides*, *Sorbus torminalis*, *Populus tremula*, *Betula pendula*
Western Iran	*Quercus aegilops* var. *brantii*, *Q. infectoria*, *Q. libani*, *Q. iberica*, *Q. persica*, *Acer*, *Pyrus*, *Prunus*
Caspian and Black seas	*Acer laetum*, *Carpinus caucasica*, *Diospyros lotus*, *Fagus orientalis*, *Platanus orientalis*, *Pterocarya fraxinifolia*, *Quercus harwissiana*, *Q. iberica*, *Q. imertiana*, *Tilia multiflora*, *T. caucasica*, *Ulmus elliptica*, *U. foliacea* and *Zelkova carpinifolia* (moist forests)
Eastern edge of temperate zone (including east of Caspian)	*Acer*, *Albizia*, *Carpinus*, *Fraxinus*, *Gleditsia*, *Parrotia*, *Populus*, *Prunus*, *Quercus*, *Ulmus*, *Zelkova*

glabra. At higher elevations, *Fagus orientalis* grows with coniferous species like *Abies bornmülleriana*, *Abies equitrojani*, *Abies nordmanniana*, *Picea orientalis* and *Pinus sylvestris*, along with some of the broadleaf species found at the lower elevations.

North and Central American

Temperate forests occupy much of North America, from north of the border between the USA and Canada south into Mexico and some Central American countries at high altitudes. The conifer-dominated forests of western North America are separated from the more deciduous forests of the east by vast areas of former prairie now dominated by agriculture where rainfall or irrigation permit.

USA and Canada. The eastern North American temperate forest is primarily deciduous, although the forests of the south-eastern USA contain large amounts of conifers, both natural and planted, as do the states bordering the Great Lakes. The forests of the region have been extensively described, especially by Braun (1950) and more recently Barnes (1991). I follow Barnes's (1991) classification scheme, which is derived from Braun's (1950) and Küchler's (1964). The forest regions described below are roughly confined to specific geographical areas, to which I refer in each section in an attempt to maintain some consistency with the other sections in this chapter.

The South-eastern Evergreen region covers the coastal plain of the south-eastern USA, stretching roughly from Virginia along the Atlantic coast, continuing south across Florida and the Gulf of Mexico to Texas. *Pinus palustris–Aristida stricta* savannahs (Fig. 2.9) formerly dominated much of the uplands but have been largely converted to other uses or cover types, and both *Pinus elliottii* and *Pinus taeda* are presently the most abundant conifers in the region. *Magnolia grandiflora* and *Fagus grandifolia* are believed to be climax dominants on mesic sites, although a number of other genera are abundantly represented (Table 2.8).

Inland of the South-eastern Evergreen region is the Oak–Hickory–Pine region (Fig. 2.10), extending from Pennsylvania to Alabama and bordered by the coastal plain towards the Atlantic Ocean and the Appalachian mountains to the west. As its name implies, the principal species include those in the genera *Quercus* and *Carya*, often in mixture with species common to adjacent regions, including *Pinus*.

The Appalachian Oak region includes the Appalachian mountains, from Georgia north and eastward into eastern New York and Massachusetts, and is the most floristically diverse in North America (see Table 2.8). Oaks are abundant at all elevations and aspects. At higher elevations (>1100 m), species are more characteristic of the beech–sugar maple and sugar maple–basswood forests to the north (see below). At the highest elevations are boreal forests composed of *Picea rubens* and the endemic *Abies fraserii* in the south, and *Picea rubens* and *Abies balsamea* to

Fig. 2.9 *Pinus palustris–Aristida stricta* savannah in Georgia, USA. (Photo by the author.)

Table 2.8 Forest regions, their geographical extent and principal tree species in eastern North America.

Forest region	Geographic extent	Principal species or genera
South-eastern Evergreen	Atlantic coastal plain (Virginia–Texas)	*Pinus palustris* (former); *Pinus elliotii, Pinus taeda* (especially plantations); *Magnolia grandiflora, Fagus grandifolia* (mesic uplands); *Acer, Carya, Fraxinus, Liquidambar styraciflua, Magnolia, Quercus, Ulmus* (various); *Taxodium, Nyssa* (wetlands)
Oak–Hickory–Pine	From Pennsylvania to Alabama, between South-eastern Evergreen to east and Appalachian to west	*Quercus* (especially *Q. alba, Q. coccinea, Q. falcata, Q. marilandica, Q. rubra, Q. stellata, Q. velutina*), *Carya* (especially *C. glabra, C. ovata, C. tomentosa*), *Pinus echinata, Pinus taeda;* also *Acer, F. grandifolia, Fraxinus, Nyssa sylvatica, Liquidambar styraciflua, Liriodendron tulipifera*
Appalachian Oak	Appalachian Mountains, from New England to Georgia	*Quercus alba, Q. coccinea, Q. prinus, Q. rubra, Q. velutina,* with *Acer, Aesculus octandra, Betula, Carya, F. grandifolia, Liriodendron tulipifera, Magnolia, Pinus pungens, Pinus rigida, Pinus virginiana, Tsuga canadensis*
Mixed Mesophytic	West of Appalachian Oak, from western Pennsylvania to northern Alabama	*Acer, F. grandifolia, Liriodendron tulipifera, Prunus serotina* (especially north), *Quercus, Tilia, Tsuga canadensis*
Western Mixed Mesophytic	S. Illinois, Indiana and Ohio, south to N. Mississippi and Alabama	*Quercus, Carya, Pinus echinata, Pinus taeda, Pinus virginiana*
Oak–Hickory	S. E. Wisconsin, Illinois southwest through Missouri, W. Arkansas and E. Oklahoma	Unglaciated south: *Quercus alba, Q. marilandica, Q. stellata, Q. velutina, Carya illinoensis, C. ovata, C. texana, C. tomentosa, Juniperus virginiana*
		Glaciated north: *Acer saccharum, F. grandifolia, Q. alba, Q. macrocarpa, Q. rubra, Q. velutina*
Maple–Beech	Ohio, Indiana, S. Michigan and S. W. Ontario	*Acer saccharum, F. grandifolia, Acer, Carya, Fraxinus, Quercus, Tilia americana, Ulmus*
Sugar Maple–Basswood	S. W. Wisconsin and E. Minnesota	*Acer saccharum, Tilia americana, Acer, Fraxinus, Quercus, Prunus serotina, Ulmus*
Northern Hardwood–Conifer	N. Minnesota east across N. Wisconsin, N. Michigan and E. central Ontario	*Acer, Betula alleghaniensis, B. papyrifera, F. grandifolia, Fraxinus, Pinus banksiana, Pinus resinosa, Pinus strobus, Populus balsamifera, Populus grandidentata, Populus tremuloides, Quercus, Tilia americana, Thuja occidentalis, Tsuga canadensis*

the north. *Pinus* spp. dominate dry exposed sites throughout, often with a dense ericaceous under-storey. *Castanea dentata* was formerly domi-nant in the region, but the fungus *Endothia parasitica* had killed virtually every tree by the 1940s. It persists mostly as small root-sprouted saplings.

The Mixed Mesophytic region lies north of the Oak–Hickory–Pine region and east of the Appalachian Oak region. It is orientated north-

Fig. 2.10 Mature forest of *Quercus* and *Carya* spp. in the Oak–Hickory region of eastern North America (Georgia, USA). (Photo by the author.)

east to south-west from western Pennsylvania to northern Alabama, and coincides physiographically with much of the unglaciated Appalachian plateaus. The vegetation comprises principally species from the regions which it borders (see Table 2.8). The Western Mixed Mesophytic region lies immediately to its west, covering central and western Tennessee and Kentucky, the southern boundary of the Wisconsinan Glaciation in southern Indiana and Ohio, and northern Mississippi and Alabama. Forests dominated by *Quercus* spp. cover most of the region, with a large component of *Carya* and *Pinus*.

The westernmost region is the Oak–Hickory, bordered to the west formerly by prairies and presently by agricultural lands. It extends from Manitoba, Canada to southern Texas. South of the glaciated portion lie the Ouachita Mountains and Ozark Plateau. *Quercus alba* is the dominant species on dry-mesic uplands, with various species of *Carya*, *Quercus* and *Pinus* characteris-

tic of drier sites. In the glaciated region to the north, *Quercus* spp. are common on drier uplands, while *Fagus grandifolia* and *Acer saccharum* grow on more mesic sites. *Quercus macrocarpa* formerly formed savannahs interspersed with prairie, most of which has been lost to agriculture and urbanization.

The Sugar Maple–Beech region occurs on glaciated till in the southern half of Michigan's lower peninsula, Indiana and Ohio; *Acer saccharum* and *Fagus grandifolia* are the dominant species in mature forests, although a number of other species are common (see Table 2.8). The Sugar Maple–Basswood region to the north of the Sugar Maple–Beech region is situated on glaciated terrain and, as the name implies, *Acer saccharum* and *Tilia americana* are dominant. Other associates are those characteristic of the Sugar Maple–Beech region (see Table 2.8).

The Northern Hardwood–Conifer region extends across the northern USA and southern Canada from Manitoba to Maine and Nova Scotia. The terrain is glaciated and supports a wide array of forest types. Species common to the Sugar Maple–Beech region predominate on glacial till, although the region extends beyond the northern range of *Carya*, *Liriodendron*, *Nyssa* and other genera common to the more southern forest regions. *Quercus ellipsoidalis* grows on dry glacial outwash plains, but fire-prone *Pinus banksiana* is the most dominant species on these soils. Much of the presettlement forest was dominated by *Pinus strobus* but virtually all of the original forest was cut and converted naturally to hardwoods, including *Betula papyrifera*, *Populus tremuloides*, *Populus grandidentata* and *Quercus* spp. Coniferous swamps on calcareous soils often consist predominantly of *Thuja occidentalis* with lesser amounts of other hardwoods.

The temperate forests of western North America differ from those in the east by a preponderance of conifers rather than hardwoods. Furthermore, much of forested western North America is mountainous and the vegetation is often distributed in rather discrete elevational bands, as opposed to the more subtle changes with landscape position associated with the eastern deciduous forests.

The Rocky Mountains extend from north-central Canada, through the USA and into Mexico, and form the eastern border of North America's temperate coniferous forests. At the lowest elevations (1500–1600 m), *Pinus edulis* and *Pinus monophylla* grow with various species of *Juniperus* and *Quercus*. At higher elevations are *Pinus ponderosa* forests, both closed canopy and open park-like stands (Fig. 2.11), above which lie forests dominated by *Pseudotsuga menziesii*. In the southern Rockies, the associates of *Pseudotsuga menziesii* are few, but in the northern Rockies it is variously found with *Abies grandis*, *Larix occidentalis*, *Pinus contorta* and *Populus tremuloides*. With increasing elevation in the northern Rockies, *Pseudotsuga menziesii* occurs with *Thuja plicata* and *Tsuga heterophylla*. These two species often form nearly pure stands. *Populus tremuloides* grows at similar elevations, colonizing after disturbances, principally fire. Just below the treeline are found *Picea engelmannii* and *Abies lasiocarpa*. In the far northern Canadian Rockies, the principal sub-

Fig. 2.11 Open *Pinus ponderosa* forest in western North America (Idaho, USA). (Photo by the author.)

alpine species are *Abies lasiocarpa*, *Picea glauca* and *Pinus contorta*.

In the Sierra Nevada Mountains of California, *Pinus sabiana* is a dominant conifer in the western foothills, growing with *Quercus* spp. On western slopes, *Pinus lambertiana* grows with *Pinus ponderosa* at higher elevations (750–1800 m), together with *Abies concolor* and *Calocedrus decurrens*. *Pinus jeffreyi* occupies lower eastern slopes and higher west-facing elevations. Further south, *Pseudotsuga macrocarpa* and *Pinus coulteri* grow in the elevations between desert scrub and the mixed conifer forests described above. *Sequoiadendron giganteum*, the world's largest trees, are distributed between 1350 and 2250 m on the western slopes of the Sierra Nevada. *Abies magnifica* and *Pinus contorta* grow between roughly 1800 and 2400 m, in regions of extremely high snowfall (4–20 m snowpack). The subalpine forests (>2400 m) consist of *Pinus albicaulis*, *Tsuga mertensiana*, *Pinus balfouriana*, *Pinus flexilis* and the long-lived (>4000 years) *Pinus aristata*.

The coastal mountain ranges of the northwestern USA and south-west Canada differ from the interior mountains principally by the abundant precipitation they receive (upwards of 2000 mm annually) and the exceptionally large size that many of the species attain. Along the Pacific coast from northern California into southeastern Alaska and the Kodiak Islands, *Picea sitchensis* forests are found in narrow bands (generally only a few kilometres wide) below 150 to 600 m elevation (Franklin & Dyrness 1973). Its associates include *Pseudotsuga menziesii*, *Thuja plicata*, *Tsuga heterophylla* and *Alnus rubra*. In southern Oregon and northern California, *Sequoia sempervirens* forests grow at similar elevations, extending as far as 15–20 km inland, and growing on slopes and sometimes at higher elevations.

Above the *Picea sitchensis* zone is the *Tsuga heterophylla* zone (550–1200 m) (Franklin & Dyrness 1973), above which *Abies amabilis* dominates the elevational bands ranging between 600 and 1500 m. The highest elevation forests are the subalpine *Tsuga mertensiana* forests that grow as high as 2000 m in southern Oregon. Principal associates include *Abies amabilis*, *Abies lasio-*

carpa and *Abies procera*. *Tsuga mertensiana* is replaced by *Abies lasiocarpa* in the interior portions of the Cascades and on the drier eastern slopes. The most extensive forests of the eastern slopes of the Cascades are those dominated by *Abies grandis*, which grow at elevations of 1000–1500 m in the north and 1000–2000 m in southern Oregon.

Central America and Mexico. Geographically speaking, the forests of the region are primarily (sub)tropical in nature, although there is a large degree of taxonomic similarity between the vegetation of the USA and Mexico and Central America. Some of the similarity is due to the continuity of species' range distributions, but it is also believed that the region provided a refuge during cool periods of the Oligocene, after which geological uplift and further climatic changes isolated the temperate forests to disjunct islands among otherwise (sub)tropical forests from Mexico southward into Nicaragua (Röhrig 1991d).

Among deciduous species, *Carpinus caroliniana*, *Liquidambar styraciflua*, *Nyssa sylvatica*, *Ostrya virginiana* and *Prunus serotina* grow in both the USA and Mexico. Species with a high degree of similarity include *Fagus mexicana* and *Fagus grandifolia*, *Carya ovata* var. *mexicana* and *Carya ovata*, and *Acer sutchii* and *Acer saccharum*. Several genera are well represented in both regions, including *Alnus*, *Carya*, *Fraxinus*, *Juglans*, *Magnolia*, *Platanus*, *Populus* and *Salix*. The genus *Quercus* is especially abundant, both deciduous and evergreen.

In montane deciduous forests, *Juglans olancha*, *Liquidambar styraciflua* and *Ostrya virginiana* are dominant, growing at 1200–1800 m in the Sierra de Chucharas in the Mexican state of Tamaulipas. Forests dominated by *Quercus* spp. and *Pinus* spp. often occupy the highest elevations (>1800 m), with (sub)tropical species prominent in the lower elevations (Röhrig 1991d). Riverine forests may be composed primarily of deciduous genera like *Carpinus*, *Fraxinus*, *Platanus*, *Populus*, *Salix* and *Ulmus*, but uplands typically contain representatives from more (sub)tropical families like Clethraceae, Fabaceae, Sabiaceae, Lauraceae, Staphylaceae, Cunoniaceae and Rutaceae.

Mexico and Central America also contain a rich coniferous flora, especially of pines. There are at least 47 *Pinus* species (Perry *et al.* 1998), which is about half the world total. Many species are five-needled pines, and they occupy niches from coastal to high montane environments. Some, e.g. *Pinus caribaea* and *P. patula*, are widely planted in the tropics.

South American

The South American temperate forests are the smallest in extent, confined primarily to Chile within and between the coastal and Andean mountain ranges, with a small extension into Argentina. Both broadleaf evergreen and deciduous species are present, often in mixture, although evergreen species are dominant. *Nothofagus* spp. are probably the most abundant trees; a total of 10 species grow in the region, of which seven are deciduous (*N. obliqua*, *N. alpina*, *N. pumilio*, *N. antarctica*, *N. glauca*, *N. leoni* and *N. alessandri*) and three evergreen (*N. betuloides*, *N. dombeyi* and *N. nitida*). The only other deciduous species present is *Acacia cavens*, which grows in a more Mediterranean climate in the northern part of the temperate region, between about 31 and 38° S.

Nothofagus glauca and *N. obliqua* grow in the wetter climate to the south of the *Acacia* forests in central Chile, the former on eastern and northern slopes and the latter on lowland sites. *Nothofagus alessandri* mixes with *N. glauca* on moister sites of the coastal mountains (Schmaltz 1991). *Nothofagus obliqua* dominates much of the lowlands and valleys south of 36–38° S, depending on proximity from the ocean. It is often found with *N. alpina* and *N. dombeyi*, the latter of which often forms pure stands in cool wet gorges (Schmaltz 1991). The conifer *Libocedrus chilensis* is often present on dry slopes opposite *N. obliqua*, and *Laurelia sempervirens* and *Persea lingue* are common, though lower-stature, associates of *N. obliqua*. At higher elevations in the Andes, *N. dombeyi* replaces *N. obliqua*. *Nothofagus alpina* forests occur in narrow elevational bands in the northern Chilean Andes and the species is widespread in mixed forests of the coastal range, often found with *N. dombeyi* and the associates of *N. obliqua* (Schmaltz 1991). *Auracaria araucana* occurs over small areas of the coastal range and is more widespread in the Andes (37–41°S) above the *N. alpina* forests (600–900 m). *Nothofagus pumilo* occupies cold snowy sites, at elevations of 1500–1800 m at about 37°S down to sea level at Cape Horn. *Nothofagus antarctica* is a high-elevation krummholz-forming species that also colonizes primary successional habitats (e.g. volcanic deposits) and other harsh or exposed sites (Schmaltz 1991). *Nothofagus nitida*, *Podocarpus nubigenus* and *Weinmannia trichosperma* occupy humid wetter sites between 600 and 900 m elevation south of about 40°S. *Fitzroya cupressoides* is a large-sized tree occurring on wetter sites between about 40° and 43° 30′S in both the Andean and coastal ranges, and at elevations from about 700 m to the timberline (Veblen *et al.* 1983).

2.5 SAVANNAHS AND OPEN WOODLANDS

Savannah is a broad term used to describe a number of ecosystem types. Generally speaking, savannahs are ecosystems dominated by a grass ground cover with an overstorey of widely spaced trees that do not form a closed canopy. Cole (1986) described five types of savannah.

1 Savannah woodlands have a canopy of large-stature deciduous or semideciduous trees (>8 m) with tall grasses (>80 cm) and grow in mesic environments.
2 Savannah parklands also consist of tall grasses but with scattered low-stature trees (<8 m).
3 Savannah grasslands lack tree or shrub cover.
4 Low tree/shrub savannahs contain widely spaced low-stature grasses (<80 cm) and trees or shrubs (<2 m).
5 Thicket/scrub savannahs consist of medium-stature trees that lack a grass understorey.

Savannahs have been substantially altered by humans on a global scale. Conversion to agriculture, especially livestock grazing, has dramatically altered community composition and structure. These problems have been most acute in less-developed nations, where the climate is often the most arid and where heavy demands for wood for fuel and shelter, combined with grazing, have led to severe resource overexploitation. However, even the North American savannahs have been subjected to heavy use. For example, the *Pinus palustris* forests of the south-eastern

USA now occupy less than 1% of their original habitat, the rest having been converted to agriculture, urban environments or plantations of other species.

Fire, caused by both lightning and humans, has long been an important force in shaping the structure and composition of savannah ecosystems. Generally, fire promotes the regeneration of the grass layer while inhibiting the establishment of trees. Humans have deliberately set fire to Africa's savannahs for at least 50 000 years (Rose Innes 1972) and for perhaps 40 000 years in Australia (Nicholson 1981), and this has undoubtedly promoted their persistence and expansion. Fires occur naturally during the dry season, but most fires in savannahs are kindled intentionally in order to clear land for ploughing or promote palatable new growth for livestock. Fire interacts with climate and soils, though, and is often regarded as a secondary causative factor in the establishment of savannahs. For example, humans have been burning savannahs for only about 5000 years in South America, an insufficient length of time to account for their broad expanse there. However, there is no question about the importance of fire in maintaining the open structure and grass understorey of pre-existing savannahs. In North America, for example, the large expanse of prairie and savannah (dominated by *Quercus macrocarpa*) that extended into the midwestern USA at the time of the European settlement is generally ascribed to the action of fire, particularly those set by humans. These areas subsequently reverted to forest after European settlement and fire suppression, and few remain today.

Savannahs are extensively grazed throughout much of the world. Prior to their exposure to domesticated livestock, the savannahs of Africa in particular had been subjected to extensive pressure from wildlife grazing for much of their 25 million year history. Grazing generally inhibits the regeneration of woody plants, but where there is a long history of intensive selective herbivore pressure on plants, such as Africa, many woody species have evolved both chemical and morphological defences against grazers and browsers. Intensive grazing by introduced domesticated livestock with which the plants have not evolved has led to the replacement of native species by exotics in some regions. As is the case with fire, though, grazing alone is insufficient to explain the presence and maintenance of savannahs.

2.5.1 Distribution and extent

Savannahs occupy about 25 million km^2, most of which (24 million km^2) are found in the tropics (Fig. 2.12). Although there are about 700000 km^2 of savannah in North America, I focus my discussion here on the tropical regions. I have described some of the most important temperate savannahs, like the *Pinus palustris* forests of the southeastern USA (see Fig. 2.9) and the pinyon–juniper woodlands of the western USA, in the section on temperate forests (see pp. 47–52). McPherson (1997) has extensively described the North American savannahs and I refer the interested reader to his text.

Africa contains most of the world's savannahs, about 15 million km^2 (Johnson & Tothill 1985), most of which occupies a broad latitudinal area between about 29°S and 16°N. A band of savannah ecosystems growing between the equator and 20°N covers, or once covered, the continent nearly from coast to coast. The miombo woodlands of eastern Africa consist of about 3 million km^2 and constitute the largest single vegetation type in the region. In Australia, savannahs occur between 11 and 30°S, stretching across much of northern Australia and curving south into southeastern Queensland. The Asian formation formerly covered much of eastern India, although it has largely been cleared for agriculture. In South-East Asia, savannah is interspersed with tropical forest, and is found in parts of Cambodia, Laos and Vietnam and to a lesser extent in Thailand and Burma (Archibold 1995).

The South American savannahs occur between 25°S and 23°N, with the largest portion composed of the Brazilian cerrado (1.8 million km^2), the pantanal (400000 km^2) in south-west Brazil adjacent to Bolivia and Paraguay, the chaco (800000 km^2) of western Paraguay, eastern Bolivia and northern Argentina, and the Llanos de Orinoco in western Colombia and central Venezuela (Medina & Silva 1991; Solbrig 1996). Lesser areas include the Llanos de Orinoco Moxos of Bolivia and the Andean foothills, and the Gran Sabana of Venezuela.

2.5.2 Climate

Savannahs are characterized by a largely tropical temperature regime and an extended dry season, typically in winter. Mean annual temperatures range from approximately 19 to 27°C, and reach their maximum at the end of the wet season (Archibold 1995). Mean warm season temperatures can exceed 30°C. Annual precipitation ranges from 500 to 1500 mm, although the dry season may last for 6–7 months, during which only a small percentage (<5%) of annual precipitation may fall. The rainy season typically lasts from late spring through autumn.

2.5.3 Soils

Savannah soils in the higher-rainfall regions are often Oxisols: old, highly weathered, high in clay

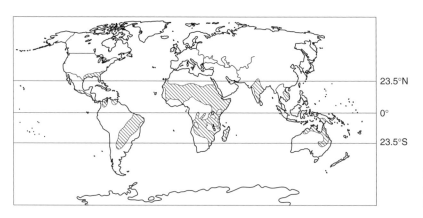

Fig. 2.12 Global extent and distribution of savannah ecosystems.

and somewhat acidic (pH 4.8–5.2). In dryer climates, Alfisols are more common, as found in North and East Africa, India and portions of Brazil. Entisols are found in the driest savannahs, such as those in southern Africa and northern Australia. They are characteristically low in phosphorus and often gravelly or stony. Vertisols are found in parts of eastern Australia. They tend to be high in montmorillonite clay, low in phosphorus, somewhat saline and have a low pH.

Many savannah soils have carbon concentrations in the range 1.5–3.0% (Lopes & Cox 1977); savannah soils as a whole contain about 14% of the world's soil carbon (Kirchmann & Eklund 1994). As stated above, however, nutrient contents are often low, and internal cycling is efficient and quantitatively important. Efficient cycling of critical nutrients like nitrogen, combined with atmospheric inputs, compensate for losses due to fire and leaching (Kellman 1989).

2.5.4 Regional floras

African

Between moist tropical forests and the shrub–steppe ecosystems that border deserts are a wide variety of dry forests, including both open woodlands and savannahs (Fig. 2.13). The miombo woodlands of southern, eastern and central Africa are dominated by the legumes *Brachystegia* spp. and *Isoberlinia* spp. (Aubréville 1956). Various species of *Baikiea*, *Cryptosepalum* and *Burkea* vary locally in their importance. Representatives of the Euphorbiaceae (*Uapaca*) and Dipterocarpaceae (*Monotes* and *Marquesia*) are of minor importance. Forests of *Baikiea plurijuga* grow on Kalahari sand formations in portions of Zimbabwe. The forests of south-western Africa are less extensive and less species-rich, the better forests being dominated by *Isoberlinia* and *Uapaca*. To the north in West Africa, *Anogeissus leiocarpus* and *Lophira lanceolata* grow with an understorey of *Andropogon*, *Hyparrhenia* and *Pennisetum* spp. (Archibold 1995). Forests of *Adansonia digitata* (see Fig. 2.13) and *Sclerocarya birrea* grow to the east. In more arid regions of southern Africa, *Colophospermum mopane* forms nearly pure stands (Fig. 2.14), and shrubby *Acacia* savannahs occur on drier soils as well. Palm savannahs occur on more poorly drained sites, principally *Borassus* spp. (Africa) and *Medemia* (Madagascar).

Australian

Eucalyptus and *Acacia* spp. dominate most of Australia's savannahs and open woodlands. In

Fig. 2.13 African baobab (*Adansonia digitata*) forest in Zimbabwe. (Photo by Roger Ruess.)

Fig. 2.14 *Colophospermum mopane* forest in Hwange National Park, Zimbabwe. (Photo by Roger Ruess.)

northern Australia, *Eucalyptus tetrodonta* is widespread where annual rainfall is 700–1500 mm (Beadle 1981). The graminoids *Aristida* spp., *Chrysopogon latifolius, Heteropogon triticeus, Sorghum* spp. and *Themeda australis* are common in the understorey. Other *Eucalyptus* species occurring in high-rainfall areas include *E. miniata* and *E. polycarpa*, which grow with *E. tetrodonta, E. tectifica, E. confertiflora* and *E. grandiflora* on finer soils, and with *E. ptychocarpa* in riparian environments. In the drier areas to the south, *E. argillacea* and *E. terminalis* grow in fine-textured soils and *E. dichromophloia* occurs on stony soil and deep sands. *Eucalyptus crebra* and *E. drepanophylla* are common in north-eastern Queensland. Low-stature *Melaleuca viridiflora* and *Melaleuca nervosa* woodlands occur in the north and in eastern Queensland (Gillison 1994), along with a number of *Eucalyptus* spp. (*E. brevifolia, E. dichromophloia, E. jensenii, E. papauana* and *E. prunosa*).

Further inland, eucalypts are replaced by low-stature *Acacia* shrubland. However, *Acacia* spp. also form open woodlands in areas of higher precipitation. In central Queensland, *Acacia harpophylla* woodlands (brigalow) reach nearly to the coast, with *Brachychiton rupestris* a common associate. In north-eastern Australia, *Acacia* woodlands and savannahs grow on both coarse, shallow, acid soils and on deep, fertile, alkaline soils (Johnson & Burrows 1994). *Acacia shirleyi* and *Acacia catenulata* often dominate the poor soils, with *Acacia harpophylla* and *Acacia cambagei* common on the better sites. *Acacia aneura* grows along the coast of Western Australia with short tussock grasses, especially *Eragrostis eriopoda* and *Monochather pardoxa*.

South American

South America's largest region of savannah is the cerrado of east-central Brazil (Fig. 2.15), occurring predominantly on tablelands at elevations of 300–1000 m. Dominant grasses include *Andropogon, Aristida, Paspalum* and *Trachypogon* (Archibold 1995). Associated tree species include *Caryocar brasiliense* and *Curatella americana*. In the transition zone between closed forests and savannah, species such as *Dipteryx alata, Dilodendron bipinnatum, Luehea paniculata, Magonia pubescens* and *Pseudobombax tomentosum* are characteristic (Ratter 1992).

The llanos occur at lower elevations in the floodplain of the Orinoco River in Venezuela and Colombia. Grasses and sedges of the genera *Panicum, Leersia, Eleocharis, Luziola* and

Fig. 2.15 Burned Brazilian campo cerrado near Brasilia. (Photo by Daniel Markewitz.)

Hymenachne dominate, with trees widely scattered and often forming small 'islands' (Blydenstein 1968; Walter 1985). Tree species include *Bowdichia virgilioides*, *Byrsonima crassifolia* and *Curatella americana*. The llanos have a pronounced seasonal wet–dry climate, often flooded during the wet season, while drought of 5 months or more is not uncommon during the dry season. The swampy pantanal lies to the south of the llanos and has many of the same species. Common grasses include *Panicum*, *Paspalum* and *Sorghastrum* spp., and the tree *Curatella americana* is common on wetter sites.

In Central America, the most abundant savannah species is probably *Pinus caribaea*, although several species of *Quercus* also occur. *Pinus caribaea* savannahs dominate much of southern Belize, with associated palms, cecropias and miconias (Kricher 1997).

Indian and Asian

Much of eastern India was formerly savannah but little remains; most has been cleared for agriculture. The small trees *Anogeissus latifolia* and *Diospyros melanoxylon* occur in small woodlands, and thorny species like *Acacia catechu* and *Zizyphus jujuba* have increased in response to heavy grazing pressure (Archibold 1995). The remaining savannahs of South-East Asia are dominated by deciduous dipterocarps. Savannahs typically occur where the dry season is 5–7 months in length and precipitation varies between 1000 and 1500 mm (Rundel & Boonpragob 1995), predominantly in hilly or mountainous terrain. Dominant species include *Shorea obtusa*, *Shorea siamensis*, *Dipterocarpus obtusifolius* and *Dipterocarpus tuberculatus*.

2.6 MEDITERRANEAN ECOSYSTEMS

The vegetation of Mediterranean ecosystems is characteristically low in stature, with a relatively open canopy. Evergreen shrubs and sclerophyllous trees are the dominant plant life forms, and frequent fires have historically played an important role in regulating community composition and structure. Most of these ecosystems have been heavily disturbed, subject to agricultural and horticultural conversion, grazing, cutting for wood and charcoal, and a long history of urbanization.

2.6.1 Distribution and extent

Mediterranean ecosystems occupy about 1.8 million km², distributed among five geographical regions (Fig. 2.16). In North America, Mediterranean ecosystems occur along the southern coast of California and portions of northern Baja California, Mexico, between roughly 30 and 40°N. In South America, they occur along the coast of Chile between 30 and 39°S. Much of the coastal region surrounding the Mediterranean Sea in Europe, the Middle East and Africa is occupied by the ecosystem type bearing its name. The northern coast, from Portugal to Turkey, is (or was) predominantly Mediterranean woodland, as are portions of Libya, Morocco, Algeria and Tunisia. Southern African Mediterranean forests include both the coastal and mountain fynbos of the Cape in South Africa. The remaining Mediterranean ecosystems occur in south-western Australia, and portions of South Australia and Victoria.

2.6.2 Climate

The Mediterranean climate is characterized by summer drought and wet cool winters. Annual precipitation ranges between about 300 and 1250 mm, of which two-thirds or more falls during the winter. Summer temperatures are warm, with July mean values of 20–30°C, and daily temperatures above 35°C are not uncommon. Winters average 10–12°C (Archibold 1995) and frosts are infrequent in all but the higher elevations. Relative humidity is often low, and in combination with strong seasonal winds and the sclerophyllous vegetation promotes hot, fast-moving fires.

2.6.3 Soils

Mediterranean ecosystems are characterized by heterogeneous topography, including mountains, steep-sided canyons and intervening valleys, plains and tablelands. Consequently, the soils are highly variable. Generally speaking, soils are thin and low in available nitrogen and phosphorus, with relatively little horizon differentiation. Surface horizons tend to be neutral, with higher pH values in deeper horizons. A proposed sequence of decreasing fertility has been proposed: Chile > Mediterranean Basin > California > South Africa > Australia. Although based on data from macronutrient sampling from a few locations, it is reasonably accurate, at least with respect to Chile, California and Australia (Lamont 1995). However, broad geographical generalities about particular regions being characteristically low in a particular nutrient (e.g. phosphorus in Australia) are not necessarily especially widely applicable.

2.6.4 Regional floras

Australian

Mediterranean *Eucalyptus* forests occur in south-western Western Australia, coastal South

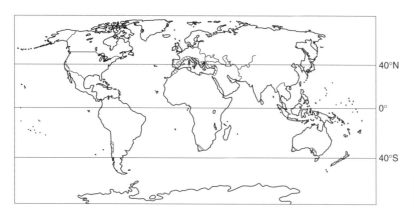

Fig. 2.16 Global extent and distribution of Mediterranean forests.

Australia and parts of Victoria. In Western Australia, Beadle (1981) recognized 14 alliances in the region, generally distributed as follows. In regions with relatively high rainfall (1000–1500 mm), *Eucalyptus diversicolor* and *E. jacksonii* dominate on fertile granitic soils, with *E. marginata* and *E. calophylla* on laterite soils and *E. gomphocephala* on calcareous soils of the coast. In drier areas, *E. wandoo* and *E. loxophleba* replace these species, with *E. dundasii*, *E. lesouefii*, *E. salmonphloia* and *E. slubris* on the most xeric soils. *Eucalyptus cornuta* and *E. occidentalis* occupy wet soils. The mallee, a region dominated by eucalypts of shrubby morphology, forms the Mediterranean forests of South Australia and Victoria. There are about 100 species of *Eucalyptus* in the region. On deeper soils are found species like *E. diversifolia*, *E. dumosa*, *E. foecunda*, *E. gracilis*, *E. incrassata*, *E. loxophleba*, *E. oleosa*, *E. socialis* and *E. viridis*.

Chilean

Mediterranean forests are confined to coastal Chile, where they occur from the lowlands up to about 2000 m elevation, at which point they merge into alpine communities. The flora is diverse, with nearly 2400 identified species. The herbaceous and shrub communities share a large degree of taxonomic affinity with Californian species, but the trees are relatively unique. *Salix*, *Prosopis* and *Rhamnus* are found in both regions, while *Gormotega*, *Jubaea*, *Legrandia* and *Pitvia* are endemic to Chile (Arroyo *et al.* 1995). The Chilean matorral comprises primarily low-stature (1–3 m) evergreen species and occurs on the interior slopes of the coastal mountains and the Andean slopes north of 36° S and along the coast south of 32° S (Arroyo *et al.* 1995). Dominants include *Colliguaja* spp., *Cryptocarya alba*, *Lithrea caustica*, *Peumus boldus* and *Quillaja saponaria*. On moister sites *Salix chilensis* and *Cryptocarya alba* assume greater stature, with a shrub understorey. In Andean valleys the matorral grades into tall dry forests, and the two share many of the same species. To the north, the shrubs *Bahjia ambrosioides*, *Fuschia lycioides* and *Adesmia* spp. appear, and the vegetation is more scrubby and open. Scattered *Acacia cavens* grow above herbs and grasses in the interior espinal (Archibold 1995).

Mediterranean Basin (European and North African)

Because of the long history of human occupation and the extensive nature of human disturbance, much of the natural Mediterranean forest in the region has been lost. The remaining ecosystems share many of the species that occur in the Submediterranean region of Europe (see p. 46) and grade into these forests and temperate forests as elevation and/or distance from the coast increase.

In Spain and Portugal, *Quercus ilex* and *Quercus suber* are the dominant hardwoods, the latter occurring on non-calcareous soils. In general, these forests are heavily disturbed and rather sparsely populated with trees. The pines *Pinus halepensis*, *Pinus pinea* and *Pinus pinaster* grow at higher elevations. *Quercus ilex* is the most abundant species in the region, accounting for over 20% of forestland (Giordano 1956). The mix of species in Italy is similar, and *Quercus ilex* is often found in mixture with *Arbutus unedo*, *Erica* spp., *Phylleria variabilis*, *Pistacia lentiscus* and *Myrtus communis*. *Pinus pinea* grows on sandy coastal sites, and with *Pinus halepensis* and *Pinus pinaster* inland. In Greece, *Quercus ilex* is common, being replaced by *Quercus coccifera* on calcareous soils. The mix of associates is similar to the rest of the region, with the addition of *Arbutus andrachne* and *Olea europaea* among others (Giordano 1956). Conifers in the region include *Pinus halepensis*, *Pinus brutia* and *Cupressus sempervirens*. In Turkey, *Quercus macrolepis* and other *Quercus* spp. occur, along with *Fagus orientalis* and the conifers *Pinus brutia* and *Pinus nigra* at lower elevations and *Cedrus libani* and *Abies cilicica* at higher elevations. In the coastal regions of the Middle East, *Quercus calliprinos* is common, along with many of the oaks and pines characteristic of forests to the west. Along the north coast of Africa, *Quercus ilex* is again often the dominant species. Associates include *Pinus halepensis* and *Abies numidica* at lower elevations and *Cedrus atlantica* above 1200 m (Archibold 1995).

North American

Mediterranean ecosystems occur primarily in the coastal region of California and at low elevations in the Sierra Nevada. The coastal and desert-margin arborescent or shrubby vegetation (chaparral) covers about 3.4 million ha, about 20% of which is dominated by small trees. The shrub *Adenostoma fasciculatum* is common between 300 and 1500 m, and *Artemisia californica*-dominated coastal sage scrublands occur at lower elevations. *Ceonothus cuneatus* occurs with *Adenostoma fasciculatum*, and *Arctostaphyllos* spp. are common at higher elevations (Archibold 1995). On more mesic sites, trees are principally low-stature *Quercus* spp., including *Q. dumasa*, *Q. agrifolia* and *Q. wilslizenii*. There are about 3 million ha of open *Quercus*-dominated woodlands in California, not all of which experience a strong Mediterranean climate, but all of which have an open structure, are fire prone and are affected by increased demand for grazing land and urban expansion (Fig. 2.17).

South African

The Mediterranean ecosystems in South Africa are represented primarily by the fynbos. At lower elevations (300–700 m) and on drier sites, these are primarily mid-stature (>2 m) shrub communities, dominated by the genus *Protea*. *Protea neriifolia* is common, as are *Protea repens*, *Protea nitida* (rocky soils) and occasionally *Olea europaea* var. *africana* on granitic soils (van Wilgren & McDonald 1992). *Erica* spp. are also dominant in some areas, including *Erica hispidula* and *Erica versicolor*. Their primary associates include *Restio inconspicuus*, *Tetraria bromoides* (with *Erica hispidula*) and *Agathosoma ovata* (with *Erica versicolor*) (McDonald 1993). In riparian areas, 3–6 m high forests composed of *Brabejum stellatifolium* occur, with an understorey of *Diospyros glabra*, *Halleria elliptica* and *Rhus angustifolia*. Tall (8–16 m) forests grow on stable boulder screes (van Wilgren & McDonald 1992). Canopy species include *Hartogiella schinoides*, *Olinia ventosa* and *Podocarpus elongatus*. On loose boulder fields, short forests (2–8 m) consisting of *Hartogiella schinoides*, *Heeria argentea*, *Olinea ventosa* and *Maytenus acuminata* grow, with a *Podocarpus elongatus* understorey.

2.7 SUMMARY

The earth supports a tremendous amount and

Fig. 2.17 Heavily grazed Mediterranean *Quercus* woodland in coastal southern California, USA. (Photo by the author.)

diversity of forests, which humans have utilized as a source of food, fibre and forage for countless millennia. In the process, we have inventoried forests based on their composition and structure, recognizing each for the particular products they supply and the organisms they contain. The inventory and classification process was informal and driven by intuition and necessity in the past, but within the past century or so it has been performed increasingly by design. By virtue of their low species diversity and limited geographical or topographical distribution, some forests (e.g. the coniferous forests of western North America) have lent themselves well to more formal and specialized classification. Others, with abundant species, many of which have highly localized distributions (e.g. the tropical rain forests), have proved difficult or impossible to assign to a particular forest type. Moreover, many economically important species have been widely planted outside their native range, so much so that some (e.g. *Pinus radiata*) are far more abundant outside their native range than within it. In spite of these difficulties, I hope that I have been able to adequately differentiate among, and generally describe, the world's forest biomes and the forest types or communities within them, and at the same time conveyed to the reader an appreciation for their rich ecological diversity.

REFERENCES

Adamson, R.S. (1956) South Africa. In: Haden-Guest, S., Wright, J.K. & Teclaff, E.M., eds. *A World Geography of Forest Resources*, pp. 385–91. Ronald Press, New York.

Ahti, T. (1977) Lichens of the boreal coniferous zone. In: Seaward, M.R.D., ed. *Lichen Ecology*, pp. 141–81. Academic Press, New York.

Ahti, T., Hämet-Ahti, L. & Jalas, J. (1968) Vegetation zones and their sections in northwestern Europe. *Annales Botanici Fennici* 5, 169–211.

Anderson, R.C. & Mori, S.A. (1967) A preliminary investigation of rapia swamps, Puerto Viejo, Costa Rica. *Turrialba* 17, 221–4.

Archibold, O.W. (1995) *Ecology of World Vegetation*. Chapman & Hall, New York.

Arroyo, M.T.K., Cavieres, L., Marticorena, C. & Muoz-Schick, M. (1995) Convergence in the Mediterranean floras in central Chile and California: insights from comparative biogeography. In: Arroyo, M.T.K., Zedler, P.H. & Fox, M.D., eds. *Ecology and Biogeography of Mediterranean Ecosystems in Chile, California and Australia*, pp. 43–88. Springer-Verlag, New York.

Ashton, D.H. & Attiwill, P.M. (1994) Tall open-forests. In: Groves, R.H., ed. *Australian Vegetation*, pp. 157–96. Cambridge University Press, Cambridge.

Aubréville, A.M.A. (1956) Tropical Africa. In: Haden-Guest, S., Wright, J.K. & Teclaff, E.M., eds. *A World Geography of Forest Resources*, pp. 353–84. Ronald Press, New York.

Bailey, R.G. (1989) Explanatory supplement to ecoregions map of the continents. *Environmental Conservation* 16, 307–9.

Barnes, B.V. (1991) Deciduous forests of North America. In: Röhrig, E. & Ulrich, B., eds. *Temperate Deciduous Forests. Ecosystems of the World Vol. 7*, pp. 219–344. Elsevier, Amsterdam.

Beadle, N.C.W. (1981) *The Vegetation of Australia*. Cambridge University Press, Cambridge.

Blydenstein, J. (1968) Tropical savanna vegetation of the llanos of Columbia. *Ecology* 48, 1–15.

Braun, E.L. (1950) *Deciduous Forests of Eastern North America*. McGraw-Hill, New York.

Busby, J.R. & Brown, M.J. (1994) Southern rainforests. In: Groves, R.H., ed. *Australian Vegetation*, pp. 131–56. Cambridge University Press, Cambridge.

Chaturvedi, M.D. (1956) India. In: Haden-Guest, S., Wright, J.K. & Teclaff, E.M., eds. *A World Geography of Forest Resources*, pp. 455–82. Ronald Press, New York.

Ching, K.K. (1991) Temperate deciduous forests in East Asia. In: Röhrig, E. & Ulrich, B., eds. *Temperate Deciduous Forests. Ecosystems of the World Vol. 7*, pp. 439–555. Elsevier, Amsterdam.

Cockayne, L. (1958) *The Vegetation of New Zealand*, 2nd edn. Engleman, Leipzig.

Cole, M.M. (1986) *The Savannas. Biogeography and Geobotany*. Academic Press, London.

Cole, M.M. (1992) Influences of physical factors on the nature and dynamics of forest–savanna boundaries. In: Furley, P.A., Proctor, J. & Ratter, J.A., eds. *Nature and Dynamics of Forest–Savanna Boundaries*, pp. 63–75. Chapman & Hall, New York.

Entrican, A.R. & Holloway, J.T. (1956) New Zealand. In: Haden-Guest, S., Wright, J.K. & Teclaff, E.M., eds. *A World Geography of Forest Resources*, pp. 591–609. Ronald Press, New York.

Esseen, P.-A., Ehnström, B., Ericsson, L. & Sjöberg, K. (1997) Boreal forests. *Ecological Bulletins* 46, 16–47.

Eyre, F.H. (1980) *Forest Cover Types of the United States and Canada*. Society of American Foresters, Washington, DC.

FAO (1997) *State of the World's Forests*. Publications Division, Food and Agriculture Organization, Rome, Italy.

Franklin, J.F. & Dyrness, C.T. (1973) *Natural Vegetation of Oregon and Washington*. General Technical Report

PNW-8 United States Department of Agriculture Forest Service, Portland, Oregon.

Frodin, D.G. (1984) *Guide to Standard Floras of the World*, pp. 146–94 Cambridge University Press, Cambridge.

Gentry, A.H. (1995) Diversity and floristic composition of Neotropical dry forests. In: Bullock, S.H., Mooney, H.A. & Medina, E., eds. *Seasonally Dry Tropical Forests*, pp. 146–94. Cambridge University Press, Cambridge.

Gillison, A.N. (1994) Woodlands. In: Groves, R.H., ed. *Australian Vegetation*, pp. 227–56. Cambridge University Press, Cambridge.

Giordano, G. (1956) The Mediterranean region. In: Haden-Guest, S., Wright, J.K. & Teclaff, E.M., eds. *A World Geography of Forest Resources*, pp. 317–52. Ronald Press, New York.

Groves, R.H. (ed.) (1994) *Australian Vegetation*. Cambridge University Press, Cambridge.

Hamilton, A. (1989) African forests. In: Lieth, H. & Werger, M.J.A., eds. *Tropical Rain Forest Ecosystems. Ecosystems of the World Vol. 14B*, pp. 155–82. Elsevier, Amsterdam.

Howard, T.M. & Ashton, D.H. (1973) The distribution of *Nothofagus cuninghamii* rainforest. *Proceedings of the Royal Society of Victoria* **86**, 47–76.

Hyun, M.S. (1956) Korea. In: Haden-Guest, S., Wright, J.K. & Teclaff, E.M., eds. *A World Geography of Forest Resources*, pp. 561–72. Ronald Press, New York.

Jahn, G. (1991) Temperate deciduous forests of Europe. In: Röhrig, E. & Ulrich, B., eds. *Temperate Deciduous Forests of Europe. Ecosystems of the World Vol. 7*, pp. 377–502. Elsevier. Amsterdam.

Johnson, R.W. & Burrows, W.H. (1994) Acacia–open forests, woodlands and shrublands. In: Groves, R.H., ed. *Australian Vegetation*, pp. 257–90. Cambridge University Press, Cambridge.

Johnson, R.W. & Tothill, J.C. (1985) Definition and broad geographic outline of savanna lands. In: Furley, P.A., Proctor, J. & Ratter, J.A., eds. *Nature and Dynamics of Forest–Savanna Boundaries*. Australian Academy of Science, Canberra.

Kellman, M. (1989) Mineral nutrient dynamics during savanna-forest transformation in Central America. In: Proctor, J., ed. *Mineral Nutrients in Tropical Forest and Savanna Ecosystems*, pp. 137–52. Blackwell Scientific Publications, Oxford.

Kellman, M. & Tackaberry, R. (1997) *Tropical Environments: the Functioning and Management of Tropical Ecosystems*. Routledge, New York.

Kirchmann, H. & Eklund, M. (1994) Microbial biomass in a savanna-woodland and an adjacent arable soil profile in Zimbabwe. *Soil Biology and Biochemistry* **26**, 1281–3.

Kricher, J. (1997) *A Neotropical Companion*. Princeton University Press, Princeton, New Jersey.

Küchler, A.W. (1964) *The Potential Natural Vegetation of the Conterminous United States*. American Geographic Society Special Publication 36. American Geographical Society, New York.

Lamont, B.B. (1995) Mineral nutrient relations in Mediterranean regions of California, Chile and Australia. In: Arroyo, M.T.K., Zedler, P.H. & Fox, M.D., eds. *Ecology and Biogeography of Mediterranean Ecosystems in Chile, California and Australia*, pp. 211–35. Springer-Verlag, New York.

Larsen, J.A. (1980) *The Boreal Ecosystem*. Academic Press, New York.

Lauer, W. (1989) Climate and weather. In: Lieth, H. & Werger, M.J.A., eds. *Tropical Rain Forest Ecosystems. Ecosystems of the World Vol. 14B*, pp. 7–53. Elsevier, Amsterdam.

Li, H.-L. (1972) Eastern Asia–Eastern North America species-pairs in wide-ranging genera. In: Graham, A., ed. *Floristics and Paleofloristics on Asia and Eastern North America*, pp. 66–78. Elsevier, Amsterdam.

Lieth, H. (1975) Primary productivity of the major vegetation units of the world. In: Lieth, H. & Whittaker, R.H., eds. *Primary Productivity of the Biosphere*, pp. 203–15. Springer-Verlag, New York.

Longman, K.A. & Jenik, J. (1987) *Tropical Forest and its Environment*, 2nd edn. John Wiley & Sons, New York.

Lopes, A.S. & Cox, F.R. (1977) A survey of the fertility status of surface soils under 'cerrado' vegetation in Brazil. *Soil Science Society of America Journal* **41**, 742–7.

Lovett, J.C. (1996) Elevational and latitudinal changes in tree associations and diversity in the Eastern Arc mountains of Tanzania. *Journal of Tropical Ecology* **12**, 629–50.

Mabberly, D.J. (1992) *Tropical Rain Forest Ecology*, 2nd edn. Blackie, Chapman & Hall, New York.

McDonald, D.J. (1993) The vegetation of the southern Langeberg, Cape Province. 3. The plant communities of the Bergfontein, Rooiwaterspruit and Phesantefontein areas. *Bothalia* **23**, 239–63.

McPherson, G.R. (1997) *Ecology and Management of North American Savannas*. University of Arizona Press, Tucson.

Medina, E. & Silva, J.F. (1991) Savannas of northern South America: a steady state regulated by fire–water interactions on a background of low nutrient availability. In: Werner, P.A., ed. *Savanna Ecology and Management*, pp. 59–69. Blackwell Scientific Publications, Boston.

Monod, T. (1957) Les grandes divisions chorologiques de l'Afrique. Conservation Science Afrique Sud Sahara. 24.

Ng, F.S.P. & Low, C.M. (1982) *Check list of endemic trees of the Malay Peninsula*. Malay Forestry Department Research Pamphlet 88. Malaysia.

Nicholson, P.H. (1981) Fire and the Australian aborigine: an enigma. In: Gill, A.M., Groves, R.H. & Noble,

I.R., eds. *Fire and the Australian Biota*, pp. 55–76. Australian Academy of Science, Canberra.

Nikolov, N. & Hemisaari, H. (1992) Silvics of the circumpolar boreal forest tree species. In: Shugart, H.H., Leemans, R. & Bonan, G.B., eds. *A Systems Analysis of the Global Boreal Forest*, pp. 13–84. Cambridge University Press, New York.

Oksanen, J. (1983) Vegetation of forested dunes in northern Karelia, eastern Finland. *Annales Botanici Fennici* **20**, 281–95.

Oksanen, J. & Ahti, T. (1982) Lichen-rich pine forest vegetation in Finland. *Annales Botanici Fennici* **19**, 275–301.

Ovington, J.D. & Pryor, L.D. (1983) Temperate broad-leaved evergreen forests of Australia. In: Ovington, J.D., ed. *Temperate Broad-leaved Evergreen Forests. Ecosystems of the World Vol. 10*, pp. 73–101. Elsevier, Amsterdam.

Perry, D.A. (1994) *Forest Ecosystems*. Johns Hopkins University Press, Baltimore, Maryland.

Perry, J.P., Graham, G. & Richardson, D.M. (1998) The history of pines in Mexico and Central America. In: Richardson D.M., ed. *Ecology and Biogeography of Pinus*, pp. 137–47. Cambridge University Press, Cambridge.

Pool, D.J., Snedaker, S.C. & Lugo, A.E. (1977) Structure of mangrove forests in Florida, Puerto Rico, Mexico and Costa Rica. *Biotropica* **9**, 195–212.

Prance, G.T. (1989) American tropical forests. In: Lieth, H. & Werger, M.J.A., eds. *Tropical Rain Forest Ecosystems. Ecosystems of the World Vol. 14B*, pp. 99–132. Elsevier, Amsterdam.

Ratter, J.A. (1992) Transitions between cerrado and forest vegetation in Brazil. In: Furley, P.A., Proctor, J. & Ratter, J.A., eds. *Nature and Dynamics of Forest–Savanna Boundaries*, pp. 417–27. Chapman & Hall, New York.

Raup, H.M. (1946) Phytogeographic studies in the Athabasca–Great Slave Lake region II. *Journal of the Arnold Arboretum, Harvard University* **27**, 1–85.

Richter, D.D. & Babbar, L.T. (1991) Soil diversity in the tropics. *Advances in Ecological Research* **21**, 161–71.

Röhrig, E. (1991a) Introduction. In: Röhrig, E. & Ulrich, B., eds. *Temperate Deciduous Forests. Ecosystems of the World Vol. 7*, pp. 1–5. Elsevier, Amsterdam.

Röhrig, E. (1991b) Deciduous forests of the Near East. In: Röhrig, E. & Ulrich, B., eds. *Temperate Deciduous Forests. Ecosystems of the World Vol. 7*. Elsevier, Amsterdam.

Röhrig, E. (1991c) Climatic conditions. In: Röhrig, E. & Ulrich, B., eds. *Temperate Deciduous Forests. Ecosystems of the World Vol. 7*. Elsevier, Amsterdam.

Röhrig, E. (1991d) Temperate deciduous forests in Mexico and Central America. In: Röhrig, E. & Ulrich, B., eds. *Temperate Deciduous Forests. Ecosystems of the World Vol. 7*. Elsevier, Amsterdam.

Rose Innes, R. (1972) Fire in West African vegetation. In: *Proceedings of the Tall Timbers Fire Ecology Conference* Vol. 11, pp. 147–73. Tall Timbers Research Station, Tallahasee, Florida.

Rumney, G.R. (1968) *Climatology and the World's Climates*. Macmillan, New York.

Rundel, P.W. & Boonpragob, K. (1995) Dry forest ecosystems of Thailand. In: Bullock, S.H., Mooney, H.A. & Medina, E., eds. *Seasonally Dry Tropical Forests*, pp. 93–123. Cambridge University Press, Cambridge.

Satoo, T. (1983) Temperate broad-leaved forests of Japan. In: Ovington, J.D., ed. *Temperate Broad-leaved Evergreen Forests. Ecosystems of the World Vol. 10*, pp. 169–89. Elsevier, Amsterdam.

Schmaltz, J. (1991) Deciduous forests of southern South America. In: Röhrig, E. & Ulrich, B., eds. *Temperate Deciduous Forests. Ecosystems of the World Vol. 7*, pp. 557–78. Elsevier, Amsterdam.

Schroeder, P.E. & Winjum, J.K. (1995) Assessing Brazil's carbon budget. I. Biotic carbon pools. *Forest Ecology and Management* **75**, 77–86.

Sewandono, R. (1956) Southeast Asia. In: Haden-Guest, S., Wright, J.K. & Teclaff, E.M., eds. *A World Geography of Forest Resources*, pp. 491–517. Ronald Press, New York.

Slaughter, C.W. & Viereck, L.A. (1986) Climatic characteristics of the taiga in interior Alaska. In: Van Cleve, K., Chapin, F.S., Flanagan, P.W., Viereck, L.A. & Dyrness, C.T., eds. *Forest Ecosystems in the Alaskan Taiga*, pp. 9–21. Springer-Verlag, New York.

Solbrig, O.T. (1996) The diversity of the savanna ecosystems. In: Solbrig, O.T., Medina, E. & Silva, J.F., eds. *Biodiversity and Savanna Ecosystem Processes: a Global Perspective*, pp. 1–27. Springer-Verlag, New York.

Thieret, J.W. (1964) Botanical survey along the Yellowknife Highway, Northwest Territories, Canada. II. *Canadian International Development Agency Contributions to Botany* **1**, 187–239.

Tomlinson, P.B. (1986) *The Botany of Mangroves*. Cambridge University Press, Cambridge.

Van Cleve, K., Viereck, L.A. & Dyrness, C.T. (1996) State factor control of soils and forest succession along the Tanana River in interior Alaska, USA. *Arctic and Alpine Research* **28**, 388–400.

van Wilgren, B.W. & McDonald, D.J. (1992) The Swartbsokloof Experimental Site. In: van Wilgren, B.W., Richardson, D.M., Kruger, F.J. & van Hensbergen, H.J., eds. *Fire in South African Mountain Fynbos*, pp. 1–20. Springer-Verlag, New York.

Veblen, T.T., Schlegel, F.M. & Oltremari, J.V. (1983) Temperate broad-leaved evergreen forests of South America. In: Ovington, J.D., ed. *Temperate Broad-leaved Evergreen Forests. Ecosystems of the World Vol. 10*, pp. 5–31. Elsevier, Amsterdam.

Viereck, L.A., Dyrness, C.T., Batten, A.R. & Wenzlick,

K.J. (1992) *The Alaska Vegetation Classification.* General Technical Report 286, US Forest Service Pacific Northwest, Portland, Oregon.

Walter, H. (1985) *Vegetation of the Earth.* Springer-Verlag, New York.

Wardle, P., Bulfin, M.J.A. & Dugdale, J. (1983) Temperate broad-leaved evergreen forests of New Zealand. In: Ovington, J.D., ed. *Temperate Broad-leaved Evergreen Forests. Ecosystems of the World Vol. 10*, pp. 33–71. Elsevier, Amsterdam.

Webb, L.J. & Tracey, J.G. (1994) The rainforests of northern Australia. In: Groves, R.H., ed. *Australian Vegetation*, 2nd edn, pp. 87–129. Cambridge University Press, Cambridge.

West, R.C. (1977) Tidal salt-marsh and mangrove formations of Middle and South America. In: Chapman, V.J., ed. *Wet Coastal Ecosystems. Ecosystems of the World Vol. 1.* Elsevier, Amsterdam.

White, F. (1983) *The Vegetation Map of Africa.* United Nations Educational, Scientific and Cultural Organization (UNSECO), Paris.

Whitmore, T.C. (1975) *Tropical Rain Forests of the Far East.* Clarendon Press, Oxford.

Whitmore TC. (1984) *Tropical Rainforests of the Far East*, 2nd edn. Clarendon Press, Oxford.

Whitmore, T.C. (1989) Southeast Asian tropical forests. In: Lieth, H. & Werger, M.J.A., eds. *Tropical Rain Forest Ecosystems. Ecosystems of the World Vol. 14B*, pp. 195–218. Elsevier, Amsterdam.

Yim, Y.J. (1977a) Distribution of forest vegetation and climate in the Korean Peninsula. III. Distribution of tree species along a thermal gradient. *Japanese Journal of Ecology* **27**, 177–89.

Yim, Y.J. (1977b) Distribution of forest vegetation and climate in the Korean Peninsula. IV. Zonal distribution of forest vegetation in relation to thermal climate. *Japanese Journal of Ecology* **27**, 269–78.

Zachrisson, O. (1985) Some evolutionary aspects of life history characteristics of broadleaved tree species found in the boreal forest. In: Hägglund, B. & Peterson, G., eds. *Broadleaves in Boreal Silviculture: an Obstacle or Asset?* Department of Silviculture, Swedish University of Agricultural Science, Uppsala, Sweden.

Part 2
Biological and Ecological Processes

Forests are complex ecosystems. They are complex not only because trees are large and long lived but because of all the interactions with other lifeforms that go on in any forest stand. In this part we have attempted to set out the influences leading to the great variety of trees and forest types, as God has been pleased to bestow, and then show how various impacts—disturbances—affect the processes that take place. Indeed, disturbance is taken as the unifying theme of Chapters 4 to 6. Understanding this will help foresters and forest ecologists, and others responsible for the well-being of the world's forest and all who benefit from them, to manage forests better and in more sustainable ways knowing what impacts different activities have. Forests are incredibly rich and intricate webs of life: here we disentangle a few of the strands.

Gene Namkoong and Matthew Koshy (Chapter 3) summarize the genetic history and interactions of forest systems that provide the foundation for much of the Earth's biodiversity. Crucially, they point out that the genetic variation within the presently existing tree species is not a fixed endowment, and the question we all face is what we do with it. George Peterken (Chapter 4), Jaboury Ghazoul and Eunice Simmons (Chapter 5) and Hugh Evans (Chapter 6) dissect the elements of disturbance in forest stands from perspectives of the tree community itself, the interrelated and integral fauna and flora, and the specific dimension of invertebrates respectively. Each perspective brings further contrasts, with Chapter 4 focusing mainly on semi-natural temperate forests, Chapter 5 drawing largely on tropical forests and Chapter 6 primarily addressing the topic with plantation forestry in mind.

3: Genetics and Speciation in the World's Forests

GENE NAMKOONG AND MATHEW P. KOSHY

3.1 INTRODUCTION

The vast majority of plant and animal diversity is contained in forests, comprising both the trees that dominate this environment as well as forest-dependent species. The genetic and species diversity of trees provide the foundation for much of the earth's biodiversity (see Chapter 2). Among the trees themselves, taxonomic diversity is exemplified by the 25 000 species and the high levels of genetic variation within most of them. While this is only around 10% of the total number of vascular plant species, the forests of the world cover over one-third of the dry land area. In addition, the functional diversity of tree species is commonly assumed to be necessary for the sustenance of forests and the organisms dependent on them. An equivalent amount of the dry land area of the earth is in pasture and agriculture, but these ecosystems are relatively low in species and genetic and functional diversity. Hence, much of the world's biodiversity at the genetic, species and ecosystem levels of organization depends on the processes that create and sustain the genetic and species diversity of forest trees.

With the loss of original forests around the world, there is concern over the robustness of this diversity, and a need to understand its origins. How did the trees arrive at their present locations and how did tree species arrive at their present distributions? Do the accidents of history determine the outcome of evolution or is there selectivity for an optimum distribution? Are some families simply more prone to diversify and to create new species? To what extent does arrival time and the capacity to migrate determine distribution and how much influence is exerted by the capacity to grow and reproduce, and to adapt to local competition for resources?

Forest composition is the result of several genetically influenced factors, including the probabilities and distributions of species establishment and their responses to the stresses and opportunities for adaptation. Genetic capacities thus influence ecological dynamics. However, once established on a site, the levels and distribution of genetic variation are partially determined by stand dynamics and the distribution of available parents for establishing the next generation. Thus, the genetic patterns of variation are partly influenced by establishment, survival and reproductive dynamics, and species dynamics are partly determined by genetic dynamics. Since these factors are particularly complex in forest ecosystems, the relationships between genetic and ecological dynamics are particularly important to understand. The evolution of forests is neither a matter of genetic factors preceding ecological dispositions nor of ecological factors determining optimal genetic composition; it is an interaction of genetics and ecology.

The joint dynamics of genetics and ecology are also influenced by historic events. We might first note that in the evolution of the modern species of forest trees, the history of distinctive species is relatively long. The fossil record indicates that recognizably similar trees have existed for much longer periods of time than for the grasses and that the range of environments over which trees grow is also much larger than for annual plants. This implies that forest trees have had a long period of time to evolve, for some species to go extinct and for others to proliferate variations. It also implies that a large variety of evolutionary histories have had the opportunity to develop over large spans of time and geography.

In a broad view of forest ecosystems, there are about 4 billion ha of forests. Approximately 32% are in the coniferous forests of Europe and North America, 27% in other temperate forests and 41% in dry and moist tropical forests. Since the tropical zones contain the greatest number of species, there is an uneven packing of species around the world in different forest ecosystems. Some families and groups of genera are more heavily represented in some parts of the globe than others. Historically contingent migration and the existence of migration routes also affect where species may have had the opportunity to exist. In addition, environmental differences would give selective preference to some species and hence allow only selected species to compete for resources. A further consequence of differential species packing is that the number of individuals per species must be reduced in any given forest if species number increases, and this would affect the density distribution and mating system of those species. Some species may evolve as environmental specialists that avoid competition and mate in small neighbourhoods.

In addition, historical contingencies as well as selective factors affect the dynamics of within-species distribution. Among species, we would therefore expect differences in how they are subdivided, how distant and how often parents can find mates, and how they adapt to the different environments that they encounter. With different climates and soils around the world, the different species that react to their environments can be expected to develop different evolutionary responses and different life histories for their growth and reproduction. Then, since species dynamics differ, we can expect that the distribution of genetic variation would also be different among the species in different ecosystems. Thus genetic variation is both the product of evolution and the engine that drives the capacity of organisms to adapt. The interplay of environment and history that affects the distribution of species and genes is affected by the genetic and evolutionary capacity of those species and has created the rich biotic dynamics which is our forest legacy. To survey this wealth of species, we first look at the origins of tree species.

3.2 THE WEALTH OF FOREST TREE SPECIES

The earliest forests of the Carboniferous Age were probably not composed of woody tree species but of ferns, cycads and other vascular plants that were neither conifer nor angiosperm. The more complex vascular and reproductive systems of coniferous species did not evolve until the Devonian Period some 350 million years ago. Conifer trees were not prominent until the Permian Period 100 million years later (Scagel 1965). It was another 70–100 million years before the pines differentiated, around 180 million years ago. The soft and hard pines diverged around 130 million years ago, during and following the break-up of Pangaea and the great continental migrations. Angiosperms evidently did not originate until the Cretaceous Period, but soon broadleaf tree species would become more prominent in forests. By around 100 million years ago they began to dominate over the ferns and gymnosperms in many forests.

The end of the Cretaceous Period, around 70 million years ago, is marked by major animal extinctions, wide climatic fluctuations, continental shifts with episodes of mountain building, and the differentiation of the major angiosperm family groups and genera. Periods of major extinctions also occurred during the Permian and during the global cooling that took place at the end of the Tertiary. Further speciation continues in both gymnosperms and angiosperms to this day, but largely on the development plans established by the end of the Cretaceous.

Since most of the major families and genera existed at least as prototypes by the end of the Cretaceous Period, there are large patterns of geographical differentiation in species, genera and families. Among the nearly 600 species of conifers, the Pinaceae (10 genera, 250 species) is a Northern Hemisphere family, as is Taxaceae (5 genera, 15 species), Cephalotaxaceae (1 genus, 5 species) and Taxodiaceae (9 genera, 12 species, with one southern genus with 3 species), while Podocarpaceae (7 genera, 131 species) and Araucareaceae (2 genera, 36 species) are predominantly Southern Hemisphere families. Cupressaceae (16 genera, 142 species) is spread over both hemispheres. All species are monoecious and wind pollinated.

Among the angiosperms, a much larger number of species (20000–25000) evolved during the Cretaceous and the development of the progenitors of many presently known species continued through the early Tertiary Period of large climate change and widespread extinction. A wide variety of reproductive behaviour evolved, including monoecy and dioecy, and all forms of sexual divergence in between (Bawa & Opler 1975; Charlesworth & Charlesworth 1978; Charlesworth 1984, 1989), which affected the rates of population and species divergence and evolution. However, the establishment of mating barriers is not the *sine qua non* of speciation for plants as it is for animals, and other forces have relatively more effect on the creation of new species (Grant 1971; Grant & Mitton 1979). To understand how the present array of tree species arose and how future diversity may evolve, several evolutionary factors must be considered.

The selective forces of the environment may favour different inherited traits and force divergence. Mutations may also generate diversity, and the reproductive system itself may erect barriers to intermating among the derived and the progenitor groups. Other events such as geographical isolation may form barriers, and further isolation could reinforce the effects of selection and mutation to generate distinct species. If further divergence occurs, then species-level differences can accumulate to generic and familial levels of divergence as later speciation events accumulate differences and sets of descendants diverge further. Such divergence may be rapid or slow depending on the strengths of the evolutionary factors, but all factors were probably strong simultaneously during those periods of rapid speciation such as occurred at the Cretaceous–Tertiary boundary.

3.2.1 Cladogenesis

The development of species (cladogenesis) is seen as a process that proceeds from the normal evolution of a species which, if no split occurs, would be called anagenesis, or the evolution of a species lineage without subdivision. Within a human lifetime and on a local scale of observation, microevolutionary changes composed of shifts in species traits can be observed, such as when tree breeding programmes change growth rates or stem form. Species seem to remain fairly constant in their general form and behaviour, though with gradual shifts in a continuing evolution of adaptation (anagenesis). Evolution may be largely dominated by events common to the whole species, if populations are often intermating and subject to similar environmental conditions. Even if subject to different and strongly selective environmental conditions, occasional interpopulation mating can force their evolutionary dynamics to be very similar to that of a single population. However, on a longer time-scale than included in a single human generation, or on a large geographical scale, species may not always evolve as a unit with gradual continuous steps.

On a global scale, it is expected that environmental differences are so large that the segregation of different adaptive capacities would often be advantageous and hence that selection on heritable traits would be a factor in speciation. Since ecologically critical factors are not distributed uniformly around the world, divergence rates among species would also not be expected to be uniform. We can see some of the broad strokes of large effects in the contrasts between the boreal and tropical forests. The obvious factors of temperature and moisture are so different among broad regions that divergent adaptations, at least to major climate extremes, would clearly be advantageous. In addition, it can also be observed that in the boreal forests most species carry both sexes on the same tree, have their pollen carried by wind and are usually found in relatively large continuous stands.

In contrast, tropical forest trees often carry males and females on different trees. These species are most often sparsely distributed throughout most of their range (Ashton 1976; O'Malley & Bawa 1987), and insects, birds or mammals carry pollen. Opportunities for inbreeding or for restricted localization of mating neighbourhoods are much greater in such species

(Bawa 1974). Tree species differ in their reproductive behaviour as well as in the environments to which they are better adapted. It can therefore be expected that forest regeneration dynamics would differ among regions and that their genetic dynamics would also differ, leading to segregation into divergent populations or subspecies.

Among the angiosperms, many reproductive systems exist, from those that are primarily self-pollinating to those which are primarily outcrossing (Murawski & Hamrick 1991). Pollination may be predominantly mediated by wind, insects, birds or mammals, and seed dispersal may be proximal or distant; both may be episodic between years or may be continuous throughout a year. The mating system of some species is conducive to restricting the pool of potential parents. The mating system may also force some species to depend on one or other type of pollinator and would lead to reproductive isolation (Bawa 1974). In addition, the possibility of the mating pool being very limited would, if coupled with a non-continuous ecological distribution, allow some species sets to have very rapid rates of evolutionary divergence.

Thus if ecological conditions act as selective agents that differentially affect survival or reproduction of some genotypes, the genetic composition of the population may be biased by selection for tolerance to competition or by sexual selection on rare genotypes. In such cases population evolution would be dependent on the combination of genetic and ecological events. The evolution of populations can then affect not only the immediate survival of that population but can also determine the future subspecies structures that may lead to further speciation or other genetic structure for the species as a whole. Thus, the mating system, environmental differences and history can all affect the rates of species creation by cladogenesis. Cladogenesis and anagenesis would be expected to often follow the same rules of evolutionary dynamics whether in single or multiple populations, and to be determined by the effects of mutation, migration, selection and population size.

The forces of evolution act on the capacity of a species to respond to challenges of the reproductive or vegetative environment, and that capacity is determined by its genetic endowment. This genetic endowment determines what it is that we call a species at any one moment in its evolution and also determines what its evolutionary trajectory may be. Therefore, discussing the species present in forests requires discussion of what their genetic endowment is, as well as what forces are acting on it. Discussion on how species are changing, or can be conserved, also requires discussion on how their genetics is changing and how their genes can be conserved. Since many of the dynamical events that affect evolution occur among individual trees, the individual is considered to be the primary unit of evolution, and the genotype (the collection of all the genes in an individual) is the unit of gene expression. The phenotype is the physical expression of the genotype within the environment of the individual's lifetime and is the unit of selection. The correspondence between individual-level events that we can observe and the long-term evolution of species groups, which takes a longer time, is the realm of evolutionary and population genetics.

However, evolution is not a unidirectional phenomenon, with trees and their genes passively exposed to a selective environment. It does not proceed solely from a given set of environments that determine which genes are present, nor solely from genes to how the trees will perform. At a large level, it is clear that the forces which affect reproductive behaviour and selection are themselves changing partly as a result of the genes that are favoured. As climates and human influences change, the ranges and density distributions of species can also be expected to change. The migration of species may thus be driven by general environmental influences, but would also be influenced by how other elements of the biota abet or hinder their movements. These are factors that are partially dependent on which genes are present. For example, in the temperate and boreal forests the postglacial movement of species would not be expected to be uniform since they differ in colonizing ability and their initial distributions (Pielou 1991).

In addition, some species would form migrational barriers to other species and hence the history of migration would also affect the distribution of species and would inhibit the existence of an optimal composition of species in communities. One pioneer species will often exclude other potential pioneer species, and secondary succession may depend on those initial events as well. Thus, tree communities may in fact be associations of species that have not coevolved to their mutual benefit but coexist due to independent adaptations to similar environmental conditions, or that jointly occur for ephemeral and historically dependent reasons. Evolution within forest tree species may therefore be largely independent of the evolution of other tree species. They are less likely to be independent of interacting species such as their pollinators.

The dynamics of forest ecology depends on the characteristics of the species that can invade and maintain themselves, and is therefore partly a result of the species endowment available. However, the evolution within species also depends on those same ecological factors. While it may be obvious that major differences in forest dynamics exist among the forest zones, all forests go through periods when some species establish themselves in new populations while others die out, at least in local areas. On a given area, the displacement of one species by another is a common occurrence that can be triggered by extrinsic forces such as fire or by autonomous factors such as competitive exclusion of one species by another (Shugart 1984). Some of these displacements are mediated by differences in species adaptations to regeneration under shade vs. open areas or to differences in competitive ability for soil, water or light resources. Some displacements may be due only to the accidents of the timing of seed arrival or other random events.

While a forest cover may exist in some areas for many centuries, at least local variations in species composition over time would be the general rule and hence the local regeneration of any one species iswould be expected to be often sporadic and episodic (Loveless & Hamrick 1984; Shugart 1984). Even for forests composed of shade-tolerant and self-replacing species, gaps frequently disrupt the continuity of forest cover and generate

sequences of displacement (Clarke & Clarke 1984; Clarke & Clarke 1991). Some species, called pioneers, are also dependent on large gaps or cleared openings for regeneration, and normally form dense colonies in the foundation of populations (Bannister 1965). The seeds of such species may be easily transported over long distances (Wright 1976) or may be stored in a soil seed bank, enabling cleared areas to regenerate a forest cover. Their mode of regeneration also implies that they suffer local extinctions, through their own effects on their environment, through displacement by superior competitors or by environmental catastrophes. Forest dynamics itself brings evolutionary pressure on the forest's constituent species.

3.2.2 Selection

The selective advantage of one individual over another is one of the main forces of evolutionary genetics, and along with mutation, migration and drift form the genetic endowment of a species and of the populations of which it is composed. Evolution from the individual level of selective and reproductive events is one of the ways that we can understand the transition from what is called microevolution to higher taxonomic levels, or macroevolution.

Competition dynamics is often invoked as one of the strong selective factors in forest tree evolution (Campbell 1979). Species that benefit from or tolerate shade as seedlings or saplings may occur in higher frequency under a canopy, or may invade areas already occupied by other vegetation. They may emerge into the upper canopy when openings occur and hence the distribution of reproductively mature individuals would follow patterns established by the size and frequency of canopy openings (Bawa 1974). Some species in tropical forests require shade or a protective canopy cover and can reproduce only under a canopy where their crowns can form a low canopy layer. However, light is not the only environmental limiting factor that affects the distribution of trees. Other factors such as soil type and soil moisture can have a strong influence on the extent and patchiness of a species distribution.

One of the basic necessities for evolution is the

capacity to adapt to new stresses in the physical or biotic environment (Loehle & Namkoong 1987). For any selection to lead to adaptive evolution, there must be genetic variation in the stress response and the trees with the favoured trait expression must leave more progeny.

Since many environmental factors are themselves changing in response to extrinsic forces, we can expect that stand composition and species distributions are perpetually changing for extrinsic reasons. Thus, selection factors are not expected to be consistent over time. In addition, historic influences on the recruitment pool for species establishment can affect the probabilities of species replacements and simple stochastic variations on the demographic distribution can also influence distribution for reasons not related to optimality of fitness. Hence, the accidents of history, stochastic variations in environmental variables, lag times in selective effects, as well as deterministic causes affect the actual distribution of species.

Virtually all species can be considered to be involved in a dynamical system, to still be migrating both geographically and ecologically, and to be evolving in response to directional evolutionary forces as well as to both deterministic and stochastic forces. Therefore, forest tree species can be expected never to be in an optimally adapted state or, if so, only transiently.

Thus, while selection can be a significant force in the evolution of species and in the evolution of diversity among species, it has effects within a complex ecosystem and on complex organisms, and its effects have a lag time. Its effects also depend on the genetic endowment and structure of variation already present in the species and on the other forces of evolution, including mutation, migration and drift. These forces can change the species characteristics and if sufficient genetic variation is maintained within or between populations, continued evolution can maintain the species and can sustain forests. *Sustainable forestry and the maintenance of species thus require a sustainable genetic resource in some form*. The wealth of genetic variation within and between populations and the combined genetic and ecological dynamics thus jointly determine the inherent structure of forests.

3.2.3 Migration

If we examine how species migrate and change their geographical distributions over a few generations, the establishment and extinction of local populations is not an unusual event. If such populations become reproductively isolated and if, either by chance or through selective events, they also genetically diverge in mating behaviour and adaptability, the basis for developing a new species exists. If populations do diverge, they would be likely to do so at many genetic loci and at many alleles at the same time. Hence, whole suits of traits and sets of genes would be expected to shift together as populations segregated into different mating groups. Not all of these changes would be ascribable to selective advantage of the alleles, but rather to chance events and to accidental associations of genes and traits.

In the relatively simple boreal ecosystems, the movements of populations escaping from glacial encroachment (Pielou 1991) into sufficiently hospitable areas and alternately advancing into retreating glacial boundaries would be expected to allow populations to carry adaptations to conditions that lag behind actual environments. Nevertheless, during episodes of environmental change or of species movement into new environments, some traits would be of adaptive significance and near optimality, while others may be of little significance or may lag far from any current optimality (Namkoong 1969). Only if the populations carry sufficient genetic variation could they respond rapidly to optimizing selection and ultimately achieve or approach an optimum condition, and then only if the environment is also stabilized.

During periods of change, populations may diverge from a base population or from each other in location or habitat (Grant 1971). For divergence to become inherent among them, strong selection on populations with genetic variation in traits of adaptive significance and an almost absolute barrier to pollen or seed migration between populations would be required. For *Pinus contorta*, strong interpopulation differences can be observed despite substantial gene migration (Yang & Yeh 1995). Most often, it can be expected that without designed or accidental interventions, separated populations would sooner or later reconverge and that periodic intermating among populations would be frequent enough that traces of different gene frequencies and linkages would be difficult to detect. If populations once diverge, they may often simply integrate upon secondary contact and re-form as a single intermating population. Without strong mating barriers for a long time, the long-term evolution of the species as a whole would be a slow approach to an average optimality by the increments of mutations to new alleles at some loci.

Occasionally however, divergence is strongly established for long enough that at least partial mating barriers are formed. Then, splitting of a single former species into two or more branches could generate new species. While the occasional speciation event may not be often repeated, incipient speciation could generate stronger and stronger divergences in genetic composition as selective differences and unique mutations accumulated. Once a coherent species is formed, any further speciation events would follow on the basis of prior divergence and there could then ensue a hierarchical series of branching species that ultimately develop into genera and family level divergence. This is the basis on which phylogenetic relationships are constructed and which underlies the power of molecular genetic data for reconstructing evolutionary trees.

The existence of large populations undergoing population isolation and of changing environments were pronounced during the Cretaceous Period, when many of the major forest tree speciation events occurred. Subsequently, this led to the generic and family level structures that we see today. Though the potential for speciation events continues to exist through modern times, the events occur at slower rates than humans are attuned to observe. However, by using experimental evolutionary approaches, including strong selection and the construction of strong mating barriers, such as among divergently selected breeding populations, it is possible to create large population divergence both genotypically and phenotypically. As long as new species arise by splitting parental species, and each accumulates random mutational changes, divergence among species would follow patterns of statistical

distributions, and by tracing backwards in time, similarities can be used to reconstruct coalescent histories.

3.2.4 Chromosomal change

Occasionally speciation events in plants also occur as a result of reorganization of the genetic material, especially by means of chromosome rearrangements and duplications. For example, it is reasonable to conjecture that chromosome rearrangements explain the number of chromosomes of Douglas fir, which has 26 chromosomes in its vegetative cells instead of the usual 24 carried in all other conifers. This is probably the result of one of the usual conifer chromosomes dividing in two and developing its own centromere. In tropical hardwood species, it is not uncommon to find irregular series of chromosome numbers (Mehra & Bawa 1969), seen in the dipterocarp species especially (Jong 1976).

Within most families, the basic haploid number of chromosomes (x) is the same for all genera and species, but for some, as among the Aceraceae and Betulaceae, there is a wide variation in chromosome number among genera and species (Grant 1971; Morgenstern 1996). There is a series of species with chromosome numbers that are multiples of the basic number (e.g. $2x$ to as much as $6x$) which form distinctive species. In aspens, occasional triploids arise following species hybridizations (Morgenstern 1996). The genes in these species may simply be replicated among the chromosome sets and all continue to add to gene expression.

However, the kinds of interactions among alleles can be affected by their pairings as well as by the numbers of alleles at the same 'locus' and among 'loci'. Such a series of chromosome multiplication is a polyploid series, and may arise as the result of a meiotic failure of the original nuclear genome to split into two separate nuclei. This may occur often enough within a population that the pollen or egg cell can mate with another that also carries the doubled set, and instead of producing a diploid would produce a tetraploid. If a diploid germ cell is paired with a normal haploid cell, a triploid would form, although these usually fail to generate viable germ cells. However, if a diploid and a tetraploid happen to mate and those cells double their chromosome set, a hexaploid could result.

The creation of polyploid series may be stimulated by hybridization events, where chromosome mechanics are sufficiently disturbed that unusual meiosis events may occur more frequently. However, hybridization events by themselves can also create new kinds of species dynamics such that new species are generated even if the chromosome numbers are the same and do not otherwise change.

3.2.5 Hybridization

Hybridizations among divergent species occur when they are not too divergent for the intermating to occur but are distinctly different in form or function. They may be frequent enough among plant species that their potential for shaping evolutionary relationships cannot be ignored (Barton & Bengtsson 1986). Hybridization occurs with much greater frequency among plant species than among animals. Examples of hybridization in fruit trees can be found with different outcomes. At some point during the divergent development of distinct species, a barrier may exist that further reinforces intermating such that no hybridizations may occur, although this seems to be at fairly wide levels of divergence. Before that time, hybridizations can intermingle the genetic endowments of the incipient species and can thereafter serve as foundation taxa for further evolution of divergence among species. Therefore, the existence of hybrids and populations of intermediate genotypes implies that the speciation process has not gone so far that new species can evolve by hybridization. In fact the processes of divergence and hybridization are continuing to this day. The large areas over which lodgepole pine and jack pine hybridize indicate a broad compatibility of distinct species that may stabilize in some areas (Wheeler & Gurries 1987), but with intermediate types developing as a new taxon.

In some cases, the population swarm that characterizes recent hybridizations may segregate the parental types as second-generation intercrosses or as backcrosses to their parental populations.

Alternatively, they may maintain intermediate types in most traits and eventually stabilize as a taxonomic unit with some intermediate trait expression and with some mixture of different parental traits.

In the evolution of a hybrid population between *Pinus taeda* and *Pinus palustris* created by the natural intercrossing of the parental species in the 1920s, later generations failed to maintain any intermediate types and, 20 and 40 years later, when their progeny reproduced they predominantly gave rise to the parental types (Namkoong 1966). Apparently, the original hybrids were successful only in the peculiar environment of abandoned farms from which fire was excluded; when a forest cover was established, with occasional brush fires, the hybrid either was not selectively stable or was chromosomally unstable, or both. In contrast, the hybrids between *Picea sitchensis* and *Picea glauca* form a large and apparently stable zone of introgression in northern British Columbia and hence can be stabilized as a hybrid with backcrossing to both parental species (Yeh & Arnott 1986). The hybrid types may disappear or may form a permanent zone for hybridization, or may eventually form a third 'species'.

If just two parental species are involved in the development of a hybrid, it might be difficult to distinguish if an intermediate type is a recent hybrid of historic species or if the intermediate is a historic source population for newly divergent taxonomic units. If historical evidence is available on prior migrations, then the origins of the units can be readily distinguished by the times of establishment. However, in the absence of historical data the intermediate may be either a hybrid progeny of two divergent parents or the parental source of divergent progeny. If genetic measures can be obtained for multiple, independent and neutral loci, it would be expected that the isolation of divergent populations would initially involve a random segregation of genes from the original population.

This is a case of single source population producing two derived branches, and instead of simply splitting into two, three branches may exist from a single origin. Then the derived progeny populations would receive an admixture of alleles such that among the progeny populations themselves there would be similarities in some traits and differences in others. If alleles or traits were different between the progeny populations, they would have diverged from the parent population but would not necessarily diverge in symmetrically opposing directions. The allelic richness of the source population would be expected to be at least as great as that of both of the derived progeny populations, each of which would contain subsets of the parent population's alleles.

In contrast, a recently derived hybrid population would share most alleles at frequencies between the parental populations and would therefore be intermediate in most traits simultaneously. The divergent populations then would seem to diverge in symmetrically opposing directions from the intermediate population at all divergent loci and in all traits. The hybrid population would also be expected to have alleles that exist in at least one of the parent populations but not to contain all of the alleles of both parents.

The eventual divergence of hybrid populations can, of course, involve backcrossing and selection such that they ultimately cannot be distinguished from the previous model of divergence from a parental stock. However, linkage disequilibrium in the hybrid population should linger and be detectable for many generations. In this case, two parental species can give rise to three species if the hybrid continues to develop and establishes mating barriers such that it can evolve independently. In the case of *Pinus rigida*, *Pinus taeda* and *Pinus clausa*, there are three two-way hybrids between the species that could indicate a close relationship between them and a splitting of an original one into two or three by a subsequent split. However, it could also imply a recent hybridization of at least two of the parental species (Smouse 1972).

These forms of speciation can lead to further diversification of lineages or to convergence of genetic lineages, if some sets of genes are transferred among several species. These forms of speciation may then form multibrachial taxonomic structures and cross-linkages of sets of genes, neither of which conforms to evolutionary trees. The existence of three-way hybrids in pines, for example, indicates that such complexities in speciation may not be unusual at least among the pines and may not have been unusual in the development of the family. The construction of genetic lineages therefore is not identical with the construction of species lineages, and local events in the evolution of populations can have significant effects on long-term phylogenies.

The events involved in speciation do not generally occur in all units of a species at the same time, but instead involve genetic events that change individuals or that occur among population isolates. If species-level events are contained within a matrix composed of other species, they may often exist as a set of reproductively isolated islands, and if the parents of new populations rarely mate with members of other islands, the species may exist in a metapopulation structure of loosely connected subpopulations. This may even lead to incipient speciation as the islands become more distinct if they further form mating barriers among populations. Thus, a species may be distributed in the form of large, indistinct, continuously intermating populations, or as discrete and distinct metapopulations and in mixtures of large and small populations with partial migration among some subsets. The amount and distribution of genetic variation will be influenced by the metapopulation structure as populations split into large or small segments and as they either advance or retreat from habitable sites.

3.3 THE WEALTH OF POPULATIONS

Within plant species, genetic variation commonly exists at a high proportion of loci and tree species commonly exhibit the highest levels of all plants (Hamrick & Murawski 1991). For the species surveyed, which are mostly boreal and temperate pioneer species, the levels of genetic variation within populations often seem to exceed those between populations. The lack of strong differentiation may be due to the wide dispersal abilities of pollen and seed, which maintain a degree of connectedness between mating neighbourhoods among trees over long distances. A few studies of tropical tree species that are not pioneers indicate considerable variation among populations of trees, which may reflect a finer subdivision of mating neighbourhoods.

Another measure of genetic variation is the heritable variation that exists in traits controlled by many gene effects simultaneously but in which individual gene loci may have small effects. These are the so-called quantitative traits that can involve many genetic loci which act indirectly on the observed trait and hence have small direct effect. Most of these loci would not change much in frequency from one generation to the next. Cumulatively, however, they could cause large changes in the trait expression. Thus, many of the traits selected for in breeding populations can shift mean expression by as much as one standard deviation in one generation even if there is little shift in gene frequencies at loci sampled by molecular means. Furthermore, since differences among populations in the frequency of molecular markers are more difficult to detect than differences measured in quantitative traits (Lewontin 1984), traits of adaptive significance may be more obviously segregated among populations than molecular genetic measures may indicate.

Some traits and the gene loci affecting them may be under strong selection pressure that is divergent between populations, while other traits may be under mild or virtually no selection pressure. Some traits may be indirectly related to strong selection events, such as caused by fire, drought, or forest practices that leave few resistant individuals, but only weakly related to resistance mechanisms. Other genes may not be presently involved in any recent selection events. Thus the genes associated with selection response may be under strong or weak selection according to the strength of the selection event and the degree of relatedness of the trait to the selection agent. There may also be multiple genes that affect a trait expression, and the combined effects of the genes may not be simply assigned to any simple action of any one of them. This kind of multigene interaction is called epistasis and it may cause multiple gene locus differences to be more pronounced at the level of trait observations than in single-locus allele frequency differences. These kinds of interactions may mask any obvious trends in gene frequency changes associated with selection trends.

Since the phenotypic traits themselves can be expected to be under a wide array of selection intensities and the genes associated with them can also have a wide range of influence on those traits, we can expect that there would be a wide range of selection pressures among the genes of any one population (Campbell 1987). Thus, if the relative divergence of populations was tracked by weakly selected traits or loci, population similarities would predominantly reflect migration effects. In the absence of selection, even very low migration rates would tend to maintain similar allele frequencies among populations. However, if divergence was tracked using traits or loci that respond strongly to divergent habitat selection, then similarities would tend to reflect habitat differences. In any one population, strong directional selection would force disfavoured alleles towards the point of extinction unless maintained by an input from migration or mutation.

Among populations exposed to divergent directional selection, different alleles would be at high and low frequencies among them and hence their frequencies would be highly dissimilar. If a large number of traits or loci are used and some loci reflect similarities where others reflect multiple patterns of divergence, then contrasting patterns of allelic distribution must coexist. Similarly, traits subject to different intensities of divergent selection would also display different degrees of divergence. Clearly, if phylogenetic relationships were estimated using selectively significant loci along an environmental gradient, then genetic similarities or distances would be correlated with that environmental variable. Alternatively, if selectively neutral traits were used, the genetic distance measures would be correlated with migration distances, and the derived phylogenies would be different. The existence of differing strengths of divergence has been observed in variations in the patterns of geographically related population traits in *Pinus armandii* (Ma 1989). Differences in the amount and geographical pattern of isozyme allele frequencies and in growth traits have also been found for *Pinus lambertiana* (Martinson 1996) and red alder (Hamann et al. 1998).

In addition to those forces of evolution, the mating system can have heterogeneous effects on the distribution of alleles. If pollen and seed have different migration distances or different direction of flow due to their migration season or the influence of different wind pollinator effect on the sexes, then the balancing effects of migration and selection would be different (Namkoong & Gregorius 1985). Even for selectively neutral alleles, differences between the migration paths of the sexes causes heterogeneity of allele distribution.

Genetic variation also exists at several levels of expression and the amounts and patterns of variation can depend on which measures of variation are used to estimate that differentiation. With currently available tools of molecular measurement, it is possible to measure very fine levels of difference so that individual trees (or humans for that matter) can be discriminated by unique changes in the fine structure of single genes. It is also possible to discriminate differences among local populations if they share maternal or paternal lineages, by using particular molecular genetic markers that distinguish among populations even though the populations are not obviously different phenotypically (Millar & Libby 1991). Furthermore, some genes are expected to be more influenced by the different forces of evolution and therefore the distribution of variation among genes may vary according to their function (Ma 1989).

Therefore, there is no one preferred measure of genetic variation but several options, and the choice of which measure to use depends on the purpose of the measure and the type of gene being measured. For forest trees it is satisfying to note that the levels of genetic variation measured are usually among the highest in the plant kingdom, although their distribution and the differences among loci have not been well studied. If the frequency of heterogeneous associations is high, the evolution of forest species may be unusually complex.

Some forest trees have relatively high levels of genetic variation between populations in several traits and at several molecular loci, while others may have low general levels (Millar & Libby 1991). Since the beginnings of interest in genetics, foresters have known about economically significant levels of heritable genetic variation among populations in traits relating to growth and soil and climatic adaptability (Morgenstern 1996). One of the earliest tests to indicate the utility of population differences is recorded for *Cryptomeria* in Japan in 1883 (Toda 1961). However, the existence of graded changes in traits is widely recognized in forest trees as clines (Teich & Holst 1974; Millar & Libby 1991; Morgenstern 1996) and with high levels of variation among populations relative to genetic variation within populations (Wheeler & Gurries 1982). The variation patterns often differ depending on which traits are measured but growth rates and climatic adaptabilities are often considered to be optimized within local populations (Rehfeldt 1989). While there is debate as to the generality of the concept that local populations are optimal (Namkoong 1969; Rehfeldt 1995), regular trends in performances are often observed.

The geographical patterns of variation among populations or provenances may vary from a single, large, contiguous and intermating population with gradations across mating and environmental neighbourhoods to one of discretely divergent populations with little relationship among subpopulations. Some species may be composed of a main, large, central population with satellites connected only to the central one and with only one-way germplasm flow, or may be a series of more or less equivalent-sized populations with equal and multidirectional flows. Some of them may exist along streamside corridors as single units with connections between drainage basins being more tenuous, while others may form island populations on the tops of separate mountain peaks or ranges.

Since the stability of the interpopulation structure depends on the stability of ecological dynamics, the level of population or provenance differences may fluctuate over time and geographical area. In turn, the response of populations to environmental divergence is contingent on the genetic dynamics that can occur within an intermating neighbourhood. For the most part, the potential for selective divergence among the populations is moderated by migration among populations. Only traits with strong divergent selection would maintain high levels of interpopulation difference.

3.4 THE WEALTH
OF INTRAPOPULATION
GENETIC VARIATION

Within populations, four forces (selection, migration, mutation and population size) determine the structure of genetic variation. The interaction of all four factors can be highly complex and their effects are determined by the mating system and the kinds of gene effects that exist. A problem in analysing even this level of genetic evolution is that the kinds of gene effects that exist depend on the effects that other forces of evolution have had on gene actions themselves. To consider how the forces can affect evolution, we can first start with each of them and then consider how their interactions can change predictions.

For selection to have an effect on speciation and the structure of forests, there must be a difference among genotypes in the traits affected by survival or reproduction events and the individuals must accordingly leave more or fewer progeny than other genotypes. The genotype is defined as the individuals with a distinctive ensemble of alleles at its gene loci that distinguishes it from other genotypes. This is a multiple gene locus definition, which is often simplified by considering every one of the loci to be independent in the occurrence of its alleles in both frequency and effect.

In this case, we can consider the genotype to be the summation of all independent loci and hence can examine gene effects and dynamics one locus at a time. This is, of course, an oversimplified view since genes are in fact linked on chromosomes and therefore have linkages that correlate their occurrences, and most genes affect processes that are combined in their net effect on traits; furthermore, most genes affect enzymatic pathways that ultimately affect several traits when measured at the level of whole organism performance. These effects are called linkage, epistasis and pleiotropy, respectively, and tie the whole organism together into an integrated set of actions and functions that can survive and reproduce. While undoubtedly true in general, there is little evidence that individuals are so bound by such constrained genetic constitutions and physiological limitations that substantial independence among loci exists. The evidence for the existence of strongly selected, tightly linked 'coadapted gene complexes' is not strong and hence the model of independence can be usefully employed, even though it is surely wrong at least in detail.

Within a gene locus, many alleles may exist and the pairwise interactions among them are called the dominance relations. For a diploid organism, if both alleles at a locus are the same, the individual is said to be a homozygote; if the two alleles are different, the individual is said to be a heterozygote. If the heterozygote has a phenotype intermediate between the two homozygotes, the dominance effect is considered to be zero and the gene action is said to be additive. If the heterozygote is other than intermediate, it may be less than the lesser of the two homozygotes. If its phenotype lies between the homozygotes but is not exactly intermediate and if it appears to be greater than the greater of the homozygotes, the gene action is said to be underdominant, partially dominant or overdominant, respectively.

The effect of selection would depend on the kind of gene action, since directional selection for greater or lesser phenotypes would favour different homozygote and heterozygote states. Selection may also be stabilizing if it favours an intermediate phenotype or disruptive if it disfavours the intermediate and its effect on the frequency of the alleles would depend on the gene actions (Namkoong & Koshy 1995). Directional selection experiments seem to indicate that most alleles have partial dominance actions, though many may be involved in multiple locus interactions. Since selection would be expected to favour survival and reproduction of certain genotypes, we expect that it would increase the population level of fitness.

Selection is therefore usually considered to have a positive effect on the evolutionary potential of a population or species in small increments every generation but may not generate an optimal condition at any one time. For alleles that have partial dominance effects, the dynamics of the rates of displacement of one allele by another are such that the frequency of the allele itself affects its own rate of displacement. When alleles are at either low or high frequency, their rates of change are slower than when they are at an intermediate frequency. Except in small populations where accidental loss can occur, selection would not generally lose alleles that are favoured for adaptation.

Another force of evolution is mutation, which is usually considered to be a random change in gene structure and function that occurs indiscriminately throughout the genome. Since most species have evolved a genetic and physiological level that is close to a sufficient capacity for continued evolution, novel changes are not expected to improve performance or fitness but rather to generally decrease adaptability. However, there are always going to be mutations and while most may be disfavoured by selection, there is likely to be a balance between the rate at which new mutants are introduced and the rate of their elimination.

Most mutations occur at low rates for any one locus in any one generation and would not be expected to affect the adaptability of a population unless they are accumulated in small populations. For most conifer species where the frequency of deleterious mutants has been estimated, there are a larger number of lethal alleles carried at one or another locus in most individual trees than in most other plant or animal species. This might indicate that there are many effective loci in trees that can mutate to a harmful state, that the mutation rate is high or that the mutants are favoured in some conditions or are of some advantage in heterozygous conditions (Bishir & Namkoong 1987). However, independent mutations and selection can generate differences among populations, which can increase fitness in each. Hybridization among populations can then generate high levels of genetic variance in traits that may then be useful in new environments (Koshy *et al.* 1998).

A third force of evolution is migration, which can occur among many different kinds of population structures. The input of pollen or seed from other populations that have different allele frequencies may bring in alleles that might otherwise be lost, and if the migration rate is high enough among all populations may induce all populations to have approximately the same frequencies. If the immigrant alleles are favoured, then further immigration would advance the rate

of its increase; if disfavoured, immigration would maintain alleles at higher frequencies than expected in a manner similar to mutation; and if neither, would tend to make frequencies homogeneous among all populations. In the presence of divergent selection among populations that exist in different environments, selection would act to eliminate migrant alleles and would maintain population divergence.

The fourth force is sampling error induced by small population sizes, which tends to increase the probability that low-frequency alleles are lost by chance. If populations remain at low sizes for several generations, the chances of random loss are accumulated; even populations of more than 20–50 adults can lose alleles that may have started with moderate frequencies. Since most forest trees seem to carry many deleterious mutants, even if at low frequency at any one locus, another problem is that small populations will suffer from inbreeding depression and may eventually suffer such debilitation from several mildly depressive mutants that the population goes extinct (Lynch *et al.* 1995). Individual populations would have to be large enough to sustain viability, and to avoid environmental accidents and catastrophes if no migrants were recruited.

There are two problems faced in persistently small populations, the loss of alleles and inbreeding depression; hence, larger population sizes are generally better for avoiding extinction. However, the effects of selection, migration and mutation are confounded with the sampling errors incurred in real populations of finite size. Thus, if populations are separated and selected for different environmental optima, migration can introduce alleles in high frequency that depress fitness. On the other hand, migration can maintain alleles in populations where they might be useful, but would be lost due to sampling accidents in small populations.

It is reasonable to expect that all species are under the influence of several or all of the forces of evolution simultaneously and that the balance between the several forces is not the same for all genes in all populations. Furthermore, since the physical and biotic environments of forests are rarely at an equilibrium, the genetic and ecological dynamics of most species would have to be considered to be in other than a stable state.

The actual distribution of the multiple functional forms of alleles at each of the several tens of thousands of gene loci in each of the individuals of a population must therefore be the resultant of the mixture of forces felt at each locus, subject to historical events. The vast possibilities of mixed distributions are overwhelming, although it is also obvious that for most of the species we can study, species are not random mixtures of an infinite number of possible combinations.

3.5 CONCLUSIONS

The wealth of genetic variation that presently exists within species can also be used to diversify populations when subject to intensive selection and breeding. The examples of one- and two-generation breeding effects are numerous and indicate that ordinary breeding in trees can mimic the rate of progress experienced in agricultural crop species. The early experience with many species indicates that differences in average performance can produce plants that lie outside the range of present performance within three generations, and that within 10 generations there would be no individuals in either population that would overlap in performance with the other type. Within 20 generations, long-term selection can generate even further differences due to the effects of unique mutations in different populations and thus entirely unique gene combinations can exist when only one variety formerly existed.

The beginnings of speciation may therefore not require very long periods of observation if strong selection in populations is coupled with a complete barrier to interpopulation migration. However, if this is possible, it is not clear that such events occurred, or could have occurred, often in the past or that future speciation can be artificially induced. It is also not clear whether the level of speciation that we observe today has or has not completely explored the space of all possible gene combinations. It may well be that the complexities of growth and development are such that newly desirable trait combinations are not possible within the constraints of development. Similarly, the complexities of linkage and epistasis may impose constraints that prohibit reaching those evolutionary states from any present state. We do not know what our limits are, either retrospectively or prospectively.

The nature of genetic variation within the presently existing tree species is not a fixed endowment, neither is the array of forest tree species that we see today. It is certainly wise to avoid losses when possible, although in a dynamical system losses are inevitable, whether caused by human interventions or not. The evolutionary processes that generate the genetic variation within and among species are in a dynamic state and the same processes are what we use for breeding and which serve as the basis for response to future evolutionary pressures. The question for conservation and management is therefore not what we must preserve but rather what our endowment is and where it will go.

REFERENCES

Ashton, P.S. (1976) An approach to the study of breeding systems, population structure and taxonomy. In: Burley, J., Hughes, C.E. & Styles, B.T., eds. *Tropical Trees*, pp. 35–42. Academic Press, London.

Bannister, M.H. (1965) Variation in the breeding system of *Pinus radiata*. In: Bakerand, H.G. & Stebbins, G.L., eds. *The Genetics of Colonizing Species*, pp. 353–72. Academic Press, New York.

Barton, N. & Bengtsson, B.O. (1986) The barrier to genetic exchange between hybridizing population. *Heredity* **57**, 357–76.

Bawa, K.S. (1974) Breeding systems of tree species of a lowland tropical community. *Evolution* **28**, 85–92.

Bawa, K.S. & Opler, P.A. (1975) Dioceism in tropical forest trees. *Evolution* **29**, 167–79.

Bishir, J. & Namkoong, G. (1987) Unsound seed in conifers. Estimation of number of lethal alleles and of the magnitude of external effects. *Silvae Genetica* **36**, 180–5.

Campbell, R.K. (1979) Genecology of Douglas-fir in a watershed in the Oregon Cascades. *Ecology* **60**, 1036–50.

Campbell, R.K. (1987) Biogeographical distribution limits of Douglas fir in southwestern Oregon. *Forest Ecology and Management* **18**, 1–34.

Charlesworth, B. (1984) Evolutionary genetics of life histories. In: Shorrocks, B., ed. *Evolutionary Ecology: 23rd Symposium of the British Ecological Society*, pp. 117–33. Blackwell Scientific Publications, Oxford.

Charlesworth, B. (1989) The evolution of sex and recombination. *Trends in Ecology and Evolution* **4**, 264–7.

Charlesworth, B. & Charlesworth, D. (1978) A model for the evolution of dioecy and gynodioecy. *American Naturalist* **112**, 975–97.

Clarke, D.A. & Clarke, D.B. (1984) Spacing dynamics of a tropical rain forest tree. Evaluation of the Janzen–Cannell model. *American Naturalist* **124**, 769–88.

Clarke, D.B. & Clarke, D.A. (1991) The impact of physical damages on canopy tree regeneration in tropical rain forests. *Journal of Ecology* **79**, 447–57.

Grant, M.C. & Mitton, J.B. (1979) Elevational gradient in adult sex ratios and sexual differentiation in vegetative growth rates of *Populus tremuloides*. *Evolution* **33**, 914–18.

Grant, V. (1971) *Plant Speciation*. Columbia University Press, New York.

Hamann, A., El-Kassaby, Y.A., Koshy, M.P. & Namkoong, G. (1998) Multivariate analysis of allozymic and quantitative trait variation in *Alnus rubra*: geographic patterns and evolutionary implications. *Canadian Journal of Forest Research* **28**, 1–9.

Hamrick, J.L. & Murawski, D.A. (1991) Levels of allozyme diversity in populations of uncommon neotropical tree species. *Journal of Tropical Ecology* **7**, 395–9.

Jong, K. (1976) Cytology of Dipterocarpaceae. In: Burley, J. & Styles, B.T., eds. *Tropical Trees: Variation, Breeding and Conservation*, pp. 79–84. Academic Press, London.

Koshy, M.P., Namkoong, G. & Roberds, J.H. (1998) Genetic variance in the F2 generation of divergently selected parents. *Theoretical and Applied Genetics* **97**, 990–3.

Lewontin, R.C. (1984) Detecting population differences in quantitative characters opposed to gene frequencies. *American Naturalist* **123**, 115–24.

Loehle, C. & Namkoong, G. (1987) Constraints on tree breeding: growth trade off, growth strategies, and defensive investment. *Forest Science* **33**, 1089–97.

Loveless, M.D. & Hamrick, J.L. (1984) Biological determinants of genetic structure in plant populations. *Annual Review of Ecology and Systematics* **15**, 65–95.

Lynch, M.J., Corney, J. & Burger, R. (1995) Mutation accumulation and extinction of small populations. *American Naturalist* **146**, 489–518.

Ma, C.G. (1989) Geographic variation in *Pinus armandii* Franch. *Silvae Genetica* **38**, 3–4.

Martinson, S.R. (1996) *Association among geographic, isozyme, and growth variables for sugar pine (Pinus lambertiana Dougl.) in southwest Oregon and throughout the species range*. PhD dissertation, North Carolina State University, Raleigh.

Mehra, P.N. & Bawa, K.S. (1969) Chromosomal evolution in tropical hardwoods. *Evolution* **23**, 466–81.

Millar, C.I. & Libby, W.J. (1991) Strategies for conserving clinal, ecotypic and disjunct population diversity in widespread species. In: Falk, D.A. and Holsinger, K.E. eds. *Genetics and Conservation of Rare Plants*, pp. 149–70. Oxford University Press, New York.

Morgenstern, E.K. (1996) *Genetic Variation in Forest Trees*. University of British Columbia Press, Vancouver.

Murawski, D.A. & Hamrick, J.L. (1991) The effect of the density of flowering individuals on the mating system of nine tropical tree species. *Heredity* **67**, 167–74.

Namkoong, G. (1966) Statistical analysis of introgression. *Biometrics* **22**, 488–502.

Namkoong, G. (1969) The nonoptimality of local races. In: *Proceedings of the 10th Southern Conference on Forest Tree Breeding*, pp. 149–53. Texas A & M University Press, College Station, Texas.

Namkoong, G. & Gregorius, H. (1985) Conditions for protected polymorphism in subdivided populations. 2. Seed vs pollen migration. *American Naturalist* **125**, 521–34.

Namkoong, G. & Koshy, M.P. (1995) Managing genetic variance simultaneously for higher gain and adaptive potential. In: Larereau, J., ed. *Proceedings of the 25th Canadian Tree Improvement Association, Victoria, BC*, pp. 17–23. Canadian Tree Improvement Association, Victoria, B.C.

O'Malley, D.M. & Bawa, K.S. (1987) Mating systems of a tropical rain forest tree species. *American Journal of Botany* **74**, 1143–9.

Pielou, E.C. (1991) *After the Ice Age: the Return of Life to Glaciated North America*. University of Chicago Press, Chicago.

Rehfeldt, G.E. (1989) Genetic variances and covariances in freezing tolerance of lodgepole pine during early winter acclimation. *Silvae Genetica* **38**, 133–7.

Rehfeldt, G.E. (1995) Genetic variation, climate models and the ecological genetics of *Larix occidentalis*. *Forest Ecology and Management*. **78**, 21–37.

Scagel, R.F. (1965) *An Evolutionary Survey of the Plant Kingdom*. Wadworth Publications, Belmont, California.

Shugart, H.H. (1984) *A Theory of Forest Dynamics. The Ecological Implication of Forest Succession Models*. Springer-Verlag, New York.

Smouse, P.E. (1972) The canonical analysis of multiple species hybridization. *Biometrics* **28**, 361–71.

Teich, A.H. & Holst, M.J. (1974) White spruce lime stone ecotypes. *Forestry Chronicle* **50**, 110–11.

Toda, R. (1961) Studies on the genetic variance in *Cryptomeria*. *Japan Forestry Experiment Station Bulletin* **132**, 1–46.

Wheeler, N.C. & Gurries, R.P. (1982) Population structure, genetic diversity, and morphological variation in *Pinus contorta* Dougl. *Canadian Journal of Forest Research* **12**, 595–606.

Wheeler, N.C. & Gurries, R.P. (1987) A quantitative measure of introgression between lodgepole pine and jack pine. *Canadian Journal of Forest Research* **65**, 1876–85.

Wright, J.W. (1976) *Introduction to Forest Genetics.* Academic Press, New York.

Yang, R.C. & Yeh, F.C. (1995) Patterns of gene flow and geographic structure in *Pinus contorta* Dougl. *Forest Genetics* **2**, 65–75.

Yeh, F.C. & Arnott, J.T. (1986) Electrophoretic and morphological differentiation of *Picea sitchensis*, *Picea glauca* and their hybrids. *Canadian Journal of Forest Research* **10**, 791–8.

4: Structural Dynamics of Forest Stands and Natural Processes

GEORGE F. PETERKEN

Forest structure is constantly changing in accordance with its internal pattern of growth and mortality and in response to external forces. Even in natural forests, stands may be 'felled' by natural events, such as storms and fires, whilst stands change between events at a measured pace as individual trees grow, compete, dominate and die. The events that disturb the apparent stability and orderly growth of the forest occur frequently enough to form a permanent influence on forest structure and composition.

This chapter reviews these disturbances and describes their influence on forests at various spatial and temporal scales using mainly European and North American examples. Inevitably it overlaps with other chapters. The 'ordinary' growth of undisturbed forest is considered in Chapters 8 and 9, while Chapter 6 discusses how pests disturb forests. The disturbance regime of any particular forest is to some extent a reflection of climate (see Chapter 11). Whilst most disturbances are localized, some act on a grand scale, influencing forest patterns over an entire landscape. Oliver and Larsen (1996) give a comprehensive account of forest growth and change. Peterken (1996) provides an illustrated summary for the temperate region that is related to conservation issues.

4.1 WHAT COUNTS AS DISTURBANCE?

When a forest that has stood, apparently unchanging, for as long as people can remember is blown down overnight there is little doubt that the forest has been 'disturbed'. The change is abrupt, the event is discrete and the effect on the forest is a complete discontinuity with the previous state of affairs. Although there is always a degree of air movement, it is possible to define, quantify and delimit storms in space and time, and thus to count and study their frequency, pattern and impacts.

When the deer in a forest become so numerous that they browse down all seedlings and basal shoots, the state of the forest is clearly being changed but it is more difficult to see this as a disturbance. The changes are gradual, and in any case herbivores are a natural component of the forest. Abrupt increases or decreases in populations may have immediately discernible effects that can be dated, but more often populations fluctuate over years. In this case the impact of the deer is chronic, a continuously operating site factor, rather than a discrete event.

When a large canopy tree in an old-growth forest becomes rotten at the core and hollowed out by fungi, it is likely to collapse in a breeze that would have offered no threat when it was sound. The chance that a breeze will disintegrate a large tree increases when the branches are covered in ice, the crown is in full leaf or the soil has been soaked by weeks of rain. If one looks at the forest as a whole there will always be a few trees collapsing in this way each year, part of the perpetual cycle of regeneration, growth, death and decay. From this perspective, such mortality can hardly be regarded as a disturbance without destroying the meaning of the word. However, from the perspective of the patch of ground under that tree, the sudden disintegration of a tree that has stood for perhaps 300 years is indeed an event; at this smaller scale it can readily be regarded as a disturbance, which presents an opportunity for regeneration.

The cause of a disturbance is less certain the closer one looks. In the previous example, the cause can be the great age of the tree, the presence of fungi, the physical instability created by the hollow trunk, the wet soil or the burden of ice, or the breeze that finally brought it crashing to the ground. There is even ambiguity in the cause of a massive blowdown: whereas the wind is an obvious cause, the fact remains that the stand would not have blown down if it had been young rather than mature, i.e. age is also a 'cause'. Thus, each disturbance has a chain or network of causation, and the kind, degree and scale of its impacts are determined by a multiplicity of factors. Put another way, a moderate gust of 50 km h^{-1} will disturb one stand but leave another unaffected, depending on the state of the stand and its circumstances.

Despite these uncertainties, the storm that blows a whole forest down overnight is obviously a disturbance, or is it? From the point of view of an individual person living near that forest there is little doubt that this is the event of a lifetime, likely to cause as much disturbance to that person's peace of mind as it does to the forest's structure. However, storms recur and the forest will again grow to maturity, but in 200–300 years time the forest may again be blown flat. Since 200–300 years is well within the natural span of many forest dominants, it can be argued that blowdowns are a continuing factor in the life of the forest and the evolution of the species it contains. Events that count as disturbances on one time-scale may be ongoing ecological factors on another.

These points are made not to confuse but to emphasize that ecological disturbances are rarely, if ever, discrete events with sharp boundaries and an obvious single cause. Rather, they are often ill-defined in space and time, and the proximal cause may be insignificant in comparison with the multitude of contributory factors. Pickett and White (1985) have provided a definition of disturbance that incidentally conveys some of the uncertainty: 'a relatively discrete event in time that disrupts ecosystem, community or population structure and changes resources, substrate availability, or the physical environment'. Attempts have been made to distinguish catastrophic dis-

turbances from ordinary disturbances; indeed there is a clear difference between a stand-destroying storm and a wind that snaps a few branches, but the difference is just a matter of degree. Whether one is considering a single kind of disturbance or the cumulative load of disturbances sustained by a single forest, we are dealing with a continuum.

4.2 SOURCES OF DISTURBANCE

Several natural agencies disturb forests and this section briefly describes their immediate impacts and interactions.

4.2.1 Wind

Wind disturbs a very wide range of forest types (e.g. Foster & Boose 1995; Pontailler *et al.* 1997) (Fig. 4.1). It breaks branches and trunks and it blows trees over. A falling tree or branch usually falls to the ground, although a small proportion comes to rest leaning on other trees. A falling tree or branch commonly hits other trees and shrubs, crushing some, stripping branches off others and knocking over a few. Large trees high on steeply sloping ground may skittle numerous trees standing immediately below them, thereby generating distinctive elongated 'avalanche' gaps.

Fallen trees and branches tend to be clustered. It is common for groups of trees to fall together and in the same direction (e.g. Bouchon *et al.* 1973). Even where single trees fall, the branchwood is clustered where the crown fell, leaving the ground clean where the tree originally grew. Fallen trunks of narrow-crowned trees occasionally embed themselves into the ground with the impact of the fall, but generally trunks lie close to the ground surface or form a low bridge supported by major branches and rootplates.

Trees that snap leave a snag. Those that break near the base leave only a ragged stump, but breakage may occur higher up the trunk, often just below the main crown branches. These higher breakages leave a snag, which usually dies within a few years, often after sprouting weakly from the shattered top. As they decay, snags lose their bark and eventually break up. The roots, which remain moist in the soil, decay faster

Fig. 4.1 Destruction of a beech stand at Slindon, southern England by the storm of October 1987. The stand had been carefully manipulated into a small-group selection forest, a silvicultural system that imitates natural structures and dynamics. Although such woods are normally disturbed only by small-scale events, leaving single-tree gaps, on rare occasions they are levelled. Note the survival of a few mature trees and a substantial stock of tall saplings. Left to itself, this stand would recover rapidly to a form of two-storied high forest, stocked principally by individuals established before the disturbance. (Photo by the author.)

than trunks, so snags usually fall when partly decayed.

Blown trees form an upraised rootplate, supported by roots on the downwind side, many of which remain unbroken and functional. Rootplates bring a ball of earth and rock with them as they fall, creating a pit on the upwind side. Trees that have been tipped over in this way often survive the fall, although the chance of this happening depends on the species and circumstances. Thus, a *Tilia* that falls into a gap will generally sprout vigorously from both the rootplate and the prostrate trunk, while a *Fraxinus* which falls into the shade of a neighbouring tree will probably die.

Wind is capable of completely levelling a stand, leaving little more than a few underwood shrubs, saplings and a few snags standing. Disturbance on this scale is achieved by tornados and hurricanes and by the stronger gusts of storms as they hit exposed slopes. Landform can concentrate wind into gulleys and can create eddies that devastate stands beyond a ridge. Generally, however, storm disturbance is patchy, leaving groups of fallen trees within a matrix of unscathed stands, or groups of standing trees in otherwise levelled forests.

4.2.2 Fire

Fires are the most easily appreciated type of natural disturbance and the one most readily controlled by foresters (Heinselman 1973) (Fig. 4.2). Under natural conditions they start with lightning strikes, which smoulder in the litter and undergrowth. They become stand-destroying infernos when fires run into stands with large fuel loads and a multilayered structure, which provides a ladder up which the fire can climb into the canopy. Once a ground fire becomes a canopy fire it is maintained more effectively by convection and spreads faster by wind and sparks. When well established in inflammable coniferous forest with the wind behind it, a fire can burn hundreds of thousands of hectares (Pyne 1982).

Ground fires burn litter, ground vegetation and saplings but leave large trees intact. Two trees standing close together act as mutual heat

Fig. 4.2 A recently burned coniferous stand at Kåtaberget, northern Sweden. Before the fire, this had been a natural old-growth Scots pine stand with an underwood of Norway spruce and Scots pine. The fire killed most of the underwood but spared the larger overstorey trees, save for old individuals with large fire scars that smouldered until they fell. In fact, this was not a natural fire but the first (courageous) prescribed burn of a long-established old-growth reserve in Sweden. Under the supervision of Per Linder of the Swedish University of Agricultural Sciences at Umeå, AssiDomän set fire to the reserve on 27 June 1995 in an attempt to arrest the succession towards spruce dominance. Under natural conditions, fires burned frequently enough to maintain pine and restrict spruce, but a century of fire suppression in favour of the commercial forests around the reserve had permitted a succession that would probably not have happened if the district had remained wholly natural. (Photo by the author.)

reflectors, so are more likely to be killed; moderate fires thus tend to leave a well-spaced stand of larger trees. Crown fires appear devastating and leave groves of blackened snags, although the capricious pattern of fires generally allows groups of trees to survive. Furthermore, heat is directed upwards, so that buried seed, rootstocks, patches of ground vegetation and small mires survive the severest conflagrations.

4.2.3 Drought

Unlike fires and storms, which occur on a particular day, droughts develop over a season or two (Hursh & Haasis 1931; Clinton *et al.* 1993). Initially trees are merely stressed and growth slows, but as the drought endures growth ceases and leaves fall prematurely. However, stress induced by heat and water shortage may occur regularly within the normal year-on-year variation and cannot be counted as a disturbance. Droughts disturb forests when they kill trees or damage them so severely that they decline and die prematurely.

Trees killed by severe droughts gradually disintegrate from the top. Twigs, small crown branches and, eventually, major branches decay and fall, until the roots rot enough for the whole snag to fall. Trees damaged by drought, and the associated excessive dry heat, lose most of their crown twigs and may effectively stop growing for several years, although vigour may be restored eventually by the growth of epicormic shoots from the trunk and crown branches. The rejuvenating crown may contain several dead crown branches for a decade or more after the drought. Strips of bark may also be killed during a drought and this admits pathogenic fungi to hitherto sound trunks, leading eventually to premature death.

4.2.4 Biotic

The interactions between plants, animals and trees are sometimes substantial and sharp enough to cross the threshold into disturbance. Perhaps the most obvious are insect epidemics, which can kill canopy trees over large areas (Blais 1965), and localized population explosions, such as bark beetle in accumulations of dead wood, which go on to kill standing trees. Fungal infections are ever present, but occasionally groups of trees are killed together.

The impact of mammals is perhaps more difficult than most to relate to disturbance. Deer and other herbivores are, or were, a natural component of the forest, although they only become a disturbance when their numbers increase or decrease beyond the range of recent fluctuations. Even when deer are clearly far more numerous than would naturally be the case, high browsing and grazing pressures become more a chronic condition than a discrete disturbance. High browsing pressures are felt most keenly by shrubs and saplings, so that the impact of herbivores on forest structure takes many years to develop. Deer browse selectively and thus influence forest composition (e.g. McInnes *et al.* 1992). If their populations depart from the natural range for no more than a few years, their long-term impact may be barely detectable in the age-class distribution, but where high deer numbers persist they influence structure and composition for decades. For example, in Bialowieza Forest, Poland, heavy browsing between 1892 and 1915 admitted a cohort of *Salix caprea* that lasted for 50–70 years (Falinski 1998) and generated changes in the balances between forest dominants that are still continuing (Bernadzki *et al.* 1998).

Small-scale disturbances are commonplace. Wild pigs and cattle disturb the soil surface and create wallows (Falinski 1986). Woodpeckers excavate holes, which may weaken trunks. Beavers fell small trees and dam streams, creating pools and killing trees whose roots become submerged. Eventually, the dam breaks and the resulting moist glade becomes a preferred feeding ground for herbivores. Vole populations may explode for one summer, during which many saplings may be killed by ring-barking at the base. All these effects can be regarded as disturbances on a small scale in time and place but remain ordinary conditions of the forest when regarded on larger scales.

4.2.5 Water

Heavy rains and floods rarely disturb forests, even those growing on floodplains. Cloudbursts occasionally cause mudflows on slopes and spates in small streams, both of which will uproot trees (Hack & Goodlett 1960). Where the currents flow strongly, water can bend and break shrubs and smaller trees. Ice and floating woody debris will abrade trunks, but rarely to the point of damage.

The main disturbances due to water are long term. Channel movement within floodplains uproots forests at one point but deposits the alluvium elsewhere and thus provides a substrate that is colonized by woodland. Channel movement takes many forms but the general pattern is of elongated patches of bottomland forest being destroyed on the outside of meanders, while elongated shoals are simultaneously built up on the inside.

4.2.6 Ice

Ice and snow can break branches and may bend saplings and small trees to the ground. Lake ice will generate a disturbed zone round islands and river-borne ice will abrade floodplain trees. In subalpine and other exposed forests, where icy winds come mainly from one direction, the ice causes 'flagging', i.e. killing twigs on the upwind side. Accumulated snow generates avalanches, which can sweep forests from the mountainside, although most avalanches follow tracks which are so predictable that avalanche-proof trees have evolved to grow on them (see also Volume 2, Chapter 3). Thus, in most circumstances these kinds of impact form a normal part of the forest.

Ice and snow become disturbing at a local scale when, for example, an unusually heavy glaze storm breaks numerous crown branches (De Steven *et al.* 1991). It becomes an event, barely a disturbance, when a late spring snowfall flattens saplings and weak basal shoots. Since some species will be killed by this whereas others will propagate themselves by layering, such an event can influence forest composition.

4.2.7 Topography and landform

The physical environment can itself generate disturbances to forests. The most spectacular, if rare, are earthquakes, volcanic eruptions and meteorite impacts, which can devastate forests over large areas. Landslips can be equally devastating on a local scale but are also rare. Rockfalls from cliffs may generate avalanche-type gaps in the upper margins of the forests below.

4.3 DISTURBANCE REGIMES

Disturbance events and agents of disturbance can be described in isolation but they rarely occur like that in practice. Rather, they interact and combine to form disturbance regimes and thus ensure that most changes in forest structure and composition have a chain of causation: one thing leads to another (e.g. Foster 1988).

Numerous interactions between disturbances have been recognized. For example, a blowdown or a pest outbreak, which generates substantial concentrations of dead wood, initiates circumstances in which a fire is more likely. Wind may break only those trees that have already been weakened by fungal infection, itself originating in fire scars (Matlack *et al.* 1993). Fungi may infect only those trees that have been damaged by drought, or when bark has been stripped by squirrels or deer. The widespread fire of 1644 in the Allegheny Mountains, Pennsylvania, was probably associated with a drought (Hough & Forbes 1943).

Vulnerability to particular disturbances changes with the state of growth (White 1987). If a forest has been free of disturbances for many years, it is likely to have a large stock of mature and weakened trees that will be particularly vulnerable to the first storm, fire or drought to materialize. Conversely, the second storm in 3 years will have little impact if all the vulnerable trees were brought down by the first storm. A particular storm may devastate a mature stand, while leaving an adjacent young stand of the same species barely ruffled. Fires are most likely to destroy young and old-growth stands, where fuel loads are high, but are less likely to do so in middle-aged stands. Each site has its unique history of disturbance, and the impact of any particular disturbance is a function of that history.

The distribution of disturbances is uneven, or patchy, at all scales from single trees up to whole countries or states (e.g. Canham & Loucks 1984). This was exemplified by the storm on 16 October 1987 in south-east England, which brought down 15% of the standing timber volume over several hundred square kilometres (Peterken 1996). Although initial perceptions were of all-embracing devastation, it quickly became obvious that some districts escaped with little damage, whereas the woods in others had been levelled, i.e. the storm appeared to be 'streaky' at a scale of several kilometres. At a much smaller scale, windthrow patches within individual woods were generally small, but the pattern was largely inexplicable. Whereas in some sites blow-downs were associated with wet ground or exposed slopes and gulleys, most windthrown groups appeared to have been no different from nearby groups that remained standing. Patchiness was evident at all scales from districts down to parts of individual woods, and at the smallest scale there appeared to be a strong element of chance in the pattern of gaps.

Despite the random element, disturbance regimes vary in a more-or-less predictable way within a catchment. Forests on the floodplain are disturbed mainly by channel movement, windthrow and perhaps the effects of prolonged unseasonal inundations. Drought is likely to affect stands on upper slopes and plateaus, but leaves stands on flushed lower slopes and the deeper soils of the river terraces untouched. Wind impacts will be determined partly by topography, particularly in regions with a marked prevailing wind direction: slopes facing the wind and stands on the plateau will be particularly vulnerable, and perhaps also stands just over the brow of a ridge, which can be brought down by eddies in the flow. Likewise, fires tend to brake on ridges and stop at lake margins, so that the pattern of the fire will be partly determined by the landform and the lake pattern. Rockfalls and landslips will be confined to the steepest ground. Grazing may be concentrated in places where the vigour of tree growth is reduced over thin or permanently wet soils. Taken as a whole, therefore, a catchment constitutes a mosaic of different disturbance regimes, which are likely to be manifested in a mosaic of different stand structures and composition. Equivalent variation occurs at a continental scale. Thus variation in climate results in variation in disturbance regimes and thereby partly determines broad-scale forest patterns.

Disturbances also vary in intensity, although to some extent this is a matter of scale. Thus, a low-intensity storm may blow down no more than

0.1% of the canopy trees in a forest, but in the gaps actually created the disturbance is no less intensive than it would have been in a greater storm. Nevertheless, at a particular scale there is real variation in the intensity of many types of disturbance. For example, a single fire can consume all the trees in a stand; or it may merely singe the underwood as it smoulders through dry litter; or it may burn the small trees and saplings, leaving dominant trees as a form of shelterwood. Similar variation in intensity can be expressed by droughts, pest outbreaks and periods of ice damage. It has become customary to distinguish between catastrophic (i.e. stand-destroying) and non-catastrophic disturbances, even though these are no more than the ends of a continuous variable.

Disturbance regimes are thus the characteristic mix of disturbances experienced by a forest. At the smallest scale, each point in a forest will have experienced a unique sequence of disturbances that go some way to explain its current composition and structure. At larger scales, disturbance regimes can be expressed in two ways.

1 Qualitatively, by listing the kinds of disturbance encountered and describing their frequency, interactions and patterns. This may be all that can be achieved in forests subjected to a variety of disturbances, none of which is dominant.

2 Quantitatively, by calculating various parameters, such as frequency, return time and time to disturb the whole of the forest. This is realistic and appropriate where one form of disturbance is dominant. It works best in fire-dominated forest types, where stand-replacing fires leave a well-defined patchwork of even-aged stands, on whose older trees the scars of past fires can be dated (Heinselman 1973; Zackrisson 1977).

Forest disturbances take two main forms: gap formation and stand destruction. The former occurs when single trees or small groups of canopy trees die or fall, leaving a scatter of gaps in a matrix of intact canopy. The latter occurs when all canopy trees are killed or blown down. Gap formation is a chronic process characteristic of most temperate forests in most decades (Runkle 1982) and is commonly caused by the interaction of wind and biotic disturbances. Stand destruction is

an event that is characteristic of forests in some extreme environments (e.g. Sprugel 1976), of seasonally dry climates and in hurricane belts. Fire is the most common agent of stand destruction, although wind and invertebrate pests can also destroy stands on a large scale. The history of many forests can be summarized as long periods of gap creation, punctuated at intervals of centuries by stand destruction and subsequent regrowth (Henry & Swan 1974; Stewart 1988). Intermediate scales and intensities of disturbance are frequent. For example, extensive fires may leave a shelterwood of large trees, or droughts may weaken or remove a common canopy species; both may leave stands in which 'gaps' can be said to form the matrix for the remaining trees. Rarely, some forests appear to escape disturbance, for example the montane bristlecone pine parklands, where the vegetation is too sparse to allow fires to spread.

4.4 TREE AND SHRUB RESPONSES

The various responses of trees and shrubs to individual disturbances combine with the overall disturbance regime to determine forest composition. Each species posesses a unique combination of responses that enable it to survive and reproduce. The key variables between species relate to the form of reproduction, dispersal, tolerance of shade, growth rate and longevity.

Species vary in the degree to which their fruits or other propagules are dispersed. Whilst most propagules fall close to parent trees, a few species achieve long-distance dispersal. The lightweight fruits of *Betula*, *Populus* and *Salix* are widely dispersed by wind. These and other species growing in regions with persistent winter snow are widely dispersed as fruits are blown across the icy surface. Species with fleshy fruits are dispersed by birds, either passing through the intestinal tract (e.g. *Sorbus*, *Malus*) or being transported as food (acorns of *Quercus*). Species that grow close to rivers are dispersed downstream if their propagules fall in the water. Efficient dispersal enables species to occupy disturbed ground rapidly and facilitates speedy long-distance movement of populations. Species with a limited dispersal ability must have a capacity to survive on-site, usually by vegetative reproduction and/or living to great ages.

Species that demand high light intensities must be able to occupy disturbed ground rapidly and maintain themselves in the upper stratum. Such shade-intolerant species tend to be either short-lived shrubs or fast-growing trees with strong leaders and widely dispersed foliage. Tolerant species survive in shade but their growth is slow and most grow faster in higher light levels. Their shade tolerance enables them to survive in undisturbed stands indefinitely and to enter established stands, where they commonly form a persistent underwood of shrubs, saplings and subcanopy trees. Shade tolerance is a continuous variable (Lorimer 1983) so that intermediate responses are widespread. Furthermore, it can vary with age, such that seedlings are shade tolerant but saplings of the same species become intolerant as they age (e.g. *Fraxinus excelsior*).

Another key variable is growth rate, specifically growth in height. Fast-growing species will maintain their position in the canopy, whereas slow-growing species remain in subordinate strata and must therefore bear shade in order to survive.

These three variables are correlated enough to allow a useful distinction to be drawn between two survival strategies.

Strategy A: widely dispersed, shade-intolerant species with rapid height growth. These species are equipped to respond rapidly to disturbance. They colonize disturbed ground and gaps in established forest and maintain their position in the canopy.

Strategy B: shade-tolerant species, capable of surviving with slow growth rates and commonly limited in their ability to disperse. They are equipped to survive on the sites they currently occupy. Whilst they are commonly found in the underwood and subcanopy below a canopy of strategy A species, they are usually capable of persisting in the canopy.

Other important variables can be related to these strategies, though there seems to be no close connection between strategy and longevity. An ability to live to a great age prolongs canopy residence time and can thus form an advantage for strategy A species, several living to 300 years or more (e.g. *Pseudotsuga menziesii*, *Picea sitchensis*, *Quercus robur*, *Liriodendron tulipifera*, *Pinus sylvestris*) and forming some of the most enduring and impressive forest stands. Great longevity is also advantageous for tolerant species, enabling them to persist long enough in the underwood to grow into large canopy trees (e.g. *Tilia cordata*) or massive subcanopy individuals (e.g. *Ilex aquifolium*, *Taxus baccata*), though in most cases these species probably live longer if they occupy the canopy or prevent inherently taller species growing through them. On the other hand, many strategy A species are short-lived, lasting no more than a century (e.g. *Betula pubescens*, *Populus tremula*, *Salix caprea*) or two centuries (*Alnus glutinosa*, *Fraxinus excelsior*, *Populus nigra*), even when they succeed in staying in the upper strata. Furthermore, strategy B species are not necessarily long-lived: beech and Norway spruce may take well over a century to reach the canopy, but then have only 100 years of residence once they get there.

Early, large and regular seed production is characteristic of some short-lived strategy A species (e.g. *Betula pendula*), which facilitates colonization whenever and wherever opportunities are created by disturbance. Otherwise there seems to be little correlation between seed production and strategies. Late, irregular seed production is characteristic of mast trees, such as *Quercus robur* and *Fagus sylvatica*, and is said to facilitate regeneration by periodically outproducing the capacity of predators.

A capacity to reproduce vegetatively is also common in both strategies. Most broadleaved species will sprout from stumps after the main stem has been destroyed or knocked down in a disturbance, and some species (e.g. *Tilia americana*) regularly reproduce in this way even in undisturbed stands. This is a classic strategy B device, which effectively restocks storm-damaged forests (Peterson & Pickett 1991) and indefinitely prolongs occupation of the site. On the other hand, several strategy A species do not sprout strongly, for example *Betula pendula*. However, the correlation with strategies is limited and there are many exceptions. For example, floodplain willows (e.g. *Salix alba*) will root from almost any fragment of branch, which enables them to take advantage of severe disturbances. Aspens (*Populus tremula, Populus tremuloides*) grow short-lived trees from persistent rootstocks, thereby combining the two strategies. *Fagus grandifolia*, a classic tolerant species, generates suckers from a persistent rootstock.

Germination rates also vary significantly. Fast germination enables species to take immediate advantage of temporarily favourable conditions. When the germination of a particular seed crop is spread over a number of years, the chance of germinating in favourable conditions is increased at the cost of a reduction in the number of germinants at any given time. Prolonged dormancy appears to be associated more with light-demanding species, for example *Cytisus scoparius* can persist in the seed bank until gaps appear. Perhaps the most extreme adaptation is serotiny, in which a proportion of the seed is retained on the tree until released by a fire into a suitable seed bed and freedom from competition (e.g. *Pinus banksiana*).

There are thus many combinations of characteristics and responses that enable species to survive in forests where disturbance is irregular. In undisturbed conditions strategy B species predominate, leaving strategy A species to a fugitive existence in gaps and localized disruptions, for example in southern Britain *Tilia* spp. were dominant until widespread prehistoric forest clearance (Greig 1982). After prolonged quiet, a massive disturbance gives strategy A species an opportunity to thrive, provided they can react faster than strategy B species regenerating vegetatively. In environments that are frequently and substantially disturbed, strategy A species tend to predominate (see example of ice-dominated forests below).

4.5 FOREST STRUCTURE AND PATTERNS

The interactions between disturbances and forest trees generate a complex range of forest structures and patterns (e.g. Elliott-Fisk *et al.* in Barbour & Billings 2000). Perhaps the best route through this complexity starts with two contrasting stand structures (for descriptive purposes a stand is taken to be a compact patch of woodland of about 2 ha).

1 *Even-aged*: stands are even-aged if the canopy trees all started growth in approximately the same year. A limited age spread of perhaps 10 years is permissible within the definition. Such stands originate after a stand-destroying disturbance or an abrupt change in land use. Colonizing trees may take longer than 10 years to fill the available space, but as the stand develops its age range narrows: latecomers will have smaller, lower crowns than those that started growth immediately and this disadvantage ensures that their mortality rate is greater.

2 *Mixed-age*: stands comprising a wide range of age-classes, from old trees to saplings, intimately intermixed. Such stands tend to form a mosaic of even-aged groups at the scale of one or two canopy trees. They function by a process of relatively constant gap formation, in which gaps formed by the loss of canopy trees or major branches are filled by groups of saplings. These groups compete amongst themselves as they grow, so that eventually only one or two trees become established in the canopy, thereby achieving a 1 : 1 replacement of the original gap-forming tree. Such stands can only survive where there has been no catastrophic disturbances within the adult lifetime of the oldest canopy trees.

Even-aged stands develop through a sequence of stages, which are explained in Chapter 9. In effect, they become steadily more mixed-age as they develop, initially through recruitment in the underwood and eventually through mortality of canopy trees and replacement with individuals initiated later (e.g. Aplet *et al.* 1988). In the absence of catastrophic disturbance, an even-aged stand eventually becomes mixed-aged, a stage described as 'shifting-mosaic steady-state' by Bormann and Likens (1981) or 'old-growth' by Oliver and Larsen (1996)

This relatively simple model is complicated by non-catastrophic disturbances. Gaps created by storms, for example, are often irregularly patchy. Canopy disturbance created by droughts may take the form of an irregular thinning. Likewise, fires of intermediate intensity commonly leave a fairly even scatter of large trees. Regeneration follows both patterns of disturbance, so that moderate disturbances eventually generate both small-scale patches of younger growth within a hitherto even-aged stand and two-storied stands in which the survivors from the previous generation persist in a matrix of younger growth.

Partial disturbances give rise to stands intermediate between the mixed-aged and even-aged stands defined above (e.g. Maissurow 1941). They are mixed in the sense that different age-classes are mixed together at a small scale, even at canopy level, but contain distinct generations, i.e. concentrations of individuals within narrow age bands. Jones (1945) recognized that strong even-aged elements within mixed-age stands were common within virgin temperate forests. At the stand scale, a plot of the age (and often also the size) of trees demonstrates pulses of recruitment, and it may be possible to partition the stand into two to four more-or-less distinct cohorts. In truly mixed-age stands, the age and size distributions take a negative exponential form (Hough 1932). In stands with strong even-aged elements, the relationship remains a negative exponential but with peaks and dips. Thus, for example, the size distribution of *Quercus mongolica* in virgin old-growth in Hokkaido showed a fairly smooth, continuous, J-shaped distribution, even though the population comprised five distinct generations spanning 500 years (Sano 1997).

A second complication arises from the tendency of gaps to enlarge once they have formed. For example, windthrow gaps within coniferous forests become foci for populations of the beetle *Ips typographus*, which then colonize and kill trees standing by the gaps. In forests of all kinds, once a gap has been opened by wind, neighbouring trees are more exposed to subsequent winds and thus more likely to be blown down (Runkle & Yetter 1987). As gaps enlarge, so they tend to coalesce, until the stand ceases to be a canopy punctuated by discrete gaps and becomes an irregular scatter of fully grown trees in a matrix of shorter, and usually younger, individuals.

European forest ecologists, led by Liebundgut (1959), have not recognized gaps as distinct features but have attempted to represent the structure of old-growth stands as a patchwork of different stages of stand development. Each stage is defined as a distinct spatial arrangement of trees of different sizes and ages and is represented as a developmental link between two or more other stages, the whole representing a cycle of stand development with several pathways (e.g. Zukrigl *et al.* 1963; Mueller-Dombois 1987). Several different versions of the classification of stages and the dynamic cycle have been published. Maps have been prepared to show the patchwork of stages in particular forests (Mayer *et al.* 1980).

In practice, neither gaps nor the Liebundgut cycle fully represents the complex structure and dynamics of undisturbed old-growth. Gaps are readily defined and delimited in the early stages of canopy break-up but become far less distinct as break-up progresses and the stand becomes multi-layered. Furthermore, most gaps are small, so that the question of what counts as a gap is crucial in any quantification, especially in stands where the crowns of canopy trees rarely touch, let alone overlap, for example *Pseudotsuga menziesii* old-growth. In the case of the Liebundgut stages, the patches so clearly defined and mapped on paper are found on the ground to be ill-defined and poorly delimited. Intermediate conditions are common. Each patch of a particular stage is seen to be internally heterogeneous, i.e. to be a micropatchwork of different structures. The stages and maps based on them represent reality as perceived at a particular scale: both the classification of stages and the maps would have been quite different if a smaller scale had been adopted.

In so far as they can be delimited, gaps have internal structure. The distribution of dead wood is generally irregular, for example a gap created by the fall of a single tree has an accumulation of crown branches and debris on one margin, leaving the rest of the gap with a clean floor. In many forest types the underwood of shrubs and groups of advance regeneration survives gap formation: its original irregularities will be increased by some flattening as the canopy tree falls. In larger gaps it is common for one or two subcanopy trees to survive, thereby retaining a light shelterwood within the gap. Irrespective of size, gaps have an internal geometry which ensures that in the Northern Hemisphere the northern side receives more direct sunlight than the southern side and that the eastern side becomes drier in afternoon sun than the western side, which receives direct sunlight during the more-humid mornings. Furthermore, light intensity is greater at the centre of gaps than at the edges. In general, larger gaps permit a larger representation of shade-intolerant species (Poulson & Platt 1989). Internal heterogeneity created by fallen trunks and branchwood generates irregularities in browsing intensity and small-scale shading patterns, and thus also contributes to the variety of regenerants (Hutnik 1952).

Both gaps and the representation of stages have been used to quantify canopy turnover. Gaps are formed, or canopy space is vacated, at about 1% per annum. The rate varies from year to year (Runkle 1982) and on a scale of decades (e.g. Poulson & Platt 1996) but where it has been measured in different forests, gaps have formed at 0.5–2% annually. This implies a residence time of about 100 years for trees that reach the canopy. Where the extent and duration of growth stages have been quantified, it has been possible to compute the duration of the growth cycle at some 300 years and to determine the degree to which the age distribution of patches departs from the theoretical steady state (Mayer & Neumann 1981).

In practice, individual trees live far longer than 100 years. Firstly, trees take several decades to reach the canopy, particularly those shade-tolerant species that remain virtually static when they are heavily shaded. Even fast-growing light-demanding species take 30 years or so to reach canopy height. Secondly, some canopy species are inherently short-lived, leaving space for other species to reside longer in the canopy. Thirdly, gap formation is not absolutely correlated with age of trees, so that some individuals live far longer than the majority. In temperate forests, individuals of oak, tulip, hemlock and lime regularly achieve ages of 300–500 years or more (Lorimer 1980; Runkle 1981).

When forest structure is considered at a landscape scale we see that the fundamental difference between even-aged and mixed-age stands is a matter of scale. A forested landscape made up entirely of mixed-age stands has the same size distribution and age-class distribution as a landscape covered in even-aged stands, each of a different age. It is only when the forest is considered at a stand scale that even-agedness is perceived. A landscape of even-aged stands is usually mixed-aged in aggregate, i.e. the various stands comprising the landscape represent a range of ages (e.g. Heinselman 1973). Thus, at a very large scale all forested landscapes are mixed-aged, although the scale of their even-agedness varies from groups no larger than a single canopy tree up to patches covering several thousands of hectares. Some historic fires have burned over whole counties, leaving patches of hundreds of thousands of hectares that would have regenerated as a massive even-aged stand, for example the widespread 500-year age-class of *Pseudotsuga menziesii* in the Cascades (Franklin & Hemstrom 1981).

Variation in the scale of the patchwork has several consequences for forest composition and structure. A very small-scale patchwork implies a fairly steady rate of gap creation when computed at a large scale, which in turn implies that forest structure will remain more or less constant. A large-scale patchwork implies that at moderate scales (say individual catchments of first- to fourth-order streams) forest structure can change widely with time according to the irregularities of disturbances, for example Yellowstone (Romme 1982). Frelich and Lorimer (1991) used growth stages and transition times between each stage to compute the state of the landscape in a region of moderate disturbance, and found that most of the forest remained old-growth for most of the time.

4.6 DEAD WOOD

Dead wood, or coarse woody debris (CWD), is an integral component of the forest (Maser *et al.* 1988). It is generated not only when trees are killed by disturbances but also when trees or individual branches die in the ordinary course of competition between individuals and change in stand structure. It also develops within the heartwood of living trees. Above-ground CWD thus takes the form of fallen wood, dead standing stems (snags and stumps), dead branches on living trees and decay columns within living trees.

The amount, distribution and rate of decay of CWD in natural temperate forests varies widely (Harmon *et al.* 1986). CWD generally increases with stand age but is also high in young-growth stands growing through the debris left by a storm or other disturbance (e.g. Spies *et al.* 1988). The amount of CWD varies through time in any particular stand: pulses of CWD input caused by disturbances are superimposed on a steady background input from stem exclusion processes. CWD distribution is also patchy, i.e. it is concentrated in recently disturbed forest, which is itself patchy, and accumulates in gulleys and debris dams in rivers.

The quantity of CWD is often measured as the volume of snags and fallen logs. In most old-growth temperate deciduous stands the values generally fall between 50 and 140 m^3 ha^{-1} (Kirby *et al.* 1998), to which should be added perhaps 20–30 m^3 ha^{-1} for snags. Values from other forms of temperate forest tend to be much higher and comparable with volumes in boreal forests. Thus values in the *Picea sitchensis–Pseudotsuga menziesii* forests of the Pacific North-west have been estimated as ranging from 600 to over 1400 m^3 ha^{-1} (Harmon *et al.* 1986), and for *Nothofagus* forests in New Zealand as 370–800 m^3 ha^{-1} (Stewart & Burrows 1994).

4.7 DYNAMIC PATTERN OF FOREST COMPOSITION

Interactions between tree species and between them and disturbance regimes generate distinctive assemblages of species, forest dynamics and landscape patchworks (White 1987). These are best illustrated by examples at the ends of the range of disturbance frequency.

4.7.1 Highly disturbed environments

Where disturbances are frequent, large scale or severe, forests tend to be dominated by intolerant and fast-colonizing species. Three examples illustrate the range of circumstances.

Floodplain forests

Floodplain forests are influenced by channel movement, which destroys mature stands but reworks the deposits into new shoals. In northern temperate regions, these are colonized mainly by *Salix*, *Alnus* and *Populus* species, which grow into even-aged, often monospecific stands. The sequence of channel movements is manifested as a pattern of elongated even-aged stands, whose age increases with distance from the channel. Tolerant species colonize beneath these pioneer stands and, given sufficient time, develop into mixed old-growth. In northern temperate deciduous regions the principal long-term dominants are *Ulmus*, *Fraxinus*, *Quercus*, *Carya*, *Sassafras* and, in the Pacific North-west, *Picea sitchensis* and *Pseudotsuga menziesii*. At any one time, the pioneer stands tend to predominate near the present channel, and the mixed old-growth tends to survive in elongated patches at some distance from the river (Hupp 1988; Cordes *et al.* 1997).

Ice-dominated forests

The classic type is the wave-regenerated *Abies* forests of the eastern USA and Japan (Sprugel 1976). Exposed mature forest degenerates when foliage is stripped by ice and wind from trees that have already lost the vitality of youth. Death of exposed trees exposes others, leading to a 'wave' of mortality, which moves steadily through the forest in the direction of the prevailing wind. Regeneration starts within the degenerating stands and grows vigorously in the lee of slightly older stands. The forest as a whole takes the form of a series of parallel waves, which move through the forest at 1–3 m annually on a return time of 60–70 years. This perpetual recycling ensures that *Abies balsamea* remains dominant and that the longer-lived *Picea rubens* is perpetually excluded.

Fire-dominated forests

The most widespread form of highly disturbed forest is dominated by fire (e.g. Heinselman 1973; Romme 1982). Most boreal forests are naturally fire-dominated, but so too are Mediterranean forests and the forests that fringe extensive grassland and desert regions. Each region has a suite of species that undergo a characteristic succession after fire, for example in Scandinavia *Betula–Pinus* develops into *Pinus* dominance, which is then succeeded by *Picea*. Exceptionally, some patches remain unburned, which allows the pioneers to be completely displaced, although generally fire returns in good time to ensure that pioneer species remain a permanent feature of the forest. In fact, there is much variation in return time, associated with variation in topography, ground vegetation and the configuration of water bodies (Zackrisson 1977). Nevertheless, fires were frequent enough to maintain most boreal forests as young or maturing stands, not old-growth. Similar fire-dominated regimes control other forests, such as the *Eucalyptus* forests of Australia and the *Pinus*-dominated forests of the coastal plain of the south-eastern USA (Christensen 2000). However, in western North America many tree species not only withstand fires once they have achieved a moderate size but also grow to great size and age, thus generating the monumental forests of *Sequoiadendron giganteum*, *Sequoia sempervirens* and *Pseudotsuga menziesii* in which old individuals may bear the scars of several fires. On the margins, the boundary between mesic- and fire-dominated forests advances and retreats according to the history of fires.

4.7.2 Relatively undisturbed environments

At the other end of the range are mesic forests growing in relatively undisturbed environments. These are not disturbance-free; instead, catastrophic disturbances are rare enough to allow most of the forest to develop into old-growth, where disturbances are small-scale events (Fig. 4.3).

Over much of the north temperate zone, a distinction can be drawn between *Fagus* forests and mixed deciduous forests (Ellenberg 1988; Barnes 1991; Jahn 1991). *Fagus*-dominated forests are found mainly in northern latitudes and submontane elevations, where *Fagus* spp. often share dominance with conifers, for example *Tsuga canadensis* in the eastern USA, *Abies alba* and *Picea abies* in central Europe. European beech, *Fagus sylvatica*, is not long-lived, but casts dense

Fig. 4.3 Old-growth mixed deciduous forest at Porter's Flat in the Great Smoky Mountains National Park, eastern USA, where white ash, buckeye, silverbell, basswood and maples are mixed with many other species. Typically, these stands take the form of high forest comprising a very wide range of tree sizes, punctuated by small gaps containing snags, fallen logs and branchwood, groups of saplings and/or groves of fast-growing poles, depending on how long ago the gap was formed. Within such mixtures, the species that develop in a particular gap will probably not be the same as the species whose death generated the gap. However, the net effect of such small-scale species transfers calculated over many hectares of forest may be close to zero, thereby maintaining the mixture. Small-gap formation sustained over 200–300 years without catastrophic disturbance generates an intimate mixture of trees of all ages. (Photo by the author.)

shade, is capable of regenerating in the small transient gaps generated in beech forests and is almost fireproof, so that it is able to both dominate the site and perpetuate this dominance. However, it is prone to disaster, in the sense that mature stands are vulnerable to drought and high winds. A few intolerant species, such as *Salix caprea*, maintain a foothold in large gaps, while the shade-tolerant *Acer pseudoplatanus*, like *Acer saccharum* in the eastern USA, fills gaps, competes in advance regeneration and can grow into mature stands. Diversity is maintained partly by the tendency of species not to regenerate under themselves. Thus, *Fagus grandifolia* and *Acer saccharum* (Fox 1977) in the USA and *Fagus sylvatica* and *Abies alba* in Europe have been reported to alternate, thereby maintaining a small-scale mosaic of different dominance, though in the former case coexistence has been ascribed to different responses to light intensity (Poulson & Platt 1996). Occasionally, such stands are destroyed by storms (Peterson *et al.* 1990), whereupon pioneer species dominate the regrowth, although beech thrives in the under-wood and eventually restores its position.

Mixed deciduous forests are those which lack the dominating influence of beech. At their greatest development, for example in the southern Appalachians, they comprise a mixture of several dozen species, each with the capacity to occupy the canopy or subcanopy (Lorimer 1980) (see also p. 49). A wide range of genera are represented, notably *Acer*, *Quercus*, *Fraxinus*, *Carya*, *Aesculus*, *Betula*, *Castanea*, *Tilia*, *Magnolia*, *Carpinus* and *Halesia*, forming rich mixtures in which no single species becomes absolutely dominant. Such forests are rarely devastated by any single disturbance and for most of the time are renewed by gap-phase regeneration on a small scale. Pioneer species regenerate in the larger gaps and may dominate the canopy after periods of enhanced gap creation. An example is *Liriodendron tulipifera*, which not only forms the tallest trees in the Appalachian forests but can also live for 500 years, enabling it to perpetuate itself through long periods lacking disturbance.

Within these complex mixtures, each species has a distinctive pattern of growth, longevity and regeneration. Furthermore, there is a tendency for individuals of one species to be replaced by another species. The forests comprise a small-scale mosaic of groups of different canopy and underwood species, each with its particular successional trend. The trends in one patch are countered by opposite trends in other patches, thereby retaining the mixture. Recruitment of particular species tends to be irregular, depending on particular combinations of mast-years and disturbances. Composition remains fairly constant overall but at a small scale changes perpetually, except where individual trees replace themselves vegetatively, e.g. *Tilia* spp.

Single-species groups are common in mixed forests. These may develop in response to small differences in site conditions and/or the chance coincidence of an episode of gap creation with heavy seed production by a particular species. Simulations have recently shown that neighbourhood effects may also play a part (Frelich *et al.* 1998). Where there is a high probability that canopy trees will be replaced by individuals of the same species, 10 generations is enough for this feedback to generate small-scale single-species patches. The scale of the patches increases substantially where minor environmental differences result in 5% alterations in recruitment probabilities.

4.7.3 Interactions and intermediate conditions

The forests described above are stereotypes that represent the ends of the range of variation. In practice, many forest types are intermediate in some respect, although their dynamics and composition can be related to the stereotypes. Some examples illustrate the range of possibilities.
• Forests in which partial stand destruction is commonplace can develop into complex mosaics of tolerant and intolerant species (e.g. Stewart 1988). An example is the *Pseudotsuga menziesii* forests of the Pacific North-west, where saplings and underwood are frequently burned but the overstorey is unaffected. Moreover, light fires tend to be patchy, burning some places but leaving other patches untouched. This allows the overstorey of *Pseudotsuga menziesii* to be infiltrated by *Abies amabilis*, which develops as patches in a mosaic with groups of younger *Pseudotsuga menziesii* in larger burned patches.
• Complex interactions can develop between disturbance regimes. For example, in the south-eastern USA *Pinus* forests were maintained

naturally by frequent fires. However, where stands happen to remain unburned, *Quercus* and other broadleaved species colonize, form an underwood and eventually dominate the canopy (Iwaschkewitsch 1929; White 1987). Since the broadleaved stands are relatively fireproof they tend to remain undisturbed, thereby generating a patchwork of broadleaved and *Pinus*-dominated stands whose pattern is determined by disturbance history. This capacity for two forest types to develop on one site type is inherent in any region where disturbance becomes less likely as succession proceeds.

• The characteristics of *Fagus sylvatica* in central Europe determines much of the forest pattern. Its capacity to coppice enables it to form distinctive low scrub woodlands with *Sorbus aucuparia* near the treeline, where wind and snow maintain a chronically disturbed environment. In this case, the *Fagus sylvatica* is effectively undisturbed. Elsewhere, in steep slopes in gorges, the stands are chronically destabilized by the fall of large trees growing on insecure rootholds, and this breaks the dominance of beech and enables a mixture of tree species to survive, including *Tilia*, *Ulmus* and *Fraxinus*.

Disturbances have long-term effects on forest composition. For example, drought rarely destroys stands entirely but confers an advantage on species that survive droughts. Ultimately this is expressed as a distinctive assemblage on drought-prone sites, such as outcrops and convex slopes with thin soils. In this instance, disturbance tends to generate adapted forests, which thus reduces the incidence of disturbance. Alternatively, species that depend on disturbance may both invite it and be ready to survive. The best examples are the fire-dependent *Pinus banksiana* and *Pinus contorta* forests, which are inherently inflammable, particularly those that exhibit serotiny, where long-lived cones release their seed only after fire and which falls into a seed bed free of litter and competing vegetation. A more extreme example is seen in the *Pinus mugo* of avalanche tracks, where repeated disturbance has provoked the evolution of a species that can withstand the disturbance.

The multitude of interactions between trees and disturbance regimes is expressed as different assemblages on different site types and as a range of successional states on each site type. Disturbance should be seen as an integral component of the forest type, not an external destructive force (White 1979). Collectively, disturbances enhance regional diversity by creating distinctive forest types, maintaining a range of successional states, and by enhancing the amount of edge in the landscape.

4.7.4 Influence of people

The relationships described so far have dealt with natural conditions. In practice, most forest types are overwhelmingly influenced by people:

• Secondary successions to forest on land previously cleared are a response to severe disturbance, i.e. the complete loss of tree cover and the alternative use of land for cultivation or pasture. Pioneer trees are almost always intolerants. The pattern of secondary succession depends on seed sources, so that colonization may be extremely patchy. Colonization often takes place in waves, for example when initial colonization by wind-dispersed trees is followed by bird-dispersed trees once perches have been created.

• Coppice systems (Rackham 1980) enable shrubs and intolerant trees to form a higher proportion of a stand than they would naturally maintain. Despite the constant disturbance of felling, individual trees can sprout again indefinitely and thus live well beyond their natural span. In terms of individuals and small-scale patterns, coppices are probably more stable than the natural forests from which they were derived.

• In woodland pastures (Rackham 1980), regeneration is inhibited by grazing and browsing, although established individuals have little competition and can expand to great sizes. Moreover, the constant lopping of branches for fuel and fodder, followed by regrowth from the pollard, enables individual trees to grow to great ages. Here, too, the effect of management disturbance is to generate unnaturally high stability.

Finally, there is the problem of *Quercus petraea* and *Quercus robur* in Europe. These are long-lived intolerant species that have been abundant in the pollen rain over very long periods (Godwin 1975), which implies that the forests were considerably disturbed, albeit at long intervals. They have been maintained at unnaturally high levels by traditional management, which favoured them as timber trees in coppices and wood-pastures (Godwin & Deacon 1974). Under natural conditions it is possible that oaks were maintained by the high level of browsing and grazing imposed by populations of large herbivores, such as deer, horses and cattle. In fact, it has been proposed that natural forest was more like savannah before prehistoric hunters reduced and domesticated the larger herbivores (Wallis de Vries 1995). However, this seems far from proven, although the possibility remains that herbivore populations maintained a controlling presence on the structure and composition of some types of temperate forest, particularly perhaps those on wet ground and heavy fertile soils (where ground vegetation would be vigorous) and steep dry slopes (where tree growth might be poor) (Bradshaw & Mitchell 1999).

4.8 STABILITY, SUCCESSION AND CLIMAX

The concepts upon which forest ecology developed for many decades centred on the notion of forests as stable places (McIntosh 1985). Change was recognized but changes formed part of a succession that would lead eventually to a stable endpoint, or climax state, which over much of the globe was a form of forest. As the name implies, climax forests were the peak of development, an ideal in which biomass, biodiversity and productivity were maximized in self-sustaining groves of stupendous arboreal giants. As the theories developed, change was recognized even in climax vegetation, allowing a cycle of structures within the climax state; further, it was accepted that under some circumstances vegetation regressed 'beyond' the climax to a postclimax state, less productive, with lower biomass and generally less impressive than the climax. Disturbances were seen and understood but they were regarded as destructive, not part of the system.

The climax concept originated as a basis for vegetation mapping (Cowles 1899) and was sustained by its simplicity, the idealism of its time and as a target for conservation measures (Raup 1981). However, it has now largely been supplanted by alternative models. 'Stability' can be defined by the capacity to return rapidly to a former state after disturbance: in this sense assemblages of arable weeds are more stable than ancient forests (Horn 1975). Vegetation is seen to develop through defined stages (i.e. succession) to no particular end-point (Bormann & Likens 1979; Oliver 1981). Succession continues predictably only until the next disturbance. The next disturbance may totally destroy the vegetation and thus reinitiate succession, or it may partially destroy vegetation and thus merely deflect succession. The composition of the early colonists, which may be determined largely by chance or transient conditions, can determine the long-term rate and direction of change (Egler 1954). Thus, in each forest region, succession must be seen not as a single highway but more as a network of possible paths along which the vegetation at any one site moves according to its particular sequence of disturbances and responses (Green 1981; Miles 1988). Even in undisturbed old-growth forests, there is no one-to-one relationship between composition and site factors, merely various probabilities that certain species will be present (e.g. McCune & Allen 1985).

Perceptions of change and stability depend on the scale at which they are considered. The natural forests of a catchment may retain the same suite of species for millennia, although the precise pattern of these species within the catchment may change over decades. It may be perpetually disturbed but retain a constant balance between young-growth and old-growth in the catchment as a whole. Even within 'stable' old-growth stands single-tree gaps will form and the replacement trees will probably not be the same species as the fallen tree (Fox 1977; Runkle 1981). Thus, each of these conditions can be regarded as either stable or unstable according to the spatial scale or time-scale considered. The concept of *minimum dynamic area* has been advanced as an expression of 'the smallest area with a natural disturbance regime, which maintains internal recolonization sources and hence minimizes extinction' (Pickett & Thompson 1978), although this does not imply absolute stability in the proportions of tree species. Estimates of such areas for the forests in the eastern USA were 10^2–10^4 ha for forests disturbed only by small-scale gap creation and 10^5 ha for forests controlled by fires (Shugart 1984).

At much larger scales of space and time, tree species migrate individualistically in response to long-term, large-scale climatic oscillations (Davis 1981; Huntley & Birks 1983), which shuffle arboreal assemblages and perpetually alter the context of competition within which each species evolves, i.e. trees migrate, spread and retreat but do not migrate as consistent communities. Individual species and assemblages of species at any particular point may never be quite in balance with their current environment (Raup 1957). At any given time, particular species may be advancing into a stand (e.g. *Fagus* in Britain), whereas others may be relicts (e.g. *Tilia* in Britain).

4.9 PEOPLE AS AGENTS OF DISTURBANCE

Much of this chapter is based on natural forests, where people and their influences are insignificant by definition. Such forests still exist, but in the real world people have long been a factor.

4.9.1 Seminatural disturbances

The varied roles of people in the disturbance regimes of even 'natural' forest reserves may be clarified with an example. After 50 years of detailed observation of Lady Park Wood in the UK, we can list the disturbances to which it has been subject and the way in which these have been influenced by people (Peterken & Mountford 1995). The wood itself is an ancient seminatural broadleaved mixture that was managed as coppice for centuries; however, in the late nineteenth century a decision was taken to promote it by periodic thinning to high forest, and then in 1940–44 part of the wood was virtually clearfelled to supply timber to the war effort. Since 1945 the wood has been allowed to grow without direct human intervention. The recorded disturbances are summarized in Box 4.1.

Evidently, the influence of people has been all-pervasive. The impact of drought and many other disturbances has depended on the age and composition of the stands, which have been determined by the history of management. Eventually this will grow out, but even so most of the biotic disturbances will probably remain seminatural in the sense that the actions of people were essential in the chain of causation. Furthermore, events around the reserve, such as felling and variation in deer control, will remain significant within the reserve. The reserve may appear to be natural but its disturbance regime is no more than seminatural.

4.9.2 Silvicultural systems

Many commentators have pointed out that felling is equivalent to natural disturbance (e.g. Whitehead 1982), although the analogy should not be pushed too far. Felling followed by timber extraction removes most of the potential volume of dead wood and generates additional disturbance to the microtopography and soil structure. Felling itself is not identical with windthrow: it leaves low flat stumps in the ground and does not generate ragged snags nor rootplates and tip-up mounds. Death standing is not part of forestry.

The structure, pattern and mode of regeneration of most natural forests are broadly equivalent to the high-forest silvicultural systems practised by foresters (Jones 1945). Within this, perhaps the greatest degree of equivalence to a particular silvicultural system is achieved by fire, which can remove timber and potential dead wood almost as effectively as timber harvesting. In boreal forests fires generate either a shelterwood structure where scattered mature trees remain or patchy retentions where groups escape fire. At a landscape scale, fires generate a mosaic of large- and small-scale patches, whose boundaries commonly coincide with ridges (watershed boundaries to Americans), rivers and wetlands, and this corresponds with large-scale forestry operations. Rotations under a natural fire regime are short (Zackrisson 1977), which is not so different from modern timber production. Prescribed burning after felling and extraction creates ground conditions that resemble those created by natural fires. Virtually all regeneration is from seed. Natural fire-dominated disturbance regimes can readily be mimicked by commercial forestry as either

---

Box 4.1 Disturbances recorded in Lady Park Wood reserve, UK

Drought
The main form of disturbance in the last 50 years, particularly severe in 1976 when many mature beech and 30–35-year-old birch were killed, generating an irregular pattern of gaps (Peterken & Mountford 1996). The pattern of impact was natural: it varied according to soil depth and site hydrology, neither of which had been altered by past management. Widespread pollution was believed by some to have intensified the impact.

Wind
Several trees have been blown down but only after their exposure had been increased by the death of elms due to disease (see below) and felling of nearby plantations. This increased exposure would not have been significant if the stands had been young.

Biotic
1 Having been introduced to Britain in the late nineteenth century, grey squirrels spread naturally to the wood. Despite prolonged attempts to control their populations, these animals continue to strip the bark of mid-sized beech. A potential long-term dominant is thereby deformed and may be restricted to the underwood.
2 Likewise, another introduced species, fallow deer, has long since spread to the wood, where it is numerous enough to prevent most regeneration. Its impact is a function of control measures in the district as a whole, which in turn reflect resources and the social acceptability of deer shooting. In this case, a moderate level of deer browsing can be accepted as natural, and control at such levels compensates for past destruction of large carnivores.
3 Dutch elm disease in its modern form arrived about 1971, having been introduced to Britain in imported wood. It largely removed wych elm from the canopy and remains a chronic factor (Peterken & Mountford 1998).
4 Bank voles, a native species, periodically reach plague proportions. On the last occasion in 1985 they killed an established cohort of small beech saplings.
5 Sycamore, an introduced tree, would spread into the wood from seed brought down from planted individuals by the river. In fact, site managers prevent this by cutting.

Ice and water
1 A late April snowfall in 1984, which lay for just 3 h, brought many weak stems to the ground. Most died, but lime remains as arched stems, pinned by falling snags (themselves the product of drought acting on birch), which would have rooted if deer browsing had not been so intense.
2 If the wood had still extended on to the floodplain, the lowest levels would still have been flooded, but flood frequency would have been reduced by water abstraction upstream and channel movement would not have been permitted.

Topography
Trees rooted on cliffs and the steepest slopes commonly fall once their crowns become too heavily weighted on the downslope side. Large trees bring down an avalanche of lesser trees as they fall.

---

irregular even-aged patches or two-storeyed high forest (Angelstam 1997).

In the wind-dominated disturbance regimes of temperate deciduous forests, the silvicultural equivalents are selection and group-selection systems combined with occasional large coupes containing scattered shelterwood retentions. A reasonably exact mimicry of natural systems would be achieved with minimal thinning, rotations allowing canopy residence times of 100 years, and indefinite retention of groups of canopy dominants. In floodplain forests, where the principle disturbance is channel movement, the equivalent silvicultural system would be long-rotation even-aged high forest.

Traditional coppice and wood-pasture systems have very few natural equivalents. Volumes of dead wood are very small in coppices, less than

10% of natural levels (Kirby *et al.* 1998), or are concentrated in the decay columns of parkland pollards. Forms of coppice can be generated by repeated leader mortality in extreme environments, such as the exposed ridges in subalpine scrub and the broadleaved underwood in fire-dominated regions (New Mexico). Wood-pasture structures can also be generated in the marginal zones between forest and grassland, notably in central California and the prairie margins. Wood-meadow structures can also be envisaged as a natural feature of the stable portions of floodplains, where luxuriant herbaceous vegetation would inhibit tree regeneration and attract herds of deer and cattle: any tree that does become established would grow rapidly and expansively. Natural circumstances of this kind have vanished from temperate floodplains, but they might have been common in the Pleistocene era of megaherbivores.

4.9.3 Traditional cultures as part of nature

Whereas we readily accept that present-day disturbance regimes and forests owe their character to both people and nature in varying degrees, we have assumed that there was a time when people had little or no effect. In northern temperate Europe this era closed about 5000 years ago, when expanding agriculture initiated widespread forest clearance. The forests that developed before this time are termed 'original-natural', 'primeval' or 'wildwood' A few survived in remote or protected areas into historic times.

In North America, the original-natural forest has been thought to be more within historic reach. In many places, old-growth forests that have not been directly altered since European settlement survive in the modern landscape, and great swathes of second-growth forest are returning to a condition similar to that which obtained before the original forests were logged. It was against this background that the concept developed of a stable climatic climax as the destination of natural succession.

Latterly, however, the profound role of native Americans in shaping the forests has increasingly been recognized (Denevan 1992). Apart from areas of cultivation and controlled pasture, vast tracts were burned sufficiently to maintain grassland where trees would grow and maintain open forests that might otherwise be closed. Much of the 'natural' forests or 'pristine wilderness' encountered by Europeans was regrowth from 1500 to 1750 after the populations of native Americans had been greatly reduced by European diseases. Relicts of these developments can still be seen in many eastern old-growth forests, which still contain a generation of large old oaks that grew up in a landscape kept open by native Americans. In fact, native Americans had maintained their influence on the forests for tens of thousands of years, long enough for people and forests to evolve together (Pyne 1982). At the present time, remaining old-growth stands are changing in response to the direct effects of visitors and the indirect effects of management and environmental modification in the surrounding landscape (e.g. Cook 1996).

So what about the perpetual debate about people as part of, or separate from, nature? Semantically, we make a distinction: in ecology and conservation, 'natural' is a separate word that must exclude people to retain its meaning; scientifically, it is useful to think in terms of natural systems as controls or reference points without people, which enable the influence of people to be measured. Historically, however, we know that people have long interacted with the environment and other species. Practically, we know that this will continue. And, ethically, we believe this continued interaction can be desirable if the balance does not shift too far towards people.

REFERENCES

Angelstam, P. (1997) Landscape analysis as a tool for the scientific management of biodiversity. *Ecological Bulletins* **46**, 140–70.

Aplet, G.H., Laven, R.D. & Smith, F.W. (1988) Patterns of community dynamics in Colorado Engelmann spruce–subalpine fir forests. *Ecology* **69**, 312–19.

Barbour, M.G. & Billings, W.D. (eds) (2000) *North American Terrestrial Vegetation*. Cambridge University Press, Cambridge.

Barnes, B.V. (1991) Deciduous forests of North America. In: Röhrig, E. & Ulrich, B., eds. *Temperate Deciduous Forests. Ecosystems of the World Vol. 7*, pp. 219–344. Elsevier, Amsterdam.

Bernadzki, E., Bolibok, L., Brzeziecki, B., Zajaczkowski, J. & Zybura, H. (1998) Compositional dynamics of natural forests in the Bialowieza National Park, northeastern Poland. *Journal of Vegetation Science* **9**, 229–38.

Blais, J.T. (1965) Spruce budworm outbreaks in the past three centuries in the Laurentide Park. *Quebec. Forest Science* **11**, 130–9.

Bormann, F.H. & Likens, G.E. (1979) Catastrophic disturbance and the steady state in northern hardwood forests. *American Scientist* **67**, 660–9.

Bormann, F.H. & Likens, G.E. (1981) *Patterns and Process in a Forested Ecosystem*. Springer-Verlag, New York.

Bouchon, J., Faille, A., Lemée, G., Robin, A.M. & Schmitt, A. (1973) *Notice sur les cartes des sols, du peuplement forestier et des groupements vegetaux de la reserve biologique de la Tillaie en foret de Fontainebleau*. Laboratory of Plant Science, University of Paris XI, Orsay.

Bradshaw, R. & Mitchell, F.J.G. (2000) The palaeoecological approach to reconstructing former grazing–regetation interactions. *Forest Ecology and Management* **120**, 3–12.

Canham, C.D. & Loucks, O.L. (1984) Catastrophic windthrow in the presettlement forests of Wisconsin. *Ecology* **65**, 803–9.

Christensen, N.L. (2000) Vegetation of the southeastern coastal plain. In: Barbour, M.G. & Billings, W.D., eds. *North American Terrestrial Vegetation*, pp. 397–448. Cambridge University Press, Cambridge.

Clinton, B.D., Boring, L.R. & Swank, W.T. (1993) Canopy gaps characteristics and drought influences in oak forests of the Coweeta Basin. *Ecology* **74**, 1551–8.

Cook, A.E. (1996) Cook Forest State Park: reflections of a preservationist. In: Davis, M.B., ed. *Eastern Old-growth Forests*, pp. 284–90. Island Press, Washington, DC.

Cordes, L.D., Hughes, F.M.R. & Getty, M. (1997) Factors affecting the regeneration and distribution of riparian woodlands along a northern prairie river: the Red Deer River, Alberta, Canada. *Journal of Biogeography* **24**, 675–95.

Cowles, H.C. (1899) The ecological relations of the vegetation on the sand dunes of Lake Michigan. *Botanical Gazette* **27**, 95–117, 167–202, 281–307, 361–91.

Davis, M.B. (1981) Quaternary history and the stability of forest communities. In: West, D.C., Shugart, H.H. & Botkin, D.B., eds. *Forest Succession: Concepts and Applications*, pp. 132–53. Springer-Verlag, New York.

Denevan, W.M. (1992) The pristine myth: the landscape of the Americas in 1482. *Annals of the Association of American Geographers* **82**, 369–85.

De Steven, D., Kilne, J. & Matthiae, P.E. (1991) Long-term changes in a Wisconsin *Fagus–Acer* forest in relation to glaze storm disturbance. *Journal of Vegetation Science* **2**, 201–8.

Egler, F.E. (1954) Vegetation science concepts. I. Initial floristic composition, a factor in old-field vegetation development. *Vegetatio* **4**, 412–17.

Ellenberg, H. (1988) *Vegetative Ecology of Central Europe*, 4th edn. Cambridge University Press, Cambridge.

Falinski, J.B. (1986) *Vegetation Dynamics in Temperate Lowland Primeval Forests*. Junk, Dordrecht.

Falinski, J.B. (1998) Dynamics of *Salix caprea* L. populations during forest regeneration after strong herbivore pressure. *Journal of Vegetation Science* **9**, 57–64.

Foster, D.R. (1988) Disturbance history, community organisation and vegetation dynamics of the old-growth Pisgah forest, south-western New Hampshire, USA. *Journal of Ecology* **76**, 105–34.

Foster, D.R. & Boose, E.R. (1995) Hurricane disturbance regimes in temperate and tropical forest ecosystems. In: Coutts, M.P. & Grace, J., eds. *Wind and Trees*, pp. 305–39. Cambridge University Press, Cambridge.

Fox, J.F. (1977) Alternation and co-existence of tree species. *American Naturalist* **111**, 69–89.

Franklin, J.F. & Hemstrom, M.A. (1981) Aspects of sucession in the coniferous forests of the Pacific Northwest. In: West, D.C., Shugart, H.H. & Botkin., eds. *Forest Sucession: Concepts and Application*, pp. 219–29. Springer-Verlag, New York.

Frelich, L.E. & Lorimer, C.G. (1991) A simulation of landscape-level stand dynamics in the northern hardwood region. *Journal of Ecology* **79**, 223–33.

Frelich, L.E., Sugita, S., Reich, P.B., Davis, M.B. & Friedman, S.K. (1998) Neighbourhood effects in forests: implications for within-stand patch structure. *Journal of Ecology* **86**, 149–62.

Godwin, H. (1975) *The History of the British Flora*, 2nd edn. Cambridge University Press, Cambridge.

Godwin, H. & Deacon, J. (1974) Flandrian history of oak in the British Isles. In: Morris, M.G. & Perring, F.H., eds. *The British Oak: its History and Natural History*, pp. 51–61. Classey, Farringdon.

Green, D.G. (1981) Time series and post-glacial forest ecology. *Quaternary Research* **15**, 265–77.

Greig, J. (1982) Past and present lime woods of Europe. In: Bell, M. & Limbrey, S., eds. *Archaeological Aspects of Woodland Ecology*, pp. 23–55. British Archaeological Society, Oxford.

Hack, J.T. & Goodlett, J.C. (1960) *Geomorphology and Forest Ecology of a Mountain Region in the Central Appalachains*. Geological Survey Professional Paper 347, United States Government Printing Office, Washington, DC.

Harmon, M.E., Franklin, J.F., Swanson, F.J. *et al.* (1986) Ecology of coarse woody debris in temperate ecosystems. *Advances in Ecological Research* **15**, 133–302.

Heinselman, M.L. (1973) Fire in the virgin forests of the Boundary Waters Canoe Area. *Minnesota. Quaternary Research* **3**, 329–82.

Henry, J.D. & Swan, J.M.A. (1974) Reconstructing forest history from live and dead plant material: an approach to the study of forest succession in southwest New Hampshire. *Ecology* **55**, 772–83.

Horn, H.S. (1975) Forest succession. *Scientific American* **232**, 90–8.

Hough, A.F. (1932) Some diameter distributions in forest stands of northwestern Pennsylvania. *Journal of Forestry* **30**, 933–43.

Hough, A.F. & Forbes, R.D. (1943) The ecology and silvics of forests in the high plateaus of Pennsylvania. *Ecology* **46**, 370–3.

Huntley, B. & Birks, H.J.B. (1983) *An Atlas of Past and Present Pollen Maps for Europe: 0–13 000 Years Ago.* Cambridge University Press, Cambridge.

Hupp, C.R. (1988) Plant ecological aspects of flood geomorphology and paleoflood history. In: Baker, V.R., Kochel, R.C. & Patton, P.C., eds. *Flood Geomorphology*, pp. 335–56. Wiley Interscience, New York.

Hursh, C.R. & Haasis, F.W. (1931) Effects of 1925 summer drought on southern Appalachian hardwoods. *Ecology* **12**, 380–6.

Hutnik, R.J. (1952) Reproduction on windfalls in a northern hardwood stand. *Journal of Forestry* **50**, 693–4.

Iwaschkewitsch, B.A. (1929) Die wichtigsten Eigenarten der Structur und der Entwicklung der Urwald bestände. In: *Proceedings of the 7th International Congress of Forestry Experimental Stations, Stockholm*, pp. 129–47.

Jahn, G. (1991) Temperate deciduous forests of Europe. In: Röhrig, E. & Ulrich, B., eds. *Temperate Deciduous Forests. Ecosystems of the World Vol. 7*, pp. 377–502. Elsevier, Amsterdam.

Jones, E.W. (1945) The structure and reproduction of the virgin forest of the north temperate zone. *New Phytologist* **44**, 130–48.

Kirby, K.J., Reid, C.M., Thomas, R.C. & Goldsmith, F.B. (1998) Preliminary estimates of fallen wood and standing dead trees in managed and unmanaged forests in Britain. *Journal of Applied Ecology* **35**, 148–55.

Liebundgut, H. (1959) Über Zweck und Methodik de Structur und Zuwachsanalyse von Urwäldern. *Schweizerische Zeitschrift für Forstwesen* **110**, 111–24.

Lorimer, C.G. (1980) Age structure and disturbance history of a southern Appalachian virgin forest. *Ecology* **61**, 1169–84.

Lorimer, C.G. (1983) A test of the accuracy of shade-tolerance classifications based on physiognomic and reproductive traits. *Canadian Journal of Botany* **61**, 1595–8.

McCune, B. & Allen, T.F.H. (1985) Will similar forests develop on similar sites? *Canadian Journal of Botany* **63**, 367–76.

McInnes, P.F., Naiman, R.J., Pastor, J. & Cohen, Y. (1992) Effects of moose on vegetation and litter of the boreal forest, Isle Royale, Michigan, USA. *Ecology* **73**, 2059–75.

McIntosh, R.P. (1985) *The Background of Ecology. Concept and Theory.* Cambridge University Press, Cambridge.

Maissurow, D.K. (1941) The role of fire in the perpetuation of virgin forests in northern Wisconsin. *Journal of Forestry* **39**, 201–7.

Maser, C., Tarrant, R.F., Trappe, J.M. & Franklin, J.F. (eds) (1988) *From the Forest to the Sea: a Story of Fallen Trees.* General Technical Report PNW-GTR-229, United States Department of Agriculture Forest Service, Washington, DC.

Matlack, G.R., Gleeson, S.K. & Good, R.E. (1993) Treefall in a mixed oak–pine coastal plain forest: immediate and historical causation. *Ecology* **74**, 1559–66.

Mayer, H. & Neumann, M. (1981) Structureller und entwicklungsdynamischer Vergleich der Fichten-Tannen-Buchen-Urwälder Rothwald/Niederösterreich und Corkova Uvala/Kroatien. *Forstwissenschaftliches Centralblatt* **100**, 111–32.

Mayer, H., Neumann, M. & Sommer, H.-G. (1980) Bestandesaufbau und Verjüngungsdynamik unter dem Einfluss nütürlicher Wilddichten im Kroatischen Urwaldreservat Corkova Uvala/Plitvicer Seen. *Schweizerische Zeitschrift für Forstwesen* **131**, 45–70.

Miles, J. (1988) Vegetation and soil changes in the uplands. In: Usher, M.B. & Thompson, D.B.A., eds. *Ecological Change in the Uplands*, pp. 57–70. Blackwell Scientific Publications, Oxford.

Mueller-Dombois, D. (1987) Natural dieback in forests. *Bioscience* **37**, 575–83.

Oliver, C.D. (1981) Forest development in North America following major disturbances. *Forest Ecology and Management* **3**, 153–68.

Oliver, C.D. & Larsen, B.C. (1996) *Forest Stand Dynamics*, update edn. McGraw-Hill, New York.

Peterken, G.F. (1996) *Natural Woodland.* Cambridge University Press, Cambridge.

Peterken, G.F. & Mountford, E. (1995) Lady Park Wood reserve: the first half century. *British Wildlife* **6**, 205–13.

Peterken, G.F. & Mountford, E.P. (1996) Effects of drought on beech in Lady Park Wood, an unmanaged mixed deciduous woodland. *Forestry* **69**, 117–28.

Peterken, G.F. & Mountford, E.P. (1998) Long-term change in an unmanaged population of wych elm subjected to Dutch elm disease. *Journal of Ecology* **86**, 205–18.

Peterson, C.J. & Pickett, S.T.A. (1991) Treefall and resprouting following catastrophic windthrow in an old-growth, hemlock–hardwoods forest. *Forest Ecology and Management* **42**, 205–17.

Peterson, C.J., Carson, W.P., McCarty, B.C. & Pickett, S.T.A. (1990) Microsite variation and soil dynamics within newly created treefall pits and mounds. *Oikos* **58**, 39–46.

Pickett, S.T.A. & Thompson, J.N. (1978) Patch dynamics and the design of nature reserves. *Biological Conservation* **13**, 27–37.

Pickett, S.T.A. & White, P.S. (1985) Patch dynamics: a synthesis. In: Pickett, S.T.A. & White, P.S., eds. *The Ecology of Natural Disturbance and Patch Dynamics*, pp. 371–84. Academic Press, Orlando.

Pontailler, J.-Y., Faille, A. & Lemeé, G. (1997) Storms drive successional dynamics in natural forests: a case study in Fontainebleau forest (France). *Forest Ecology and Management* **98**, 1–15.

Poulson, T.L. & Platt, W.J. (1989) Gap light regimes influence canopy tree density. *Ecology* **70**, 553–5.

Poulson, T.L. & Platt, W.J. (1996) Replacement patterns of beech and sugar maple in Warren Woods. *Michigan. Ecology* **77**, 1234–53.

Pyne, S.J. (1982) *Fire in America. A Cultural History of Wildland and Rural Fire.* Princeton University Press, Princeton, New Jersey.

Rackham, O. (1980) *Ancient Woodland.* Arnold, London.

Raup, H.M. (1957) Vegetation adjustments to the instability of the site. In: *Proceedings of the 6th Technical Meeting, International Union for the Conservation of Nature and Natural Resources*, IUCN, Morges, pp. 36–48.

Raup, H.M. (1981). Reflections on American forestry. In: Stout, B.B., ed. *Forests in the Here and Now.* Montana Forest and Conservation Station, Missoula, Montana.

Romme, W.H. (1982) Fire and landscape diversity in subalpine forests of Yellowstone National Park. *Ecological Monographs* **52**, 199–221.

Runkle, J.R. (1981) Gap regeneration in some old-growth forests of eastern United States. *Ecology* **62**, 1041–51.

Runkle, J.R. (1982) Patterns of disturbance in some old-growth mesic forests of eastern North America. *Ecology* **63**, 1533–46.

Runkle, J.R. & Yetter, T.C. (1987) Treefalls revisited: gap dynamics in the southern Appalachians. *Ecology* **68**, 417–24.

Sano, J. (1997) Age and size distribution in a long-term forest dynamics. *Forest Ecology and Management* **92**, 39–44.

Shugart, H.H. (1984) *A Theory of Forest Dynamics.* Springer-Verlag, New York.

Spies, T.A., Franklin, J.F. & Thomas, T.B. (1988) Coarse woody debris in Douglas-fir forests of western Oregon and Washington. *Ecology* **69**, 1689–702.

Sprugel, D.G. (1976) Dynamic structure of wave-regenerated *Abies balsamea* forests in the north-eastern United States. *Journal of Ecology* **64**, 889–911.

Stewart, G.H. (1988) The influence of canopy cover on understorey development in forests of the western Cascade Range, Oregon, USA. *Vegetatio* **76**, 79–88.

Stewart, G.H. & Burrows, L.E. (1994) Coarse woody debris in old-growth temperate beech (*Nothofagus*) forests of New Zealand. *Canadian Journal of Forest Research* **24**, 1989–96.

Wallis de Vries, M.F. (1995) Large herbivores and the design of large-scale nature reserves in western Europe. *Conservation Biology* **9**, 25–33.

White, P.S. (1979) Pattern, process and natural disturbance in vegetation. *Botanical Review* **45**, 229–99.

White, P.S. (1987) Natural disturbance, patch dynamics and landscape pattern in natural areas. *Natural Areas Journal* **7**, 14–22.

Whitehead, D. (1982) Ecological aspects of natural and plantation forests. *Forestry Abstracts* **43**, 615–24.

Zackrisson, O. (1977) Influence of forest fires on the North Swedish boreal forest. *Oikos* **29**, 22–32.

Zukrigl, K., Eckhardt, G. & Nather, J. (1963) Standortskundliche und waldbauliche Untersuchungen in Urwaldresten der niederösterreichischen Kalkalpen. *Mitteilungen Forstwesen. Bundesversuchsanstalt Wien* **62**.

5: Biological Interactions and Disturbance: Plants and Animals

JABOURY GHAZOUL AND EUNICE SIMMONS

5.1 INTRODUCTION

Anthropogenic disturbance of forest environments in recent decades is an increasing cause for concern with regard to conservation of global biodiversity. Indeed, the rate of forest clearance is greatest in areas where the highest endemism of birds (Balmford & Long 1994) and plants (Mittermeier *et al.* 1998) is recorded. The distribution of biodiversity is generally well known but our knowledge of the ecological roles of forest species remains poor, as emphasis has been placed on documenting biodiversity solely by the number and abundance of species and individuals. The 'biodiversity' usually monitored by research ignores biotic interactions and differs from conceptual definitions of biodiversity that include variety at several hierarchical levels (genes, species, habitats, landscapes). Additionally, species are not equal in the way they contribute to ecosystem functions (such as turnover and productivity) or processes (e.g. pollination, decomposition and herbivory), and there is some degree of ecological redundancy where the functions previously performed by extinct species can be taken over by one or more extant species (Walker 1992; Lawton & Brown 1993). Although vertebrates are dwarfed by invertebrates in terms of species richness and biomass, they remain extremely important in shaping the temporal dynamics of forest structure and composition by mediating ecological processes through a variety of plant–animal interactions. For this reason the impact of anthropogenic disturbances on vertebrate populations in forests, particularly tropical forests, requires careful consideration by forest managers as well as conservation biologists.

Selective logging in the tropics refers to the extraction of commercially valuable trees exceeding a specified minimum size. This accounts for about 3–7% of forest trees, although incidental damage by mechanized logging may result in the destruction of 50% or more trees (Pinard & Putz 1996). In boreal forests where tree species richness is limited to a few species, clearcut harvesting is undertaken and decisions on the size and distribution of clearcut patches have great relevance to the persistence of biotic forest communities (Franklin & Forman 1987). The natural disturbance regime in tropical forests tends to be smaller but more frequent (fallen branches and trees, landslides), although hurricanes can be a major force in some tropical forests, while in temperate and boreal forests natural disturbances are less frequent but more extensive and severe (fire, pest and disease outbreaks, windthrow, etc.). *Consequently, approaches to the ecological management of production forests differ between tropical and temperate regions, with greater emphasis placed on local management in tropical forests, while landscape-scale approaches are adopted in temperate regions.* Disturbances on both local and landscape scales affect vertebrates by altering the abundance and quality of resources such as food or breeding sites, while the composition and stability of whole communities, sometimes encompassing the entire range of species (e.g. the spotted owl in north-western USA and Canada), respond to landscape-scale disturbances such as habitat fragmentation and conversion. All forest habitats are also subject to natural disturbance regimes that act on spatial and temporal scales, ranging from the local and brief (trampling, grazing, treefalls and gap formation) to regional and prolonged (fires and hurricanes) (see Chapter 4). Understanding the role of vertebrates in shaping the forest community dynamic, and the response of vertebrates to the natural as well as the his-toric and increasing anthropogenic disturbance regimes at the whole range of spatial and temporal scales, is crucial to the sustainable management of natural forests and the biodiversity they contain.

The purpose of this chapter is to outline the importance of terrestrial vertebrates in plant–animal interactions in tropical and temperate forests, and to describe the impact of disturbances on vertebrates and on the ecological processes they mediate. Although this chapter is limited to a consideration of forest vertebrates, we recognize the important role invertebrates have in mediating forest processes (for a recent review see Ghazoul & Hill, in press; see also Chapter 6). We concentrate on anthropogenic disturbances that are the cause of much concern with regard to the persistence of forest vertebrates and also focus on tropical forests where vertebrate species diversity is greatest; however, we also draw upon studies from temperate forests (for a more thorough discussion of natural disturbances in temperate forests see Chapter 4). Most studies are conducted at local scales and deal with one or a few species, but in reviewing these studies we attempt to describe emerging regional patterns of responses of vertebrates and vertebrate–plant interactions to local- and landscape-level disturbances.

5.2 DIVERSITY AND DISTRIBUTION OF FOREST VERTEBRATE RESOURCES

5.2.1 Habitat structure and some generalities of vertebrate responses to resource variability

Vegetational diversity and structural complexity have long been correlated with animal diversity (MacArthur & MacArthur 1961; Karr & Roth 1971; Urban & Smith 1989; Hansen & Hounihan 1996). Vertical habitat complexity and heterogeneity of regeneration phases in old-growth forests, compared with structurally simpler early successional forests, provide a variety of foraging, nesting and roosting sites and diverse microcli-

matic conditions that support high vertebrate diversity by niche segregation (MacArthur & MacArthur 1961). In northern temperate old-growth forests, elevated light levels in canopy gaps provide vertebrates with highly productive forage patches compared with younger forests that have continuous dense canopies (Alaback 1982). Large standing and fallen dead trees are far more common in old-growth than young regenerating forest (Spies *et al.* 1988) and also provide a range of foraging and nesting opportunities. Foliage volume may be related to avian diversity by affecting the abundance of food resources or nest sites (Mills *et al.* 1991; see also Willson 1993). However, correlation of forest structural complexity with bird species diversity has proved to be highly variable among regions (Hansen & Hounihan 1996) and causal mechanisms linking forest structural complexity with bird diversity remain to be found.

Mobility allows vertebrates to exploit temporally variable food resources and many, particularly mammals, are relatively generalist in their food requirements (Karr *et al.* 1990; Willson 1993). Consequently, vertebrates are able to respond behaviourally to changes in the distribution of ephemeral resources by localized habitat shifts (Levey 1988; Loiselle & Blake 1991), expansion of the home range (Fleming 1992) or switching to alternative food sources (Worthington 1982; Willson 1993). Behavioural responses to regional changes in resource availability can affect local abundance of migratory birds (Stiles & Clarke 1989) and mammals (Ashton 1988; Tutin *et al.* 1997). High mobility buffers forest vertebrates (with some exceptions) from resource-associated breeding failure and extreme fluctuations of population size. However, extensive and prolonged anthropogenic or natural disturbances can affect vertebrate population dynamics by reducing clutch size or causing temporary suspension of breeding (Wiens 1989; Holmes 1990). Some resources have importance disproportionate to their abundance, often because they are available at times of food scarcity, and these may be particularly relevant to breeding success and population viability. Fig trees provide such 'keystone' resources to frugivorous birds and mammals in some tropical forests, and loss of fig trees is predicted to have wide-ranging impacts on the vertebrate communities of these forests (Terborgh 1986; Lambert & Marshall 1991).

5.2.2 Issues of scale

Vertebrates are highly mobile and use resources on much larger spatial scales than plants or invertebrates. Foraging ranges of vertebrates may encompass several hundred hectares or more (Table 5.1), and habitat requirements for nesting and feeding may be in quite different locations. Local disturbances may therefore be less important than large-scale disturbance phenomena in affecting vertebrate communities, and habitat patterns across landscapes should be more relevant to management strategies directed to forest vertebrates. Thus vertebrates are more likely to respond to habitat fragmentation and the heterogeneity and relative abundance of different habitat types than to forest structure or local complexity.

Many vertebrates use resources from a variety of habitat types that occur in natural forested landscapes. These habitat types include a range of successional stages that are topographically, edaphically and hydrologically distinct vegetation zones. Human land uses tend to simplify or homogenize landscapes (Krummel *et al.* 1987). Forest fragmentation leads to habitat isolation, reduced patch size and increased importance of edge effects as well as, more obviously, habitat loss. Single forest patches in anthropogenically fragmented landscapes are unlikely to contain the variety of habitat types and resources previously represented in the continuous forest landscape. Vertebrates that use resources from several forest habitat types are therefore susceptible to changes in resource availability at the landscape scale. Single species operate at a variety of spatial scales during their life history. Thus many small birds defend nesting territories measured in square metres, use foraging ranges extending to several hectares, yet cover many kilometres during seasonal migrations. Even non-migratory species use resources, with varying strengths of interaction, at a range of spatial scales, and incorporating habitat features at multiple scales has been an important planning requirement for conservation of vertebrates such as the spotted owl *Strix occidentalis* in the forests of the Pacific North-West of the USA (King *et al.* 1997) (Fig. 5.1).

In fragmented landscapes the extent of disruption depends on the scale of fragmentation relative to the mobility of the organisms and on the

Table 5.1 Spatial scales relevant to some Neotropical species. (Adapted from Terborgh 1992 with permission.)

Organism unit	Area occupied by one reproductive unit (ha)
Ant colony	<0.01
Butterfly food web	2–10
Troop of howler monkeys	25
Bird specialist on treefall openings	30–100
Community of spider monkeys	150
Troop of squirrel monkeys	500
Scarlet macaw	>1000
Jaguar	5000
Herd of white-lipped peccaries	>10000
Fruit crow	>10000

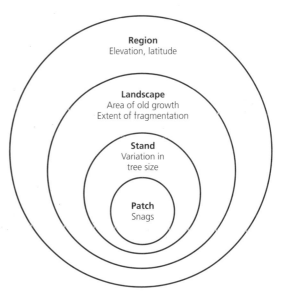

Fig. 5.1 The associations of spotted owls with habitat features occurring at several spatial scales. (Adapted from Hansen *et al.* 1993 with permission.)

nature of the habitat matrix in which the fragments are located. Vertebrates that feed on unpredictable patchy and ephemeral resources, such as nectarivores and frugivores, operate on large spatial scales and tend to be more mobile than folivores or insectivores of similar size (e.g. Oates 1987; Fleming 1992). Mobility facilitates movement between patches and consequently offers a degree of protection from the effects of habitat fragmentation, provided the matrix in which resource patches are located is favourable for movement and relatively risk-free. The nature of the habitat matrix and persistence of the resource is particularly important for vertebrates that use temporally and spatially predictable but ephemeral resources that require seasonal migrations between habitat zones (e.g. wild pigs, Ashton 1988; primates, Tutin *et al.* 1997). Seasonal habitat shifts are widespread among birds and mammals in response to climate, the availability of food resources, and nesting requirements (reviewed in Fleming 1992). For example, bears in North America feed on berries in forest habitats but move to streams to feed on salmon during the annual spawning runs (Willson 1993). Sparse berry crops can increase the movement of bears to streams, while a good fruit crop reduces the size of their home range.

Forest patches surrounded by a hostile habitat matrix (or internal fragmentation features such as fences and roads) can impede vertebrate movement that, in addition to limiting access to resources, isolates subpopulations and prevents genetic mixing. By isolating subpopulations fragmentation increases local extinction probabilities by reducing the size of effective breeding population below thresholds of genetic and demographic viability (e.g. Kinnaird & O'Brien 1991). Provision of habitat corridors and careful planning of the land-use regime can increase landscape connectivity and facilitate greater movement of vertebrates between patches. Recent years have seen an increase in studies describing the responses of vertebrates to habitat fragmentation. These have been accompanied by theoretical developments, most notably describing thresholds of fragmentation beyond which the effects of isolation and small patch size increase rapidly (Andrén 1994; With *et al.* 1997).

The concepts of scale and heterogeneity in natural ecosystems are highly relevant to the integration of ecological principles with land management. The size and distribution of habitat patches need to be considered relative to the management objectives, be they conservation of particular species, continued sustainability of ecological processes, or both. Land managers will be aided by remote sensing techniques and geographic information systems (GIS) in the assessment of the suitability of landscapes to support high biodiversity (Short *et al.* 1996), although these approaches need to be combined with studies of life history traits (Hansen *et al.* 1993) and species requirements. Variation in the frequency and size of natural disturbance events should be accounted for in management planning by incorporating or mimicking natural disturbance regimes at several spatial and temporal scales. Such planning needs to take a hierarchical perspective to match the mobility and requirements of biological entities that vary in size from small arthropods to herds of large mammals.

5.3 VERTEBRATE RICHNESS AND DIVERSITY IN PRIMARY AND DISTURBED FORESTS

Many studies have examined how habitat transformation affects the richness and abundance of forest animal communities (Thiollay 1992; Andrade & Rubio 1994; Laurance 1994; Malcolm 1994). Results from these studies are highly variable, reflecting the complex interplay of abiotic and biotic factors acting at different spatial and temporal scales on community structure and composition. Disturbance itself is qualitatively and quantitatively highly variable; deforestation, the most widespread anthropogenic disturbance, causes habitat fragmentation (with associated impacts of edge effects and isolation), microclimatic changes and alteration of habitat structure (see Chapter 4). These changes may further lead to increased susceptibility to fire (Holdsworth & Uhl 1997; Uhl 1998), windthrow (Laurance 1997), pests and pathogens, or invasive exotic species. This interplay of biotic and abiotic complexity, coupled with a scarcity of long-term datasets, makes it difficult to distinguish patterns describing the effects of habitat alteration on biodiversity and ecological processes (Kruess & Tscharntke 1994; Pimm & Sugden 1994; Tilman & Downing 1995).

Vertebrates themselves may be instrumental in structuring the physical nature of the habitat, which in turn has important positive and negative effects for other species occurring within the environment. Particular vertebrate species may have disproportionate roles as agents of habitat structural manipulation. Jones *et al.* (1994, 1997) term such organisms 'physical ecosystem engineers'. Ecosystem engineering plays a major role in the structure and function of ecosystems by controlling the availability of resources through the physical modification, maintenance or creation of habitats (Jones *et al.* 1997). In temperate forests, beavers have an important function as ecosystem engineers through the damming of forest streams, which increases habitat and resource heterogeneity (Naiman *et al.* 1988). Squirrels can also be important through the damage they cause to trees. In tropical systems, elephants and herds of ungulates change habitat physical structure by damaging trees and trampling the undergrowth (e.g. Lewis 1991). The effects of some engineers may be dramatic in proportion to their size or abundance. Woodpeckers excavate nesting cavities in dead wood that are later used by a wide variety of birds, mammals and invertebrates unable to excavate cavities themselves (Harris & Silva-Lopez 1992). Recognizing the role of these species as ecosystem engineers in forest systems is crucial for ecologically sustainable forest management.

Particular organisms can also have key roles in the food web. The role of predation in structuring vertebrate communities has largely been unexplored, though recent work indicates that predation may indeed be important (Terborgh 1988; Brosset 1990; Fonseca & Robinson 1990; Karr *et al.* 1992). As top predators, large carnivores are susceptible to habitat fragmentation or conversion (as well as hunting) owing to their small population sizes, large foraging range requirements and relative inflexibility in their food preferences (Terborgh & Winter 1980; Fonseca & Robinson 1990). The effects of the loss of top predators are not clear, although island communities or recently fragmented habitats that cannot, or no longer, support populations of large predators indicate that densities of herbivores and seed-predators increase, with further effects cascading through the community. The El Verde forest of Puerto Rico, for example, has unusually high abundances of lizards and amphibians. High densities of small mammals on Barro Colorado Island in Panama are thought to be due to the absence of regulation by large predators (Terborgh & Winter 1980). Changes in trophic structure can shape plant communities by altering patterns of herbivory and seed predation. Thus on Barro Colorado Island, recruitment of *Dipteryx panamensis*, a large canopy tree, is much reduced compared with the mainland due to unusually high seed predation pressure by small mammals (De Steven & Putz 1984).

The existence of forests themselves may be a temporal dynamic regulated by the interaction of abiotic and biotic disturbances. These disturbances can, in combination, cause major environmental changes that each on its own cannot achieve. The interaction of disease, fire and grazing by large mammals appears to be responsible for temporal changes in savannah woodland cover in Africa (Pellew 1983; Dublin *et al.* 1990). For example, woodland growth in the Mara region of the Serengeti is facilitated by wildebeest, which maintain fires at low frequency and intensity by grazing the grass fuels. Increased fire following reduction of wildebeest populations by disease can lead to a contraction of forested land, although this is exacerbated by population increases of elephants (due to a policy of elephant conservation) that inhibit forest regeneration by browsing of tree seedlings (Dublin *et al.* 1990).

The rest of this section reviews vertebrate diversity and abundance in anthropogenically disturbed forest. The response of bird and, to a lesser extent, mammal communities to disturbance has been well studied throughout temperate and tropical regions, permitting some generalizations to be drawn. In contrast, very few studies have investigated impacts on amphibians and we could find only one referring specifically to reptiles.

5.3.1 Birds

Forest degradation and fragmentation has been repeatedly shown to reduce bird diversity and richness, increase the dominance of the commonest species, and favour gap and forest edge specialists at the expense of forest interior species, which often decline in density or become locally extinct (Johns 1991; Thiollay 1992; Bierregaard & Stouffer 1997; Canaday 1997; Marsden 1998). Insectivores (Johns 1991; Thiollay 1992; Bierregaard & Stouffer 1997; Canaday 1997; Marsden 1998) and to a lesser extent large frugivores (Thiollay 1992; Bierregaard & Stouffer 1997) appear to be most sensitive to disturbance. Among the insectivorous birds, leaf litter insectivores are often the most severely affected (Bierregaard & Stouffer 1997; Canaday 1997), probably reflecting changes in invertebrate abundance and distribution following drying of the leaf litter in fragmented forests (Wolda 1978; Kapos 1989; Canaday 1997). Nectarivores appear to be more resilient to forest degradation and fragmentation, and little change in behaviour or abundance of hummingbirds has been noted in Central or South America (Feinsinger 1976; Bierregaard & Stouffer 1997; Thiollay 1997; see section 5.4.1). Some large tropical birds, such as hornbills and parrots, are thought to be tolerant of recent logging (Lambert & Marshall 1991; Marsden 1998), but these species are nomadic, long-lived and require large trees for nesting, and their long-term viability in logged forests has yet to be assessed (Marsden 1998).

In Amazonian forest fragments common frugivores were heavily affected while uncommon species showed no apparent response (though small sample sizes of rare species limit interpretation) (Bierregaard & Stouffer 1997). Such results offer some promise for conservation of avian biodiversity, but there may be serious implications for seed dispersal and plant regeneration (see also Silva *et al.* 1996). Recolonization of Amazonian forest fragments by insectivores appears to be a function of the quality of the surrounding habitat matrix (Bierregaard & Stouffer 1997), which may have great importance in maintaining the long-term biodiversity and resilience of insectivore communities. In contrast, little recolonization of

fragments by frugivores occurred regardless of the nature of the surrounding matrix, neither was there any compensation of the seed dispersal function in fragments by frugivores associated with secondary growth.

Retention of living and dead residual trees in clearcuts or large forest gaps has been frequently proposed as a strategy that benefits wildlife (Raphael & White 1984; Hunter 1990; Payne & Bryant 1994). Retention of standing trees can increase the use of clearcuts by birds (Dickson *et al.* 1983; Marcot 1983; Niemi & Hanowski 1984) and increase local avian diversity (Dickson *et al.* 1992). However, bird species differ in their habitat requirements and there is concern that management for high diversity may be detrimental to habitat specialists or rare species (Harris 1984; Yahner 1988). Merrill *et al.* (1998) addressed this concern and found that residual trees in Minnesota aspen clearcuts do benefit several avian species of high conservation priority but cautioned that the value of residual stands is relatively minor compared with large intact forests for supporting endangered forest bird species.

Fragmentation of the northern boreal forests of Canada, which has one of the richest avian breeding sites in North America (Robbins *et al.* 1989), can reduce the quantity and quality of the available breeding habitat (Saunders *et al.* 1991) with consequent effects on the avian community. Predation and brood parasitism of forest birds have been shown to increase in anthropogenically disturbed or fragmented forests (Andrén *et al.* 1985; Wilcove 1985; Andrén & Angelstam 1988; Yahner 1988, 1996; Burkey 1993; Andrén 1995; Robinson *et al.* 1995; Bayne & Hobson 1997) and along forest edges (Andrén *et al.* 1985; Santos & Telleria 1992; Paton 1994; Andrén 1995). In a quantitative review of these studies Hartley and Hunter (1998) concluded that predation rates are much higher in deforested areas but that the strength of edge effects is a function of landscape composition, notably the extent of remaining forest cover and the quality of the non-forested matrix. Nest parasitism by brown-headed cowbirds, *Molothrus ater*, which proliferated following human settlement and forest fragmentation, has severely affected the avifauna of North American forests (Brittingham & Temple 1983; Robinson & Wilcove 1994). However, birds in boreal forests are more likely to be affected by fragmentation caused by agricultural development than fragmentation due to logging (Hartley & Hunter 1998), as the latter is more likely to mimic natural disturbance regimes caused by fire or insect tree-pest outbreaks (Hansson & Angelstam 1991; Hansson 1992; Bayne & Hobson 1997; see also Santos & Telleria 1992). Finally, fragmentation or decline of temperate forests can affect the populations of tropical migrants (Robbins *et al.* 1989; Robinson & Wilcove 1994; Robinson *et al.* 1995) and coordinated forest management across widely separated regions is needed to prevent the decline of migratory species.

5.3.2 Mammals

At intensities of 70–90 m^3 ha^{-1} (and sometimes reaching 120 m^3 ha^{-1}), selective logging in South-East Asian dipterocarp forests is among the highest in the tropics. Despite this, most mammals studied are able to persist in logged forest (civets, Heydon & Bulloh 1996; primates, Johns 1992; ungulates, Heydon 1994—quoted in Heydon & Bulloh 1997), although carnivores may be more adversely affected than herbivores and omnivores. Diet flexibility may contribute to the persistence of mammal populations in logged forest, and the ability of frugivorous primates to include leaves in the diet appears to be correlated with their survival (Johns & Skorupa 1987). High-energy requirements, associated with small size, may restrict mammalian diets to foods that are easily digestible and have high calorific content. Thus small mammals (e.g. mousedeer, Heydon & Bulloh 1997) are thought to be more susceptible to changes in resource quality or abundance. There are contrasting responses to disturbance between insectivorous and frugivorous bats, presumably a reflection of their feeding guilds. Insectivorous bats are not affected by losses of canopy cover as abundance of their aerial insect prey does not vary across disturbed and undisturbed forest habitats or even in surrounding pastureland (Fenton *et al.* 1998). Flying foxes feed on forest fruits and are more susceptible to forest degradation (Cox *et al.*

1991), presumably due to the loss of fruiting trees in logged forest (Johns 1992).

The resilience (i.e. rate of population recovery) of mammal communities in logged forests appears to be high. Studies in peninsular Malaysia suggest that the initial effects of extraction of more than 50% of trees on primates and other mammals may be considerable but that most species present prior to disturbance had persisted or recolonized 12 years later (Johns 1992). However, as with birds, the nature of the surrounding habitat matrix has a large influence in determining the composition of forest fragments and Malcolm (1997) showed that primary forest fragments in Amazonia surrounded by secondary forest maintained much higher densities and diversities of small mammals than fragments surrounded by pastureland. Some studies have reported marsupials (Fonseca & Robinson 1990) and rodents (Malcolm 1997) to be even more common in secondary than primary forest. The structural and microclimatic conditions of secondary forest may resemble closely that of natural treefall gaps where rodents preferentially forage (Forget 1997).

Hunting pressure can be high in logged forests, the preferred species being large birds and mammals (Redford 1992; Bennett & Dahaban 1995; Rumiz *et al.* in press), although rodents are also targeted (Forget & Milleron 1991). These are often the animals that have important roles as seed predators and dispersers (see below) and there is some evidence that tree recruitment may be negatively affected by changes in the composition of mammal communities through hunting pressures (see Guariguata & Pinard 1998).

5.3.3 Amphibians

Strong site fidelity and limited dispersal capabilities appear to make amphibians particularly sensitive to the effects of logging and habitat fragmentation (Bradford *et al.* 1993; Blaustein *et al.* 1994; Marsh & Pearman 1997), although there is evidence suggesting that observed declines in amphibian richness immediately after a disturbance event is reversed in following years and richness may ultimately even exceed predisturbance levels (Tocher *et al.* 1997). Changes in forest structure can lead to the drying of the habitat and increased daily temperature variation (Bastable *et al.* 1993). Most terrestrial amphibians are highly sensitive to these changes due to physiological and developmental dependence on high atmospheric humidity and abundant water. However, some species (e.g. *Eleutherodactylus* spp.) deposit their eggs on land and undergo direct development to the adult form within the egg. Such life-history features allow these species to achieve very high densities and become distributed widely throughout the forests in which they occur (Reagan *et al.* 1996). More generally, amphibian species richness has been correlated with habitat characteristics such as humidity, leaf litter thickness (Fauth *et al.* 1989; Woinarski & Gambold 1992) and understorey density (Pough *et al.* 1987; Dupuis *et al.* 1995), and amphibian abundance positively correlated with forest fragment size and proximity to larger forest tracts (Marsh & Pearman 1997). The availability of a deep litter layer that maintains high moisture levels appears to be particularly important for terrestrial nesting amphibians (Hodl 1990). Thus shady dense second growth appears to be more accessible to amphibians than exposed pasture and, consequently, the nature of the surrounding matrix affects rates of amphibian colonization of disturbed areas. Like other animal groups, amphibians exhibit highly species-specific responses to disturbance (Tocher *et al.* 1997).

5.3.4 Reptiles

In the almost total absence of published studies we make some tentative predictions about the response of reptiles to anthropogenic disturbances based on physiological and ecological attributes. Reptiles are poikilothermic and therefore depend on external sources of heat to raise body temperature and, consequently, they have low metabolic rates and energy requirements. Canopy opening by disturbance changes the understorey microclimate by raising daytime temperatures in canopy gaps. This is likely to benefit heliothermic reptiles by extending their activity periods (Vitt *et al.* 1998). As with insectivorous birds, drying of the leaf litter may affect smaller lizards due to reduced arthropod avail-

ability. However, compared with homeothermic birds and mammals, reptiles are not likely to be limited by food or water availability as many are able to survive long periods between feeding events. Increased abundance of rodents in gaps (Forget 1997) and secondary forest (Malcolm 1997) may even increase food availability for large reptiles such as snakes and teiid lizards (Vitt *et al.* 1998). This could increase the abundance of predaceous reptiles by immigration and population growth, which may have a cascade effect on other forest species (Vitt *et al.* 1998). Furthermore, fragmentation can concentrate forest birds into smaller areas, thereby raising the density of prey such as eggs and chicks. Smaller patch sizes and increased edge effects may facilitate the movement of reptiles normally associated with open pasture into forested areas. Fragmentation could affect populations of forest turtles if forest patches become too small to include streams or areas of seasonally available standing water.

Thus we predict that reptiles are less likely than birds or mammals to be affected by conversion of primary forest to secondary forest or to a series of forest fragments. Food and water (except perhaps for turtles) are unlikely to be limiting as their metabolic requirements are relatively low. Predation by birds and mammals may be more relevant in structuring reptile communities, but little is known about the changes in the foraging preferences of predators in disturbed habitats.

5.4 VERTEBRATE-MEDIATED ECOLOGICAL PROCESSES AND IMPACTS OF DISTURBANCE

Ecological interactions have critical importance in the maintenance of biodiversity and ecological functions such as productivity and turnover. Mutualistic interactions are extremely common and include interactions between all major organismal groups (Table 5.2), they have been largely neglected in the ecological literature, with greater emphasis placed on competition and predation. Many interactions may be quite unspecific with a great degree of ecological redundancy, such that the loss of a single species member of a mutual-

Table 5.2 Some direct mutualisms involving vertebrates in forested environments. (Adapted from Willson 1996 with permission.)

Plants and animals
Flowering plants and animal pollinators:
  all communities, all latitudes, many orders of plants, chiefly insects, birds and mammals
Seed plants and animal dispersal agents: most communities, most latitudes, many orders of plants, chiefly vertebrates and ants

Animals and other animals
Interspecific flocks and herds: best known in birds and animals, all latitudes, any communities
Specific cases at local distribution including hornbills and mongooses, honeyguides and mammals, tickbirds and mammals in Africa

Animals and fungi
Spore dispersal of mycorrhizal fungi by small mammals: many communities, many latitudes, many taxonomic groups of plants, several genera of mammals

ism results in little change in the overall outcome of the interaction. The extent of redundancy and the resistance of ecological interactions to anthropogenic disturbances are poorly known, yet of critical importance to forest managers in determining extractive mechanisms and limits. Vertebrate–plant interactions should be of interest to both forest managers and conservation biologists owing to the roles of vertebrates in plant recruitment and in regulating forest structure and composition through pollination, seed dispersal, and seed predation and herbivory. This section reviews these four plant–vertebrate interactions in forest habitats and considers disruption of these interactions by anthropogenic disturbance.

5.4.1 Pollination

Pollination by vertebrates is almost non-existent in northern temperate forests where most species are wind-pollinated. Indeed there is a trend of increasing wind pollination with increasing latitude (Regal 1982). In tropical forests, high tree diversity and wide spacing between conspecifics limit the efficiency of wind pollination and tree species in lowland tropical forests are almost

all pollinated by animals (including insects). To attract vertebrate pollinators plants need to produce nectar in quantities much greater than that required for insect pollination. The energy costs of nectar production to plants can therefore be considerable. Thus nectar produced by vertebrate-pollinated plants tends to be relatively dilute. Vertebrate-pollinated plants are therefore most abundant in wet tropical climates where water does not limit nectar production, and are rarely found in dry forest or savannah habitats. Among vertebrates, bats, birds and a few non-flying mammals are the only known pollinators (Bawa 1990); although they pollinate only a minority of plants (6% of canopy tree species in tropical moist forest at La Selva, Costa Rica; Bawa 1990), they remain crucial for the successful reproduction of many large forest trees and a much greater variety of forest understorey plants (Hammond *et al.* 1996; Corlett 1998; Brewer & Rejmanek 1999).

Well-documented cases of non-flying mammal pollination involve 59 mammal species distributed among 19 families and six orders (Carthew & Goldingay 1997). The marsupials of Australia (12 species) and primates of Madagascar (9 species) and South America (10 species) are the best-known examples. However, the importance to plants of non-flying mammals as pollinators is debatable, as many of these plants are often served by other vertebrate and insect pollinators. Only in Madagascar do non-flying mammals (lemurs) act as sole pollinators of the plants they visit (Nilsson 1993; Kress *et al.* 1994), and the precipitous decline of some lemur species may cause an associated decline of the plants they pollinate (e.g. the traveller's tree *Ravenala madagascariensis* pollinated by the endangered black-and-white ruffed lemur). Selective exclusion experiments have demonstrated that marsupials are at least as effective as birds in pollinating *Banksia* in Australia (Goldingay *et al.* 1991; Carthew 1993) and *Protea* spp. in South Africa (Wiens *et al.* 1983); indeed, declining small-mammal activity at *Banksia* inflorescences due to habitat fragmentation resulted in a sharp decrease in seed set of *Banksia goodii* (Lamont *et al.* 1993). However, documented cases of reduced seed set due to declines in mammal pollinators are rare. In most cases, evidence for effective pollination by non-flying mammals is largely circumstantial and, while non-flying mammals may feed at flowers and occasionally pollinate them, this service must be balanced with the damage they cause (Janson *et al.* 1981).

Bats pollinate species belonging to several tropical plant families including Bombacaceae (*Pseudobombax*), Mimosaceae (*Parkia*), Passifloraceae (*Passiflora*) and Caesalpiniaceae (*Bauhinia*); on oceanic islands, such as Samoa, the majority of the dominant canopy tree species may be pollinated by bats (Cox *et al.* 1991). A number of economically important tropical trees rely on bats for pollination, including durian, banana, neem and timber species of *Eucalyptus* and several species of palm (Fujita 1991—quoted in Allen-Wardell *et al.* 1998). Although it is energetically expensive to produce bat-attracting flowers rich in nectar, plants benefit from extensive pollen transfer as bats forage over large areas (Heithaus *et al.* 1975). Habitat change, hunting and the introduction of exotic species is causing the decline of bats across the tropics and several species are threatened with extinction. This is of particular concern on oceanic islands, where alternative pollinators for endemic tree species are rare or absent (Cox *et al.* 1991).

In the Neotropics hummingbird pollination is widespread among several plant families (Regal 1982), while old-world pollinators include a much greater variety of birds, primarily sunbirds but also leafbirds, lorikeets, white-eyes and honey-eaters. However, birds are rarely the exclusive pollinators of plants (but see Ford & Paton 1985) and their importance relative to insect pollinators is not well resolved. Declines in fruit set due to temporary absence of hummingbirds have been noted, even where other invertebrate pollinators were present (Waser 1979), but obligate pollinator dependency by plants on hummingbirds or sunbirds has not been documented. Rather, these birds may be important as biological 'insurance' where temporal fluctuation of invertebrate pollinator populations may be severe (Aizen & Feinsinger 1994).

Loss of vertebrate pollinators resulting from forest conversion or fragmentation is only likely to affect plant seed production if the plant–pollinator interaction is highly specialized. Most plants are pollinated by a suite of mainly generalist pollinators (Waser *et al.* 1996) and are unlikely to suffer a large decline with the loss of one or a few closely related pollinator species. Nevertheless, in tropical floras specialist interactions among animals and plants do exist. On the small isolated Caribbean island of Tobago the absence of hermit hummingbirds resulted in reduced seed production of *Mandevilla hirsuta* (Apocynaceae) compared with the neighbouring island of Trinidad where the birds were present (Linhart & Feinsinger 1980). In contrast, there was no difference across islands in seed production of *Justica secunda* (Acanthaceae) that is pollinated by a more general array of hummingbirds. In fact most hummingbirds appear to be quite plastic in their habitat requirements and regularly use edges and move between primary and secondary habitats to take advantage of a wide variety of floral resources (Feinsinger 1976; Bierregaard & Stouffer 1997); there is also evidence that sunbirds, which in the Palaeotropics fill a similar ecological niche to hummingbirds, are more abundant at flowering trees in disturbed secondary forests (Marsden 1998). However, predicting the impacts of habitat and landscape modification on plant–pollinator interactions is difficult as both the behaviour and abundance of pollinators may change as a result of changing resource distribution (Ghazoul *et al.* 1998). Pollinators may also depend on resources that are far from the plants they pollinate. In peninsular Malaysia, for example, the fruit bat *Eonycteris spelea* is the exclusive pollinator of durian trees but feeds primarily in coastal mangroves. The bats visit, and pollinate, durian *en route* to the mangroves from their limestone-cave roosts (Start & Marshall 1976). Destruction of either the caves or the mangroves could affect the bat population and hence the pollination of durian.

5.4.2 Seed predation

Predation of seeds by animals has been proposed as a mechanism for the maintenance of the high tree species diversity in tropical forests (Janzen 1970; Connell 1971). This theory suggests that density-dependent consumption of seeds by mammalian seed predators prevents domination of the vegetation by species that have competitively superior seeds or greatest seed production. Terborgh (1988) suggested that vertebrate seed predators selectively prey upon the largest tree seeds, and where carnivores are eliminated following perturbations forest composition might shift to smaller-seeded species. There is some evidence to support this prediction from forests in Panama and Mexico (De Steven & Putz 1984; Dirzo & Miranda 1990). In Panama, high densities of small mammals on Barro Colorado Island due to an absence of large predators (Terborgh & Winter 1980) is thought to be the cause of recruitment failure among the large canopy tree *Dipteryx panamensis* due to the seed predation pressure exerted by the small mammals (De Steven & Putz 1984).

Some studies show seed predation by vertebrates (most often rodents) to be higher in mature forest compared with early successional forest (Janzen 1985a; Santos & Tellería 1992; Aide & Cavelier 1994; Osunkoya 1994), while other studies found the opposite (Uhl 1987; Hammond 1995). Rodents tend to avoid large areas of open forest or pasture probably due to the lack of protective cover and the scarcity of food resources (Price & Jenkins 1986). However, seed predation intensity under different disturbance regimes varies greatly with the species of seed plant (Willson & Whelan 1990; Terborgh *et al.* 1993; Osunkoya 1994; Chapman & Chapman 1996; Holl & Lulow 1997) making it difficult to draw generalizations to guide forest restoration programmes. However, where seed predator densities are high, commercially available low-value grains might be used to satiate the seed predators in order to minimize mortality of the desired species. Use of these methods should be based on a sound knowledge of the factors that limit the population dynamics of the predator species, otherwise the availability of superabundant food resources leads to rapid population growth and subsequent heavy seed mortality in later years.

5.4.3 Seed dispersal

Seed dispersal has been favoured by evolution

because it decreases the probability of seed predation, which is often disproportionately high close to parent trees, and increases the probability that seeds find a suitable establishment site (Howe & Smallwood 1982). Thus seeds have become adapted for dispersal by a variety of vectors, the most common of which are wind and animals, and the role of animals in the dispersal of seed and fruit has received much attention in the biological literature (see review by Sallabanks & Courtney 1992). Seeds carried by animals often have edible appendages (e.g. *Acacia* spp., *Anacardium* spp.) or are enclosed within fleshy fruits that are consumed and the seeds ejected later. In some cases the seeds themselves are harvested but dropped, or stored and later forgotten or abandoned.

In Neotropical forests, 50–90% of the canopy trees and nearly all shrubs and subcanopy trees bear fruit adapted for animal dispersal, while a smaller but still substantial proportion of species in Palaeotropical forests are also dispersed by animals (35–48% and 70–80%, respectively) (Howe & Smallwood 1982; Howe 1986). Birds, bats and scatter-hoarding rodents are the major dispersers of tropical tree seeds (Hammond *et al.* 1996; Corlett 1998; Brewer & Rejmanek 1999), while other seed dispersers include primates, fish in Amazonian forest, tortoises, some herbivorous lizards and even a frog (Stiles 1989, 1992). In temperate coniferous and deciduous forests, seeds may be more often dispersed by wind or by cache-hoarding mammals and birds (Fenner 1985). However, this dichotomy between temperate and tropical forest dispersal mechanisms has been challenged by Wheelwright (1988), who asserts that interactions between plants and their fruit-eating bird dispersers are as complex and ecologically important in temperate regions as they are in the tropics. Howe and Smallwood (1982) state that more than 60% of temperate trees have animal-dispersed seeds and record a significant negative correlation between annual precipitation and the percentage of wind dispersal. Nevertheless, there are more species of fruit-eating birds in the tropics and they comprise a larger fraction of the avian biomass (Terborgh 1986). Only in the tropics are there bird species that rely exclusively on a diet of fruit (e.g. fruit pigeons *Ptilinopus* spp. in Australia and *Treron* spp. in

South-East Asia), and most birds with a predominantly fruit diet are tropical species.

Seed dispersal syndromes are thought to reflect coevolutionary trends between plants and vertebrate (bird and mammal) seed dispersers. Bird-dispersed fruit are small and red, black, blue or purple (van der Pijl 1969), are conspicuous by sight rather than smell, and persist on the tree until removed by a disperser (Janson 1983; Howe 1986). Mammal-dispersed fleshy fruits are large, green or brown (van der Pijl 1969; Willson *et al.* 1989), emit odours, and frequently abscise shortly after ripening (Janson 1983; Howe 1986). However, the existence of clearly defined seed dispersal syndromes has recently been questioned (Tamboia *et al.* 1996) and consumption of a particular type of fruit by an animal does not imply that the fruit is adapted for dispersal by that animal (Janzen 1985b). The emerging view is that most interactions between plants and avian seed dispersers, in both tropical and temperate regions, are generalist, opportunistic and largely inefficient (from the plant's perspective), and there are few examples of tightly coupled coevolved mutualistic interactions. One well-documented exception is the commercially important brazilnut *Bertholletia excelsa*, which is entirely dependent on agoutis to gnaw open the seed capsules and release the seeds (Peres & Baider 1997).

Animal behaviour can be crucial to the success of plant dispersal and recruitment. Animal seed dispersers often place seeds in sites suitable for germination, and seeds of some species (e.g. oak) will not germinate unless they are buried (Griffin 1971). Burial further reduces the probability that the seeds will be found by seed predators (Vander Wall 1993). Bats may reduce the suitability of fruit to insect seed predator attack, leading to increased survival of seeds dropped at a bat roost (Janzen 1982). Animal digging, burrowing and tunnelling (by armadillos, coatis, porcupines, pigs, gophers) can bury seeds or unearth buried seeds. At Mount St Helens, for example, the activity of surviving gophers in bringing soil and propagules to the surface of the sterile ash deposits facilitated plant establishment and the start of forest regeneration (Andersen & Mac-Mahon 1985). Secondary dispersal of wind-dispersed seeds of *Pinus jeffreyi*, *Pinus ponder-osa* and *Pinus contorta* by chipmunks and other rodents in the Sierra Nevada is instrumental in moving seeds up to 69 m away from the parent tree compared with only 12 m by wind dispersal alone (Vander Wall 1992). Mockingbirds, doves and land iguanas have been essential in facilitating regeneration of the tree *Bursera graveolans* (Burseraceae) on the Galapagos Islands after the elimination of introduced goats. The *Bursera* seeds fail to germinate directly beneath adult trees, and recovery of the population to levels prior to the introduction of goats is entirely dependent on the native seed dispersers (Clarke & Clarke 1981). Vertebrate dispersal agents may further benefit plants by dispersing mycorrhizal fungal spores in addition to seeds, as in the Pacific North-west where squirrels disperse the obligate symbiotic mycorrhizal fungi of conifer seedlings (Maser & Maser 1988).

Dependence on animals for seed dispersal exposes the plants to the risk of dispersal failure should seed vectors become rare or extinct. The population structure of the tambalocoque tree *Sideroxylon sessiliflorum* (Sapotaceaea) on Mauritius is thought to have been affected by the extinction of the dodo (Temple 1977, 1979; but see Witmer & Cheke 1991), which dispersed the seeds and whose elimination doomed the tree to extinction as one of the 'living dead' (Janzen 1988). The likely cause of the dramatic decline of colonization rates of many fleshy fruited species in the remnant rain forests of the Mascarene Archipelago, in which Mauritius is located, is thought to be due to the extinction of over 50% of its land vertebrates in recent decades (Cheke 1987; Thebaud & Strasberg 1997). Cox *et al.* (1991) describe preliminary observations of pollination and seed dispersal failure for several bat-pollinated and -dispersed plants on the island of Guam where bats have suffered severe declines in recent years. Indeed, Fujita and Tuttle (1991) note that at least 289 plant species (according to current published information) depend to varying degrees on large populations of flying foxes for their propagation, and the fact that flying foxes are increasingly threatened is a cause for broad concern.

Selective logging and fragmentation of forest habitats have a large impact on vertebrate abundance and behaviour, which in turn is likely to affect seed dispersal of many animal-dispersed tree species. Forget and Sabatier (1997) reported that large frugivores avoid recently created gaps, and many bird species do not move into central regions of large gaps (Gorchov *et al.* 1993). However the responses of species to fragmentation can be highly variable and the impact on forest tree regeneration is often difficult to predict. Frugivorous birds may decline sharply after fragmentation (Terborgh & Winter 1980) leading to decreased seed dispersal (Santos & Tellería 1994), while small mammal populations remain stable or even increase in abundance (Santos & Tellería 1994; Malcolm 1997). However, the behaviour of small mammals may alter as the landscape is transformed into a primary and secondary forest mosaic. Terrestrial mammals readily move between primary and secondary forests (Malcolm 1997) and may promote regeneration by dispersing primary forest tree seeds into degraded secondary forest (Silva *et al.* 1996; Lamb *et al.* 1997), although seeds can also be transported in the opposite direction and alter the nature of primary forest succession.

Management of animal seed-dispersers may be necessary to accelerate regenerations of degraded areas (Silva *et al.* 1996; Lamb *et al.* 1997) and to promote the establishment of native plant species. Availability of hollows, snags, perches and nest boxes increases the abundance and diversity of birds (McClanahan & Wolfe 1993) and bats (Fenton *et al.* 1998), thereby preserving dispersal functions and accelerating forest regeneration (McClanahan 1986). Where vertebrates choose to sit while processing seeds is critical to the reproductive success of plants and appropriate provision of perches and nest boxes can encourage frugivorous birds to process (and defecate) consumed seeds in degraded sites. However, if tightly coupled mutualisms exist between a plant and animal, then specific management strategies may be required to ensure the continued existence of both.

5.4.4 Herbivory

Vertebrate herbivores are instrumental in determining the vegetational composition of forest systems. The role of the African elephant in maintaining the balance between savannah and woodland in Africa is well documented (Pellew 1983; Dublin *et al.* 1990; Ruess & Halter 1990) and the impact of grazing and browsing mammals on temperate forests has been reviewed by Gill (1992a,b). The loss of large herbivores from Europe in the Pleistocene is thought to have led to the transformation of short grasslands to less productive forests with concomitant extinctions of many small mammals dependent on the more nutrient-rich vegetation (Owen-Smith 1987, 1989). More recently, the elimination of rabbits by the disease myxomatosis in Britain has resulted in many grassland areas reverting to scrub woodland. Overgrazing by deer in European and North American woodlands causes a decline in native woodland species, facilitates the spread of invasive weeds (Koh *et al.* 1996), prevents regeneration of woody species (Ross *et al.* 1970; Tilghman 1989; Trumball *et al.* 1989; Cooke & Lakhani 1996) and may result in local extirpation of herbaceous plants (Augustine & Frelich 1998). Moderate grazing can lead to plant species replacements. In North America, preferential feeding of white-tailed deer on pine *Pinus strobus* results in its replacement by birch *Betula papyrifera*, which has a greater resistance to browsing (Ross *et al.* 1970). Moderate or low levels of grazing can also result in greater vegetational diversity than either overgrazing or an absence of grazing (Mitchell & Kirby 1990). Thus, in tropical Mexican forests the defaunation of large mammal herbivores such as tapirs, peccari and deer by hunting and illegal trading led to increased density but decreased diversity of seedlings and saplings (Dirzo & Miranda 1991).

Grazing represents a nutritional drain on plants and the ability to survive sustained grazing pressure may depend on soil quality and efficiency of nutrient capture. The response of *Colophospermum mopane* trees to high levels of elephant herbivory in *C. mopane*-dominated woodland in Zambia was related to the availability of soil nutrients; poor soil quality combined with elephant herbivory increased plant species diversity and reduced dominance by *C. mopane* (Lewis 1991). Periodic large-scale diebacks were also suggested to be a function of soil nutrient depletion from sustained grazing by elephants in these woodlands.

By browsing and debarking seedlings, saplings and trees, large mammalian herbivores can influence forest plant diversity and abundance, which in turn affect the resource and habitat quality and quantity for a range of other plant and animal species. Woodland degradation in Botswana by African elephants resulted in substantial changes in bird species composition, although bird diversity was unchanged (Herremans 1995), while culling of hippopotamus in Uganda allowed regeneration of plant diversity and consequential increases in buffalo, elephant and waterbuck densities (Hunter 1992). In tropical wet forests, large mammal densities are considerably lower than in African savannah forests and vertebrate herbivores do not have a major impact on habitat structure or composition.

5.5 VERTEBRATE INVADERS AS MEDIATORS OF CHANGE

Vertebrates themselves can be agents of anthropogenic disturbance where exotic species are released into new environments. Although most introductions fail, some succeed; of these, a few can cause large changes in the composition and structure of native forests. A familiar example in the UK is the introduction and subsequent spread of the grey squirrel *Sciurus carolinensis* at the end of the nineteenth century. Aside from eliminating the native red squirrel *Sciurus vulgaris* from much of its former range in the UK, the grey squirrel attacks hardwood trees by stripping bark, which leads to scars in the wood (and even death by ring-barking), and by damaging crowns.

Invasions of alien predators can devastate naïve prey species. The introduced brown tree snake *Boiga irregularis* has been responsible for the extinction or decline of 10 species of native birds on the island of Guam, including the Philippine turtle dove *Streptopelia bitorquata*, an important disperser of seeds into degraded vegetation remnants (Savidge 1987). Diseases spread by the introduction of exotic vertebrates is cited as a cause of the extinction of native Hawaiian birds, although habitat changes and predation by introduced grazing mammals and rats may be more impor-

tant causes (Pimm 1991). The spread of alien plants may be facilitated by native or alien generalist seed dispersers or pollinators. Thus the spread of the Brazilian pepper *Schinus teribinthifolius* in south Florida was facilitated by the American robin *Turdus migratorius* and the introduced red-whiskered bulbul *Pycnonotus jocosus* (Ewel 1986). North American mammals readily incorporate the fruit of exotic *Prunus*, *Malus*, *Pyrus* and other tree species into their diets (Willson 1993). Likewise, goats and pigs in Hawaii have been responsible for destruction of native vegetation, the creation of open areas and the dispersal of exotic plant seeds through their dung, all factors that facilitate further invasion by exotic plants (Loope & Stone 1996).

The importance of scale and historical association in determining ecological outcomes of species introductions is illustrated by the impact of red squirrels *Tamiasciurus hudsonicus* introduced to Newfoundland on the native subspecies of red crossbill *Loxia curvirostra percna*. The squirrels, introduced as prey for pine martens, quickly reached high densities by outcompeting Newfoundland red crossbills for the seeds of black spruce. The Newfoundland red crossbill is now close to extinction, though the mainland subspecies continues to coexist with the squirrels because of evolutionarily derived differences in habitat and food preferences, a result of a history of coexistence, and because mainland populations are able to travel widely over a largely forested landscape in search of new food resources (Pimm 1990). Ecologically sensitive forest management must therefore consider broad spatial (i.e. landscape) and historical patterns in order to minimize the impacts of introduced predators and competitors and the spread of generalists into forested areas at the expense of obligate forest-interior species.

5.6 ECONOMIC VALUE, HUNTING AND TOURISM

Vertebrate conservation by landscape management and protected area design has been discussed in this chapter and is elaborated further in Volume 2, Chapter 4. Successful conservation depends on integration of human needs and perceptions with those of conservation at local and regional scales. In tropical and temperate regions vertebrates represent a valuable source of income and are widely exploited as such. Exploitation may be destructive, such as hunting and collection for trade, or unobtrusive when income is generated merely through aesthetic or existence value. Vertebrate populations are susceptible to rapid decline following destructive exploitation as initial population sizes and population growth rates are often low. While attentive management of exploited populations is needed, incentives for sustainable use of vertebrates are also required. Here we briefly highlight two approaches particularly relevant to vertebrates where economic incentives are used to promote conservation.

As logging enterprises make large areas of previously remote forest accessible, vertebrates become increasingly subject to hunting by the influx of people. The extensive trade in wildlife from tropical forests places direct pressure on populations of many vertebrates despite attempted regulation by government-imposed control (Wilkie *et al.* 1998). Illegal harvesting and trade continues unabated principally because wildlife is considered to be common property. The financial burden of conservation is therefore placed on the state and there is little incentive for resource management by self-interested individuals for financial gain through sustainable harvest. Thus conservationists and economists often argue that wildlife resources need to generate income that can be appropriated by private resource owners or managers to create incentives for preservation. However, there are problems associated with commercialization for the purpose of conservation. Resource owners may have difficulty in regulating the harvest of migratory species or of species that forage over areas that exceed single-owner control. Some vertebrate populations may fluctuate widely due to conditions that are beyond the control of the resource manager (e.g. disease or climatic fluctuations), leading to uncertain income generation. There may also be costs in maintaining high densities of wildlife through damage to crops and property (Naughton-Treves 1998; Norton-Griffiths 1998). Furthermore, it is often not clear that sufficient

income can be generated from an ecologically sustainable harvest.

A popular alternative to hunting to promote wildlife conservation by income generation is wildlife tourism (Weber 1993; McNeilage 1996). Tourism has the capacity to promote support for conservation by generating employment, increasing foreign exchange earnings and attracting investment capital. For example, gorilla tourism in Democratic Republic of Congo (Zaire) and Rwanda was correlated with an increase in revenue and a decline in poaching, though establishing causal links with the latter is difficult (Butynski & Kalina 1998). It is now apparent that there are many conservation problems associated with nature tourism, which has been linked to the loss of species and habitat degradation (Butler 1991; Duffus 1993). The success of tourism depends on political stability and if this is not assured then tourism can be a highly uncertain source of revenue, as illustrated by the cessation of revenues from gorilla tourism in Rwanda following the civil war in 1994. For conservation benefits to be realized there must be positive economic gains accruing to the people living close to the wildlife populations; however, in most cases revenues flow directly to central government and little is reallocated to the local people. At the same time tourism can have an inflationary effect on local economies. Tourism may also cause changes in the behaviour and fitness of target populations, and the potential risk of disease spreading from humans to habituated primate groups is a major concern that is rarely addressed (Butynski & Kalina 1998).

A community-based approach that advocates giving people an economic stake to ensure sustainable management of their wildlife resources is epitomized by the CAMPFIRE programme in Zimbabwe (Metcalfe 1994). Focusing on communal lands it provides local people with the authority to manage wildlife (Cumming 1990). The principles behind communal wildlife management are elaborated by Murphree (1991). Central to the debate are issues of rights and responsibilities of those who stand to benefit from the commercial exploitation of wildlife. Zimbabwe has gone some way towards exploring the hierarchy of accountability arising from the integration of local interests with those at regional and national levels.

Logging, habitat degradation and conversion are likely to be the primary causes of vertebrate declines and extinctions for some time to come, but as forests become accessible to ever-greater numbers of people the pressures of hunting and tourism will become increasingly important. Privately regulated hunting and tourism are unlikely to ensure the long-term sustainability of vertebrate populations. Care has to be taken that economic incentives are not promoted to the extent where the conservation principle is forgotten. Rather, integration of economic gain with conservation objectives requires that revenue be obtained in the context of a conservation-orientated policy framework based on objective ecological science.

5.7 CONCLUSIONS

Forest managers face a difficult task in managing forests for both timber production and wildlife conservation. Information is lacking on critical issues such as the distribution and ecology of component species, the contribution of biodiversity to the resilience and resistance of ecosystems, and the appropriate spatial and temporal scales to which management must be directed. Even monitoring biological diversity to assess the success of management strategies is subject to difficulties, reflected in the lack of universally acceptable monitoring protocols. Objectives defining desired production outputs and conservation priorities need to be more clearly defined, and terms such as 'biodiversity', 'ecosystem health' and 'ecosystem sustainability' are not amenable to universal definition and too broad to be useful in evaluating management strategies. A more effective strategy for forest management, which combines the twin aims of production and conservation, is to focus on conservation strategies for species that contribute to the maintenance of ecological interactions. Thus it may be more effective to focus on species groups known to have a large impact on community structure and dynamics, i.e. the 'ecosystem engineers' and 'keystone species', as well as the suite of organisms that have important roles as pollinators,

dispersers, decomposers, etc. Management for the maintenance of these species has to consider all scales at which the species operate, defined by their resource requirements, mobility, body size, metabolic rates and life-history and demographic attributes (Hansen & Urban 1992). Evaluation of the extent of ecological redundancy (Walker 1992) in forest systems will help to set conservation targets that are integrated with production objectives.

Forest managers must proceed with management activities despite limited availability of ecological information. Spatiotemporal attributes, such as topography, seasonality, roads and water courses, and the natural disturbance regime affect both land-use decisions and habitat use by animals. In the short term, managers must make more effective use of the spatiotemporal information available and, particularly, need to recognize the importance of planning at several nested spatial and temporal scales. Likewise, knowledge of the ecological processes affecting community composition, productivity and turnover is needed to predict environmental responses to disturbances. In the longer term, field studies and monitoring programmes implemented by managers will provide this information to be used for evaluation and improved adaptive forest management, and the costs involved in undertaking such research is likely to be offset by the social, economic and ecological benefits gained.

REFERENCES

Aide, T.M. & Cavelier, J. (1994) Barriers to lowland tropical forest restoration in the Sierra Nevada de Santa Marta, Colombia. *Restoration Ecology* **2**, 219–29.

Aizen, M.A. & Feinsinger, P.S. (1994) Habitat fragmentation, pollination, and plant reproduction in a Chaco Dry Forest, Argentina. *Ecology* **75**, 330–51.

Alaback, P.B. (1982) Dynamics of understory biomass in Sitka spruce–western hemlock forests of southeastern Alaska. *Ecology* **63**, 1932–48.

Allen-Wardell, G., Bernhardt, P., Bitner, R. *et al.* (1998) The potential consequences of pollinator declines on the conservation of biodiversity and stability of food crop yields. *Conservation Biology* **12**, 8–17.

Andersen, D.C. & MacMahon, J.A. (1985) Plant succession following the Mount St Helens volcanic eruption: facilitation by a burrowing rodent, *Thomomys talpoides*. *American Midland Naturalist* **114**, 62–9.

Andrade, G.I. & Rubio, T.H. (1994) Sustainable use of the tropical rainforest: evidence from the avifauna in a shifting-cultivation habitat mosaic in the Columbian Amazon. *Conservation Biology* **8**, 545–54.

Andrén, H. (1994) Effects of habitat fragmentation on birds and mammals in landscapes with different proportions of suitable habitat: a review. *Oikos* **71**, 355–66.

Andrén, H. (1995) Effects of habitat edge and patch size on bird-nest predation. In: Hanson, L., Fahrig, L. & Merriam, G., eds. *Mosaic Landscapes and Ecological Processes*, pp. 225–55. Chapman & Hall, New York.

Andrén, H. & Angelstam, P. (1988) Elevated predation rates as an edge effect in habitat islands: experimental evidence. *Ecology* **69**, 544–7.

Andrén, H., Angelstam, P., Lindstrom, E. & Widen, P. (1985) Differences in predation pressure in relation to habitat fragmentation: an experiment. *Oikos* **45**, 273–7.

Ashton, P.S. (1988) Dipterocarp biology as a window to the understanding of tropical forest structure. *Annual Review of Ecology and Systematics* **19**, 347–70.

Augustine, D.J. & Frelich, L.E. (1998) Effects of white-tailed deer on populations of an understory forb in fragmented deciduous forests. *Conservation Biology* **12**, 995–1004.

Balmford, A. & Long, A. (1994) Avian endemism and forest loss. *Nature* **372**, 623–4.

Bastable, H.G., Shuttleworth, W.J., Dallarosa, R.L.G. & Nobre, C.A. (1993) Observations of climate, albedo, and surface radiation over cleared and undisturbed Amazonian forest. *International Journal of Climatology* **13**, 783–96.

Bawa, K.S. (1990) Plant–pollinator interactions in tropical rain forests. *Annual Review of Ecology and Systematics* **21**, 399–422.

Bayne, E.M. & Hobson, K.A. (1997) Comparing the effects of landscape fragmentation by forestry and agriculture on predation of artificial nests. *Conservation Biology* **11**, 1418–29.

Bennett, E.L. & Dahaban, Z. (1995) Wildlife responses to disturbances in Sarawak and their implications for forest management. In: Primack, R.B. & Lovejoy, T.E., eds. *Ecology, Conservation and Forest Management of Southeast Asian Rainforests*, pp. 66–86. Yale University Press, New Haven, Connecticut.

Bierregaard, R.O. & Stouffer, P.C. (1997) Understorey birds and dynamic habitat mosaics in Amazonian rainforests. In: Laurance, W.F. & Bierregaard, R.O., eds. *Tropical Forest Remnants: Ecology, Management and Conservation of Fragmented Communities*, pp. 138–55. University of Chicago Press, London.

Blaustein, A.R., Wake, D.B. & Sousa, W.P. (1994) Amphibian declines: judging stability, persistence, and susceptibility of populations to local and global extinctions. *Conservation Biology* **8**, 60–71.

Bradford, D.F., Tabatabai, F. & Graber, D.M. (1993) Isolation of remaining populations of the native frog, *Rana muscosa*, by introduced fishes in Sequoia and Kings Canyon National Parks, California. *Conservation Biology* **7**, 882–8.

Brewer, S.W. & Rejmanek, M. (1999) Small rodents as significant dispersers of tree seeds in a Neotropical forest. *Journal of Vegetation Science* **10**, 165–74.

Brittingham, M. & Temple, S. (1983) Have cowbirds caused forest songbirds to decline? *Bioscience* **33**, 31–5.

Brosset, A. (1990) A long-term study of the rainforest birds in M'Passa (Gabon). In: Keast, A., ed. *Biogeography and Ecology of Forest Bird Communities*, pp. 259–74. SPB Academic Publishing, The Hague.

Burkey, T.V. (1993) Edge effects in seed and egg predation at two neotropical rainforest sites. *Biological Conservation* **66**, 139–43.

Butler, R.W. (1991) Tourism, environment, and sustainable development. *Environmental Conservation* **18**, 201–9.

Butynski, T.M. & Kalina, J. (1998) Gorilla tourism: a critical look. In: Milner-Gulland, E.J. & Mace, R., eds. *Conservation of Biological Resources*, pp. 294–313. Blackwell Science, Oxford.

Canaday, C. (1997) Loss of insectivorous birds along a gradient of human impact in Amazonia. *Biological Conservation* **77**, 63–77.

Carthew, S.M. (1993) An assessment of pollinator visitation to *Banksia spinulosa*. *Australian Journal of Ecology* **18**, 257–68.

Carthew, S.M. & Goldingay, R.L. (1997) Non-flying mammals as pollinators. *Trends in Ecology and Evolution* **12**, 104–8.

Chapman, C.A. & Chapman, L.J. (1996) Frugivory and the fate of dispersed and non-dispersed seeds of six African tree species. *Journal of Tropical Ecology* **12**, 491–504.

Cheke, A.S. (1987) An ecological history of the Mascarene Islands, with particular reference to extinctions and introductions of land vertebrates. In: Diamond, J.W., ed. *Studies of Mascarene Island Birds*, pp. 5–89. Cambridge University Press, Cambridge.

Clarke, D.A. & Clarke, D.B. (1981) Effects of seed dispersal by animals on the regeneration of *Bursera graveolens* (Burseraceae) on Santa Fe Island, Galapagos. *Oecologia* **49**, 73–5.

Connell, J.H. (1971) On the role of natural enemies in preventing competitive exclusion in some marine animals and in rain forests. In: den Boer, P.J. & Gradwell, G.R., eds. *Dynamics of Populations*, pp. 298–312. PUDOC, Wageningen.

Cooke, A.S. & Lakhani, K.H. (1996) Damage to coppice regrowth by Muntjac deer *Muntiacus reevesi* and protection with electric fencing. *Biological Conservation* **75**, 231–8.

Corlett, R.T. (1998) Frugivory and seed dispersal by vertebrates in the Oriental (Indomalayan) Region. *Biological Reviews of the Cambridge Philosophical Society* **73**, 413–48.

Cox, P.A., Elmqvist, T., Pierson, E.D. & Rainey, W.E. (1991) Flying foxes as strong interactors in South Pacific island ecosystems: a conservation hypothesis. *Conservation Biology* **5**, 448–53.

Cumming, D.H.M. (1990) *Wildlife Products and the Market Place: a View from Southern Africa.* Multispecies Animal Production Systems Project Paper no. 12, World Wide Fund for Nature, Zimbabwe.

De Steven, D. & Putz, F.E. (1984) Impact of mammals on the early recruitment of a tropical canopy tree, *Dipteryx panamensis*, in Panama. *Oikos* **43**, 207–16.

Dickson, J.G., Conner, R.N. & Williamson, J.H. (1983) Snag retention increases bird use of a clear-cut. *Journal of Wildlife Management* **47**, 799–804.

Dickson, J.G., Thompson, F.R., Conner, R.N. & Franzreb, K.E. (1992) *Status and Management of Neotropical Migrant Birds*, pp. 375–85. General technical report, US Forest Service, Estes Park, Colorado.

Dirzo, R. & Miranda, A. (1990) Contemporary neotropical defaunation and forest structure, function and diversity: a sequel to John Terborgh. *Conservation Biology* **4**, 444–7.

Dirzo, R. & Miranda, A. (1991) Altered patterns of herbivory and diversity in the forest understory: a case study of the possible consequences of contemporary defaunation. In: Price, P.W., Lewinsohn, T.M., Fernandes, G.W. & Benson, W.W., eds. *Plant–Animal Interactions: Evolutionary Ecology in Tropical and Temperate Regions*, pp. 273–87. John Wiley & Sons, New York.

Dublin, H.T., Sinclair, A.R.E. & McGlade, J. (1990) Elephants and fire as causes of multiple stable states in the Serengeti–Mara woodlands. *Journal of Animal Ecology* **59**, 1147–64.

Duffus, D. (1993) Tsitika to Baram: the myth of sustainability. *Conservation Biology* **7**, 440–2.

Dupuis, L.A., Smith, J.N.M. & Bunnell, F. (1995) Relation of terrestrial-breeding amphibian abundance to tree-stand age. *Conservation Biology* **9**, 645–53.

Ewel, J.J. (1986) Invasibility: lessons from South Florida. In: Mooney, H.A. & Drake, J.A., eds. *Ecology of Biological Invasions of North America and Hawaii*, pp. 214–30. Springer-Verlag, New York.

Fauth, J.E., Crother, B.I. & Slovinski, J.B. (1989) Elevational patterns of species richness, evenness and abundance in the Costa Rican leaf-litter herpetofauna. *Biotropica* **21**, 178–85.

Feinsinger, P. (1976) Organization of a tropical guild of nectarivorous birds. *Ecological Monographs* **46**, 257–91.

Fenner, M. (1985). *Seed Ecology*. Chapman & Hall, New York.

Fenton, M.B., Cumming, D.H.M., Rautenbach, I.L.N. et al. (1998) Bats and the loss of tree canopy in African woodlands. *Conservation Biology* **12**, 399–407.

Fleming, T.H. (1992) How do fruit- and nectar-feeding birds and mammals track their food resources? In: Hunter, M.D., Ohgushi, T. & Price, P.W., eds. *Effects of Resource Distribution on Animal–Plant Interactions*, pp. 355–91. Academic Press, London.

Fonseca, G.A.B. & Robinson, J.G. (1990) Forest size and structure: competitive and predatory effects on small mammal communities. *Biological Conservation* **53**, 265–94.

Ford, H.A. & Paton, D.C. (1985) *The Dynamic Partnership: Birds and Plants in Southern Australia*. Government Printer, South Australia.

Forget, P.-M. (1997) Effect of microhabitat on seed fate and seedling performance in two rodent-dispersed tree species in rainforest in French Guiana. *Journal of Ecology* **85**, 693–703.

Forget, P.-M. & Milleron, T. (1991) Evidence for secondary seed dispersal by rodents in Panama. *Oecologia* **87**, 596–9.

Forget, P.-M. & Sabatier, D. (1997) Dynamics of the seedling shadow of a frugivore-dispersed tree species in French Guiana. *Journal of Tropical Ecology* **13**, 767–73.

Franklin, J.F. & Forman, R.T.T. (1987) Creating landscape patterns by forest cutting: ecological consequences and principles. *Landscape Ecology* **1**, 5–18.

Fujita, M.S. & Tuttle, M.D. (1991) Flying foxes (Chiroptera: Pteropodidae): threatened animals of key ecological and economic importance. *Conservation Biology* **5**, 455–63.

Ghazoul, J. & Hill, J.K. (in press) Impacts of selective logging on tropical forest invertebrates. In: Fimbel, R.A., Grajal, A. & Robinson, J.G., eds. *Conserving Wildlife in Managed Tropical Forests*. Columbia University Press, New York.

Ghazoul, J., Liston, K.A. & Boyle, T.J.B. (1998) Disturbance-induced density-dependent reproductive success in a tropical forest tree. *Journal of Ecology* **86**, 462–73.

Gill, R.M.A. (1992a) A review of damage by mammals in north temperate forests. *Forestry* **65**, 145–69.

Gill, R.M.A. (1992b) A review of damage by mammals in north temperate forests. 3. Impact on trees and forests. *Forestry* **65**, 363–88.

Goldingay, R.L., Carthew, S.M. & Whelan, R.J. (1991) The importance of non-flying mammals in pollination. *Oikos* **61**, 79–81.

Gorchov, D.L., Cornejo, F., Ascorra, C. & Jaramillo, M. (1993) The role of seed dispersal in the natural regeneration of rain forest after strip-cutting in the Peruvian Amazon. *Vegetatio* **107/108**, 339–49.

Griffin, J.R. (1971) Oak regeneration in the upper Carmel Valley, California. *Ecology* **52**, 862–8.

Guariguata, M.R. & Pinard, M.A. (1998) Ecological knowledge of regeneration from seed in neotropical forest trees: implications for natural forest management. *Forest Ecology and Management* **112**, 87–99.

Hammond, D.S. (1995) Post-dispersal seed and seedling mortality of tropical dry forest trees after shifting agriculture, Chiapas, Mexico. *Journal of Tropical Ecology* **11**, 293–313.

Hammond, D.S., Gourlet-Fleury, S., van der Hout, P., ter Steege, H. & Brown, V.K. (1996) A compilation of known Guianan timber trees and the significance of their dispersal mode, seed size and taxonomic affinity to tropical rain forest management. *Forest Ecology and Management* **83**, 99–116.

Hansen, A.J. & Hounihan, P. (1996) Canopy tree retention and avian diversity in the Oregon Cascades. In: Szaro, R.C. & Johnston, D.W., eds. *Biodiversity in Managed Landscapes: Theory and Practice*, pp. 401–21. Oxford University Press, Oxford.

Hansen, A.J. & Urban, D.L. (1992) Avian responses to landscape pattern: the role of avian life histories. *Landscape Ecology* **7**, 163–80.

Hansen, A.J., Garman, S.L. & Marks, B. (1993) An approach for managing vertebrate diversity across multiple-use landscapes. *Ecological Applications* **3**, 481–96.

Hansson, L. (1992) Landscape ecology of boreal forests. *Trends in Ecology and Evolution* **7**, 299–302.

Hansson, L. & Angelstam, P. (1991) Landscape ecology as a theoretical basis for nature conservation. *Landscape Ecology* **5**, 191–201.

Harris, L.D. (1984) *The Fragmented Forest*. University of Chicago Press, Chicago.

Harris, L.D. & Silva-Lopez, G. (1992) Forest fragmentation and the conservation of biodiversity. In: Fiedler, P.L. & Jain, S.K., eds. *Conservation Biology: the Theory and Practice of Nature Conservation. Preservation and Management*, pp. 197–237. Chapman & Hall, London.

Hartley, M.J. & Hunter, M.L. (1998) A meta-analysis of forest cover, edge effects, and artificial nest predation rates. *Conservation Biology* **12**, 465–9.

Heithaus, J.L., Fleming, T.H. & Opler, P.A. (1975) Foraging patterns and resource utilization in seven species of bats in a seasonal forest. *Ecology* **56**, 841–54.

Herremans, M. (1995) Effects of woodland modification by African elephant *Loxodonta africana* on bird diversity in northern Botswana. *Ecography* **18**, 440–54.

Heydon, M.J. & Bulloh, P. (1996) The impact of selective logging on sympatric civet species in Borneo. *Oryx* **30**, 31–6.

Heydon, M.J. & Bulloh, P. (1997) Mousedeer densities in a tropical rainforest: the impact of selective logging. *Journal of Applied Ecology* **34**, 484–96.

Hodl, W. (1990) An analysis of foam nest construction in the neotropical frog *Physalaemus ephippifer* (Leptodactylidae). *Copeia* **2**, 547–54.

Holdsworth, A.R. & Uhl, C. (1997) Fire in the Amazonian selectively logged rain forest and the potential for fire reduction. *Ecological Applications* **7**, 713–25.

Holl, K.D. & Lulow, M.E. (1997) Effects of species, habitat, and distance from edge on post-dispersal seed predation in a tropical rainforest. *Biotropica* **29**, 459–68.

Holmes, R.T. (1990) Ecology and evolutionary impacts of bird predation on forest insects: an overview. *Studies of Avian Biology* **13**, 6–13.

Howe, H.F. (1986) Seed dispersal by fruit-eating birds and mammals. In: Murray, D.R., ed. *Seed Dispersal*, pp. 123–89. Academic Press, New York.

Howe, H.F. & Smallwood, J. (1982) Ecology of seed dispersal. *Annual Review of Ecology and Systematics* **13**, 201–28.

Hunter, M.D. (1992) Interactions within herbivore communities mediated by the host plant: the keystone herbivore concept. In: Hunter, M.D., Ohgushi, T. & Price, P.W., eds. *Effects of Resource Distribution on Animal–Plant Interactions*, pp. 287–325. Academic Press, London.

Hunter, M.L. Jr (1990) *Wildlife, Forests, and Forestry: Principles of Managing Forests for Biological Diversity*. Prentice-Hall, Englewood Cliffs, New Jersey.

Janson, C.H. (1983) Adaptation of fruit morphology to dispersal agents in a neotropical forest. *Science* **219**, 187–9.

Janson, C.H., Terborgh, J. & Emmons, L.H. (1981) Non-flying mammals as pollinating agents in the Amazonian forest. *Biotropica* **13** (Suppl.), 1–6.

Janzen, D.H. (1970) Herbivores and the number of tree species in tropical forests. *American Naturalist* **104**, 501–28.

Janzen, D.H. (1982) Simulation of *Andira* fruit pulp removal by bats reduces seed predation by *Cleogonus* weevils. *Brenesia* **20**, 165–70.

Janzen, D.H. (1985a) *Spondias mombin* is culturally deprived in megafauna-free forest. *Journal of Tropical Ecology* **1**, 131–55.

Janzen, D.H. (1985b) On ecological fitting. *Oikos* **45**, 308–10.

Janzen, D.H. (1988) Management of habitat fragments in a tropical dry forest? *Annals of the Missouri Botanical Garden* **75**, 105–16.

Johns, A.D. (1991) Responses of Amazonian forest birds to habitat modification. *Journal of Tropical Ecology* **7**, 417–37.

Johns, A. (1992) Vertebrate responses to selective logging: implication for the design of logging systems. *Philosophical Transactions of the Royal Society of London B* **335**, 437–42.

Johns, A.D. & Skorupa, J.P. (1987) Responses of rainforest primates to habitat disturbance: a review. *International Journal of Primatology* **6**, 157–91.

Jones, C.G., Lawton, J.H. & Shachak, M. (1994) Organisms as ecosystem engineers. *Oikos* **69**, 373–86.

Jones, C.G., Lawton, J.H. & Shachak, M. (1997) Positive and negative effects of organisms as physical ecosystem engineers. *Ecology* **78**, 1946–57.

Kapos, V. (1989) Effects of isolation on the water status of forest patches in the Brazilian Amazon. *Journal of Tropical Ecology* **5**, 173–85.

Karr, J.R. & Roth, R.R. (1971) Vegetation structure and avian diversity in several New World areas. *American Naturalist* **115**, 423–35.

Karr, J.R., Robinson, S.K., Blake, J.G. & Bierregaard, R.O. Jr (1990) Birds of four neotropical forests. In: Gentry, A., ed. *Four Neotropical Rainforests*, pp. 237–69. Yale University Press, New Haven, Connecticut.

Karr, J.R., Dionne, M. & Schlosser, I.J. (1992) Bottom-up versus top-down regulation of vertebrate populations: lessons from birds and fish. In: Hunter, M.D., Ohgushi, T. & Price, P.W., eds. *Effects of Resource Distribution on Animal–Plant Interactions*, pp. 243–86. Academic Press, London.

King, G.M., Bevis, K.R., Hanson, E.E. & Vitello, J.R. (1997) Northern spotted owl management: mixing landscape and site-based approaches. *Journal of Forestry* **95**, 21–5.

Kinnaird, M.F. & O'Brien, T.G. (1991) Viable populations for an endangered forest primate, the Tana river crested mangabey (*Cercocebus galeritus galeritus*). *Conservation Biology* **5**, 203–13.

Koh, S., Watt, T.A., Bazely, D.R., Pearl, D.L., Tang, M. & Carleton, T.J. (1996) Impact of herbivory on white-tailed deer (*Odocoileus virginianus*) on plant community composition. *Aspects of Applied Biology* **44**, 445–50.

Kress, W.J., Schatz, G.E., Andrianifahanana, M. & Morland, H.S. (1994) Pollination of *Ravenala madagascariensis* (Strelitziaceae) by lemurs in Madagascar: evidence for an archaic pollination system? *American Journal of Botany* **81**, 542–51.

Kruess, A. & Tscharntke, T. (1994) Habitat fragmentation, species loss, and biological control. *Science* **264**, 161–5.

Krummel, J.R., Gardner, R.H., Sugihara, G., O'Neill, R.V. & Coleman, P.R. (1987) Landscape patterns in a disturbed environment. *Oikos* **48**, 321–4.

Lamb, D., Parrotta, J., Keenan, R. & Tucker, N. (1997) Rejoining habitat remnants: restoring degraded rainforest lands. In: Laurance, W.F. & Bierregaard, R.O., eds. *Tropical Forest Remnants: Ecology, Management and Conservation of Fragmented Communities*, pp. 367–85. University of Chicago Press, London.

Lambert, F.R. & Marshall, A.G. (1991) Keystone characteristics of bird-dispersed *Ficus* in a Malaysian lowland rain forest. *Journal of Ecology* **79**, 793–809.

Lamont, B.B., Klinkhamer, P.G.L. & Witkowski, E.T.F. (1993) Population fragmentation may reduce seed set to zero in *Banksia goodii*: a demonstration of the Allee effect. *Oecologia* **94**, 446–50.

Laurance, W.F. (1994) Rainforest fragmentation and the structure of small mammal communities in tropical Queensland. *Biological Conservation* **69**, 23–32.

Laurance, W.F. (1997) Hyper-disturbed parks: edge effects and the ecology of isolated rainforest reserves in tropical Australia. In: Laurance, W.F. & Bierregaard, R.O., eds. *Tropical Forest Remnants: Ecology, Management and Conservation of Fragmented Communities*, pp. 71–83. University of Chicago Press, London.

Lawton, J.H. & Brown, V.K. (1993) Redundancy in ecosystems. In: Schulze, E.-D. & Mooney, H.A., eds. *Biodiversity and Ecosystem Function*, pp. 255–70. Springer-Verlag, Berlin.

Levey, D.J. (1988) Spatial and temporal variation in Costa Rican fruit and fruit-eating bird communities. *Ecological Monographs* **58**, 251–69.

Lewis, D.M. (1991) Observations of tree growth, woodland structure and elephant damage on *Colophospermum mopane* in Luangwa Valley, Zambia. *African Journal of Ecology* **29**, 207–21.

Linhart, Y.B. & Feinsinger, P. (1980) Plant–hummingbird interactions: effects of island size and degree of specialization on pollination. *Ecology* **68**, 745–60.

Loiselle, B.A. & Blake, J.G. (1991) Resource abundance and temporal variation in fruit-eating birds along a wet forest elevational gradient in Costa Rica. *Ecology* **72**, 180–93.

Loope, L.L. & Stone, C.P. (1996) Strategies to reduce erosion of biodiversity by exotic terrestrial species. In: Szaro, R.C. & Johnston, D.W., eds. *Biodiversity in Managed Landscapes: Theory and Practice*, pp. 261–79. Oxford University Press, Oxford.

MacArthur, R.H. & MacArthur, J.W. (1961) On bird species diversity. *Ecology* **42**, 357–74.

McClanahan, T.R. (1986) The effect of a seed source on primary succession in a forest ecosystem. *Vegetatio* **65**, 175–8.

McClanahan, T.R. & Wolfe, R.W. (1993) Accelerating forest succession in a fragmented landscape: the role of birds and perches. *Conservation Biology* **7**, 271–8.

McNeilage, A. (1996) Ecotourism and mountain gorillas in the Virunga Volcanoes. In: Taylor, V.J. & Dunstone, N., eds. *The Exploitation of Mammal Populations*, pp. 334–44. Chapman & Hall, London.

Malcolm, J.R. (1994) Edge effects in Central Amazonian forest fragments. *Ecology* **75**, 2438–45.

Malcolm, J.R. (1997) Biomass and diversity of small mammals in Amazonian forest fragments. In:

Laurance, W.F. & Bierregaard, R.O., eds. *Tropical Forest Remnants: Ecology, Management and Conservation of Fragmented Communities*, pp. 207–21. University of Chicago Press, London.

Marcot, B.G. (1983) Snag-use by birds in Douglas-fir clearcuts. In: Davis, J.W., Goodwin, G.A. & Ockenfels, R.A., eds. *Snag Habitat Management*, pp. 134–9. General Technical Report RM-99, US Forest Service, St Paul, Minnesota.

Marsden, S.J. (1998) Changes in bird abundance following selective logging on Seram, Indonesia. *Conservation Biology* **12**, 605–11.

Marsh, D.M. & Pearman, P.B. (1997) Effects of habitat fragmentation on the abundance of two species of leptodactylid frogs in an Andean montane forest. *Conservation Biology* **11**, 1323–8.

Maser, C. & Maser, Z. (1988) Interactions among squirrels, mycorrhizal fungi, and coniferous forests in Oregon. *Great Basin Naturalist* **48**, 358–69.

Merrill, S.B., Cuthbert, F.J. & Oehlert, G. (1998) Residual patches and their contribution to forest-bird diversity on northern Minnesota aspen clearcuts. *Conservation Biology* **12**, 190–9.

Metcalfe, S. (1994) CAMPFIRE: Zimbabwe's Communal Areas Management Programme For Indigenous Resources. In: Western, D. & Wright, M., eds. *Natural Connections: Perspectives in Community-based Conservation*, pp. 161–91. Island Press, Washington, DC.

Mills, G.S., Dunning, J.B. & Bates, J.M. (1991) The relationship between breeding bird density and vegetation. *Wilson Bulletin* **103**, 468–79.

Mitchell, F.J.G. & Kirby, K.J. (1990) The impact of large herbivores on the conservation of semi-natural woods in British uplands. *Forestry* **63**, 333–53.

Mittermeier, R.A., Myers, N. & Thomsen, J.B. (1998) Biodiversity hotspots and major tropical wilderness areas: approaches to setting conservation priorities. *Conservation Biology* **12**, 516–20.

Murphree, M.W. (1991) *Communities as Institutions for Resource Management*. CASS Occasional Paper Series, University of Zimbabwe.

Naiman, R.J., Johnston, C.A. & Kelly, J.C. (1988) Alteration of North American streams by beaver. *Bioscience* **38**, 753–61.

Naughton-Treves, L. (1998) Predicting patterns of crop damage by wildlife around Kibale National Park, Uganda. *Conservation Biology* **12**, 156–68.

Niemi, G.J. & Hanowski, J.M. (1984) Relationships of breeding birds to habitat characteristics in logged patches. *Journal of Wildlife Management* **48**, 438–43.

Nilsson, L.A. (1993) Lemur pollination in the Malagasy rainforest liana *Strongylodon craveniae* (Leguminosae). *Evolutionary Trends in Plants* **7**, 49–56.

Norton-Griffiths, M. (1998) The economics of wildlife conservation policy in Kenya. In: Milner-Gulland, E.J. & Mace, R., eds. *Conservation of Biological Resources*, pp. 279–93. Blackwell Science, Oxford.

Oates, J.F. (1987) Food distribution and foraging behavior. In: Smuts, B.B., Cheney, D.L., Seyfarth, R.M., Wrangham, R.W. & Struhsaker, T.T., eds. *Primate Societies*, pp. 197–209. University of Chicago Press, Chicago.

Osunkoya, O.O. (1994) Postdispersal survivorship of North Queensland rainforest seeds and fruits: effects of forest, habitat, and species. *Australian Journal of Ecology* **19**, 52–64.

Owen-Smith, N. (1987) Pleistocene extinctions: the pivotal role of megaherbivores. *Paleobiology* **13**, 351–62.

Owen-Smith, N. (1989) Megafaunal extinctions: the conservation message from 11 000 years b.p. *Conservation Biology* **3**, 405–12.

Paton, P.W.C. (1994) The effect of edge on avian nest success: how strong is the evidence? *Conservation Biology* **8**, 17–26.

Payne, N.F. & Bryant, F.C. (1994) *Techniques of Wildlife Habitat Management of Uplands*. McGraw-Hill, New York.

Pellew, R.A.P. (1983) The impacts of elephant, giraffe, and fire upon the *Acacia tortilis* woodlands of the Serengeti. *African Journal of Ecology* **21**, 41–74.

Peres, C.A. & Baider, C. (1997) Seed dispersal, spatial distribution and population structure of Brazilnut trees (*Bertholletia excelsa*) in southeastern Amazonia. *Journal of Tropical Ecology* **13**, 595–616.

Pimm, S.L. (1990) The decline of the Newfoundland crossbill. *Trends in Ecology and Evolution* **5**, 350–1.

Pimm, S.L. (1991) *The Balance of Nature*. Chicago University Press, Chicago.

Pimm, S.L. & Sugden, A.M. (1994) Tropical diversity and global change. *Science* **263**, 933–4.

Pinard, M.A. & Putz, F.E. (1996) Retaining forest biomass by reducing logging damage. *Biotropica* **28**, 278–95.

Pough, F.H., Smith, E.M., Rhodes, D.H. & Collazo, A. (1987) The abundance of salamanders in forest stands with different histories of disturbance. *Forest Ecology and Management* **20**, 1–9.

Price, M.V. & Jenkins, S.H. (1986) Rodents as seed consumers and dispersers. In: Murray, D.R., ed. *Seed Dispersal*, pp. 191–225. Academic Press, Orlando, Florida.

Raphael, M.G. & White, M. (1984) Use of snags by cavity nesting birds in the Sierra Nevada. *Wildlife Monographs* **86**, 1–66.

Reagan, D.P., Camilo, G.R. & Waide, R.B. (1996) The community food web: major properties and patterns of organization. In: Reagan, D.P. & Waide, R.B., eds. *The Food Web of a Tropical Rain Forest*, pp. 461–89. University of Chicago Press, Chicago.

Redford, K.H. (1992) The empty forest. *Bioscience* **42**, 412–22.

Regal (1982) Pollination by wind and animals: ecology of geographic patterns. *Annual Review of Ecology and Systematics* **13**, 497–524.

Robbins, C.S., Sauer, J.R., Greenberg, R.S. & Droege, S. (1989) Population declines in North American birds that migrate to the Neotropics. *Proceedings of the National Academy of Sciences USA* **86**, 7658–62.

Robinson, S.K. & Wilcove, D.S. (1994) Forest fragmentation in the temperate zone and its effects on migratory songbirds. *Bird Conservation International* **4**, 233–49.

Robinson, S.K., Thompson, F.R. III, Donovan, T.M., Whitehead, D.R. & Faaborg, J. (1995) Regional forest fragmentation and the nesting success of migratory birds. *Science* **267**, 1987–90.

Ross, B.A., Bray, J.R. & Marshall, W.H. (1970) Effects of long-term deer exclusion on a *Pinus resinosa* forest in north-central Minnesota. *Ecology* **51**, 1088–93.

Ruess, R.W. & Halter, F.L. (1990) The impacts of large herbivores on the Serona woodlands, Serengeti National Park, Tanzania. *African Journal of Ecology* **28**, 259–75.

Rumiz, D., Guinart, D. & Herrera, J.C. (in press) Logging in timber concessions in the Department of Santa Cruz, Bolivia. In: Fimbel, R.A., Grajal, A. & Robinson, J.G., eds. *Conserving Wildlife in Managed Tropical Forests*. Columbia University Press, New York.

Sallabanks, R. & Courtney, S.P. (1992) Frugivory, seed predation, and insect–vertebrate interactions. *Annual Review of Entomology* **37**, 377–400.

Santos, T. & Tellería, J.L. (1992) Edge effects on nest predation in Mediterranean fragmented forests. *Conservation Biology* **60**, 1–5.

Santos, T. & Tellería, J.L. (1994) Influence of forest fragmentation on seed consumption and dispersal of Spanish juniper *Juniperus thurifera*. *Biological Conservation* **70**, 129–34.

Saunders, D.A., Hobbs, R.J. & Margules, C.R. (1991) Biological consequences of ecosystem fragmentation: a review. *Conservation Biology* **5**, 18–32.

Savidge, J. (1987) Extinction of an island forest avifauna by an introduced snake. *Ecology* **68**, 660–8.

Short, H.L., Hestbeck, J.B. & Tiner, R.W. (1996) Ecosearch: a new paradigm for evaluating the utility of wildlife habitat. In: DeGraaf, R.M. & Miller, R.I., eds. *Conservation of Faunal Diversity in Forested Landscapes*, pp. 569–94. Chapman & Hall, London.

Silva, J.M.C., Uhl, C. & Murray, G. (1996) Plant succession, land management, and the ecology of fruit-eating birds in abandoned Amazonian pastures. *Conservation Biology* **10**, 491–503.

Spies, T.A., Franklin, J.F. & Thomas, T.B. (1988) Coarse woody-debris in Douglas-fir forests of western Oregon and Washington. *Ecology* **69**, 1689–702.

Start, A.N. & Marshall, A.G. (1976) Nectarivorous bats as pollinators of trees in West Malaysia. In: Burley, J. & Styles, B.T., eds. *Tropical Trees: Variation, Breeding and Conservation*, pp. 141–50. Academic Press, London.

Stiles, E.W. (1989) Fruits, seeds, and dispersal agents. In: Abrahamson, W.G., ed. *Plant–Animal Interactions*, pp. 87–122. McGraw-Hill, New York.

Stiles, E.W. (1992) Animals as seed dispersers. In: Fenner, M., ed. *Seeds: the Ecology of Regeneration in Plant Communities*, pp. 105–56. CAB International, Wallingford, UK.

Stiles, F.G. & Clarke, D.A. (1989) Conservation of tropical rain forest birds: a case study from Costa Rica. *American Birds* **43**, 420–8.

Tamboia, T., Cipollini, M.L. & Levey, D.J. (1996) An evaluation of vertebrate seed dispersal syndromes in four species of black nightshade (*Solanum* sect. *Solanum*). *Oecologia* **107**, 522–32.

Temple, S.A. (1977) Plant–animal mutualism: coevolution with dodo leads to near extinction of plant. *Science* **197**, 885–6.

Temple, S.A. (1979) The dodo and the tambolocoque tree. *Science* **203**, 1364.

Terborgh, J. (1986) Keystone plant resources in the tropical forest. In: Soulé, M.E., ed. *Conservation Biology: the Science of Scarcity and Diversity*, pp. 330–44. Sinauer Associates, Sunderland, Massachusetts.

Terborgh, J. (1988) The big things that run the world: a sequel to E.O. Wilson. *Conservation Biology* **2**, 402–3.

Terborgh, J. (1992) Maintenance of diversity in tropical forests. *Biotropica* **24**, 283–92.

Terborgh, J. & Winter, B. (1980) Some causes of extinction. In: Soulé, M. & Wilcox, B., eds. *Conservation Biology: an Evolutionary Ecological Perspective*, pp. 119–33. Sinauer Associates, Sunderland, Massachusetts.

Terborgh, J., Losos, E., Riley, M.P. & Bolaños Riley, M. (1993) Predation by vertebrates and invertebrates on the seeds of five canopy tree species of an Amazonian forest. *Vegetatio* **107/108**, 375–86.

Thebaud, C. & Strasberg, D. (1997) Plant dispersal in fragmented landscapes: a field study of wood colonisation in rainforest remnants of the Mascarene Archipelago. In: Laurance, W.F. & Bierregaard, R.O., eds. *Tropical Forest Remnants: Ecology, Management and Conservation of Fragmented Communities*, pp. 321–32. University of Chicago Press, London.

Thiollay, J.-M. (1992) Influence of selective logging on bird species diversity in a Guianan rain forest. *Conservation Biology* **6**, 47–63.

Thiollay, J.-M. (1997) Disturbance, selective logging and bird diversity: a Neotropical forest study. *Biodiversity and Conservation* **6**, 1155–73.

Tilghman, N.G. (1989) Impacts of white-tailed deer on

forest regeneration in North-western Pennsylvania. *Journal of Wildlife Management* **53**, 524–32.

Tilman, D. & Downing, J.A. (1995) Biodiversity and stability in grasslands. *Nature* **367**, 363–5.

Tocher, M.D., Gascon, C. & Zimmerman, B.L. (1997) Fragmentation effects on a central Amazonian frog community: a ten-year study. In: Laurance, W.F. & Bierregaard, R.O., eds. *Tropical Forest Remnants: Ecology, Management and Conservation of Fragmented Communities*, pp. 124–37. University of Chicago Press, London.

Trumball, V.L., Zielinski, E.J. & Aharrah, E.C. (1989) The impact of deer browsing on the Allegheny forest type. *Northern Journal of Applied Forestry* **6**, 162–5.

Tutin, C.E.G., White, L.J.T. & Mackanga-Missandzou, A. (1997) The use by rain forest mammals of natural forest fragments in an equatorial African savanna. *Conservation Biology* **11**, 1190–203.

Uhl, C. (1987) Factors controlling succession following slash-and-burn agriculture. *Journal of Ecology* **76**, 633–81.

Uhl, C. (1998) Perspectives on wildfire in the humid tropics. *Conservation Biology* **12**, 942–3.

Urban, D.L. & Smith, T.M. (1989) Microhabitat pattern and the structure of forest bird communities. *American Naturalist* **133**, 811–29.

van der Pijl, L. (1969) *Principles of Dispersal in Higher Plants*. Springer-Verlag, Berlin.

Vander Wall, S.B. (1992) The role of animals in dispersing a 'wind-dispersed' pine. *Ecology* **73**, 614–21.

Vander Wall, S.B. (1993) A model of caching depth: implications for scatter hoarders and plant dispersal. *American Naturalist* **141**, 217–32.

Vitt, L.J., Avila Pires, T.C.S., Caldwell, J.P. & Oliveira, V.R.L. (1998) The impact of individual tree harvesting on thermal environments of lizards in Amazonian rain forest. *Conservation Biology* **12**, 654–64.

Walker, B.H. (1992) Biodiversity and ecological redundancy. *Conservation Biology* **6**, 18–23.

Waser, N.M. (1979) Pollinator availability as a determinant of flowering time in ocotillo (*Fouquieria splendens*). *Oecologia* **39**, 107–21.

Waser, N.M., Chittka, L., Price, M.V., Williams, N.M. & Ollerton, J. (1996) Generalization in pollination systems, and why it matters. *Ecology* **77**, 1043–60.

Weber, W. (1993) Primate conservation and ecotourism in Africa. In: Potter, C.S., Cohen, J.I. & Janczewski, D., eds. *Perspectives on Biodiversity: Case Studies of Genetic Resource Conservation and Development*, pp. 129–50. AAAS Press, Washington, DC.

Wheelwright, N.T. (1988) Fruit-eating birds and bird-dispersed plants in the tropics and temperate zone. *Trends in Ecology and Evolution* **3**, 270–4.

Wiens, D., Rourke, J.P., Casper, B.B. *et al.* (1983) Non-flying mammal pollination of southern African proteas: a non-coevolved system. *Annals of the Missouri Botanical Garden* **70**, 1–31.

Wiens, J.A. (1989) *The Ecology of Bird Communities. Vol. 2. Processes and Variations.* Cambridge University Press, Cambridge.

Wilcove, D. (1985) Nest predation in forest tracts and the decline of migratory songbirds. *Ecology* **66**, 1211–14.

Wilkie, D.S., Curran, B., Tshombe, R. & Morell, G.A. (1998) Modelling the sustainability of subsistence farming and hunting in the Ituri forest of Zaire. *Conservation Biology* **12**, 137–47.

Willson, M.F. (1993) Mammals as seed-dispersal mutualists in North America. *Oikos* **67**, 159–76.

Willson, M.F. (1996) Biodiversity and ecological processes. In: Szaro, R.C. & Johnston, D.W., eds. *Biodiversity in Managed Landscapes: Theory and Practice*, pp. 96–107. Oxford University Press, Oxford.

Willson, M.F. & Whelan, C.J. (1990) Variation in escape from predation of vertebrate dispersed seeds: effects of density, habitat, location, season, and species. *Oikos* **57**, 191–8.

Willson, M.F., Irvine, A.K. & Walsh, N.G. (1989) Vertebrate dispersal syndromes in some Australian and New Zealand plant communities, with some geographical comparisons. *Biotropica* **21**, 133–47.

With, K.A., Gardiner, R.H. & Turner, M.G. (1997) Landscape connectivity and population distributions in heterogeneous environments. *Oikos* **78**, 151–69.

Witmer, M.C. & Cheke, A.S. (1991) The dodo and the tambolocoque tree: an obligate mutualism reconsidered. *Oikos* **61**, 133–7.

Woinarski, J.C.L. & Gambold, N. (1992) Gradient analysis of a tropical herpetofauna: distribution patterns of terrestrial reptiles and amphibians in stage III of Kakadu National Park, Australia. *Wildlife Research* **19**, 105–27.

Wolda, H. (1978) Seasonal fluctuations in rainfall, food and abundance of tropical insects. *Journal of Animal Ecology* **47**, 369–81.

Worthington, A. (1982) Population sizes and breeding rhythms in two species of manakins in relation to food supply. In: Leigh, E.G., Rand, A.S. & Windsor, D., eds. *Ecology of a Tropical Forest: Seasonal Rhythms and Long-Term Changes*, pp. 213–26. Smithsonian Institution Press, Washington, DC.

Yahner, R.H. (1988) Changes in wildlife communities near edges. *Conservation Biology* **2**, 333–9.

Yahner, R.H. (1996) Forest fragmentation, artificial nest studies, and predator abundance. *Conservation Biology* **10**, 672–3.

6: Biological Interactions and Disturbance: Invertebrates

HUGH F. EVANS

6.1 DESCRIPTION OF PROCESSES OR CHARACTERISTICS

6.1.1 Ecological attributes that determine the diversity and abundance of invertebrate species in forest ecosystems

This introductory section deals with basic ecological principles that apply in most forest ecosystems, regardless of the region in which they are found. The theme expands on a few key concepts that are assessed in relation to dynamics in section 6.2.

The diversity and abundance of invertebrate species in forest ecosystems are determined by the interaction of complex ecological and environmental factors. Research into these factors is gradually providing data that can be incorporated into models, both descriptive and mathematical, of the processes involved. At the core of such models is the description of the availability of resources that enables a particular invertebrate species to reproduce, whereas the rate of reproduction is dependent on multiple factors, including temperature, competition, etc., that determine utilization of available resources. In relation to exploitation of resources within a forest ecosystem, the primary focus for forest managers is the potential of invertebrates to cause damage and to reduce yields of the trees themselves. This is certainly the case in managed forests where the primary aim is to grow trees for commercial gain, often with associated value from recreational, landscape or environmental considerations. Any invertebrates that feed directly on the trees, or have an influence of how the trees grow, may be classified as pests and

hence may require some form of pest management to reduce the threat (Speight & Wainhouse 1989). Methods of pest management are discussed in Volume 2, Chapter 8, but it is pertinent to consider how ecological theory and practice can be used both to understand why a pest problem has arisen and to provide guidance on how to combat that problem. The same principles can be applied to understanding and enhancing the biodiversity of forests, including conservation of rare and beneficial invertebrates. Indeed, the two extremes of pest outbreak and invertebrate rarity are part of the same continuum of interaction between invertebrates and their host trees.

In general, considering that trees are known to harbour the greatest numbers of insect species (diversity) (Crawley 1983; Stork et al. 1997a), it is surprising that serious pest status is reached only relatively infrequently. For example, although oak (Quercus spp.) is known to support at least 450 species of insect in Britain (Kennedy & Southwood 1984), Bevan (1987) lists only 35 species as pests, of which five were classified as important or severe (XXX–XXXXX on Bevan's scale). Wider listing of pest species has been noted for forest insects in continental Europe, although even here the number of consistently serious pests is small relative to the total diversity of insects present (Klimetzek 1993). Naturally, the severity of a pest outbreak will be classified differently depending on the purposes of the forest manager, and thus it is not always possible to classify insects consistently as pests across geographical ranges or, indeed, between years. Further, an insect may normally be classified as uncommon or rare but, given an abundance of resources, may be described later as a pest. This was certainly the case for oak pinhole borer, *Platypus cylindrus* (Coleoptera: Platypodidae), which prior to the 1987 great storm in southern England was listed in the British Red Data Book as rare (Shirt 1987). Winter (1988) warned that the great increase in availability of freshly dead oak following the storm could result in an increase in numbers of *P. cylindrus* and, subsequently, it has been noted that the insect is now causing damage to high-value oak logs as a result of its habit of boring directly into the heartwood. Therefore, consideration of why invertebrate abundance might increase above a given threshold level can provide a more consistent appraisal of status and also potential approaches for reducing populations below that threshold or indicating measures that can be taken to encourage rare or endangered species.

Among the many interacting processes that determine insect diversity and abundance, a few attributes that are particularly well linked to the environmental status of trees can be distinguished. These provide a means to describe intrinsic processes but also allow comparison with other plants, thus offering the potential for a more consistent approach to descriptive population ecology of invertebrates irrespective of the host plant being considered (Strong *et al.* 1984). Some of these factors are described below.

Longevity

Trees are the longest-lived components of an ecosystem and, in the majority of natural situations, approach or represent the climax stage of vegetation succession. This provides opportunities for continuity of insect population growth and over time this will give a particular set of insect associates for a given tree species (Southwood *et al.* 1979). Continuing the theme of resource availability, the fact that trees may be present in the same location for tens or hundreds of years ensures that both migrant and, particularly, relatively sedentary invertebrates can exploit them at some stage during their growth. However, longevity *per se* is not a quantifiable attribute but does provide the means for the other components, described below, to influence invertebrate diversity and abundance. Increasing sta-

bility in the environment, as represented under forest conditions, leads to stability in both microclimate and resource availability, particularly favouring exploitation by specialist insect feeders and their associates.

Biogeography

Tree species richness in a given region or locality varies considerably with geographical location and thus the climax community structure and longevity of individual species may vary accordingly. Although subject to uncertainty and lack of unequivocal evidence of the underlying processes, there is certainly a gradient of species richness of plant cover, especially trees, from the tropics to the poles (Begon *et al.* 1996). Not surprisingly, the same applies to a wide range of animals and plants; as an extreme example, the species richness of orchids ranges from 15 to 2500 in a latitudinal range from 0 to 66°S (Dressler 1985). Further examples are provided by Brown (1988). More specifically, in an analysis of species richness of deciduous forests both in the tropics and, latitudinally, within North America, Brown and Gibson (1983) contrasted the high diversity of tree species in the tropics (up to 100 ha^{-1}) with the low diversity from north to south within North America (1–30 ha^{-1}). However, even within such a generalization there are exceptions, so that for example conifers are more diverse in temperate latitudes.

Geological time

The ecological structure of forests and of their associated invertebrate faunas is determined by both present and recent characteristics, particularly area, and by their history in geological time. The latter characteristic is driven as much by time *per se* as by the history of major disturbances (glaciations, volcanic eruptions, land shifts, etc.) within geological time. Thus, present flora and fauna may well be determined by recolonization since the last major disturbance, the diversity of species being a reflection of both colonization and speciation relative to more recent ecological conditions. The gradient of species from polar to tropical regions could therefore be explained by the high level of disturbance closer to the polar masses compared with the relative stability of climate in the tropics (Brown 1988). Evidence from habitats that have become isolated since the last major disturbance indicates that current species diversity may reflect the historical link to the original major habitat. For example, islands that were once joined by land bridges to the larger land masses of New Guinea have species diversities of birds greater than expected compared with other islands that were never connected (Diamond 1975). However, although local ecological communities can be explained by historical patterns, the gradients over wide geographical scales are not so easily attributed to time as the major determinant. The concept that major disturbances drive change undoubtedly carries weight, but closer analysis of tropical rain forests reveals that they too have been subject to disturbances of various sorts and are, in many cases, much changed over geological time. Based on this knowledge Haffer (1969, 1982) proposed a theory that disturbances in the tropics during the Pleistocene (alternating wet and dry periods giving rise to forest blocks surrounded by grassland) gave rise to refugia in which high levels of speciation took place, ultimately leading to the current high diversity characteristic of the tropics. While this theory is plausible and has been invoked by other authors (Connor 1986), tests of the hypothesis have cast doubt over its wide applicability (Lynch 1988; Flenley 1999).

Irrespective of the mechanistic bases for variation in species richness over wide latitudinal ranges, time is a significant determinant at a more local scale. Birks (1980) analysed the insect faunas of trees in Britain during the period since the last glaciation 13000 years ago and concluded that there was a significant correlation with the time that a given tree species had been continuously present in Britain during that period. Taking this further, Kennedy and Southwood (1984) included Birks' radiocarbon estimates of time in Britain as a variable in a multiple regression analysis of insects on British trees. They showed that time was a significant predictor, both in its own right and as a co-predictor with log abundance (area). This is logical considering that over time there will be a tendency for well-established tree genera

to increase their abundance, despite the recent proliferation in areas arising from commercial afforestation.

Size

A consequence of basic structure and longevity is that trees tend to be the largest components of the phylloplane, thus providing a much wider number and range of ecological niches compared with herb and shrub layer plants (Crawley 1983). Such attributes enable more insect species to colonize the plants without facing competition, both interspecific and intraspecific (Strong *et al.* 1984). Size has many interrelated characteristics that help to explain the presence of both herbivores and their natural enemies.

• A larger physical presence is more easily detected and colonized by invertebrates. This concept, termed 'apparency', reflects a higher visual profile for colonizing adults or passively migrating immature life stages (e.g. mean height was a good predictor in models of the distribution of Scolytidae in Finland; Heliovaara & Vaisanen 1995) and a greater production of olfactory cues (e.g. the bark beetle *Tomicus piniperda* can recognize pine tree volatiles and assess suitability for breeding while in flight; Byers *et al.* 1985).

• Complex plant architecture is a consequence of greater size and for trees is particularly linked to the provision of many niches capable of supporting invertebrates with a wide range of feeding strategies (guilds). Although the concept of the guild is not always clearly defined (Hawkins & MacMahon 1989), it is nevertheless a useful means of describing the trophic interactions of invertebrates and trees without having to consider their precise taxonomic status. Thus, seed, bud, leaf, bark and wood feeders can be distinguished, each occupying a different part of the tree and therefore potentially avoiding competition.

• Increased biomass provides greater food resources per individual, thus encouraging growth and breeding success whilst reducing potential intraspecific competition. Price (1992) has reviewed the role of plant resources in insect population dynamics and concludes that, in part, the carrying capacity of an ecosystem depends on plant succession in time and space. Carrying capacity is therefore greater for trees, which generally appear late in succession, compared with herbs or shrubs. Thus biomass is available for longer (see below) and in greater quantities on trees compared with other plants.

• The concept of enemy-free space applies to the likelihood of herbivores being overlooked by searching parasitoids or predators (Jeffries & Lawton 1984). One aspect of size is that there is more opportunity for herbivores to occupy a part of the plant that offers protection against natural enemies (and competition from other herbivores). The complexity and number of niches present on a plant (plant architecture; Lawton 1983) therefore has an effect on enemy-free space. The greater the architectural complexity, the greater the time that natural enemies may have to spend in searching for prey, thus reducing their overall effectiveness and enhancing the survival probability of herbivores.

Area

Forests tend to occupy greater areas than other plants. There is evidence to support the view that the greater the area occupied by the host plant, the greater the number of insect species associated with that host. This is called the species–area relationship and was originally linked to the island biogeography theories of species isolation and colonization (MacArthur & Wilson 1967). Species–area relationships have been demonstrated for many plants and animals, including both highly mobile and sedentary species. With regard to the numbers of insect species on trees, data on the British insect fauna, one of the best characterized in the world, have been analysed by a number of authors. Arising from earlier tentative analyses (Southwood 1961), Kennedy and Southwood (1984) analysed the relationships between insect species and characteristics of tree species in Britain. They concluded that there was a significant relationship between the area occupied by British tree species and the numbers of insect species associated with them (Fig. 6.1). The area occupancy data were based on county and subcounty information of plant records and were thus potentially overestimates of actual areas

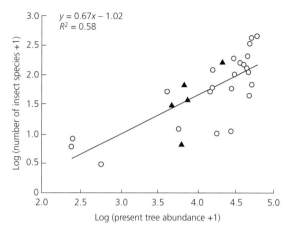

Fig. 6.1 A double logarithmic plot of the relationship between area, measured as present tree abundance in quadrats, and numbers of insect species on British trees. ○ broadleaved tree species; ▲ conifer species. (Adapted from Kennedy & Southwood 1984.)

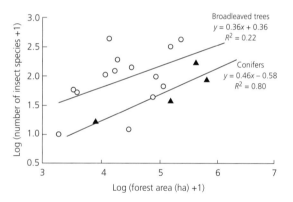

Fig. 6.2 A double logarithmic plot of the relationship between area, as total hectares derived from Forestry Commission census data, and numbers of insect species on British trees. ○ broadleaved tree species; ▲ conifer species, with separate regression lines. (Adapted from Claridge & Evans 1990.)

occupied. Claridge and Evans (1990) reanalysed the species–area relationships using the same insect data as Kennedy and Southwood but employing actual area information derived from Forestry Commission census data (Locke 1987). They concluded that area was not a good descriptor for insect species on deciduous trees but was just significant for conifers (Fig. 6.2).

However, further unpublished analysis of the data broken down into numbers and areas of habitat patches occupied by the tree genera indicated that whereas total area was not a good descriptor, the degree of fragmentation of that area provided a better explanation for the observed insect species numbers (Fig. 6.3). A similar conclusion, concentrating on Lepidoptera that have moved hosts to include conifers, was reached by Fraser and Lawton (1994) in an analysis of new associations on British trees. The overall finding that area provides a broad descriptor of insect diversity remains valid, although this does not explain the mechanisms that lead to these associations. This is discussed in more detail in section 6.2.

Taxonomic relatedness

In a given ecosystem, the closer the taxonomic relatedness of host plants, the more similar the numbers of insect species associated with those plants. This has been documented for tree species

that have a long history in a particular location and thus provide rich sources of invertebrate species that may subsequently colonize other tree species, especially if they are taxonomically related (Claridge & Wilson 1981; Kennedy & Southwood 1984). A particularly good example of the process in action is the colonization of *Nothofagus*, a member of the Fagaceae newly introduced to Britain, which has been colonized rapidly by insects associated with closely related members of the native Fagaceae (Welch & Greatorex-Davies 1993).

6.1.2 Geographical isolation as a critical factor in determining invertebrate diversity and implications for international movement of pest organisms

Geographic isolation drives the acceleration of speciation and provides the means to derive principles on the processes involved. In addition, the fact of isolation means that some inverte-

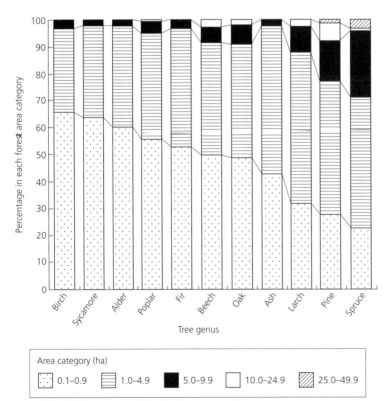

Fig. 6.3 Degree of fragmentation of the major tree genera in British forests showing proportion of area occupancy in a range of area categories. Area data from Forestry Commission Subcompartment Database.

brates abundant in one region will be absent elsewhere, despite conditions being suitable for their establishment. In such situations, it is pertinent to examine the international plant health implications of geographical isolation and how modern patterns of trade influence the degree of isolation and later colonization by new insect species.

Geographical isolation as a factor in determining insect diversity

In general, the basic principles in section 6.1.1 apply across ecosystem boundaries and geographical zones extending from the poles to the tropics. In addition, the isolation afforded by relatively small islands provides further evidence of the principles of species colonization over space and time. Great Britain provides a particularly good example, because of its island position in the Atlantic Offshore climatic zone and the fact that it has been subjected to glaciation in recent geological time. Among the ecological characteristics that arise from this set of circumstances are a relatively depauperate mix of both animal and plant species that have arisen from the combination of losses resulting from the last ice age and the slow rate of recolonization arising from island status. British flora and fauna are therefore impoverished relative to continental Europe (Rackham 1990). Further influences arise from the preponderance of exotic tree species in commercial forestry in Britain, which although dominated by exotic conifer species also has increasing numbers of exotic broadleaved species (Locke 1987).

Expanding on the principle that colonization of tree species is determined by the features described in section 6.1, the ranges of numbers of insect species associated with trees in Britain reflect both the area (with provisos) and length of time that trees have been continuously present in Britain (Kennedy & Southwood 1984). However, there are exceptions, so that Sitka spruce in particular would appear to support far fewer species than predicted on the basis of area, although this probably also reflects the relatively short history of the species and recent expansions in areas planted in Britain

Implications for international movement of pest insects

Trading patterns between countries have had a considerable influence on both the flora and fauna of most of the countries in the world. This reflects deliberate introductions of plants and animals for agricultural, forestry and amenity purposes and also accidental introductions by association with other commodities. In the majority of cases, the introduced species have had a neutral to beneficial effect, reflecting their 'domesticity' within managed crop or animal husbandry systems. However, there are many exceptions to this generalization and in most cases the results of introduction of new species were not foreseen, either through lack of knowledge of the potential impacts or because the introduction was accidental and undetected. The literature on international biological control is dominated by examples where attempts have been made to rectify the effects of new pest introductions (both animal and plant) (Waage & Greathead 1988). A classic example is the control of the woodwasp *Sirex noctilio* (Hymenoptera: Siricidae) in exotic pine forests in Australia with the introduced nematode *Deladenus siricidicola* (Nematoda: Neotylenchidae) isolated from *Sirex* in New Zealand. This nematode has a life cycle that includes a free-living form that feeds on the fungus *Amylostereum areolatum* and a parasitic form that invades *S. noctilio* larvae and subsequently migrates to the reproductive organs of the female wasp, causing sterility (Madden 1988). In this situation, all elements of the interaction are exotic to the new location, with little prospect of natural control being realized.

6.2 PRINCIPLES AND DYNAMICS THAT GOVERN EVENTS AND OCCURRENCE

Expanding on the basic principles in the introductory section, the remainder of this chapter deals with more specific consideration of dynamic processes. This is illustrated by reference to factors that determine the interface between endemic and epidemic insect populations in forests and woodlands. The emphasis is on feedback processes that tend to keep insect populations 'in balance' with their food sources, particularly taking account of the attributes in section 6.1. Departure from this 'balance' can result in either development of pest status or reductions in populations of rare insects, possibly leading to extinctions.

6.2.1 Interactions with the food plant: phytophagous invertebrates (herbivores)

It has already been mentioned that insect densities generally do not reach damaging levels, thus implying that there are constraints to population growth. Important determinants of population growth are palatability and nutritional suitability of the host tree. Both adult and larval stages may be affected by these parameters. In the majority of cases, adult insects locate suitable host trees for their progeny and lay eggs accordingly. The evolution of oviposition behaviour and host preferences in Lepidoptera has been discussed by Thompson and Pellmyr (1991), who considered host chemistry, plant distribution and suitability of resources as well as the evolution of the wingless forms of many lepidopterous species. They concluded that no single factor governs the evolution of host specificity and that many studies of the mechanisms of selection fail to separate the many possible host selection processes. In the case of flying females, selection of oviposition sites by females is determined by choice, either directed by the cues given off by the host plant, or by landing and 'tasting' the substrate for suitability. An interesting example of such a phenomenon is the teak defoliator moth *Hyblaea puera* (Lepidoptera: Hyblaeidae) in tropical teak forests, especially in India, which swarms as both males and females and then disperses *en masse* to eventually locate and oviposit in the crowns of teak trees that may be considerable distances from the site of origin. Considering that in its natural habitat teak is a relatively localized tree species, growing within mixed forests with many other tree species present, this implies that the moth has well-developed host location capabilities even when the tree is intimately mixed with other tree genera (Nair 1988). Once the tree is located, eggs are laid on the expanding 'tender'

leaves, thus providing the newly hatched larvae with the only food suitable within the teak canopy. This and other examples indicate that active host choice by female invertebrates is often linked to the suitability of the oviposition site, or its immediate vicinity, for consumption by the more vulnerable first instar larvae (Trewhella *et al.* 1997). By contrast, those insects having flightless females have more limited opportunities to place their eggs in suitable locations on the 'ideal' host plant. Fraser and Lawton (1994) considered this phenomenon in assessing expansions in the host ranges of British moths and concluded that a greater than expected proportion of 'host-shifting' moths were unable to fly actively to place their eggs. Moths with flightless females, such as winter moth *Operophtera brumata* (Geometridae), European gypsy moth *Lymantria dispar* (Lymantriidae), fall cankerworm *Alsophila pometaria* (Psychidae) and wattle bagworm *Kotochalia junodi* (Psychidae), tend to lay eggs either close to the pupation site (*O. brumata* and *K. junodi*) or more indiscriminately on both the original plant and on other substrates in the vicinity (*L. dispar*). Dispersal in these moths tends to be through the action of undirected larval movements, often by wind carriage of early instar larvae on threads of silk. Interestingly, Hunter (1991) included flightless females among the characteristics inherent in outbreak species of Lepidoptera, although other features such as polyphagy and high fecundity played equally important roles.

A further aspect of host selection and success in establishment is the coincidence of egg hatch, which determines when the vulnerable first larval stages commence feeding, and host phenology. This is particularly crucial for spring-feeding larvae that rely on newly expanding buds and shoots for initial establishment of young larvae on the most succulent tissues. This is most apparent in those species of insect that overwinter in the egg stage and where early instar larvae must feed on newly expanding foliage in the spring to survive. Among such species, oak leaf roller moth *Tortrix viridana* (Lepidoptera: Tortricidae) is particularly vulnerable to lack of synchrony between egg hatch and budburst of oak. Studies in France have demonstrated that populations of the moth adapted to specific budburst periods of *Quercus pubescens* (early budburst) and *Quercus ilex* (budburst up to 3 weeks later) could be distinguished in terms of egg hatch and hence early larval establishment (Merle 1983). Hunter (1992) also confirmed the dependent relationship between oak budburst phenology and establishment success of neonate *T. viridana* and *O. brumata* larvae. Similar results were obtained in tests of clonal poplars in relation to timing of budburst and foliar suitability (nutrition and phenolic defences) for larvae of forest tent caterpillar *Malacosoma disstria* (Lymantriidae) (Robison & Raffa 1997).

The development of tissues with differing degrees of palatability to potential herbivores is an intrinsic part of phenological coincidence between herbivore and host. Palatability may vary widely as plant parts develop during the growth season and also between seasons and is linked to both physical structure and chemical composition of the different tissues of the host plant (Price 1992). For example, the suitability of oak foliage to a range of herbivores is determined in part by reactions to the presence of tannins (negative) (Feeny 1976), total phenolics (negative) (Rossiter *et al.* 1988), foliar astringency (positive) (Hunter 1997) and nutritional content (Hunter & Schultz 1995).

In general, there is a feedback relationship between the degree of defoliation suffered by a tree and the absolute and relative quantities of phenolic and nutritional components present in the leaves. This apparent trade-off between growth and defence is the subject of considerable research and provides some interesting insights into how plant quality can act on population dynamics of herbivores. It is also an intrinsic component of multitrophic interactions in which other factors, particularly the roles of natural enemies (discussed in section 6.2.2), affect the dynamics of herbivores.

Plants that are long-lived, such as trees, could be expected to develop more effective defence mechanisms than transient plants because of the longer time period over which they would be subjected to potential herbivore attack and their greater apparency in the environment (Rhoades & Cates 1976). While it is true to say that trees do

have a wide range of defensive strategies, it has not been possible to confirm that overall they are better defended than less apparent plants. The hypothesis was tested specifically by Reader and Southwood (1981) who used generalist herbivores to assess the palatability of plants with turnover rates ranging from rapid to infrequent, and showed an overall decrease in palatability with increases in successional stage, i.e. towards longer-lived plants. However, results were very variable and indicated that there were clearly a wide range of factors involved, plant chemistry being one of the more important.

The mechanistic basis of plant resistance is not universally agreed and indeed it is apparent that the actual mechanism involved will differ, depending on a combination of host plant genotype, environment and herbivore pressure. At its most simplistic level, plant defences involve both the presence of chemicals that deter herbivores and the presence of nutrients that enable them to grow and ultimately reproduce. An attractive theory proposed by several authors is that plant growth is a trade-off between carbon and nitrogen such that if nitrogen is limiting, then plant growth slows down, leading to excess carbon being diverted to secondary chemicals and thus improved plant defence (Bryant *et al.* 1983, 1993). Ruohomäki *et al.* (1996) questioned this assumption in their studies of nitrogen fertilization and its effects on mountain birch. They concluded that a resource-based hypothesis did not provide an adequate explanation for observed events and that a source–sink hypothesis was a better descriptor. The latter hypothesis assumes that the plant is split into a number of interlinked components, some of which act as sources of nutrients and carbon (e.g. photosynthesizing leaves act as sources for amino acids and provide the necessary components for future resource allocation to buds) and others as sinks (e.g. newly expanding buds take nutrition from a general 'pool' in the tree). Haukioja and Honkanen (1996, 1997) analysed various forms of induced defensive reactions in relation to the source–sink hypothesis. For birches and pines, they showed that foliar damage and the immediate and delayed effects on foliage suitability for herbivores could be explained by impacts on physiological sinks, such as newly expanding buds. This had the effect of reducing plant growth in the following year and that impact was local to the site of damage, reflecting the damage to the particular meristems affected (Honkanen & Haukioja 1994). Other studies, notably those of Bryant *et al.* (1993) on paper birch *Betula papyrifera* and Hunter and Schultz (1995) on *Quercus prinus*, showed that addition of nitrogen fertilizers reduced the ability of trees to produce secondary defensive compounds. This apparent contradiction between nutrient balance and source–sink hypotheses may be a result of specific differences in reactions depending on tree species or may represent artefacts of experimental design. For example, is there a threshold effect such that addition of excess nitrogen can override internal feedback mechanisms in the plant? The studies of Bryant *et al.* (1993) may well support the contention that excess nitrogen can mitigate any defensive characteristics of the host plant. Clearly, further work needs to be done to elucidate the precise mechanisms involved.

Irrespective of the mechanisms, there is evidence to indicate that the suite of defences produced by trees can be effective in reducing attack levels by different herbivores. Such defences vary in their effects on different feeding guilds so that it is possible to distinguish between the strategies developed by both the herbivores themselves and the plants on which they feed. This has been reviewed comprehensively by Koricheva *et al.* (1998) who carried out a meta-analysis of published experiments in which the effects of stress were analysed against a series of parameters, including feeding guild, plant phenology, plant structure, etc. The authors concluded that, overall, plant stress had little effect on insect relative growth rate, fecundity, survival and ability to colonize a plant. However, when data were analysed at the level of the insect feeding guild, there were large and significant differences in responses, both positive and negative, to plant stress (Table 6.1). Thus, insects that colonize trees by boring into the bark and/or wood (bark beetles, longhorn beetles, woodwasps, etc.) were more successful in colonizing stressed trees (there was no information on the other parameters, but it would be reasonable to assume that successful

Table 6.1 Plant stress and its impact on the performance of different insect feeding guilds. (From Koricheva *et al.* 1998.)

Performance parameter	Guild	Positive or negative effect on performance
Relative growth rate	Chewers	Minor negative
	Suckers	Positive
Fecundity	Chewers	Negative
	Suckers	Positive
	Miners	Positive
Survival	Chewers	Negative
	Suckers	Minor positive
	Miners	Positive
	Gall formers	Large negative
Colonization potential	Chewers	Minor negative
	Suckers	Negative
	Miners	Positive
	Gall formers	Large negative
	Wood borers	Large positive

colonization leads to a higher probability of success in the other parameters). Sucking insects, which use stylets to access nutrients in the phloem, also performed better on stressed plants. In contrast, gall-forming and chewing insects tended to be negatively affected, the latter being particularly suppressed on slow-growing plants such as trees. This is discussed further in Volume 2, Chapter 8 in relation to pest outbreaks and risk-rating systems.

6.2.2 Interactions between herbivores and their natural enemies

Discussion has so far been concerned with the trophic interactions between herbivores and their host trees, providing examples of the so-called bottom-up approach to understanding the dynamics of the two elements. However, a further trophic level that interacts with, and may influence critically, herbivore dynamics is the role of biotic mortality factors, collectively termed 'natural enemies'. Such factors can impinge on any stage of the herbivore life cycle and may be linked in a quantitative way to the population size of the host or prey. There is considerable debate concerning the relative importance of top-down impacts of natural enemies compared with the resource-driven, bottom-up effects of the plants themselves (Dempster & Pollard 1986; Harrison & Cappuccino 1995). However, recent literature is tending to take a balanced view that includes both resource and natural enemy impacts in determining the dynamics of herbivore interactions with their host plants (Kidd & Jervis 1997; Schultz 1997). There is no doubt that natural enemies have a role to play in determining the numbers of potential prey within a given ecosystem. Regulation of prey densities within predictable limits requires that the numbers of natural enemies are linked quantitatively to those of their prey (density-dependent regulation) (Hassell & Waage 1984). Although many models have been developed in an attempt to predict the relative dynamics of natural enemies and their prey, the most pertinent evidence in relation to management of forest insect pests comes from their demonstrated use as biological control agents (Waage & Greathead 1988). Although biological control has ranged from highly successful, requiring no further intervention, to only partial, overall the data lend strong support to the supposition that natural enemies can be important regulatory factors in herbivore dynamics. Kidd and Jervis (1997) considered the roles of parasitoids and predators on forest insect populations in some detail. They concluded that natural enemies were significant in regulation of forest pests but that they could also give rise to population increases as a result of delayed density-dependent processes. Interestingly, Kidd and Jervis considered the bottom-up vs. top-down debate by assessing the dynamics of the natural enemy–herbivore–plant interaction within the context of the synoptic population model proposed by Southwood and Comins (1976). Herbivores were classified into groups according to their growth rates and ability to disperse.

• *r*-strategists have high growth rates and effective dispersal powers, making them capable of colonizing and utilizing unstable or transient habitats. These species tend to have few effective natural enemies and are usually regulated by resource availability.

• *K*-strategists represent the opposite extreme, having slow growth rates and limited ability to disperse and are associated with stable habitats. Population size tends to be tightly linked to availability of resources and natural enemy activity has little influence on total numbers.

• *r-K* strategists have some of the characteristics of both groups and tend to lie in the 'natural enemy ravine' where the range of natural enemies is greatest and have pronounced numerical responses to changes in herbivore densities. Forest pests tend to be classified as *r-K* strategists but, because they tend towards the endemic in most cases, will be skewed to the *K* end of the continuum.

Evidence that natural enemies have a significant influence on forest insect numbers comes from examples of successful biological control in many countries of the world and from studies of the dynamics of some of the most damaging forest pests. Table 6.2, adapted from data collated by Kidd and Jervis (1997), indicates both the numbers of successful biological control introductions and also examples where natural suppression has been observed. The Lepidoptera and Homoptera dominate, reflecting the importance of these orders as damaging agents of forest trees, although in virtually all cases there are contributory influences of host plant resources, especially secondary chemicals. The Coleoptera, particularly the bark beetles, tend to have a dynamic interaction with the host plant, which may act in concert with the activities of natural enemies. This is certainly the case for *Dendroctonus micans*, which is regulated

Table 6.2 Biological control introductions and natural impacts of biological control agents. (Based on Kidd & Jervis 1997.)

Pest insect order	Numbers of examples		
	Biological control	Natural control	Tree resource limited
Homoptera	9	2	5
Lepidoptera	7	9	1
Coleoptera	4	2	4
Hymenoptera	5	1	–
Diptera	1	–	2
Total	26	14	12

by both the action of its specific predator, *Rhizophagus grandis*, and the defensive characteristics of its host spruce trees (Wainhouse & Beech-Garwood 1994; Evans & Fielding 1996).

The characteristics of successful natural enemies include a well-developed ability to locate suitable hosts (non-random search), high reproductive capacity, quantitative linkage to changes in host density and dispersal capacity to exploit spatially separated host populations. These clearly apply to many of the predators and parasitoids that have been studied but are less easy to apply to pathogenic organisms equally important as mortality factors.

Predators

Invertebrate predators search for and consume their prey directly, usually in both the adult and the larval stages. The best-known examples tend to be among the ladybirds (Coleoptera: Coccinellidae) familiar to many in urban and woodland situations. However, among the predators of forest pests it is the beetle predators in the families Cleridae, Trogostidae and Rhizophagidae that have been most studied in relation to their impacts on bark beetles (Dahlsten & Berisford 1995). Predators are not restricted to the insects; mammals, birds, reptiles, spiders and mites have all been shown to feed on tree-dwelling herbivores. In most cases the predators are generalists and will feed on the most abundant prey available. However, within this generalization, some apparently polyphagous predators act quite selectively and respond directly to the pheromones (kairomones) produced by their prey. Thus, the clerid beetle predator *Thanasimus undatulus* is attracted to the pheromones of spruce beetle *Dendroctonus rufipennis* and the secondary bark beetles *Dryocoetes affaber* and *Ips tridens* (Poland & Borden 1997). In this case, use of the *I. tridens* pheromone produced a dual effect on the primary pest *D. rufipennis* by both competitively excluding the spruce beetle and by attracting the predator to the tree. Other generalist predators have significant effects at low but not at high densities; the white-footed mouse, *Peromyscus leucopus*, has been shown to regulate low-density gypsy moth, *L. dispar*, populations through predation on late larval and pupal stages (Campbell & Sloan 1977). However, this effect was not apparent at high gypsy moth densities, presumably because the predator numerical response was insufficient to maintain density-dependent regulation (Campbell & Sloan 1978).

Parasitoids

Insects that parasitize their hosts through laying eggs in, on or near host individuals and in which their larval stages develop are known as parasitoids. This group of natural enemies is extremely common and probably makes up over 10% of insect species in the world (Godfray 1994). The majority of parasitoids are in the order Hymenoptera, notably the families Ichneumonidae and Braconidae. Dipterous parasitoids, in the family Tachinidae, have also been shown to be effective in regulating prey, for example *Cyzenis albicans*, a tachinid parasitoid, is a regulator of winter moth, *O. brumata*, populations in Canada where it was introduced from Europe (Roland & Embree 1995). Interestingly, *C. albicans* is a density-independent mortality factor in winter moth populations in Europe but is strongly dependent on larval density in Canada (Hassell 1980). The interactions between natural enemies, particularly parasitoids, and their hosts have been the subject of much research into the dynamics of the interactions. Mathematical models have been produced that explain some of the variability in responses between parasitoids and their hosts and are being increasingly refined to take account of both temporal and spatial parameters (Pacala *et al.* 1990). Models that take account of dynamics in a single isolated population tend to be unstable and therefore the dynamics in neighbouring populations also need to be taken into account. Comins *et al.* (1992) simulated predator–prey interactions as a series of metapopulations in which both organisms were allowed to disperse between patches. Dynamics of individual patches were unstable but stability was observed in the system as a whole, thus demonstrating that persistence of the whole population of either predator or prey could be achieved despite instability within its subcomponents. Such simulations show that it is necessary to have not only a detailed knowledge of natural enemy behaviour and dynamics but also an appreciation of the spatial elements in the interaction with its prey. As indicated in section 6.2.1, the influences of the host plant must also be included, again on a temporal and spatial scale.

Overall, therefore, there is evidence to show that parasitoids, by virtue of their directed search behaviour and relatively high specificity, can have a regulatory influence on host dynamics. This does not necessarily imply that the host will be regulated below an acceptable economic threshold and thus it may be necessary to take account of other mortality factors to explain fluctuations in the host population.

Pathogens

Invertebrate herbivores are subject to a wide range of naturally occurring pathogenic organisms, many of which will not be recognized as such in assessing mortality rates in the field. Table 6.3 lists the principal pathogenic organisms and their main characteristics in relation to their status as biotic mortality agents for forest-dwelling herbivores. Pathogens differ from other natural enemies in not having the ability to search actively for their hosts. Dispersal is therefore a passive event that relies on biotic and abiotic factors to ensure that the pathogen is able to encounter suitable hosts. Replication and infec-

Table 6.3 Pathogens of invertebrates and their potential roles in forest pest management.

Pathogen group	Principal representatives	Characteristics	Potential for pest management
Bacteria	*Bacillus thuringiensis*	Spore and toxic crystal. Effect on host mainly through toxin action on midgut. Must be ingested	The principal microbial control agent in forestry. Extensive use in Europe and North America (Evans 1997)
Fungi	Deuteromycetes, especially *Beauveria*, *Metarhizium* and *Verticillium*. Also Entomophthorales	Spore germinates on integument and penetrates to replicate in haemocoel and other organs	Relatively low use in forestry. Recent great promise of *Entomophthora maimaiga* as natural control in North America (Hajek *et al.* 1993)
Protozoa	Mainly Microsporidia in the genus *Nosema*	Ingestion of spores or transovarial transmission. Replication in fat body. Debilitates and sometimes kills hosts	Some use of spore preparations for pest management, especially against gypsy moth (Jeffords *et al.* 1989)
Viruses	A number of families, including Baculoviridae, Reoviridae, Poxviridae, Iridoviridae, Parvoviridae, Rhabdoviridae, etc.	Most have to be ingested to induce infection. First three families listed are occluded viruses where virus particles are bound in crystalline protein	Baculoviridae have been used successfully against Lepidoptera and Hymenoptera (sawflies). Natural epizootics occur in many high-density insect populations
Insect pathogenic nematodes	Mainly in the Steinernematidae	Nematodes penetrate cuticle and carry pathogenic bacteria, causing septicaemia and death of hosts	Commercially available and used successfully in horticulture. Prospects for use against restocking pests (Brixey 1997)

tion arise only when the infectious unit is ingested (bacteria, Protozoa and viruses) or comes into contact with the integument of the potential host (fungi) (Fuxa & Tanada 1987). Although this introduces a high element of chance in determining encounter frequency, pathogens have a number of strategies that help to overcome this potential problem.

High reproductive rate. Replication results in massive increases in the quantity of inoculum produced on death of the infected host. For example, the baculovirus of gypsy moth, *L. dispar*, increases 10 000 fold compared with the initial lethal dose to yield over 10^9 inclusion bodies per larva (Entwistle & Evans 1985). Massive increases of this nature ensure that large quantities of inoculum are released in the immediate area where other hosts are likely to be feeding, thus partially overcoming the inability to search actively for hosts.

Ability to persist. Although pathogens are subject to rapid inactivation from environmental factors such as ultraviolet light (Ignoffo 1992), they are still able to persist in the environment between host generations. This arises from a number of pathways, the most significant being the protection of the infectious unit within an outer coat (spores for bacteria, fungi and Protozoa; proteinaceous inclusion bodies for several groups of insect pathogenic viruses) and the tendency for inoculum to be deposited in protected situations such as soil, bark crevices, etc. (Evans & Harrap 1982). Infection in the following generation depends on encounter by early larval stages, which almost invariably are the most susceptible and thus require only very small quantities of inoculum to induce infection. The massive increase in secondary inoculum arising from primary infection can then lead to a local pool of inoculum available for encounter by other individuals in the same host population, thus leading to accelerating levels of infection and further increases in inoculum (Anderson & May 1981).

Transovarial transmission. In some cases, such as the Protozoa and some non-occluded viruses,

the nucleic acids of the pathogen are incorporated directly within the germplasm of the host insect and thus can persist in a latent form to the next generation of the host (e.g. the picornaviruses; Moore & Eley 1991). However, the evidence for true transovarial transmission within the host cells is often equivocal and difficult to separate from transovum transmission through external contamination with infective stages of the pathogen (Andreadis 1987). In addition, expression of the pathogen in the next generation is an uncertain event and thus in terms of overall dynamics of pathogen infections it would appear that transovarial transmission is a 'failsafe', rather than a primary route of transmission.

Transmission through other biotic agencies. Many of the pathogen groups listed in Table 6.3 have been shown to be transmitted by biotic agents such as predators or parasitoids. For example, birds eat baculovirus-infected pine beauty moth larvae and void infective virus in their faeces, thus contaminating foliage and inducing infection in other moth larvae (Entwistle *et al.* 1993). Similarly various parasitoids, including Braconidae, Encyrtidae, Ichneumonidae (all Hymenoptera) and Tachinidae (Diptera), have been shown to transmit baculoviruses between hosts (Entwistle & Evans 1985).

Not surprisingly, in view of the debate on bottom-up vs. top-down processes, there is an interaction between the infectivity of various pathogens and the food substrate consumed while the host is ingesting the pathogen. The presence of tannins and other phenolics can reduce the apparent effectiveness of *Bacillus thuringiensis* (Appel & Schultz 1994) and baculoviruses (Foster *et al.* 1992) against gypsy moth on various species of oak tree. In these cases there appears to be significant binding between the inclusion body of the pathogen and the hydrolysable tannins within the oak leaves. Understanding the dynamics of pathogen–host interactions clearly requires in-depth information on the subtle effects at various trophic levels.

6.3 INTERACTIONS WITH OTHER PROCESSES, WITH SCALE AND WITH FOREST TYPE

Many of the processes that govern the dynamics of insect associations with trees were discussed in the previous sections. In the main, these factors combine to keep insect populations at levels that do not appear to pose a threat to the health of trees or their commercial value. As Crawley (1989) has pointed out, there is a common perception that because the terrestrial environment is mainly green, then neither the plants nor their associated herbivores are limited in their resource requirements. However, the previous discussion has shown clearly that there are complex interactions between the different trophic levels associated with trees so that the end-result may give rise to major fluctuations in the relative performances of the different components. In both their natural and managed states, forests are affected by factors of scale such that regional site characteristics and local climate will influence the species mix and the growth rates of trees. Such effects will also influence invertebrate associates, which will respond both to resource availability (at both the plant and higher trophic levels) and to temperature and other abiotic conditions that will drive the rate of response. Some of the influences of scale are implicit in the species–area relationships that, at least in part, provide descriptors of species accumulation of both animals and plants in a particular region (Williamson 1988).

What, then, are the effects of scale on communities of invertebrates in forests and how are they related to forest management in a rapidly changing world? It has already been stated that the area occupied by a particular tree species is an important determinant of insect diversity, particularly if that area contains species that have been present for long periods. Such a situation applies to the natural oak populations in the north-western USA, where two of the most convincing demonstrations of the species–area effect and interactions with other local and landscape features apply. The natural distribution of oak genera and species in this region covers a wide latitudinal and altitudinal range, including the Californian region (southern Oregon to northern Baja Califor-

nia bounded by the Rocky Mountain and Sierran regions, thus isolating the zone) and Atlantic region (larger, ranging from southern Canada to the Gulf of Mexico, west to the Great Plains). The oak species in the two regions are distinct from each other, those of the Californian region being closely related to Asian rather than eastern US oaks (Cornell & Washburn 1979). Two separate studies of herbivores within specific feeding guilds have been carried out in these regions. Cornell and Washburn (1979) studied the Cynipinae (Hymenoptera: Cynipidae), a subfamily of gall-forming wasps that are extremely specialized to the genus *Quercus*. They have been extensively studied and the data were used by Cornell and Washburn to investigate relationships to the areas occupied by oaks in the Atlantic and Californian regions. An earlier study by Opler (1974) also assessed relationships to oak distribution but was concerned with leaf-mining Lepidoptera (Lepidoptera: Gracillariidae) within California only. Both studies demonstrated significant species–area relationships against total range (expressed in square miles). However, the pattern of response was quite variable depending on the subsets of data examined; some of these are reproduced in Table 6.4. The steeper slope and higher coefficient of variation in the Californian subset was attributed to the greater ecosystem variability in the latter location, taking account of the larger number of oak species, the wider altitudinal range and the greater variety of topographical features compared with the Atlantic region (Cornell & Washburn 1979).

Interestingly, the leaf-miner data for the same geographical range of oaks suggest that new moth species colonize sympatric oak species within the same area and that there is no strong link with taxonomic relatedness between oak species, thus demonstrating a response more to total resource availability than to oak species *per se* (Opler 1974). Such data provide support for habitat heterogeneity as one of the principal factors driving species–area relationships. Further studies by Cornell (1986) indicated that both the size of oak trees and, particularly, the size of the pool of cynipine wasp species in a given area were the major determinants of species richness on a given oak species. He attributed this to several of the

Table 6.4 Species–area relationships for leaf-mining Lepidoptera and cynipid gall wasps on oaks in the Pacific North-west of the USA. All insect data for cynipines, unless otherwise stated. (Based on Opler 1974; Cornell & Washburn 1979.)

Region	Oak–insect combinations	Slope of species–area relationship	Variability explained (r^2, expressed as percentage)
Atlantic region	All oaks except two species with poor faunal lists	0.25	41
	Red oaks	0.15	28
	White oaks	0.24	17
California region	All oaks	0.63	33
	Quercus oaks only	0.72	72
	All oaks minus leaf-miners	0.47	90
	Leaf-miners plus cynipines	0.56	51
	Quercus minus leaf-miners	0.46	91
	Leaf-miners plus cynipines	0.62	80

factors discussed in section 6.1.1, namely size (smaller trees are 'riskier' environments for colonization by specialists), resource availability and apparency (bigger trees are easier to find, especially if the herbivore has relatively poor dispersal capacity). The above studies therefore provide examples of how both local and regional characteristics influence species richness of a given host genus. Manipulation of these characteristics by removing or planting tree species could, over time, have a marked influence on total species diversity in a habitat.

A factor not amenable to manipulation but which may nevertheless have a major effect on both trees and their associated invertebrate faunas is climate. Much of the wide range in plant and animal diversity from the poles to the tropics is due to climatic variation, such that combinations of temperature, rainfall and sunlight suitable for the widest range of species tend to increase towards the equator (Whittaker 1975). Although this pattern holds in its widest sense, there are regional differences in climate suitability arising from altitude, local topography and proximity to major water courses. Thus, climate exerts its effects through determining the growing season of an organism, which in turn determines the reproductive rate and potential of that organism. Insects associated with trees will therefore have an intrinsic relative growth rate (RGR), usually quantified in day-degrees of tem-

perature above a metabolic threshold. For many invertebrates, the relationship between rate of development and temperature is linear above the developmental threshold and thus provides a useful predictor of activity, especially in relation to predicting when a pest will appear and potentially cause damage.

Climate change due to atmospheric pollution, including increased levels of CO_2 and the presence of sulphur and nitrogen compounds in the atmosphere, is predicted to have significant effects on both the host trees and the herbivores on those trees (see also Chapters 8, 10 and 11 in this volume). Various predictive models have been produced for both regional and world scales; for example, a detailed model for western Europe predicts that elevated CO_2 will lead to a rise of 0.2°C per decade in the Atlantic Offshore region and up to 0.35°C per decade in eastern Europe, with accompanying changes in rainfall and wind patterns (Hadley Centre 1992). Effects on host trees are likely to be manifested by increased uptake and sequestration of carbon, possibly at the expense of nitrogen concentration in foliage (Docherty *et al.* 1997). In the case of gypsy moth feeding on aspen that had been exposed to elevated CO_2, levels of phenolic glycosides in foliage increased while those of total nitrogen decreased, leading to lower growth rates of larvae and increased consumption of foliage (Lindroth *et al.* 1997). These findings mirror those of a number of studies (reviewed in Docherty *et al.* 1997), although the wide variation in responses indicates that the effects may be tree and site specific. Elevated temperatures will also act on performance of both plant and invertebrate, changing both absolute growth rates and, perhaps more importantly, the synchrony between host plant suitability for feeding and presence of the appropriate stage of the insect. Those insects with overwintering stages that require quite precise coincidence of emergence with budburst of their host trees may be particularly vulnerable to changes in the incidence of frosts, especially if emergence is driven more by daylength than by temperature *per se* (Dewar & Watt 1992). Diapause termination, triggered by a specific period of chilling below a subzero threshold temperature, may also be affected so that emergence in the spring may be reduced, thus affecting overall population size. However, some insects presently constrained by low winter temperatures may show enhanced survival and increased reproduction with an increase in warmer winters. This is likely to be the case for green spruce aphid, *Elatobium abietinum*, which is normally killed by winter temperatures below $-8°C$ (Straw 1995).

Pollutants, such as ozone, SO_2 and NO_x, may have direct effects on tree performance and both indirect and direct effects on insect growth (Brown 1995). Ozone results in amino acid changes within trees that can provide a richer food source for phloem-feeding insects in particular, potentially leading to higher growth and reproductive rates (Bolsinger & Fluckiger 1989). However, recent studies by Holopainen *et al.* (1997) indicated that ozone episodes were less important than increased levels of nitrogen fertilizers in determining the performance of the aphids *Schizolachnus pineti* and *Cinara pinea* on Scots pine and *Cinara pilicornis* on Norway spruce. These authors concluded that occasional episodes of ozone may increase amino acid availability but, generally, the effects from predicted increases in atmospheric ozone were unlikely to be significant. In considering effects on herbivores as a whole, several studies have demonstrated the high sensitivity of sap-sucking insects to the effects of atmospheric pollutants such that elevated levels of SO_2 and NO_x give rise to greater

RGR and hence reproductive success, for example *E. abietinum* responded to increased concentrations of SO_2 with increased RGR when exposed on Sitka spruce (Warrington & Whittaker 1990). Foliage-chewing insects have also been shown to respond to atmospheric pollutants but with rather inconsistent trends. Scots pine seedling trees subjected to elevated SO_2 had higher sulphur, decreased calcium and higher levels of free amino acids and were shown to reduce the larval performance of pine looper moth, *Bupalus piniaria* (Lepidoptera: Geometridae), but had no effect on pine lappet moth, *Dendrolimus pini* (Lepidoptera: Lasiocampidae) (Katzel & Moller 1993). The authors attributed the different responses to major reductions in glucose and fructose, which adversely affected *B. piniaria* whereas *D. pini* was able to tolerate the lower carbohydrate levels. The foliage of mountain birch, *Betula pubescens* ssp. *czercpanova*, subjected to experimentally enhanced levels of sulphur and nitrogen over a 10-year period had no effect on autumnal moth, *Epirrita autumnata* (Lepidoptera: Geometridae), feeding on the foliage (Suomela & Neuvonen 1997).

Overall, despite high variability, it would appear that atmospheric pollutants have a neutral to positive effect on insect performance, although there are interactions with temperature, nutrient levels, etc. that can induce different effects even within the same insect species. Even when insect performance is impaired (e.g. elevated CO_2), the net effect may be increased damage as feeding rate increases to compensate for reduced food quality. This complex area of herbivore–plant dynamics clearly requires further work to determine patterns and causes.

6.4 FACTORS INFLUENCING PROCESSES AND OPPORTUNITIES FOR MANIPULATION

The array of processes considered in the early part of this chapter range from those reflecting different niches within a single tree, through local scales to the ultimate position of trees as major landscape components. It is also apparent that there are complex multitrophic interactions between trees and their herbivores that make it difficult to generalize concerning potential to manipulate processes. Management at the tree or stand level will involve local processes that reduce some of the complexity, such that gross climate effects may be removed as driving variables, although even here there may be microclimate effects at very small scales. Effects at this scale are manifested through the intrinsic suitability of the local population of trees, thus bringing in structural, nutritional and defensive qualities (see section 6.2.1) as well as the local impacts of natural enemies (see section 6.2.2). The interactions with wider-scale attributes are illustrated schematically in Fig. 6.4. Extrinsic effects act at a range of scales, from local (within the forest), through regional to geographical, such that the effects on a given tree population become less but the wider attributes of climate, topography and site characteristics come into play. This will have the effect of changing the absolute and relative abundances of tree species and hence their associated invertebrate species over wide areas. The examples of leaf-miners and gall-wasps from the full latitudinal range of oak species in California described in section 6.3 serve to illustrate this principle. In summary, therefore, any population of herbivores associated with trees

is actually influenced by the herbivore–tree interface so that, provided the herbivore is able to find the host tree, its ultimate performance is driven by local factors that may in turn be influenced by the interaction parameters shown in Fig. 6.4. Management at this local scale will reflect the initial components of tree species choice and location (i.e. the suitability of the site for tree growth and ultimate aim of the crop). This, of course, applies particularly to artificial forests rather than natural forests, where tree species mix and stage are entirely driven by site and local climate characteristics. Once trees are established, the elements of further management that can influence susceptibility to herbivore colonization are addition of fertilizers, particularly nitrogen, and thinning regimes that will affect proximity to other trees and plants and generate open spaces and habitat corridors. All other factors will be extrinsic (proximity to other forests, topography, climate, etc.), with decreasing influence with both distance and time.

The influence of given factors is therefore a question of scale. If these are at the landscape scale, then it is possible to view forest management at a more remote level so that a high degree

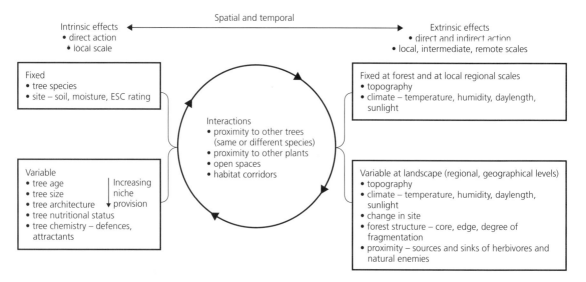

Fig. 6.4 A conceptual framework for the effects of scale and degree of fragmentation on interactions between herbivores and trees. ESC, Ecological Site Classification.

of fragmentation at a number of area scales can be considered and thus their relative influences on local herbivore–tree interactions measured at any fixed point within the wider landscape. This conceptual approach is now being used to increase understanding of the wider processes affecting invertebrate interactions with trees, making use of geographic information systems (GIS), new geostatistical methods and improved invertebrate sampling regimes to assess populations over relatively wide areas (see Stork *et al.* 1997a for a number of review articles focusing on canopy arthropods, especially in relation to tropical communities). Further impetus to such approaches has been provided by the various biodiversity conventions (UNCED, Rio, 1992; Ministerial Conference on the Protection of European Forests, Helsinki, 1993) and the implementation of biodiversity action plans at individual country levels (e.g. Sustainable Forestry, The UK Programme, delivered to Parliament in January 1994). The principles described in this chapter apply equally to conservation of invertebrate diversity and abundance, thus embracing enhancement of biodiversity and avoidance of invertebrate outbreaks within the same conceptual framework.

6.4.1 Forest fragmentation at local and landscape scales

Figure 6.4 provides a framework for considering invertebrate presence at a range of scales, acting through fixed and variable parameters. Among the parameters, the degree of fragmentation of a forest is one of the key variables that can be subject to human influence, both deliberate and accidental. It is also one of the variables most subject to natural disturbances over recent and geological time-scales. There is increasing interest in fragmentation in relation to tropical forests because of the extremely rapid loss of habitat through intensive logging. For example, models have been constructed to take account of the degree of fragmentation of tropical forests in relation to core area, edge effects and the shape of the forest area (Laurance & Yensen 1991). Further refinement, including provision of spatiotemporal components and the degree of connectivity

between fragments, has been provided by Didham (1997). This synopsis included:

• area effects, taking account of total area and interior core area which is not influenced by any edge effects;
• edge effects, especially the distance over which the edge has an influence relative to the core;
• shape of the fragment, which clearly affects both edge and core elements;
• degree of spatiotemporal isolation so that the time since fragmentation is taken into account;
• degree of habitat connectivity in the landscape, thus providing corridors for movement of species through otherwise unsuitable habitats.

Studies of the relative influences of edge to core in forests have been carried out in both temperate and tropical forest ecosystems, using a range of sampling methods. In general, it has been noted that invertebrate densities, of both individuals and species, tend to decline with distance from a forest edge to the core. Thus various beetle families show greater species richness or abundance at the edges of forests: Carabidae (Duelli *et al.* 1990), Cerambycidae (McCann & Harman 1990), Staphylinidae (Buse & Good 1993) and mixed families of beetles (Baldi & Kisbenedek 1994). Although not sampled in relation to direct interactions with host trees, the bark beetle *Ips typographus* tends to colonize edges more frequently than forest interiors (Sanders 1987). Patterns of abundance from edges to core woodland have been assumed to be a smooth gradation but recent studies on arthropod communities in Amazonian forests by Didham (1997) have demonstrated a bimodal form to relationships of edge to core abundances of leaf-litter invertebrates. This effect was compounded with the area of the forest fragment being studied, leading to three-dimensional relationships between fragment area, distance from forest edge and abundance of selected beetle species. Within the Amazonian forest system, the effect of fragmentation was felt even within large remnant forest blocks so that an area considerably greater than 100 ha, possibly approaching 1000 ha, was required before the invertebrate fauna matched that of undisturbed forests (Didham 1997).

The implications of results such as these are difficult to extrapolate to the heavily managed forest landscapes of Europe and, increasingly, North America and elsewhere. Forest fragmentation can now be regarded as the norm in such situations and therefore the invertebrate faunas in such fragments may already be greatly impoverished compared with those in original undisturbed forests. However, the general principles of maintaining the ratio of edge to core to ensure that the habitat requirements of forest specialists as well as edge colonists are maintained is a sound strategy. Unfortunately, the biological attributes of edges and cores are not clearly defined and thus the type of edge may be as important as the edge itself. For example, studies by Ozanne *et al.* (1997) in canopies of managed conifer forests in Britain showed a lower species richness of various invertebrate taxa at the edges compared with the cores of forests studied. This apparent contradiction to the studies cited earlier may reflect the type of edge, which in managed conifer plantations is often a very abrupt transition from the surrounding open ground area. The authors suggested that a gradation of the edge, with scrubby transition zones, could give rise to greater species richness at the edges but this would probably be at the expense of forest canopy specialists. Such uncertainty reinforces the need for more detailed studies of the mechanistic basis for observed effects of forest fragmentation, particularly taking account of the wide range of ecological factors that have been discussed in this chapter. Studies such as those of Waltz and Whitham (1997) provide useful data on the characteristics of individual trees that may influence edge and core populations of particular herbivores. They studied the effects of canopy structure in clonal poplars in relation to arthropod communities dominated by the gall-forming aphid *Pemphigus betae* and the leaf-feeding beetle *Chrysomela confluens*. There were direct correlations between the mature ramets of the trees and invertebrate species richness and this varied with the clone studied. The authors concluded that such within-tree variability in resource availability explained the higher biodiversity in mixed-aged stands and at edges compared with interiors of even-aged stands.

As Stork *et al.* (1997b) have pointed out, the understanding of forest arthropod dynamics, especially canopy-dwelling arthropods, is hampered by the tendency to gather observational data rather than by rigorous hypothesis testing. They also correctly point out the sheer difficulty of carrying out such studies, graphically illustrated by the 2-year effort by over 20 taxonomists at the Natural History Museum, London to sort and partially identify samples collected over a mere 12-day sampling period in Borneo. The underlying ecological principles that govern invertebrate communities of forests are clearly complex and yet unless they are incorporated into studies of forest fragmentation, there is little prospect of developing a sound science of both pest management and invertebrate conservation for the future.

6.4.2 Forest disturbance as a factor in invertebrate dynamics

The processes described above emphasize that size and scale matter in relation to communities of forest invertebrates. It was also stated that the areas of natural forest remnants are progressively shrinking and this is likely to have an impact on associated invertebrates. Much of the disturbance that leads to this extensive fragmentation is a result of human activity, directly or indirectly (Vitousek 1994). Anthropogenic influences such as aerial pollution and climate change are having an increasingly global effect that will take many years to influence trees and their invertebrate communities. Prediction of these effects is imprecise because of the relatively poor understanding of the processes involved and the difficulties, mentioned above, of tracking changes against an imperfect baseline.

Some indication of the effects of habitat fragmentation have been provided in studies of the outbreaks of some of the more damaging forest pests and these serve to illustrate the complexity of managing change. Edge effects are well known in relation to initial infestations of several damaging bark beetles, particularly in relation to felling programmes that can provide breeding resources as well as opening the remaining forest to the risks of windthrow, etc. *Ips typographus*, the most damaging bark beetle in western Europe, is sustained at low densities in most forest locations by reliance on sporadic availability of weakened trees for initial breeding. However, if a sudden increase in breeding material is brought about by felling programmes or wind damage, populations can build up rapidly and attack surrounding living trees (Christiansen & Bakke 1988). Such eruptive patterns of attack are common in different species of bark beetle and the triggering of the infestation is frequently linked to large-scale provision of suitable breeding material on-site, either through human intervention or via natural effects such as fire or severe winds. A good example of such an effect was the major fire in Yellowstone Park, USA during 1988 which left a mosaic of fire-killed, fire-damaged and undamaged areas in the area (Rasmussen *et al.* 1996). Surveys showed that 12.7% of Douglas fir (*Pseudotsuga menziesii*), 17.9% of lodgepole pine (*Pinus contorta*), 6.6% of Engelmann spruce (*Picea engelmannii*), 7.5% of subalpine fir (*Abies lasiocarpa*) and 2.8% of whitebark pine (*Pinus albicaulis*) were killed following the build-up of bark- and wood-boring beetles in fire-damaged and stressed trees. The unexpected and severe nature of the Yellowstone fires also illustrates one of the adverse consequences of fire-management regimes that have been in place in the USA since the early 1900s. Fire was previously a normal component of forest ecosystems in the USA and Canada and had a considerable influence on species composition, succession and forest stability (McCullough *et al.* 1998). Regimes of fire suppression have changed the structure of many forests and, in some cases, have led to insect outbreaks. A particularly good example of such an effect is the balsam fir and white spruce forest ecosystem of eastern Canada and the north-eastern USA. Fire suppression was started in 1920 and has been accompanied by increased frequency, intensity and scale of spruce budworm, *Choristoneura fumiferana* (Lepidoptera: Tortricidae), outbreaks (Blais 1983). Apart from any direct effects on the budworm, fire suppression altered succession by reducing pioneer species such as aspen, birch, jack pine and black spruce, thus decreasing the species mix that previously broke up the highly susceptible fir and spruce–fir stands. Recent strategies have been to return to selective use of prescribed burns to help restore the previous forest mosaics and thus to

help combat spruce budworm outbreaks (McRae 1994). McCullough *et al.* (1998) cite several other examples of linkage of insect outbreaks to fire suppression regimes and conclude that an improved understanding of the ecology of fire–tree–insect interactions will help to identify strategies to manage pests and to maintain biodiversity in northern forest ecosystems.

6.5 UNDERSTANDING THE FOREST AS A BASIS FOR MANAGEMENT

There have been few published attempts to bring together the various elements that have formed the basis of this chapter in developing rational forest management regimes to provide a balance between pest infestations and enhancement of biodiversity. The need for such balanced approaches is now acute as multipurpose forestry becomes the norm in both developed and developing countries. The interactions summarized in Fig. 6.4 serve to emphasize that it is not sufficient to deal with the relative simplicity of the forest block or stand and that a perception of the effects of landscape scales must also be considered. It is fortunate that the tools to allow such approaches to be attempted are now available and are becoming increasingly sophisticated. GIS, often combined with global positioning systems (GPS), and a range of remote sensing techniques have revolutionized the ways in which landscape and geographical features can be incorporated into spatially explicit models of insect–tree interactions (Liebhold *et al.* 1993). The STS (Slow The Spread) process for gypsy moth in the USA (see Volume 2, Chapter 8) provides a good example of the use of spatiotemporal modelling to assess and predict the effects of management regimes on the spread of the pest into new areas (Sharov & Liebhold 1998). Host plant availability and rates of natural spread as well as incorporation of landscape features (Liebhold *et al.* 1994) have enabled refinements in prediction and management for the gypsy moth STS process. Optimization has led to reduction in the costs of the programme by lowering the numbers of pheromone traps and associated control measures at increasing distances from the advancing front.

Aspects of stand and forest management

approaches can also be applied to conservation strategies. The basic tenets of interactions between insects and their forest environment remain the same, although the aims will be very different and will require a balanced approach in order to avoid creating conditions for insect outbreaks while attempting to enhance a particular species or mix of invertebrate species. Such problems are particularly acute in management of old-growth forests, such as the Sitka spruce forests in the Pacific North-west of North America (Winchester 1997). Using various trapping methods, abundance and species richness of arthropods were sampled in forest canopies, forest floors, transition zones and in clearcut areas, with emphasis on the beetle family Staphylinidae. Results from Malaise trapping of flying staphylinid beetles are summarized in Fig. 6.5.

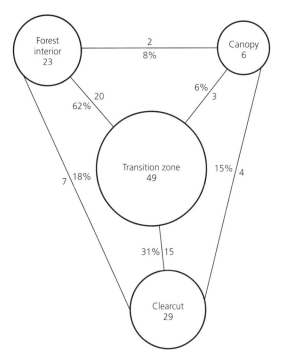

Fig. 6.5 Numbers of staphylinid beetle species in old-growth Sitka spruce forests in British Columbia in relation to disturbance. Numbers within circles are staphylinid species while those on the connecting lines are numbers and percentages of species in common between the forest categories. (Adapted from Winchester 1997.)

Although the greatest number of species was found in the disturbed transition zone, the fact that 62% of these were linked to forest interior species indicates the importance of this zone as a source of species. Principal components analysis confirmed the separation of the various zones and also confirmed the unique nature of the old-growth forest floor staphylinid fauna. In addition, canopy-dwelling oribatid mites were highly specific to old-growth canopies and it was concluded that following clearcut second-growth canopies were unlikely to provide the specialized habitat required for these species within the rotation period (up to 120 years) (Winchester & Ring 1996).

Another important habitat in old-growth forests is dead wood, which provides a specialized resource for saproxylic invertebrates. Comparison of the saproxylic Coleoptera of Scots pine and Norway spruce in old-growth and managed forests in central Finland provided some interesting contrasts between species numbers and rarity value (Vaisanen *et al.* 1993). Although numbers of both species and individuals were higher in managed forests (78 species and 1302 individuals compared with 55 species and 512 individuals), this was due mainly to the presence of bark beetles (Scolytidae). However, when species classified as rare were considered, the proportion in the samples was only 1% in managed forests compared with 12% in the old-growth forests. The authors concluded that dead-wood faunas in managed forests, although providing high biodiversity value, were virtually completely different from those of old-growth forests. This was almost entirely at the expense of relatively rare old-growth dead-wood specialists that are not likely to colonize managed forests, particularly if the degree of fragmentation of old-growth forests is large. Such findings are repeated in other studies of old-growth and managed forests, confirming that it is not just the absolute amount of dead wood that is important but also the resource pool in the surrounding area, especially for specialist saproxylic species with limited dispersal potential (Speight 1989; Albrecht 1991).

The processes that determine the interactions between forests and their associated invertebrate faunas are therefore complex and it is important that the implications of forest management practices, for whatever purpose, are taken into account in assessing potential impacts on invertebrate species richness and abundance. Some of the management options that can be considered at a range of scales have been alluded to in this chapter. It is not always apparent whether forest pest management in practice has taken account of the lessons that can be learned from existing ecological knowledge and principles. However, the fact that tools such as GIS, GPS and portable computers are now available should provide the impetus to develop more context-sensitive and realistic decision support systems for future management of forest invertebrates.

REFERENCES

Albrecht, L. (1991) Die Bedeutung des toten Holzes im Wald. *Forstwissenschaftliches Centralblatt* **110**, 106–13.

Anderson, R.M. & May, R.M. (1981) The population dynamics of microparasites and their invertebrate hosts. *Philosophical Transactions of the Royal Society of London B* **291**, 451–524.

Andreadis, T.G. (1987) Transmission. In: Fuxa, J.R. & Tanada, Y., eds. *Epizootiology of Insect Diseases*, pp. 159–76. John Wiley & Sons, New York.

Appel, H.M. & Schultz, J.C. (1994) Oak tannins reduce effectiveness of thuricide (*Bacillus thuringiensis*) in the gypsy moth (Lepidoptera: Lymantriidae). *Journal of Economic Entomology* **87**, 1736–42.

Baldi, A. & Kisbenedek, T. (1994) Comparative analysis of edge effects on bird and beetle communities. *Acta Zoologica Hungarica* **40**, 1–14.

Begon, M., Harper, J.L. & Townsend, C.R. (1996) *Ecology. Individuals, Populations and Communities*. Blackwell Science, Oxford.

Bevan, D. (1987) *Handbook 1. Forest Insects. A Guide to Insects Feeding on Trees in Britain*. Forestry Commission, Edinburgh.

Birks, H.J.B. (1980) British trees and insects: a test of the time hypothesis over the last 13 000 years. *American Naturalist* **115**, 600–5.

Blais, J.R. (1983) Trends in the frequency, extent and severity of spruce budworm outbreaks in eastern Canada. *Canadian Journal of Forest Research* **13**, 539–47.

Bolsinger, M. & Fluckiger, W. (1989) Ambient air pollution induced changes in amino acid pattern of phloem sap in host plants: relevance to aphid infestation. *Environmental Pollution* **56**, 209–16.

Brixey, J. (1997) *The potential for biological control to reduce* Hylobius abietis *damage*. Forestry Commission Research Information Note 273, pp. 1–6.

Brown, J.H. (1988) Species diversity. In: Myers, A.A. & Giller, P.S., eds. *Analytical Biogeography*, pp. 57–89. Chapman & Hall, London.

Brown, J.H. & Gibson, A.C. (1983). *Biogeography*. C.V. Mosby, St Louis.

Brown, V.C. (1995) Insect herbivores and gaseous air pollutants: current knowledge and predictions. In: Harrington, R. & Stork, N., eds. *Insects in a Changing Environment: 17th Symposium of the Royal Entomological Society*, pp. 219–49. Academic Press, London.

Bryant, J.P., Chapin, F.S. & Klein, D.R. (1983) Carbon/nutrient balance of boreal plants in relation to vertebrate herbivory. *Oikos* **40**, 357–68.

Bryant, J.P., Reichardt, P.B., Clausen, T.P. & Werner, R.A. (1993) Effects of mineral nutrition on delayed induced resistance in Alaska paper birch. *Ecology* **74**, 2072–84.

Buse, A. & Good, J.E.G. (1993) The effects of conifer forest design and management on abundance and diversity of rove beetles (Coleoptera: Staphylinidae): implications for conservation. *Biological Conservation* **64**, 67–76.

Byers, J.A., Lanne, B.S., Lofqvist, J., Schlyter, F. & Bergstrom, G. (1985) Olfactory recognition of host-tree susceptibility by pine shoot beetles. *Naturwissenschaften* **72**, 324–6.

Campbell, R.W. & Sloan, R.J. (1977) Natural regulation of innocuous gypsy moth populations. *Environmental Entomology* **6**, 315–22.

Campbell, R.W. & Sloan, R.J. (1978) Natural maintenance and decline of gypsy moth outbreaks. *Environmental Entomology* **7**, 389–95.

Christiansen, E. & Bakke, A. (1988) The spruce bark beetle of Eurasia. In: Berryman, A.A., ed. *Dynamics of Forest Insect Populations: Patterns, Causes, Implications*, pp. 479–503. Plenum Press, New York.

Claridge, M.F. & Evans, H.F. (1990) Species–area relationships: relevance to pest problems of British trees? In: Watt, A.D., Leather, S.R., Hunter, M.D. & Kidd, N.A.C., eds. *Population Dynamics of Forest Insects*, pp. 59–69. Intercept, Andover, UK.

Claridge, M.F. & Wilson, M.R. (1981) Host plant associations, diversity and species–area relationships of mesophyll-feeding leafhoppers of trees and shrubs in Britain. *Ecological Entomology* **6**, 217–38.

Comins, H.N., Hassell, M.P. & May, R.M. (1992) The spatial dynamics of host–parasitoid systems. *Journal of Animal Ecology* **61**, 735–48.

Connor, E.F. (1986) The role of Pleistocene forest refugia in the evolution and biogeography of tropical biotas. *Trends in Ecology and Evolution* **1**, 165–9.

Cornell, H.V. (1986) Oak species attributes and host size

influence cynipine wasp species richness. *Ecology* **67**, 1582–92.

Cornell, H.V. & Washburn, J.O. (1979) Evolution of the richness–area correlation for cynipid gall wasps on oak trees: a comparison of two geographic areas. *Evolution* **33**, 257–74.

Crawley, M.J. (1983) *Herbivory: the Dynamics of Animal–Plant Interactions*. Blackwell Scientific Publications, Oxford.

Crawley, M.J. (1989) Insect herbivores and plant population dynamics. *Annual Review of Entomology* **34**, 531–64.

Dahlsten, D.L. & Berisford, C.W. (1995) Diversity of bark beetle natural enemies. In: Hain, F.P., Salom, S.M., Ravlin, W.F., Payne, T.L. & Raffa, K.F., eds. *Behavior, Population Dynamics and Control of Forest Insects*, pp. 184–201. Ohio State University & USDA Forest Service, Radnor, Pennsylvania.

Dempster, J.P. & Pollard, E. (1986) Spatial heterogeneity, stochasticity and the detection of density dependence in animal populations. *Oikos* **46**, 413–16.

Dewar, R.C. & Watt, A.D. (1992) Predicted changes in the synchrony of larval emergence and budburst under climatic warming. *Oecologia* **89**, 557–9.

Diamond, J.M. (1975) Distributional ecology of New Guinea birds. *Science* **179**, 759–69.

Didham, R.K. (1997) An overview of invertebrate responses to forest fragmentation. In: Watt, A.D., Stork, N. & Hunter, M.D., eds. *Forests and Insects*, pp. 303–20. Chapman & Hall, London.

Docherty, M., Salt, D.T. & Holopainen, J.K. (1997) The impacts of climate change and pollution on forest pests. In: Watt, A.D., Stork, N. & Hunter, M.D., eds. *Forests and Insects*, pp. 229–47. Chapman & Hall, London.

Dressler, R.L. (1985). *The Orchids: Natural History and Classification*. Harvard University Press, Cambridge, Massachussetts.

Duelli, P., Studer, M., Marchand, I. & Jakob, S. (1990) Population movements of arthropods between natural and cultivated areas. *Biological Conservation* **54**, 193–207.

Entwistle, P.F. & Evans, H.F. (1985) Viral control. In: Kerkut, I. & Gilbert, L.I., eds. *Comprehensive Insect Physiology, Biochemistry, and Pharmacology*, pp. 347–412. Pergamon Press, Oxford.

Entwistle, P.F., Forkner, A.C., Green, B.M. & Cory, J.S. (1993) Avian dispersal of nuclear polyhedrosis viruses after induced epizootics in the pine beauty moth, *Panolis flammea* (Lepidoptera: Noctuidae). *Biological Control* **3**, 61–9.

Evans, H.F. (1997) The role of microbial insecticides in forest pest management. In: Evans, H.F., ed. *Microbial Insecticides: Novelty or Necessity?*, pp. 29–40. British Crop Protection Council, Farnham.

Evans, H.F. & Fielding, N.J. (1996) Restoring the natural balance: biological control of *Dendroctonus micans* in Great Britain. In: Waage, J.K., ed. *Biological Control Introductions: Opportunities for Improved Crop Production*, pp. 45–57. British Crop Protection Council, Farnham.

Evans, H.F. & Harrap, K.A. (1982) Persistence of insect viruses. In: Mahy, W.J., Minson, A.C. & Darby, G.K., eds. *Virus Persistence*, pp. 57–96. Cambridge University Press, Cambridge.

Feeny, P. (1976) Plant apparency and chemical defence. *Recent Advances in Phytochemistry* **10**, 1–40.

Flenley, J. (1999) The origins of diversity in tropical rain forests. *Trends in Ecology and Evolution* **8**, 119–20.

Foster, M.A., Schultz, J.C. & Hunter, M.D. (1992) Modelling gypsy moth-virus–leaf chemistry interactions: implications of plant quality for pest and pathogen dynamics. *Journal of Animal Ecology* **61**, 509–20.

Fraser, S.M. & Lawton, J.H. (1994) Host range expansion by British moths introduced onto conifers. *Ecological Entomology* **19**, 127–37.

Fuxa, J.R. & Tanada, Y. (eds) (1987) *Epizootiology of Insect Diseases*. John Wiley & Sons, New York.

Godfray, H.C. (1994). *Parasitoids: Behavioral and Evolutionary Ecology*. Princeton University Press, Princeton, New Jersey.

Hadley Centre (1992) *The Hadley Centre Transient Climate Change Experiment 1. A numerical experiment in which atmospheric concentrations of carbon dioxide were increased at 1% per annum (compound) for 75 years*. Meteorological Office, Bracknell.

Haffer, J. (1969) Speciation in Amazonian forest birds. *Science* **165**, 131–137.

Haffer, J. (1982) General aspects of the refuge theory. In: Prance, G.T., ed. *Biological Differentiation in the Tropics*, pp. 6–24. Columbia University Press, New York.

Hajek, A.E., Larkin, T.S., Carruthers, R.I. & Soper, R.S. (1993) Modeling the dynamics of *Entomophaga maimaiga* (Zygomycetes, Entomophthorales) epizootics in gypsy moth (Lepidoptera, Lymantriidae) populations. *Environmental Entomology* **22**, 1172–87.

Harrison, S. & Cappuccino, N. (1995) Using density manipulation experiments to study population regulation. In: Cappuccino, N. & Price, P.W., eds. *Population Dynamics: New Approaches and Synthesis*, pp. 131–47. Academic Press, San Diego.

Hassell, M.P. (1980) Foraging strategies, population models and biological control: a case study. *Journal of Animal Ecology* **49**, 603–28.

Hassell, M.P. & Waage, J.K. (1984) Host–parasitoid population interactions. *Annual Review of Entomology* **29**, 89–114.

Haukioja, E. & Honkanen, T. (1996) Why are tree responses to herbivory so variable? Mattson, W.J., Niemela, P. & Rousi, M., eds. *Dynamics of Forest*

Herbivory: Quest for Pattern and Principle, pp. 1–10. USDA Forest Service, St Paul, Minnesota.

Haukioja, E. & Honkanen, T. (1997) Herbivore-induced responses in trees: internal vs. external explanations. In: Watt, A.D., Stork, N. & Hunter, M.D., eds. *Forests and Insects*, pp. 69–80. Chapman & Hall, London.

Hawkins, C.P. & MacMahon, J.A. (1989) Guilds: the multiple meanings of a concept. *Annual Review of Entomology* **34**, 423–51.

Heliovaara, K. & Vaisanen, R. (1995) Bark beetles, forest fragmentation and satellite imagery. In: Hain, F.P., Salom, S.M., Ravlin, W.F., Payne, T.L. & Raffa, K.F., eds. *Behavior, Population Dynamics and Control of Forest Insects*, pp. 154–63. Ohio State University & USDA Forest Service, Radnor, Pennsylvania.

Holopainen, J.K., Kainulainen, P. & Oksanen, J. (1997) Growth and reproduction of aphids and levels of free amino acids in Scots pine and Norway spruce in an open-air fumigation with ozone. *Global Change Biology* **3**, 139–47.

Honkanen, T. & Haukioja, E. (1994) Why does a branch suffer more after branch-wide than after tree-wide defoliation? *Oikos* **71**, 441–50.

Hunter, A.F. (1991) Traits that distinguish outbreaking and nonoutbreaking Macrolepidoptera feeding on northern hardwood trees. *Oikos* **60**, 275–82.

Hunter, M.D. (1992) A variable insect–plant interaction: the relationship between tree budburst phenology and population levels of insect herbivores among trees. *Ecological Entomology* **17**, 91–5.

Hunter, M.D. (1997) Incorporating variation in plant chemistry into a spatially explicit ecology of phytophagous insects. In: Watt, A.D., Stork, N. & Hunter, M.D., eds. *Forests and Insects*, pp. 81–96. Chapman & Hall, London.

Hunter, M.D. & Schultz, J.C. (1995) Fertilization mitigates chemical induction and herbivore responses within damaged oak trees. *Ecology* **76**, 1226–32.

Ignoffo, C.M. (1992) Environmental factors affecting persistence of entomopathogens. *Florida Entomologist* **75**, 516–25.

Jeffords, M.R., Maddox, J.V., McManus, M.L., Webb, R.E. & Wieber, A. (1989) Evaluation of the overwintering success of two European Microsporidia inoculatively released into gypsy moth populations in Maryland. *Journal of Invertebrate Pathology* **53**, 235–40.

Jeffries, M.J. & Lawton, J.H. (1984) Enemy-free space and the structure of ecological communities. *Biological Journal of the Linnean Society* **23**, 269–86.

Katzel, R.V. & Moller, K. (1993) The influence of SO_2-stressed host plants on the development of *Bupalus piniarius* L. (Lep., Geometridae) and *Dendrolimus pini* L. (Lep., Lasiocampidae). *Journal of Applied Entomology* **116**, 50–61.

Kennedy, C.E.J. & Southwood, T.R.E. (1984) The number of species of insects associated with British trees: a re-analysis. *Journal of Animal Ecology* **53**, 455–78.

Kidd, N.A.C. & Jervis, M.A. (1997) The impact of parasitoids and predators on forest insect populations. In: Watt, A.D., Stork, N. & Hunter, M.D., eds. *Forests and Insects*, pp. 49–68. Chapman & Hall, London.

Klimetzek, D. (1993) Baumarten und ihre Schadinsekten auf der Nordhalbkugel. *Mitteilungen der Deutschen Gesellschaft Fur Allgemeine und Angewandte Entomologie* **8**, 505–9.

Koricheva, J., Larsson, S. & Haukioja, E. (1998) Insect performance on experimentally stressed woody plants: a meta-analysis. *Annual Review of Entomology* **43**, 195–216.

Laurance, W.F. & Yensen, E. (1991) Predicting the impacts of edge effects in fragmented habitats. *Biological Conservation* **55**, 77–92.

Lawton, J.H. (1983) Plant architecture and the diversity of phytophagous insects. *Annual Review of Entomology* **28**, 23–39.

Liebhold, A.M., Rossi, R.E. & Kemp, W.P. (1993) Geostatistics and geographic information systems in applied insect ecology. *Annual Review of Entomology* **38**, 303–27.

Liebhold, A.M., Elmes, G.A., Halverson, J.A. & Quimby, J. (1994) Landscape characterization of forest susceptibility to gypsy moth defoliation. *Forest Science* **40**, 18–29.

Lindroth, R.L., Roth, S., Kruger, E.L., Volin, J.C. & Koss, P.A. (1997) CO_2-mediated changes in aspen chemistry: effects on gypsy moth performance and susceptibility to virus. *Global Change Biology* **3**, 279–89.

Locke, G.M. (1987) Census of woodlands and trees 1979–82. *Forestry Commission Bulletin* **63**, 1–123.

Lynch, J.D. (1988) Refugia. In: Myers, A.A. & Giller, P.S., eds. *Analytical Biogeography: an Integrated Approach to the Study of Animal and Plant Distributions*, pp. 311–42. Chapman & Hall, London.

MacArthur, R.H. & Wilson, E.O. (1967) *The Theory of Island Biogeography*. Princeton University Press, Princeton, New Jersey.

McCann, J.M. & Harman, D.M. (1990) Influence of the intrastand position of Black Locust trees on attack rate of the Locust Borer (Coleoptera: Cerambycidae). *Annals of the Entomological Society of America* **83**, 705–11.

McCullough, D.G., Werner, R.A. & Neuman, D. (1998) Fire and insects in northern and boreal forest ecosystems of North America. *Annual Review of Entomology* **43**, 107–27.

McRae, D.J. (1994) Prescribed fire converts spruce budworm-damaged forest. *Journal of Forestry* **92**, 38–40.

Madden, J.L. (1988) *Sirex* in Australasia. In: Berryman, A.A., ed. *Dynamics of Forest Insect Populations:*

Patterns, Causes, Implications, pp. 407–29. Plenum Press, New York.

Merle, P.D. (1983) Phenologies comparees du chene pubescent, du chene vert et de *Tortrix viridana* L. (Lep., Tortricidae). Mise en evidence chez l'insecte de deux populations sympatriques adaptees chacune a l'un des chenes. *Acta Oecologica, Oecologia Applicata* **4**, 55–74.

Moore, N.F. & Eley, S.M. (1991) Picornaviridae: picornaviruses of invertebrates. In: Adams, J.R. & Bonami, J.R., eds. *Atlas of Invertebrate Viruses*, pp. 371–86. CRC Press, Boca Raton, Florida.

Nair, K.S.S. (1988) The teak defoliator in Kerala, India. In: Berryman, A.A., ed. *Dynamics of Forest Insect Populations: Patterns, Causes, Implications*, pp. 267–89. Plenum Press, New York.

Opler, P.A. (1974) Oaks as evolutionary islands for leafmining insects. *American Scientist* **62**, 67–73.

Ozanne, C.M.P., Hambler, C., Foggo, A. & Speight, M.R. (1997) The significance of edge effects in the management of forests for invertebrate biodiversity. In: Stork, N., Adis, J. & Didham, R.K., eds. *Canopy Arthropods*, pp. 534–50. Chapman & Hall, London.

Pacala, S.W., Hassell, M.P. & May, R.M. (1990) Host–parasitoid associations in patchy environments. *Nature* **344**, 150–3.

Poland, T.M. & Borden, J.H. (1997) Attraction of a bark beetle predator, *Thanasimus undatulus* (Coleoptera: Cleridae), to pheromones of the spruce beetle and two secondary bark beetles (Coleoptera: Scolytidae). *Journal of the Entomological Society of British Columbia* **94**, 35–41.

Price, P.W. (1992) Plant resources as the mechanistic basis for insect herbivore population dynamics. In: Hunter, M.D., Ohgushi, T. & Price, P.W., eds. *Effects of Resource Distribution on Animal–Plant Interactions*, pp. 139–73. Academic Press, San Diego.

Rackham, O. (1990). *Trees and Woodland in the British Landscape*. Dent, London.

Rasmussen, L.A., Amman, G.D., Vandygriff, J.C., Oakes, R.D., Munson, A.S. & Gibson, K.E. (1996) *Bark beetle and wood borer infestation in the Greater Yellowstone area during four postfire years*. Research Paper no. INT-RP-487, Intermountain Research Station, USDA Forest Service, Logan, Utah.

Reader, P.M. & Southwood, T.R.E. (1981) The relationship between palatability to invertebrates and the successional status of a plant. *Oecologia* **51**, 271–5.

Rhoades, D.F. & Cates, R.G. (1976) Toward a general theory of plant antiherbivore chemistry. *Recent Advances in Phytochemistry* **10**, 168–213.

Robison, D.J. & Raffa, K.F. (1997) Effects of constitutive and inducible traits of hybrid poplars on forest tent caterpillar feeding and population ecology. *Forest Science* **43**, 252–67.

Roland, J. & Embree, D.G. (1995) Biological control of the winter moth. *Annual Review of Entomology* **40**, 475–92.

Rossiter, M.C., Schultz, J.C. & Baldwin, I.T. (1988) Relationships among defoliation, red oak phenolics, and gypsy moth growth and reproduction. *Ecology* **69**, 267–77.

Ruohomäki, K., Chapin, F.S., Haukioja, E., Neuvonen, S. & Suomela, J. (1996) Delayed inducible resistance in mountain birch in response to fertilization and shade. *Ecology* **77**, 2302–11.

Sanders, W. (1987) Studies on the activity of the engraver beetle *Ips typographus* in deciduous forests and in the open field. *Journal of Applied Entomology* **103**, 240–9.

Schultz, J.C. (1997) Factoring natural enemies into plant tissue availability to herbivores. In: Hunter, M.D., Ohgushi, T. & Price, P.W., eds. *Effects of Resource Distribution on Animal–Plant Interactions*, pp. 175–97. Academic Press, San Diego.

Sharov, A.A. & Liebhold, A.M. (1998) Model of slowing the spread of gypsy moth (Lepidoptera: Lymantriidae) with a barrier zone. *Ecological Applications* **8**, 1170–9.

Shirt, D.B. (1987) *British Red Data Books. 2. Insects*. Nature Conservancy Council, Peterborough.

Southwood, T.R.E. (1961) The number of species of insect associated with various trees. *Journal of Animal Ecology* **30**, 1–8.

Southwood, T.R.E. & Comins, H.N. (1976) A synoptic population model. *Journal of Animal Ecology* **45**, 949–65.

Southwood, T.R.E., Brown, V.K. & Reader, P.M. (1979) The relationships of plant and insect diversities in succession. *Biological Journal of the Linnean Society* **12**, 327–48.

Speight, M.C.D. (1989) *Saproxylic Invertebrates and their Conservation*. Council of Europe, Strasbourg.

Speight, M.R. & Wainhouse, D. (1989). *Ecology and Management of Forest Insects*. Oxford University Press, Oxford.

Stork, N.E., Adis, J. & Didham, R.K. (eds) (1997a) *Canopy Arthropods*. Chapman & Hall, London.

Stork, N., Didham, R.K. & Adis, J. (1997b) Canopy arthropod studies for the future. In: Stork, N., Adis, J. & Didham, R.K., eds. *Canopy Arthropods*, pp. 551–61. Chapman & Hall, London.

Straw, N.A. (1995) Climate change and the impact of the green spruce aphid, *Elatobium abietinum* (Walker), in the UK. *Scottish Forestry* **49**, 134–45.

Strong, D.R., Lawton, J.H. & Southwood, T.R.E. (1984) *Insects on Plants. Community Patterns and Mechanisms*. Blackwell Scientific Publications, Oxford.

Suomela, J. & Neuvonen, S. (1997) Effects of long-term simulated acid rain on suitability of mountain birch for *Epirrita autumnata* (Geometridae). *Canadian Journal of Forest Research* **27**, 248–56.

Thompson, J.N. & Pellmyr, O. (1991) Evolution of

oviposition behavior and host preference in Lepidoptera. *Annual Review of Entomology* **36**, 65–89.

Trewhella, K.E., Leather, S.R. & Day, K.R. (1997) The effect of constitutive resistance in lodgepole pine (*Pinus contorta*) and Scots pine (*P. sylvestris*) on oviposition by three pine feeding herbivores. *Bulletin of Entomological Research* **87**, 81–8.

Vaisanen, R., Bistrom, O. & Heliovaara, K. (1993) Subcortical Coleoptera in dead pines and spruces: is primeval species composition maintained in managed forests? *Biodiversity and Conservation* **2**, 95–113.

Vitousek, P.M. (1994) Beyond global warming: ecology and global change. *Ecology* **75**, 1861–76.

Waage, J.K. & Greathead, D.J. (1988) Biological control: challenges and opportunities. *Philosophical Transactions of the Royal Society of London B* **318**, 111–28.

Wainhouse, D. & Beech-Garwood, P. (1994) Growth and survival of *Dendroctonus micans* larvae on six species of conifer. *Journal of Applied Entomology* **117**, 393–9.

Waltz, A.M. & Whitham, T.G. (1997) Plant development affects arthropod communities: opposing impacts of species removal. *Ecology* **78**, 2133–44.

Warrington, S. & Whittaker, J.B. (1990) Interactions between Sitka spruce, the green spruce aphid, sulphur dioxide pollution and drought. *Environmental Pollution* **4**, 363–70.

Welch, R.C. & Greatorex-Davies, J.N. (1993) Colonization of two *Nothofagus* species by Lepidoptera in southern Britain. *Forestry* **66**, 181–203.

Whittaker, R.H. (1975). *Communities and Ecosystems.* Macmillan, London.

Williamson, M. (1988) Relationship of species number to area, distance and other variables. In: Myers, A.A. & Giller, P.S., eds. *Analytical Biogeography: an Integrated Approach to the Study of Animal and Plant Distributions*, pp. 91–115. Chapman & Hall, London.

Winchester, N.N. (1997) Arthropods of coastal old-growth Sitka spruce forests: conservation of biodiversity with special reference to the Staphylinidae. In: Watt, A.D., Stork, N. & Hunter, M.D., eds. *Forests and Insects*, pp. 365–79. Chapman & Hall, London.

Winchester, N.N. & Ring, R.A. (1996) Centinelan extinctions: extirpation of Northern Temperate old-growth rainforest arthropod communities. *Selbyana* **17**, 50–7.

Winter, T.G. (1988) *Insects and storm-damaged broadleaved trees.* Forestry Commission Research Information Note 133, Forestry Commission, Edinburgh.

Part 3
Environmental Interactions

Forests are both impacted by and have an impact on the environment in which they occur. Great advances in our understanding have occurred in recent decades, and the six chapters in this section attempt to present overviews of where the science has reached. As with other chapters, these analyses are presented by world authorities and are designed both to introduce a subject and reduce its complexity to make it accessible to the general reader.

While the emphasis here concerns the impact of different facets of the environment on forest growth and development, it usefully allows fuller consideration in Volume 2 of what effects changes in forests will have and what the impacts are of forest operations.

Peter Attiwill and Christopher Weston (Chapter 7) develop the theme of forest soils by focusing on the integrating role of soil organic matter, or the lack of it, as a key determinant. Their concluding remarks about harvesting and sustainability set the scene for Bob Powers's and Larry Morris's chapters in Volume 2 (5 and 7 respectively).

Dick Waring and Tony Ludlow (Chapter 8—Ecophysiology of Forests), Paul Jarvis and David Fowler (Chapter 10—Forests and The Atmosphere) and Peter Freer-Smith (Chapter 11—Environmental Stresses of Forests) together review the science of how trees and forests interact with the environment at different scales and from different perspectives. They are masterly overviews in fields of forest science which have seen enormous advance in recent years. Among these chapters, Hans Pretzsch's account of forest models (Chapter 9) is included immediately after Ecophysiology of Forests (Chapter 8) as a summary of the way in which foresters and forest scientists integrate the impacts of environmental influences at a stand scale, namely how is the growth of trees assessed in ways useful to management. In this sense this chapter bridges the two volumes of *The Forests Handbook*, as indeed does Freer-Smith's (Chapter 11).

Sampurno Bruijnzeel's (Chapter 12) overview of forest hydrology is a stand-alone précis of an enormous subject that very effectively concludes this large section on environmental interactions.

7: Forest Soils

PETER M. ATTIWILL AND CHRISTOPHER J. WESTON

7.1 INTRODUCTION

To write a chapter on forest soils in detail at a global level is an impossible task. To give the general reader an appreciation of the scope of the subject and indeed to acknowledge its historical roots, we list some major publications that have dominated the field. In the USA, an enormously successful book on soil science has been *The Nature and Properties of Soils*, a text that has evolved with the science over 11 editions (Brady 1996), starting from the earliest by T.L. Lyon and H.O. Buckman in the 1920s. Similarly in the UK, E.J. Russell's *Soil Conditions and Plant Growth*, first published as a series of monographs in 1912, is also in its 11th edition (Wild 1988). Both are concerned essentially with soils and agriculture. We also cite Jenny's (1941, 1980) pioneering works on soil formation and soil behaviour that have been of enormous influence in ecological research, including conceptual models of the rate of accumulation of soil carbon.

Texts on forest soils have been smaller and more regional. For example, in the USA there is a distinguished series of texts by Lutz and Chandler (1946), Wilde (1946), Pritchett (1979) and Binkley (1986); in Canada by Armson (1977); in Russia by Remezov and Pogrebnyak (1969); in Australia by Attiwill and Leeper (1987); to name a selective few. Again, if we were asked to name but a handful of publications in addition to texts such as these that have been seminal in the historical development of the modern science of forest soils and nutrient cycling we would include the works of Ebermayer (1876), Rennie (1955) and Likens *et al.* (1977).

7.1.1 Scope of this chapter

If we were asked 'In what ways do forest soils differ from other soils?', our first response would be that soil processes in forest soils are dominated by the biological search for C, the result of continued input to the soil of C from litterfall and root turnover. We have therefore chosen this as a dominant theme in discussing soil processes. We then address some specifically forest-related ecological and management issues. However, in both of these attempts we can only be selective rather than comprehensive in our treatment, and no doubt our Anglo-Saxon origins and our scientific biases are all too apparent.

From the 1950s through the 1980s, most of the development in forest soils centred around soil chemistry and nutrient cycling. Throughout much of that era, the concentration was on above-ground pools and processes. More recently, this imbalance has begun to be redressed and the contribution of fine-root turnover to nutrient cycling and of the rhizosphere to nutrient uptake are now major areas of research in forest soils. Nutrient cycling, roots and root turnover, and the rhizosphere are the first themes of this chapter.

There is probably general agreement that the two elements of most immediate concern are N and P, with N receiving greater attention, particularly in the Northern Hemisphere. Mineralization is the process by which the nutrients 'trapped' in organic matter, cycled from plant to soil, are made mineral again and available as simple ion for uptake. Mineralization is the gateway that controls the rate of nutrient cycling. Thus mineralization, litter decomposition and soil organic matter are inextricably linked. Where the oxidation of C compounds and the availability of nutrients are 'driven by nature' (Cadisch & Giller 1997), these linkages are fundamental to our understanding of the rate of nutrient cycling, of the forms, storage and turnover of C in forest soils, and of the response of soil C to changing environments. These are the secondary themes of this chapter.

A great deal of research in forest soils has focused on the effects on forests of increasing acidic inputs from highly industrialized societies. Acidic inputs have been associated with both forest decline (mainly through S deposition) and forest enrichment (through N deposition). Forest decline and 'N-saturation' are the third themes of the chapter.

Our last themes address the conservative and sustainable management of our forests for timber. Do our silvicultural and logging systems result in a loss of soil fertility and is that loss significant for sustained productivity? Do these systems compact the soil and is compaction significant for sustained productivity?

Our final limitation in setting the scope of this chapter is that forests grow on all soil types from the deepest sands to the heaviest clays. Soils are classified in numerous ways, using various features, at various scales, for various purposes. Soil classification is therefore too large a topic for us to cover. For an international audience, the type of soil is generally described by either the USA Department of Agriculture comprehensive classification (Soil Survey Staff 1996) or the World Soil Classification (FAO–UNESCO 1988). There are many other classifications that may be used at national, regional, local and management scales, examples of which are given in Table 7.1.

7.2 NUTRIENT CYCLING IN FORESTS

7.2.1 Soil chemistry and nutrient cycling

By and large, the study of soil chemistry has concentrated on agricultural soils. The immediate practical questions to be answered have centred on the availability of plant nutrients, the need to develop chemical tests to define those soils suitable for particular crops and the prescription of

Table 7.1 Classification of the same forest soil from Tasmania, Australia, according to various systems that have been used at different scales. (From Grant *et al.* 1995.)

Scale	Description
Management (Grant *et al.* 1995)	16.1 Red clayey soil under wet forest
Management (Hill *et al.* 1995)	Yolla soil profile class
State (Loveday 1955)	Red brown and dark red brown krasnozems on basalt
State (Stephens 1941)	Red-brown soils
National (Stace *et al.* 1968)	Krasnozem
National (Northcote 1971)	Gn3.11
National (Isbell 1996)	Haplic, dystrophic, red ferrosol
International (Soil Survey Staff 1996)	Hapludox

the quantity of fertilizer that must be added to increase the productivity of the crop. With the exclusion of N (the availability of which is driven by biological processes in all soils), we might summarize the essential difference between the availability of plant nutrients in agricultural and forest soils:
• the availability of nutrients in agricultural soils is an immediate concern that mostly addresses inorganic equilibria over weeks or months;
• the availability of nutrients in forest soils is a long-term concern that should address biological processes and inorganic equilibria over years and centuries.

The growth of forests is a long-term process in which nutrients are cycled from plant to soil in litterfall and root turnover. Nutrients are withdrawn from tissues as they age and are translocated to actively growing tissues. Timber harvesting takes away the aged tissues with relatively low nutrient concentrations, leaving the tissues with relatively high nutrient concentrations on site. In contrast, all of the nutrient supply for an annual crop is taken up from the soil within the few months the crop takes to reach maturity.

There is little cycling of nutrients from plant to soil. At maturity, the most nutrient-rich parts of the crop (seeds, leaves, storage organs) are harvested, leaving only the tissues with relatively low nutrient concentrations behind.

Thus, from the pioneering work of Ebermayer (1876) more than a century ago, much of the work on forest soils has been directed towards nutrient cycling, in particular:
• quantifying the cycle of nutrients (uptake and return between plant and soil and retranslocation within the plant);
• quantifying inputs and outputs, including nutrient removals in harvested timber.

While all of this is obvious, the science required is not obvious. Twenty years ago Stone (1979) concluded that despite a wealth of data on nutrient cycling processes for many forests, we do not have a sound experimental basis on which to assess the sustained productivity of forest ecosystems. Ten years later, Landsberg *et al.* (1991) restated that conclusion: 'Conventional chemical analyses to determine the nutritional status of the soil yield information that may have little relevance in calculating the capacity of the soil to provide nutrients to trees, and the complexity of uptake processes through complex, dynamic and poorly defined root systems is tremendous.'

Rather than attempting to review ongoing research on the experimental and analytical basis for assessing sustainability of nutrient supply, we give only a factual account of what is known of the chemistry and cycling of the major nutrients. We do this in general terms, using data for a temperate forest of above-average productivity (Fig. 7.1, Table 7.2) and we include a summary of the form, function, concentration and cycling of the major nutrients N, P, Ca, Mg and K in Table 7.3.

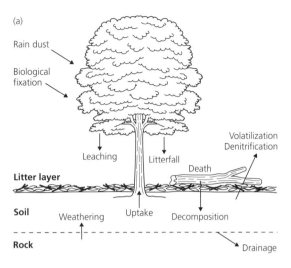

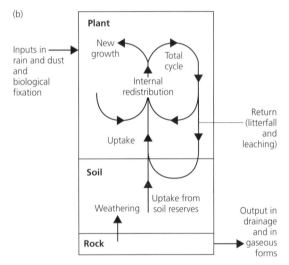

Fig. 7.1 Diagrammatic (a) and schematic (b) illustrations of the cycling of nutrients in a forest. 'Total cycle' is the sum of all fluxes between plant and soil *plus* that amount needed for new growth. (From Attiwill & Leeper 1987; Attiwill & Adams 1993.)

Biomass

The above-ground mass of the forest (trees, understorey and shrub layer) may reach $500\,t\,ha^{-1}$. Litterfall (A) is $8\,t\,ha^{-1}\,year^{-1}$ but may reach $10\,t\,ha^{-1}\,year^{-1}$. If the litter layers (Q) weigh $30\,t\,ha^{-1}$, the decomposition rate of litter $k = A/(A + Q) = 0.21\,year^{-1}$, and the half-life $t_{0.5} = 0.693/$ $k = 3.3$ years. This and subsequent analyses are based only on the above-ground stand. Data for root mass and root turnover are both more limited and variable, and have been intensively summarized (Vogt *et al.* 1986, 1996). We have previously estimated a rate of root turnover for our temperate forest of above-average productivity of $2–3\,t\,ha^{-1}\,year^{-1}$ (Attiwill & Adams 1993).

Table 7.2 Mass of organic matter (t ha^{-1}) and amounts of nutrients (kg ha^{-1}) in pools and transfers in a typical eucalypt forest of high productivity aged 50 years. 'Total cycle' is the sum of all the fluxes between plant and soil *plus* that needed for new growth. (Adapted from Attiwill & Adams 1993.)

	Organic matter	N	P	Ca	Mg	K
Pools						
Above-ground stand	500	500	50	500	250	400
Litter layer	30	200	10	110	30	25
Annual return from plant to soil						
Litterfall	8	50	2	60	20	10
Leaching	0.05	5	<0.1	5	2	15
Annual inputs and outputs						
Rainfall	0.01	5	<0.1	5	5	2
Streamflow		5	0.01	0.2	3	2
N$_2$ fixation, asymbiotic plus symbiotic		7				
Rock weathering		0	0.1	3	1	8
Annual total cycle		102	6	75	32	43
Return, plant to soil		55	2	65	22	25
Internal redistribution		30	3	0	5	10
Inputs minus outputs		7	<0.1	5	2	0
Annual net supply from soil reserves		10	1	5	3	8

Nitrogen

The concentration of N in leaves is greater than the concentration of any other nutrient, but much of this N is withdrawn from leaves before they die and fall as litter. Internal redistribution of N accounts for almost 30% of the total annual above-ground cycle. The half-life of N during litter decomposition is 3.5 years, a little longer than that for dry weight, so that N is immobilized during decomposition of the litter. The input of N in rainfall of 5 kg ha^{-1} year^{-1} is more or less balanced by the output of N in streamwater. More than 95% or so of N in the soil is in organic form; organic N is mineralized to NH$_4^+$-N and then under certain conditions to NO$_3^-$-N, a form freely available for uptake by plants. NH$_4^+$-N is held on negatively charged exchange sites of soils, but NO$_3^-$-N is mobile within the soil and is the form in which N moves with drainage waters into streams (see Table 7.3).

In soils of natural forests of the world, the generally high C/N ratio of litter and soil organic matter precludes mineralization proceeding to NO$_3^-$-N, so that the dominant form of N is NH$_4^+$-N (Jansson 1958; Wollum & Davey 1975; Raven *et al.* 1978; Attiwill *et al.* 1996). The cycling of N is therefore conservative, and N may be immobilized in litter of older forests (particularly in cooler-temperate areas) to the extent that the trees become N-deficient (Tamm 1964). Disturbance (treefall and the creation of gaps, fire, harvesting for timber) provides the conditions that result in a decrease in C/N ratio and the production of NO$_3^-$-N, in which form N may be leached from the soil into streamwater. Rapid regeneration of the forest, including rapid uptake of NO$_3^-$-N, following disturbance is therefore the key to the conservation of N (Weston & Attiwill 1990, 1996).

Phosphorus

Phosphorus occurs in highly mobile forms within plants and, like N, is withdrawn from plant parts during senescence; internal redistribution of P accounts for 50% of the total annual above-ground cycle. The half-life of P during decomposition

Table 7.3 Summary of the cycling of major elements in temperate forests. (from Attiwill 1995)

Element	Role in nutrition	Form taken up by plants	Typical concentration in tree leaves (mg kg⁻¹ dry weight)	Mobility in plants	Typical amount in above-ground parts of trees (kg ha⁻¹)	Main cycling pathway	Release during litter decomposition	Mobility in soil
Nitrogen	Constituent of amino acids, proteins. nucleic acids, etc.	NO_3^- NH_4^+	10–20	Moderate	200–500	Litterfall > redistribution > leaching	Immobilized until late stages	NO_3^-: in solution and highly mobile. NH_4^+: cation exchange and relatively immobile
Phosphorus	Constituent of sugar phosphates, nucleic acids, key role in ATP–ADP reactions	$H_2PO_4^-$	0.7–2.0	Highly mobile	10–50	Redistribution > litterfall > leaching	Immobilized until late stages. Can be leached in some forest types	Adsorbed: highly immobile
Calcium	Structural, especially in middle lamella of cell walls	Ca^{2+}	2–10	Mostly immobile	100–400	Litterfall > leaching > redistribution	Released at rate of litter decomposition	Cation exchange and relatively immobile
Magnesium	Central atom in chlorophyll molecule	Mg^{2+}	2–5	Moderate	20–200	Litterfall > redistribution > leaching	Released faster than litter decomposition	Cation exchange and relatively immobile
Potassium	Enzyme reactions, osmotic relations	K^+	2–5	Mobile	100–300	Leaching > redistribution > litterfall	Leached early during decomposition	Cation exchange and relatively immobile

is 4.1 years, i.e. P is immobilized throughout decomposition to a greater degree than N. A negligible amount of P comes in with the rain. P as $H_2PO_4^-$ is adsorbed by clay colloids and is then highly immobile in the soils so that losses of P to streamwater are negligible (see Table 7.3). However, in forest soils more than 50% of the total P in surface soils may be in organic form. Mineralization of organic P is therefore fundamental to ecosystem function, and P availability is 'determined by competition between biological and geochemical sinks for phosphate anions' (Attiwill & Adams 1993). Whereas methods for estimating rates of N mineralization have been developed and are in general application (e.g. Raison *et al.* 1987; Adams *et al.* 1989), there are no methods for routinely estimating the rate of P mineralization. The difficulty is that whenever $H_2PO_4^-$ comes into solution, there is a surface to adsorb it. This competitiveness between biological and chemical sinks for P remains a priority area of research in quantifying the long-term sustainability of P supply for forests.

Calcium, magnesium and potassium

Ca, Mg and K, and the non-essential element Na, are generally grouped as the major exchangeable cations in soil. An exchangeable cation is held adsorbed on a negatively charged surface of a colloid (both inorganic and organic). In this exchangeable form, the cations are not soluble but are brought quickly into solution and made available for uptake by exchange of protons from plant roots.

Ca is highly immobile within the plant, where it is incorporated within cell walls. Thus 80% of the total annual above-ground cycle of Ca is in litterfall. In contrast, K is highly mobile and K^+ easily leached from tree crowns by rain; leaching accounts for about 30% of the total annual above-ground cycle of K.

Trace elements (micronutrients)

The trace elements Fe, Mn, Cu, Zn, Mo and B have all played spectacular parts in agriculture and in tree plantations where plants have been introduced into new land. Applications of trace amounts (e.g. as little as $0.5\,kg\,ha^{-1}$ of Mo) have corrected deficiencies that would otherwise have made agriculture or plantation forestry unproductive. We know of no such experience in natural forests, although there is little doubt that the distribution of many species may be determined by their ability to increase or restrict the supply of one or more of the trace elements. For example, concentrations of Mn in leaves and bark differ consistently and significantly between groups of the genus *Eucalyptus* (Lambert 1981; Judd *et al.* 1996) but whether these differences are of significance in determining the distribution of species is unknown.

7.2.2 Roots and the rhizosphere

Roots supply plants with water and nutrients and provide anchorage in soil. Most studies of roots in forests have been directed to the surface horizons, where fine roots (< 2 mm in diameter) are concentrated and where water and the bulk of nutrients are sourced. There are few studies of roots below 1 m, with only nine reported in the review of Jackson *et al.* (1996). In general our knowledge of root characteristics, including biomass, distribution and fine-root area, is poorly developed compared with the above-ground components of forests. Much of the recent interest in root distribution and turnover in forests has arisen primarily from concern for increased C flux from soils due to climate change (Jackson *et al.* 1996, 1997; Silver 1998).

Distribution, mass and turnover of roots

A global analysis of root distribution in terrestrial biomes (Jackson *et al.* 1996) showed that boreal forests have the shallowest rooting profiles (80–90% of roots in the upper 30 cm of forest soils) and that temperate coniferous forests have the deepest rooting profiles (52% of roots in the upper 30 cm). When considered together, all temperate and tropical trees have 26% of roots in the top 10 cm and 60% in the top 30 cm (Jackson *et al.* 1996). Root biomass of forests has been estimated at $2–5\,kg\,m^{-2}$, with the highest values in tropical evergreen forests (Jackson *et al.* 1996). However, discrepancies between model estimates and measured characteristics of fine-root length and surface area point to the need for more data on root biomass and distribution, and for more careful measurement of fine root biomass in particular.

These analyses of root distributions and characteristics (biomass, surface area and nutrient content) among forests will enable improved modelling of water and nutrient uptake on a global scale (Jackson *et al.* 1997). They also provide the basis for improving global models of the impact of climate change on vegetation distribution and C sequestration in soil. When linked with models that predict the change in vegetation distribution and type, they will allow better prediction of the consequences of global environmental change (Jackson *et al.* 1996). The need for this type of analysis seems all the more urgent when it is considered that roots can account for more than 50% of net primary productivity (NPP) in forests (Vogt *et al.* 1986; Raich & Nadelhoffer 1989; Jackson *et al.* 1997) and that the global pool of C in fine roots is 5% of the size of the atmospheric C pool. Because about half the global pool of C in fine roots is in forests, further study of factors influencing fine-root turnover is critical to the understanding of CO_2 flux between terrestrial and atmospheric C pools.

While roots are concentrated in the upper metre of soil in most forests, deep roots become important in forests subject to periodic drought because they provide access to water held deep in the soil (Stone & Kalisz 1991; Nepstad *et al.* 1994). Carbon from these deep roots may contribute over decades to net C release from soil following conversion of forest to pasture, as shown by Nepstad *et al.* (1994) in an Amazonian rain forest. These sources of deep soil C have not been generally considered in estimates of C flux from forest soils, especially those of tropical regions where forests continue to be cleared for agriculture.

The turnover of fine roots has been investigated in relation to forest disturbance, including clearcutting, in the Northern Hardwood ecosystem (Fahey & Hughes 1994). Biomass of fine roots (<2 mm) in mature hardwood forest was 471 g m^{-2}, with average lifespan ranging from 8 to 10 months. Biomass of fine roots recovered to 71% of that in the mature forest within 3 years after clearcutting (Fahey & Hughes 1994).

Rhizosphere

The rhizosphere is the area of soil immediately surrounding the root where biotic and abiotic processes interact to create an environment distinct from bulk soil further from the root. Compared with bulk soil, rhizosphere soil is usually more intensively weathered, has a lower pH and has greater concentrations of cations and P (Griffiths *et al.* 1994; Gobran *et al.* 1998). Exudates released from plant roots and microorganisms benefit plant growth by increasing nutrient acquisition and metal detoxification, by alleviation of anaerobic stress in roots and by their action in mineral weathering (Illmer *et al.* 1995; Jones 1998).

There is no doubt that some forest species can modify the rhizosphere to access previously unavailable reserves of soil P by releasing organic acids that solubilize inorganic P. For example, Grierson (1992) demonstrated that in P-deficient soils, *Banksia integrifolia* induced the development of short-branched, tertiary, lateral roots (proteoid or cluster roots) that released citric and malic acids (the main organic acids induced under P deficiency). Organic acids probably increase the availability of P in soils through a combination of decreased adsorption of P, increased solubilization of P compounds and chelation of metals such as Fe (Bolan *et al.* 1994). Other rhizosphere processes that impact on the availability of soil P reserves include increase in root hair length and density, release of C that enhances mycorrhizal exploitation of bulk soil, and the release of phosphatases that solubilize organically bound P (Jones 1998).

The role and function of mycorrhizas in forest soils has been reviewed by Brundrett (1991) and Vogt *et al.* (1991) and there is no doubt that mycorrhizal associations benefit trees by enhancing nutrient acquisition, especially in relation to P. The evidence for N and P transfer to forest trees via mycorrhizal associations has been demonstrated in tracer studies run over relatively short time-courses. However, it is only recently that new techniques have allowed longer-term studies; these have confirmed the benefits of mycorrhizas to nutrient acquisition and have also demonstrated increased plant growth as a result (Brandes *et al.* 1998). Further work of this kind is required to establish the relative contribution of mycorrhizal uptake mechanisms to nutrient acquisition in mature forests over the long term.

The activity of microorganisms in the rhizosphere is driven by the release of low-molecular-weight organic compounds by the root and there is some evidence to suggest that the growth of certain microorganisms is favoured while that of others is suppressed. For example, Gonzalez *et al.* (1995) reported that while microbial numbers were greater in the rhizosphere of *Alnus*, roots differentially favoured colonization by proteolytic and ammonifying organisms and inhibited nitrifying organisms. Norton and Firestone (1996) reported a greater than 50% increase in N turnover (mineralization and immobilization) in the rhizosphere relative to bulk soils in *Pinus ponderosa* microcosms. Further evidence of the complex interactions in the rhizosphere come from reports of ectomycorrhizal fungi stimulating bacterial growth associated with the release of citric acid by the fungus and, conversely, of inhibition of bacterial growth (Olsson & Wallander 1998).

Recent work has aimed at characterizing the chemical and biotic nature of the rhizospere, especially as it is influenced by soil liming (Clemensson-Lindell & Persson 1993; Hüttl & Schneider 1998) and nutrient addition (Majdi & Bergholm 1995; Teng & Timmer 1995). Most work has been focused in the vicinity of industrialized areas of the Northern Hemisphere, where there are concerns for the unimpeded functioning of beneficial root–rhizosphere reactions under both controlled (fertilizer inputs) and uncontrolled (atmospheric deposition) inputs of acidity, N and S (Gorissen *et al.* 1994; Majdi & Rosengrenbrinck 1994). Reported effects of ammonium sulphate application on rhizosphere chemistry vary and many of the changes, such as increased acidity and Al (Majdi & Rosengrenbrinck 1994; Rosengrenbrinck *et al.* 1995), are found in bulk soils as well (Majdi & Persson 1995). Other effects include depletion of P and K from the rhizosphere

of Norway spruce in southern Sweden induced by ammonium deposition and the subsequent stimulation of P and K uptake (Clegg & Gobran 1997).

The benefits of rhizosphere organisms, including mycorrhizal fungi, in maintaining the growth of tree seedlings in nurseries has long been known. This concept has been extended recently to the development of soil inoculation techniques to improve the growth of seedlings in the reforestation of degraded or clearcut sites (Chanway 1997), where inoculation with forest soil has increased the number of ectomycorrhizas (Colinas *et al.* 1994a). Differential effects on survival and growth of seedlings, where planting holes were inoculated with soil transferred from forestry or plantation sites, have been reported (Colinas *et al.* 1994b). Increases in seedling growth have been associated with increased nutrient mineralization brought about by soil animals stimulating microbial turnover in transferred soil, while increases in survival probably resulted from the introduction of beneficial rhizosphere organisms (Colinas *et al.* 1994b).

Another dimension in the use of mycorrhizas to benefit tree growth has been the recognition by Garbaye (1994) of bacteria that selectively promote mycorrhizal development (the so-called 'mycorrhization helper' bacteria, MHB). In anticipation of the applications discussed by Chanway (1997) the process of selecting and using MHB cultures for controlled 'mycorrhization' has been patented by Garbaye and Duponnis (1991).

7.3 LITTER AND SOIL ORGANIC MATTER

7.3.1 Litter and litter decomposition

There have been several syntheses of rates of accession and decomposition of litter in forests over the last two decades (Schlesinger 1977; Vogt *et al.* 1986; Mathews 1997). Large amounts of litter fall from the above-ground parts of forests to the soil annually; this litter is a major determinant of nutrient cycles in forests (Attiwill & Adams 1993). On a global scale, annual above-ground litter production in forests, woodland and wooded grassland is estimated at about 39 Pg of dry matter ($1\,Pg = 10^{15}\,g$), which is a little less than 50% of total global litter production (Mathews 1997). Annual above-ground litter accession is highest in tropical, broadleaved, evergreen forests at $15\,300\,kg\,ha^{-1}$ and as low as $130\,kg\,ha^{-1}$ in boreal, needle-leaved, evergreen forests (Vogt *et al.* 1986).

On a global scale, average litter turnover time is about 5 years for litter and about 13 years for coarse woody debris (usually classified as materials $>7\,cm$ diameter) (Mathews 1997). As decomposition of forest litter is an important component of global C cycling, the potential impact of climate change on C release from litter has attracted considerable attention over the last decade (McHale *et al.* 1998; Moore *et al.* 1999). In a recent review on a global scale of litter decomposition (measured as decomposition in the first year), Aerts (1997) concluded that climate (expressed as actual evapotranspiration) is a better predictor of the rate of litter decomposition than litter chemistry, although climate indirectly influences litter chemistry. However, litter chemistry became the best predictor of decomposition rates within discrete climatic regions, especially in the tropics and in the Mediterranean region (see also Gallardo & Merino 1999) where the ratio of concentration of lignin to concentration of total N was the best indicator of litter decomposition (Aerts 1997). The strong correlation between decomposition rates and the lignin:N ratio of litter has been corroborated in a wide range of Canadian forests (Moore *et al.* 1999) where 73% of the variance in decomposition was accounted for by a multiple regression of mean annual temperature, mean annual precipitation and the lignin:N ratio of the litter.

On a regional or global scale, the rate of litter decomposition in forests is predominantly controlled by climate (temperature and moisture, as shown in general terms by a strong negative correlation with latitude) and by the quality of litter in terms of how beneficial it is to the microbial community as a source of energy or nutrients (Meentemeyer 1978; Meentemeyer *et al.* 1982; Stohlgren 1988; Currie & Aber 1997). High concentrations of nutrients relative to stored energy promote more rapid decomposition rates, so that C:N and lignin:N concentration ratios are commonly used as variables for litter quality (Harmon *et al.* 1990; Parton *et al.* 1994). Across the generally mesic forested biomes, climate and litter quality seem to be the best indicators of the rate of litter decomposition. However, the value of litter quality as a determinant of the rate of litter decomposition does not hold across all biomes, and is poorly correlated with the rate of decomposition in desert regions (Schaefer *et al.* 1985).

While accession of exogenous N to forested land has raised concern for increased decomposition of litter, reports of effects on decomposition rates vary. For example, Kuperman (1999) reported increased rates of litter decomposition (and of N mineralization) in white oak stands of the lower midwestern USA following increased N deposition. However, in a study encompassing a wide range of sites, Prescott (1995) found that increased N availability following fertilizer addition or deposition did not alter rates of litter decomposition in forests.

It seems clear that while exogenous N may increase decomposition in some forests, especially over the first year following litterfall, it may also increase the lignin content of litter and reduce long-term decomposition rates in other forests (Magill & Aber 1998). The reduced decomposition of N-rich litter may be due to the chemical formation of stable nitrogenous compounds from lignin by-products (Berg *et al.* 1995). These N compounds are thought to form by chemical condensation reactions and may be highly resistant to biological degradation (Aerts 1997). Current evidence therefore suggests that an increased rate of N deposition decreases the rate of C release in the latter stages of decomposition and stimulates C storage, both in increased woody biomass and through soil humus formation and, thereby, sequestration of C as soil organic matter (Berg 1986; Magill & Aber 1998).

Below-ground inputs of C and nutrients in forests may become the dominant pathway of nutrient return, especially in cold-temperate climates (Vogt *et al.* 1986). Berg *et al.* (1998) studied root litter decomposition in coniferous forests from northern Scandinavia to north-eastern Germany and found little correlation between the loss of mass in the first year and climatic variables when different forest types were considered together. When the *Pinus* and *Picea* stands were treated separately, stronger correlations emerged with initial concentration of P and July temperatures explaining 71% of the variation in the rate of decomposition of root litter of *Picea* stands along the transect.

Soil-warming experiments in a northern hardwood forest, where the forest floor was heated with a cable, showed that the CO_2 flux from leaf litter increased exponentially with increasing soil temperature (McHale *et al.* 1998). These increased rates of CO_2 flux from litter decomposition may represent a short-term depletion of labile C with increasing temperature, so that the response diminishes over time (Peterjohn *et al.* 1994). However, if global warming causes increases in the rate of biomass production and in the rate of litterfall, the rate of litter decomposition will increase and increases in the rate of CO_2 emissions from forest litter will be sustained (McHale *et al.* 1998).

7.3.2 Carbon in forest soils

Forest soils and carbon storage

The global cycling of C involves fluxes between the fossil C reservoir, the atmosphere, the oceans and the terrestrial biosphere (Schimel 1995). The role of the terrestrial biosphere in buffering or contributing to increases in CO_2 of the global C cycle is uncertain, partly because of the difficulty in predicting the likely impact of climate change, associated with rising atmospheric CO_2, on C flux (Smith *et al.* 1993). Because forests and wooded lands cover about one-third (4100 million ha; Table 7.4) of the land area of the earth, they are a significant component of that C stored in the terrestrial biosphere which is exchangeable with the atmospheric pool of C (Sedjo 1992; Dixon 1994; Dixon *et al.* 1994). The estimated 787 Pg of C in forest soils and associated peats (Table 7.4) is more than two-thirds of the total of 1146 Pg of C held in forests (Dixon 1994; Dixon *et al.* 1994).

About 33% of forest soils are covered by coniferous forests in the high latitudes of Russia,

Table 7.4 Global estimates of forest and woodland area, soil carbon content and turnover time of soil organic matter. The estimate of mid-latitude forest areas includes Nordic nations. (After Dixon *et al.* 1994; Mathews 1997; Schimel *et al.* 1994.)

Parameter	Latitudinal belt			Total
	High (50–75°)	Mid (25–50°)	Low (0–25°)	
Forest area (10^6 ha)	1372 (33%)	1038 (25%)	1755 (42%)	4165
Soil carbon to 1 m (Pg)	471 (60%)	100 (13%)	216 (27%)	787
Litter turnover (years)	≈ 15	≈ 8	<1	
Soil carbon turnover (years)	>70 ⟶		<20	

Canada and Alaska. About 25% of forest soils are covered by forests in the mid-latitudes of continental USA, Europe and Australia, while the low-latitude (tropical) forests of Asia, Africa and the Americas account for 42% of forested land. More than half of the low-latitude forests are in tropical America (see Table 7.4). The amount of C stored in forest soils increases with northern latitude, so that low-latitude (tropical) forests contain about 27% of the pool of soil C, increasing to about 60% in soils of high-latitude northern boreal forests (Dixon 1994; Dixon *et al.* 1994). Thus the soils of high-latitude northern forests are an important C pool and any change in the climate or the management of these forests that diminishes the soil C pool will significantly affect global C storage (Dixon 1994; Dixon *et al.* 1994; Bird *et al.* 1996). This applies equally to the tropical forests of lower latitudes, where soil C stocks are less than half those of the boreal forests, but where small increases in temperature will have a large proportional effect on metabolic activity (McKane *et al.* 1995).

The balance between net production or consumption of CO_2 in forests is largely determined by decomposition rate and the rate of soil organic C formation. This balance can be estimated reasonably well for undisturbed forests (Van

Breemen & Feijtel 1990). However, following clearing of forests the relatively low-level sinks of CO_2 in forest soils become high-level sources. Conversely, reforestation may restore the sink capacity of the soil, thereby off-setting to some extent the increase in atmospheric CO_2 due to land clearing and the oxidation of fossil fuels. Extensive tree planting and the development of policies for 'carbon credits' has therefore focused research efforts on the amount and stability of soil C and on processes that control the inputs and outputs of C.

Because soils contain the largest pool of C in terrestrial ecosystems, small changes in soil temperature or moisture or both could lead to a substantial increase in atmospheric CO_2 concentrations (Anderson 1992). Current general circulation models predict that the temperature of the earth's atmosphere will increase by 1°C over the next 35 years (Houghton *et al.* 1992), altering global climate patterns and hence impacting on many ecosystem processes. Climate warming is predicted to be greatest nearest the poles so that boreal and tundra biomes will be influenced the most; this has generated considerable interest in the interaction of climate with the biogeochemistry of these biomes (Rapalee *et al.* 1998).

Carbon accumulates in forest soil as a result of litterfall and root input, while outputs result from microbial degradation of organic matter, eluviation, solution losses and erosion. The carrying capacity for soil C increases as forests mature, and the maximum carrying capacity is controlled by climate, topography, soil type and vegetation (Dewar 1991; Van Cleve & Powers 1995). For example, Niklinska *et al.* (1999) measured the effect of increasing temperature on the rate of respiration of humus from forests of Scots pine and predicted an increase in the annual rate of soil respiration for the high latitudes of the Northern Hemisphere of 70 Tg $(1 \, Tg = 10^{12} \, g)$ CO_2 with a 2°C increase in temperature. Niklinska *et al.* (1999) therefore suggested that northern forests may become a net C source with increased global temperatures.

Much of the global store of soil C is sequestered in peats and peaty soils developed in cold and wet climates (Harrison *et al.* 1995; Hart & Perry 1999). Drying of these organic soils as a result of climate

change will significantly increase the rate of loss of soil C (Schimel *et al.* 1994), although the magnitude of change of C-flux following the drainage of peatlands and their conversion to conifer plantations has been difficult to measure (Cannell *et al.* 1993).

In general, the state of current knowledge is insufficient to predict whether rising temperatures will lead to a net release or storage of C in the terrestrial biosphere (Post & Pastor 1996; Houghton *et al.* 1998). The potential for forest soils to store or release C under changing climate and land use has been modelled on a large scale by Schimel *et al.* (1994). The CENTURY model of the turnover of soil organic matter applied to global forest and grassland ecosystems showed that soil texture and lignin concentration of foliage are key variables influencing soil C storage. Feedback between the N cycle and NPP influences model predictions of the terrestrial C balance (Schimel *et al.* 1994). All projections, based on a range of ecosystem models, suggest that increasing CO_2 will increase C inputs to soil, which will in turn widen the detrital C:N ratio, thereby decreasing the rate of decomposition. Simulations using the CENTURY model show that N cycles more rapidly under warmer temperatures and that as soil organic matter is lost more N becomes available for plant growth, resulting in more soil organic matter formation, and thereby providing negative feedback (Schimel *et al.* 1994). It seems that the efflux of organic C from soil with global warming would be much greater, perhaps double the current efflux, without this so-called 'negative feedback' (Schimel *et al.* 1990, 1994). However, speculation on the role of increased N mineralization from soil organic matter in increasing the vegetation biomass has not been substantiated with field data (Houghton *et al.* 1998).

The Intergovernmental Panel Assessment on Climate Change (IPCC) assessed the role of forests in the terrestrial C cycle over the decade 1980–90. The assessment concluded that regrowth forests of the Northern Hemisphere may have been a net sink of 0.5 Gt year[-1] of C, while deforestation in the tropics released 1.6 Gt year[-1] of C (Harrison *et al.* 1995). Climate change in tropical regions also has the potential to alter C storage significantly, as shown by Tian *et al.* (1998) in a study of the effect of interannual climate variability on C storage in Amazonian ecosystems. The hot dry weather of El Niño years in Amazonia decreased soil moisture, leading to a net release of 0.2 Pg of C in 1987 and 1992. In other years, the forests and savannahs of the Amazon basin are a sink of up to 0.7 Pg annually of C (Tian *et al.* 1998). These findings are supported by Silver (1998) who predicted that small increases in temperature and CO_2 will result in more rapid decomposition and therefore an increased CO_2 flux from moist tropical soils.

Deep soils of tropical forests contain a significant proportion of the ecosystem C, with estimates of the mass of C at depths below 1 m equal to or greater than the mass of C in the aboveground forest (Nepstad *et al.* 1994). This C, held deep in the soil, represents a significant source of CO_2 emissions following deforestation and conversion to pasture.

The impacts of forest regrowth, of increasing atmospheric concentrations of CO_2, of N deposition and of global warming on C storage in forests remain critical areas for future research. Holland *et al.* (1997) modelled the effects of atmospheric NH_x-N and NO_y-N deposition on global terrestrial C and estimated an increased rate of storage of C of 1.5–2.0 Gt year^{-1}, an amount similar to that of the missing terrestrial sink. Makipaa (1995) found increases in C storage in both humus (14–87%) and mineral soil (15–167%) following long-term N addition experiments in boreal forest. In temperate forests, N deposition is expected to both alleviate N deficiency and increase C sequestration in biomass and soil organic matter (Nilsson 1995).

Houghton (1996) claimed that the predicted increase in mid-latitude forest sequestration of C has occurred (data from Canada, USA, Europe and the former USSR), with a greater than expected accumulation of C (0.8 Pg year^{-1}) in biomass and soil pools during the 1980s. These gains of C in forests of the mid-latitudes have been offset by deforestation in the tropics, which contributed 1.6 Pg year^{-1} of C during the 1980s (Dixon *et al.* 1994; Houghton & Hackler 1995; Schimel *et al.* 1995; Houghton 1996).

Forest management to conserve and sequester C has been identified as one of three main strategies to stabilize greenhouse gases (Dixon *et al.* 1996). Johnson *et al.* (1995) likewise identified significant global-scale opportunities for both reducing C emissions from soil and increasing C sequestration in soil. The viability of such strategies has already been questioned on both sociological and scientific grounds (Binkley *et al.* 1997). Management to increase C storage in forests has been investigated at regional, national and global levels (e.g. Dixon 1995; Carter *et al.* 1998; Dick *et al.* 1998; Liski *et al.* 1998). In a global survey of 94 nations, Dixon *et al.* (1994) found that the establishment of agroforestry and alternative land-use systems (fuelwood and fibre plantations, bioreserves, intercropping systems, shelterbelts and windbreaks) on marginal and degraded lands could sequester 0.82–2.2 Pg annually. The estimated cost of these activities was $US1–69 per tonne of C, which compared favourably with other options.

Deforestation, especially of low-latitude forests, remains one of the main causes of the global net release of CO_2 from forest soils to the atmosphere, yet the magnitude of emissions is uncertain due to limited studies of changes in soil C following forest clearing (Veldkamp 1994). Dixon (1994) and Dixon *et al.* (1994) estimated that 15.4×10^9 ha of forest were lost in the two decades between 1971 and 1990; the release of C from these soils will continue for decades as the C content of the soil gradually stabilizes following the change in land use (Veldkamp 1994).

Torn *et al.* (1997) correlated the content of soil organic C with soil age in a forested chronosequence in Hawaii and concluded that the accumulation and subsequent loss of organic matter was largely driven by changes in the cycling of mineral-stabilized C over millennia rather than by changes in either the amount of fast-cycling organic matter or NPP. Soil mineralogy is therefore important in determining the quantity of organic C stored in soil, its turnover time, and C exchange between ecosystems and the atmosphere during long-term soil development. The C storage capacity of the 25% of the world's mineral soils that began developing after the last major glaciation will change as they undergo further development, and they may eventually become long-term sources of atmospheric CO_2 (Torn *et al.* 1997).

Forms and turnover of carbon in soil

Organic matter includes most of the litter and humic layer in forest soils and generally forms 1–10% of surface mineral horizons (Attiwill & Leeper 1987). The comminution and decomposition of above- and below-ground litter in forests releases CO_2, and nutrient elements are mineralized and immobilized. While all organic matter is ultimately oxidized in forest soils, compounds that decompose over centuries and millennia are synthesized in the process of humification, which involves processes of both degradation and synthesis. This humus is amorphous and is intimately mixed with soil mineral components; it is difficult to describe in any precise chemical manner. The inert organic matter of forest soils is protected from decomposition in a number of ways, including association with mineral particles, which is the basis for longer turnover times for organic matter in clay-textured soils relative to sandy soils. Most of the organic matter in soils is very old, and decomposition to inorganic constituents will take centuries to millennia (Jenkinson & Rayner 1977; Attiwill & Leeper 1987; Schimel *et al.* 1994).

Both chemical and physical methods have been used to fractionate soil organic matter; however, it has not been possible to separate organic matter into discrete components of varying decomposability (Sanchez *et al.* 1989). Traditional chemical methods date from the classical empirical fractionation based on sequential acid and alkali extractions; this recognizes fulvic acids (soluble in both acid and alkali), humic acids (soluble in alkali and precipitated by acid) and humin (insoluble in both alkali and acid). The structure of humic compounds is not accurately known but humus formation involves the polymerization of aromatic compounds, many of which are derived from lignin and polyphenolic pigments. The application of analytical techniques such as nuclear magnetic resonance, Fourier transform infrared spectroscopy and pyrolysis–mass spectroscopy are refining our understanding of the structure and reactiveness of humic compounds in soil (Schnitzer & Schulten 1995). Recent studies of humic compounds in forest soils have shown that the stability of the inert or occluded organic matter appears to be inversely related to the content of *O*-alkyl C and directly related to the content of aromatic C (Golchin *et al.* 1995).

Physical methods based on density usually separate soil according to free particulate organic matter (macro-organic matter), occluded particulate organic matter (light fraction) and colloidal or clay-associated organic matter (heavy or humified fraction) (e.g. Golchin *et al.* 1994). Density fractionation, based on dispersal of soil in NaI solution, has proved useful in separating organic matter into components that are physically and chemically distinct (Strickland & Sollins 1987), and which are mineralized at different rates by soil microorganisms (Christensen 1992; Boone 1994). Generally the light fraction extracted in NaI comprises partially decomposed roots and litter and is relatively labile; it has been reported to represent 25–50% of total soil carbon in forests of the United States (Strickland & Sollins 1987). By contrast, the heavy fraction is regarded as more resistant humic material that is adsorbed on to mineral surfaces (Theodorou 1990). The heavy fraction can be further separated into proximate fractions such as soluble fats and oils, celluloses and lignin, and phenolic compounds. Studies employing these density-fractionation techniques (e.g. Entry & Emmingham 1998) have established that old-growth forests store more C than young-growth forests and that soils of old-growth forest contain greater percentages of recalcitrant organic matter typically richer in lignin and tannic compounds. Further application of these methods, which identify the most biologically active forms of C in forest soils, are critical to developing a better understanding of soil organic matter turnover under a range of influences.

Turnover time $(1/k)$ is defined as the time that C is held in organic forms per unit volume of forest floor and soil; it is therefore the time taken for all of the organic C in the soil to be replaced. Turnover times range from hours and days for freshly added and easily decomposed organic matter to centuries and millennia for highly stabilized organo-mineral complexes in the soil.

As apparent from the discussion of C storage in forest soils, much of the recent interest in forms of organic matter in forest soils and their turnover time is in relation to the impacts of climate change and forest management on pools and fluxes of C. Models of soil C turnover have been used to predict the effects of climate change and of different land uses on soil C fluxes (e.g. Kelly *et al.* 1997; Bolker *et al.* 1998; Peng & Apps 1998; Peng *et al.* 1998). While a number of models have been developed, the most popular are those based on the CENTURY model (Parton *et al.* 1988) and the ROTH-26.3 model (Coleman & Jenkinson 1996). These models have been used to simulate changes in organic C of arable and grassland soils with accuracy (e.g. Coleman & Jenkinson 1996).

However, studies in forest soils must encompass organic matter with a wide range of decomposition constants. Although the C models were originally developed for agricultural soils (Jenkinson & Rayner 1977; Oades 1988), they have recently been adapted to model the more complex environment of forest soils by inclusion of modules to accommodate the forest floor and litter layer (Parton *et al.* 1994; Vitousek *et al.* 1994). An adapted CENTURY model predicted the pattern or ranking of decomposition of surface litter in a matrix of *Metrosideros* sites on Mauna Loa but did not predict the magnitude of differences among sites (Vitousek *et al.* 1994). The refinement of the CENTURY model as it applies to forest soils will doubtless demand a better understanding of the reactivity of organic matter on the forest floor.

Studies of the proportion of the large store of soil C that exchanges with the atmosphere within centennial and shorter time-scales (compared with time-scales of millennia) are critical to our understanding of the global C cycle (Torn *et al.* 1997). At the scale of landscapes and over geological time, Torn *et al.* (1997) demonstrated that the turnover time for the large, passive pool of soil C is controlled primarily by soil mineralogy.

7.4 FOREST SOILS AND ACIDIC INPUTS

Sulphur emissions have increased since the start of the Industrial Revolution to the extent that 75% of global emissions of S (total about 105 Mt year^{-1}) come from anthropogenic sources (Fig. 7.2). While emissions from natural sources are about evenly distributed between Northern and Southern hemispheres, 90% of anthropogenic sulphur emissions come from sources in the Northern Hemisphere (about 70 Mt year^{-1} in the Northern Hemisphere and 7.7 Mt year^{-1} in the Southern Hemisphere; Fig. 7.2).

7.4.1 Forest decline

Forest decline in the Northern Hemisphere was increasingly observed during the 1970s and 1980s and was initially linked with the increasing acidity of rain, the latter caused by nitric and sulphuric acid formed from emissions from industry, cars, etc. However, there are other sources of acidity such as dry deposition (including both direct gaseous absorption and particulate deposition), and the rates of dry deposition of SO_2 and NH_4^+ may exceed rates of deposition in rainfall. There are also many causes of forest decline other than acid deposition (see also Chapters 10 and 11).

There is little doubt that acid rain (used here in a generic sense) is a cause of some severe forest declines, particularly when close to emission sources. The most extensive studies have been in Germany (e.g. Ulrich 1983; Schulze 1989; Schulze & Ulrich 1991) and the results can be simply summarized. With increasing acidity, cation exchange capacity is reduced (leading to the loss of Ca^{2+} and Mg^{2+}) and the concentration of Al ions increases. Ca : Al and Mg : Al ratios in needles of *Abies alba* from healthy stands are greater than those in needles from stands that show symptoms of forest decline. Lindberg *et al.* (1986) estimated that atmospheric deposition supplied 40% of the annual requirement for wood production of S and 100% of that for N. Continuing input of acid rain therefore produces imbalances in the ratio of N to cations in tree crowns that result in chlorosis and, if serious enough, death. However, for many

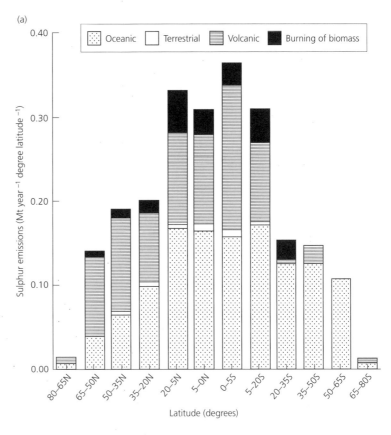

(a)

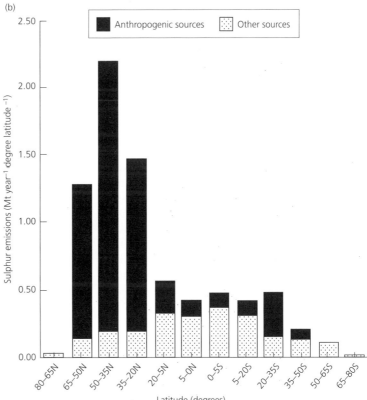

(b)

Fig. 7.2 Global emissions of sulphur from natural sources (a) and a comparison of sulphur emissions from natural and anthropogenic sources (b). (From Ayers & Granek 1995.)

forests cause and effect are not easy to link (Binkley 1986; Binkley *et al.* 1989; Markewitz *et al.* 1998). Most forest soils are acidic, so there is the overriding problem of assessing increasing acidification of an already acidic medium; added to this is the difficulty of determining and assessing the sources of acid input.

Two themes show the complexity of the acid rain problem. Markewitz *et al.* (1998) analysed soil samples from the Calhoun Experimental Forest that had been archived over three decades. They concluded that natural processes of acidification (increase in biomass and uptake of $[cations]_c > [anions]_c$ by plants, loss of $[cations]_c > [anions]_c$ by leaching, increase in soil organic matter, mineralization of N to NO_3^-, microbial and root respiration, production of organic acids) accounted for 68% of acidification over the 30 years, with acid deposition accounting for the remaining 32%. With a threefold increase in SO_2 emissions for the eastern USA from 1960 to 1980, the capacity of the surface soil to adsorb SO_4^{2-} was saturated but there was strong adsorption below 60 cm. A result of major interest is that, from 1982 to 1990, the concentration of extractable SO_4^{2-} in soils at 35–60 cm decreased significantly. This decrease was equivalent to a loss of charge of $2.2\,kmol_c\,ha^{-1}\,year^{-1}$, and Markewitz *et al.* (1998) suggest that the decrease in concentration of extractable SO_4^{2-} shows that the soils at Calhoun are responding (or recovering) to the decrease since 1980 in rates of both SO_2 emission and SO_4^{2-} deposition.

For the intensively studied Hubbard Brook Experimental Forest, north-eastern USA, inputs of acid rain have caused large losses of cations (particularly Ca^{2+} and Mg^{2+}, together with K^+ and Na^+) in drainage waters (Likens *et al.* 1996). Since 1963, atmospheric emissions have decreased, resulting in decreases in concentrations of SO_4^{2-} and the cations Ca^{2+}, Mg^{2+}, Na^+ and K^+ in bulk precipitation. The loss of cations in drainage and the net uptake of cations by the forest together represent an 'ecosystem requirement' that must be balanced by inputs in bulk precipitation, from weathering and from release of cations from exchange sites in the soil. For Ca^{2+}, precipitation supplied 29% and net release from soil 63% of this required balance during the period 1940–55. From 1976 to 1993, precipitation supplied only 12% of demand and net release from soil increased to 79%. Likens *et al.* (1996) therefore concluded that even with major reductions in emissions, the response of soil and streamwater chemistry (particularly the acid–base balance) will be significantly delayed relative to that expected from studies of inputs of acids (the biogeochemistry of sulphur) alone.

7.4.2 Nitrogen saturation

The second theme is called N saturation, the hypothesis that forests may become 'N saturated' as a result of too much N from acid deposition. Again, this is a concern for the Northern Hemisphere forests, for example in Europe (Skeffington 1990), in North America (McNulty *et al.* 1996) and in Japan (Mitchell *et al.* 1997).

Binkley and Högberg (1997) examined the hypothesis of N saturation in considerable analytical detail for forests in Sweden. While S deposition rates in Sweden have decreased by as much as 50% since 1980, the annual rates are still as high as $25\,kg\,ha^{-1}$ along the south-west coast, decreasing to $<12\,kg\,ha^{-1}$ in the north. These authors also estimated the annual rates of input of N to the south-west coast as $8–12\,kg\,ha^{-1}$ for NO_3^--N and $5–15\,kg\,ha^{-1}$ for NH_4^+-N, decreasing in the north of Sweden to $<2\,kg\,ha^{-1}$ for both NO_3^--N and NH_4^+-N. Despite these sometimes high inputs of N, the N is apparently accumulated since in only a few stands near the south-west coast do losses of N by leaching approach inputs of N in deposition. Thus the evidence is that Swedish forests are not N saturated, and forest decline is not linked with any condition of N saturation. The evidence from the application of N fertilizers is uncertain to say the least; some stands respond positively, some negatively and some not at all. Nevertheless, Binkley and Högberg (1997) concluded that 'the vast majority of conifer forests in Sweden still suffer from chronic N deficiency'.

Because of concerns about the acidification of forest soils a number of countries have initiated liming experiments, such as the Höglwald experiment in Germany, where $4\,t\,ha^{-1}$ of dolomitic limestone was applied (Kreutzer & Weiss 1998). In reviewing these experiments, Binkley and Högberg (1997) concluded that 'the overall effect of liming of forest soils in the Nordic countries has been a general decrease in forest growth (of perhaps 10%)'. A piece of jargon that has emerged from soil acidification and the N-saturation hypothesis is 'vitality fertilization' (or, even worse, 'vitalization fertilization'), a term that really means no more than the application of a fertilizer to ameliorate a deficiency (or, more ornately, to correct a nutrient imbalance). We agree strongly with Binkley and Högberg (1997): 'vitality fertilization' should not be used because it implies lack of vitality where, as in most forests, there is no symptom of decline.

Acid deposition is a problem for the Northern Hemisphere, simply as a consequence of the very much greater density of land area and population in the Northern than in the Southern Hemisphere. While forest decline is caused by acid rain in many forests, in many other forests in both hemispheres decline cannot be linked with high inputs of acid. Wild (1993) summarized a number of causes of decline other than acid deposition.

• Natural acidification processes lead to accumulation of Al and a deficiency of Mg in old trees. This problem is exacerbated by an increase in N supply that causes the development of large crowns, placing demands on Mg. Where Mg is in short supply, Mg is translated from older foliage to developing foliage.

• The increased input of N produces softer or more palatable tissues, resulting in increased insect attack.

• Ozone is directly absorbed by leaves and is extremely damaging to structure and function; decline may therefore be caused by increased inputs of ozone.

• The 1970s and 1980s in Europe were characterized by hot dry summers and cold winters; these conditions might place stress on some species in some areas.

• Poor forest management.

In summary, the 1980s saw the widespread prediction of a general decline of European forests (the concept of *Waldsterben*). By the 1990s this view was greatly modified, particularly by evidence (Kauppi *et al.* 1992, 1995) that the growing stock of European forests increased between 1971 and 1990 by 25% and that growth of the forests increased by 30%. Skelly and Innes (1994) concluded that 'the concept of a general forest decline is untenable', and that the evidence for air pollution causing forest decline 'away from known sources of pollution is extremely limited'.

7.5 TIMBER HARVESTING AND SUSTAINABILITY[1]

7.5.1 Timber harvesting, nutrient removal and sustained productivity

There have been many studies of the mass of nutrients in the above-ground parts of a wide range of forests, so nutrient removal in harvested timber can be assessed with a degree of certainty. For a range of forests of the eastern USA, Federer *et al.* (1989) calculated that a whole-tree harvest of a mature forest removes about $250\,kg\,ha^{-1}$ of N, $20\,kg\,ha^{-1}$ of P, $350\,kg\,ha^{-1}$ of Ca (up to $600\,kg\,ha^{-1}$ for oak–hickory forests), $40\,kg\,ha^{-1}$ of Mg and $150\,kg\,ha^{-1}$ of K. Losses of P, Mg and K in nutrient removal and harvest-induced leaching for a single harvest were less than 3% of the total pools, and the loss of N and Ca was less than 8%, except for oak–hickory forests where the loss of Ca was 13–19% (Federer *et al.* 1989). Data for eucalypt forests in Australia (Table 7.5) are similar; losses of N and P due to timber harvesting ranging from 2% when only the stems are harvested up to 5% for whole-tree harvesting.

The magnitude of nutrient loss varies with nutrient, species and harvesting practice, including rotation length, intensity of harvest (debarking or not, stems or whole trees removed) and site preparation for regeneration (particularly whether the logging slash is burned). Judd (1996) simulated nutrient loss due to timber harvesting for a range of harvesting regimes and for two

---

1. See also Volume 2, Chapters 5, 7 and 18.

Table 7.5 Estimates of the removal of nutrients in timber harvesting relative to pools and inputs and outputs of nutrients for a typical eucalypt forest of high productivity. (From Table 7.2 and Attiwill & Leeper 1987.)

Pool or process	N	P
Pools		
Above-ground stand (kg ha^{-1})	500	50
Litter layers (kg ha^{-1})	200	10
Soil (0–30 cm) (kg ha^{-1})	15 000	900
Outputs (kg ha^{-1}) per 100 years		
Timber harvesting		
Timber	180	20
Bark	120	12
Regeneration burning	330	10
Drainage to streamwater	500	1
Inputs (kg ha^{-1}) per 100 years		
Rain	500	5
Asymbiotic N$_2$ fixation	150	0
Symbiotic N$_2$ fixation	?	0
Weathering	0	10
Inputs – outputs (kg ha^{-1}) per 100 years		
Stem harvested	−360	−16
Stem plus bark harvested	−480	−28
Whole trees harvested	−680	−46
Inputs – outputs, change in concentration (ppm year^{-1}) in surface 30 cm of soil		
Stem harvested	−1.2	−0.05
Stem plus bark harvested	−1.6	−0.09
Whole trees harvested	−2.3	−0.15
Inputs – outputs (% of total pools)		
Stem harvested	−2.3	−1.6
Stem plus bark harvested	−3.1	−2.9
Whole trees harvested	−4.3	−4.8

subgenera of the genus *Eucalyptus*: *Monocalyptus*, which includes the fibrous-barked stringy-barks; and *Symphyomyrtus*, which includes the smooth-barked gums (Table 7.6). Judd's model excludes leaching losses. For natural forests with an 80-year rotation, Mg and K are always in positive balance and P always negative. N becomes

negative when fire is used to prepare the site for regeneration. Losses of all nutrients increase rapidly as the rotation time decreases. In the extreme comparison, the Ca balance changes over 80 years from +557 kg ha^{-1} (fibrous-barked natural forest, 80-year rotation, logs debarked on site) to −2168 kg ha^{-1} (smooth-barked plantation, 10-year rotation, logs not debarked), a fivefold increase.

This emphasis on the loss of Ca as a potential problem for sustained productivity was also raised by Turner and Lambert (1996) for plantations of *Eucalyptus grandis*, a smooth-barked eucalypt in the subgenus *Symphyomyrtus*. About 60% of the total pool of Ca in a 27-year-old stand was held above ground. Hopmans *et al.* (1993) concluded that Ca and B were the nutrients at greatest risk in managed, natural eucalypt forests in south-eastern Australia. In the study of forests of the eastern USA previously cited, Federer *et al.* (1989) calculated that the present rate of leaching will remove all the Ca in less than 1000 years. If losses due to short-rotation, whole-tree harvesting are added to these leaching losses, 50% of the Ca will be removed in only 120 years.

Clearly this rate of leaching of Ca in the forests of the eastern USA must be recent, otherwise the ecosystem would be entirely depleted. Federer *et al.* (1989) suggest that it has been caused by acid rain (Table 7.7). The H$^+$ in solution is replaced by the metallic cations Ca^{2+}, Mg^{2+}, K$^+$ and Na$^+$. If the input of H$^+$ in acid rain as H$_2$SO$_4$ and HNO$_3$ was one-tenth the current rate, the metallic cations would be in long-term balance (Table 7.7), and this is likely to have been the pre-industrial condition.

By comparison with losses of Ca due to timber harvesting, the losses of N and P (the elements previously and generally considered to be the more limiting and therefore of most likely concern) are relatively small. To put these losses of N and P into a different perspective, we have calculated loss in terms of a change in concentration in the surface 30 cm of soil (see Table 7.5). For N, the change in concentration of N in the soil over 100 years is a mere 1–2 ppm annually (compared with total N about 5000 ppm) and is an order of magnitude less for P, 0.05–0.15 ppm annually (total P about 300 ppm).

Table 7.6 Nutrient budgets under various management practices in eucalypt forests (either smooth-barked or fibrous-barked) and plantations. (Adapted from Judd 1996.)

| | Percentage of wood taken | Burn? | Debark? | Net loss or gain ($kg\,ha^{-1}$) per 80 years | | | | |
				N	P	Ca	Mg	K
Natural forest, 80-year rotation								
Fibrous-barked	0.5	No	Yes	136	−8	557	274	330
	0.5	No	No	84	−12	472	259	291
	0.5	Yes	No	−395	−31	−71	134	103
Smooth-barked	0.5	Yes	No	−395	−35	−969	88	46
Plantation, smooth-barked								
20-year rotation	0.5	No	Yes	−264	−32	−36	136	−212
	0.7	No	Yes	−468	−40	−208	76	−416
	0.7	No	No	−716	−76	−1756	−64	−644
10-year rotation	0.7	No	Yes	−576	−56	−472	−432	−472
	0.7	No	No	−776	−96	−2168	−512	−720

Table 7.7 Ionic inputs in precipitation and outputs in streamflow at the Hubbard Brook Experimental Forest, 1963–74. (From Federer *et al.* 1989.)

	Input	Output
Cations ($kmol\,(+)\,ha^{-1}\,year^{-1}$)		
H^+	0.97	0.10
NH_4^+	0.16	0.02
Ca^{2+}	0.11	0.68
Na^+	0.07	0.32
Mg^{2+}	0.05	0.26
K^+	0.02	0.05
Al^{3+}	0.00	0.21
Total	1.38	1.64
Anions ($kmol\,(-)\,ha^{-1}\,year^{-1}$)		
SO_4^{2-}	0.80	1.12
NO_3^-	0.32	0.28
Cl^-	0.20	0.14
HCO_3^-	0.00	0.13
Total	1.32	1.67

Replacing the loss of N: N_2 fixation

In our example of harvesting a eucalypt forest of high productivity on a rotation of 100 years (see Table 7.5), N balance is restored if the rate of symbiotic N_2 fixation is 3.6–6.8 $kg\,ha^{-1}\,year^{-1}$, depending on the intensity of harvesting. *Acacia* species are a natural component of both understorey and shrub layers in eucalypt forests, and where they grow densely rates of N_2 fixation of 20–30 $kg\,ha^{-1}\,year^{-1}$ have been recorded (Adams & Attiwill 1984a,b; Turvey *et al.* 1984). For most forests, however, the distribution of N_2-fixing species is irregular, and that makes the estimation of an areal rate of N_2 fixation very difficult. If we are to replace N losses by managing N_2-fixing species, we need much more research on managing the rate of N_2 fixation by these species relative to the long-term economic value of their N input.

Replacing the loss of P: weathering and pools of P

Resolution of a decrease in P of 0.05–0.15 ppm annually due to harvesting our eucalypt forest on a 100-year rotation is a futile quantitative exercise. Apart from analytical uncertainties in assessing the pool of plant-available P and the rate at which it is sustained over 100 years, there are basic questions that remain unanswered.
1 The rate of rock weathering (see Table 7.5) is a 'ball-park' estimate; how can it be improved?
2 Does P move from lower-class (or less available) reserves, in both surface soil and subsoil, to

higher class at the rate of 0.05–0.15 ppm annually?

3 Are tree roots chemically active in obtaining P from lower class reserves? For example, the roots of some forest tree species exude low-molecular-weight organic acids that may play a role in solubilizing P (Smith 1976; Fox & Comerford 1990; Fox *et al.* 1990).

Replacing nutrient losses in timber harvesting with fertilizers

The correction of a nutrient deficiency by the addition of an appropriate fertilizer is a simple and relatively economic business. However, compared with agriculture the use of fertilizers in forestry is relatively recent. White and Leaf's (1956) comprehensive bibliography of fertilizer research in forestry contained a mere 700 entries. About 25% of the papers were concerned with nurseries and nursery practice, and some 10% gave attention to 'degraded' forests (also described as 'stunted', 'degenerated', impoverished', etc.) in central Europe (particularly Germany and the former Czechoslovakia) and to the reclamation or 'improvement' of 'useless' land (swamps, bogs, peats, heathlands, moorlands, *Calluna* soils) in the UK, Sweden, Norway, Denmark and Finland and the subsequent establishment of tree plantations.

The 1960s saw the start of extensive use of fertilizers in forests. Tamm's (1968) historical view of the development of the use of fertilizers in Europe is worth recalling.

• 1865–1900: the recognition, starting with Ebermayer (1876), of nutrient cycling and nutrient demand in forests; almost no experimentation.

• 1900–25: trial-and-error field trials, with little analytical basis.

• 1925–60: scientific foundation for tree nutrition established; many experiments but limited practical application.

• 1960 onward: the start of large-scale application of fertilizers in some forests.

And, indeed, this was a large-scale expansion. For example, the area of fertilized coniferous forest in the north-western and south-eastern USA increased from zero in 1967 to 850 000 ha in 1978 (Bengtson 1979). Baule (1973) estimated that the worldwide area of forest to which fertilizer had been applied doubled from 2 million ha in 1970 to 4 million ha in 1973.

While these areas are impressive, we should remember that they are less than 0.1% of the world's forested land (Baule 1973). In short, fertilizers are not used on most of the world's forests. The traditional view of tree nutrition in silviculture of natural forests has always been that 'the loss of nutrient elements through the exploitation of timber is not very serious if the leaves and twigs are left behind in the forest' (Köstler 1956) and that 'while mineral nutrients are as indispensable for trees as for agricultural crops, the cycle [of nutrients] rarely breaks down under rational forest management, and so mineral elements "take care of themselves"' (Baker 1934).

Evidence for a decline in productivity due to nutrient removal

The assertion that the loss of nutrients associated with timber harvesting in natural forests causes a *loss* of productivity is regularly made in the ongoing debate over forests and logging. However, we have not found any definitive quantitative evidence in the literature in support of this assertion, nor are fertilizers used regularly in most of the world's forests used for timber production.

In contrast to the natural forest, fertilizers are used regularly and increasingly in plantation forests around the world. We do not attempt here to cover the many and varied fertilizer prescriptions that have been developed but point to some relatively recent reviews of the use of fertilizer in plantation forests, including Miller (1990) for European forests, Allen *et al.* (1990) for southern pine forests of the USA, Boomsma and Hunter (1990) for *Pinus radiata* plantations in Australia and New Zealand, and Attiwill and Adams (1996) for eucalypt plantations in Australia, New Zealand, Argentina, Chile, Brazil, India, China, Portugal and South Africa.

Plantation forestry very often involves species planted well outside their natural range (e.g.

Pinus in Australia and New Zealand; *Eucalyptus* in Brazil, South Africa, India, Spain and Portugal), often on soils that are viewed as marginal for most other uses. Nutrient deficiencies and imbalances are usually obvious and can be readily diagnosed and remedied. Furthermore, plantation forestry is more easily managed as a farming process than is the management of a natural forest. In plantation forestry, we manage on rotations of 10 years or so for fibre and on rotations of 15–50 years or so for timber. In natural forests we deal with rotations of 80–150 years. We establish and manage plantations for a specific purpose, whereas we manage a natural forest for all its benefits, such as water production, recreation, aesthetic well-being, conservation of fauna and flora, and timber. The costs of establishing and maintaining plantations are more readily related to a commercial benefit than are the costs of maintaining a natural forest.

Maintaining the productivity of natural forest requires very different scales of perspective, both in time and space, from those that apply in plantation forestry which, in turn, are very different from those that apply to agriculture. In agriculture, we can compare this year's crop with last year's, or with mean yields over the past 10, 20 or 50 years. In plantation forestry, we can compare the productivity of a given compartment with the productivity of the previous rotation. For most natural forests, however, we do not have the records that allow a comparison of the productivity of today's forests with those of 100 years ago.

An often-quoted example of nutrient depletion is the 'second-rotation decline' of *Pinus radiata* plantations in south-eastern Australia. Plantations of pines were established on deep, white, podsolized dune sands that previously supported natural woodlands of no commercial value for timber production. After the first rotation, the litter layer that had accumulated under the pines was heaped and burned along with the logging slash. A serious loss of production between the first and second rotations was clearly recognized by detailed assessment as early as 9.5 years into the second rotation (Keeves 1966). Remedies included both the replacement of nutrients by the addition of fertilizers (Woods 1976) and the conservation of nutrients by the conservation of organic matter between rotations (Squire *et al.* 1985).

We argue strongly against the generalization of productivity decline due to timber harvesting based on the experience in the pine plantations of South Australia, a conclusion also reached in the recent review by Evans (1999), and we argue in particular against the extrapolation of this experience to the natural forest. There are many considerations for the natural forest, including rotation length, intensity of harvest, the appropriateness of harvesting methods and silvicultural treatments relative to the ecology of natural regeneration, and the influence of other factors (e.g. Ca depletion due to increasing acid rain in some forests of eastern USA; Federer *et al.* 1989).

Nutrient demand in the natural forest

The total cycle of a nutrient in a forest (see Fig. 7.1 & Table 7.2) is the sum of all the fluxes of the nutrient between plant and soil *plus* that amount of nutrient needed for new growth. The total cycle is calculated as the sum of the amounts of nutrient: (i) cycled from plant to soil in litterfall; (ii) recycled within the plant; (iii) in inputs less that in outputs; and (iv) taken up from soil reserves. For a mature forest, the percentage contribution of litterfall to the total annual cycle decreases in the order $Ca > Mg > N > K > P$, canopy leaching $K > Ca = Mg > N > P$, internal redistribution $P > N > K > Mg > Ca$, inputs–outputs $N = Ca > Mg > K = P$, and net supply from soil reserves $K > P > N > Mg > Ca$ (Fig. 7.3). For all the nutrients discussed here, annual uptake by the mature forest from soil reserves accounts for less than 20% of the total annual demand (Fig. 7.3).

Of course, the relative contributions of nutrient cycling pathways to the total cycle must change with age of the forest. Attiwill (1980) defined three stages of forest growth and nutrient cycling, and Miller (1981) related the response of forests to fertilizers within essentially the same three stages.

1 Up to 20 years or so, the forest builds its productive system; the major proportion of NPP is used

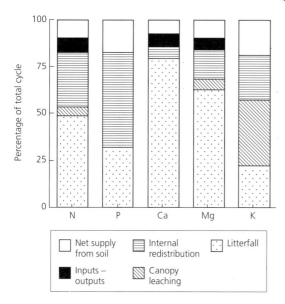

Fig. 7.3 Mass of nutrients in annual cycles ($kg\,ha^{-1}\,year^{-1}$) as percentages of the total annual cycle (uptake) of nutrients (from Table 7.2).

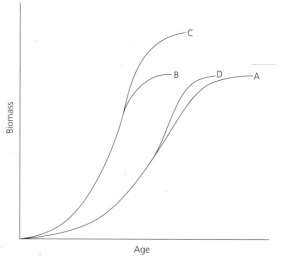

Fig. 7.4 Responses of forest growth to fertilizer additons: (A) unfertilized forest, (B) early addition of fertilizer and (D) later additon of fertilizer (probably together with thinning) increase the rate of development of the forest so that the same volume of wood is produced in less time; (C) early addition of fertilizer increases both the carrying capacity and the rate of development of the forest so that a greater volume of wood is produced in less time.

in building the photosynthetic crown and the systems for transport of water, nutrients and assimilates. This period is dominated by uptake from soil reserves, and Miller writes that this is the stage 'when responses to a wide range of fertilizer [additions] can be expected' (curve B compared with curve A, Fig. 7.4).

2 The rate of increase of the productive system is slowed. An increasing proportion of NPP is stored as physiologically dead heartwood, internal recycling accounts for an increasing proportion of the total cycle of the physiologically mobile nutrients, and the dependence on uptake from soil reserves decreases. A response to fertilizer might only be expected if the canopy is thinned (curve D compared with curve A, Fig. 7.4).

3 The trees reach maturity and the rate of biomass increase approaches zero. The major proportion of NPP is now shed as litter. Nutrients (particularly N and P) may be immobilized in the litter; additions of N and P may speed up decomposition but will have little effect on growth of the trees (e.g. Gholz & Fisher 1984).

Does the addition of fertilizer increase the carrying capacity of a forest (curve C, Fig. 7.4)? For agricultural crops, forestry where productivity has been restored on degraded land, and plantations that have been established on previously unproductive (in terms of timber production) land, limiting factors are recognizable and some of them can be easily remedied. For example, on some of the sandy soils in Australia, the success of agricultural and tree crops depends on additions of superphosphate and trace elements such as Cu and Zn. However, the productivity of ecosystems (including agricultural and forest crops) is up to an order of magnitude less than so-called 'potential' production (e.g. Jarvis 1981; Axelsson 1983). Remedies such as draining, ripping the subsoil, mounding, adding fertilizers and irrigating overcome some, but by no means all, the limitations.

However, for the natural forest and many plantations, it seems that an increase in carrying capacity due solely to the addition of fertilizers is unlikely. Nor is it likely that fertilizers will be used extensively to increase the rate of production (curves B and D, Fig. 7.4). The efficiency with which trees use nutrients added as fertilizer is low, and growth responses are short-lived (Cole *et al.* 1990). It is more likely that fertilizers will be used at relatively low rates to maintain productivity and essential ecosystem processes such as mineralization (Gessel & Atkinson 1984; Gholz & Fisher 1984).

7.5.2 Timber harvesting, soil compaction and sustainability

Timber harvesting and logging equipment compact the soil and therefore change the physical properties of soil. Compaction is due to a decrease in macropores to a depth of 30 cm or so, resulting in increases in bulk density, water-filled porosity, available water and soil strength, and in decreases in sorptivity, aeration porosity and hydraulic conductivity (Rab 1994, 1996; Herbauts *et al.* 1996; Huang *et al.* 1996; Woodward 1996; Page-Dumroese *et al.* 1998; Stone & Elioff 1998; Merino & Edeso 1999).

Despite the magnitude of some of these changes (e.g. an increase in bulk density in the surface 0–10 cm of soil of 20–40%; Rab 1996), the effects of compaction on plant growth are varied. Where disturbance by machinery is greatest (e.g. skid trails and log landings) the effects of compaction are compounded by the removal of organic layers and topsoil and by the exposure of subsoil. Among tropical forests, for example, height growth of mahogany was unaffected by soil compaction (Whitman *et al.* 1997), whereas growth of a number of species in Sabah, Malaysia, and in Amazonian Ecuador was reduced by compaction but responded more to the addition of fertilizers than to a range of treatments designed to ameliorate the effects of compaction (Nussbaum *et al.* 1995; Woodward 1996). Soil compaction may affect root growth more than shoot growth (Page-Dumroese *et al.* 1998).

Given that the snig tracks and log landings are properly revegetated, the physical properties of soils recover to precompaction conditions within 5–40 years, depending on soil type, organic content and intensity of compaction (Dickerson 1976; Greacen & Sands 1980; Wingate-Hill &

Jakobsen 1982). Thus one could conclude that where recovery time is less than one rotation, the effects of compaction *per se* will not be cumulative (Worrell & Hampson 1997). However, the problem of loss of topsoil through bad forestry practice seems to be of greater and longer-term significance than the problem of compaction alone. Restoring the fertility and physical structure of topsoil will take centuries rather than decades.

As well as direct effects of compaction and removal or erosion of topsoil on the subsequent growth of trees, there may be direct effects on diversity. The topsoil is a store of seed of some plant species and of vegetative propagules for others. Significant decreases in number or abundance of plant species have been observed after logging in aspen forests in the northern Lake States region, USA (Stone & Elioff 1998), in rain forest in Sabah, Malaysia (Nussbaum *et al.* 1995), and in mountain ash forest in Victoria, Australia (Ough & Ross 1992; Ough & Murphy 1997).

In summary, conservative management of the soil to minimize erosion and compaction is fundamental to sound land management; it is as essential in forestry as it is in agriculture. However, the nature of soil compaction in forests differs from that in agriculture (Greacen & Sands 1980). The mass of the trees, the forces generated by wind and the increasing size of roots all lead to soil compaction. Equipment used in timber harvesting is heavy, and together with the movement of large logs exerts great pressure on soils. Finally, equipment used in timber harvesting does not traverse the land as uniformly as that used in agriculture, so that heterogeneity of compaction and soil disturbance is greater in forestry than in agriculture (Greacen & Sands 1980).

Furthermore, timber harvesting generally requires the construction of roads and logging tracks to reach new logging areas. These new roads and tracks are particularly prone to erosion. Siltation of streams due to roading plus logging may be orders of magnitude greater than siltation due to logging alone (Grayson *et al.* 1993; Dargavel *et al.* 1995). Conservative management of soil in forest operations must therefore be all-embracing. Roads, skid trails and log landings must be carefully planned: steep slopes must be avoided, river crossings minimized, cross-drains created to disperse water, buffer zones established and maintained, and revegetation hastened (Lamb 1986). All these requirements can be met through strict adherence to, and monitoring of, codes of forest practice.

Assessment of the physical condition of a coupe after logging, especially the amount of soil disturbance, is targeted through sustainability indicators developed by the Working Group (1995) for the Montreal Process. Where fire is an essential component of forest ecosystems, these sustainability indicators must be sensibly interpreted. For example, *Eucalyptus regnans* (mountain ash) in Victoria, Australia, regenerates in nature only after bushfire, and the ash-bed effect (the burning of all shrub, ground and litter layers and heating of surface soil) is essential to germination, survival and growth (Ashton & Attiwill 1994; Chambers & Attiwill 1994; Ashton & Kelliher 1996). In commercial forestry operations, the trees are clearfelled and the logging slash is burned to prepare the seed bed. Subsequent seedling growth is always significantly greater on burned soils than on unburned soils, including those where a seed bed has been mechanically prepared (Van Der Meer *et al.* 1999). Sustainability indicators that emphasize retention of litter layers to protect topsoil (e.g. Rab 1999) will therefore have little basis in ecological processes when fire is involved.

REFERENCES

Adams, M.A. & Attiwill, P.M. (1984a) Role of *Acacia* spp. in nutrient balance and cycling in regenerating *Eucalyptus regnans* F. Muell. forests. I. Temporal changes in biomass and nutrient content. *Australian Journal of Botany* **32**, 205–15.

Adams, M.A. & Attiwill, P.M. (1984b) Role of *Acacia* spp. in nutrient balance and cycling in regenerating *Eucalyptus regnans* F. Muell. forests. II. Field studies of acetylene reduction. *Australian Journal of Botany* **32**, 217–23.

Adams, M.A., Polglase, P.J., Attiwill, P.M. & Weston, C.J. (1989) *In situ* studies of nitrogen mineralization and uptake in forest soils; some comments on methodology. *Soil Biology and Biochemistry* **21**, 423–9.

Aerts, R. (1997) Climate, leaf litter chemistry and leaf litter decomposition in terrestrial ecosystems: a triangular relationship. *Oikos* **79**, 439–49.

Allen, H.L., Dougherty, P.M. & Campbell, R.G. (1990) Manipulation of water and nutrients: practice and opportunity in southern U.S. pine forests. *Forest Ecology and Management* **30**, 437–53.

Anderson, J.M. (1992) Responses of soil to climate change. *Advances in Ecological Research* **22**, 163–210.

Armson, K.A. (1977) *Forest Soils: Properties and Processes*. University of Toronto Press, Toronto.

Ashton, D.H. & Attiwill, P.M. (1994) Tall open-forests. In: Groves, R.H., ed. *Australian Vegetation*, pp. 157–96. Cambridge University Press, Cambridge.

Ashton, D.H. & Kelliher, K.J. (1996) The effect of soil desiccation on the nutrient status of *Eucalyptus regnans* F. Muell. seedlings. *Plant and Soil* **179**, 45–56.

Attiwill, P.M. (1980) Nutrient cycling in a *Eucalyptus obliqua* (L'Hérit.) forest. IV. Nutrient uptake and nutrient return. *Australian Journal of Botany* **27**, 439–58.

Attiwill, P.M. (1995) Forest nutrient cycling. In: Nierenberg, W.A., ed. *Encyclopedia of Environmental Biology*, pp. 625–39. Academic Press, San Diego.

Attiwill, P.M. & Adams, M.A. (1993) Tansley review no. 50: nutrient cycling in forests. *New Phytologist* **124**, 561–82.

Attiwill, P.M. & Adams, M.A. (eds) (1996) *Nutrition of Eucalypts*. CSIRO, Melbourne.

Attiwill, P.M. & Leeper, G.W. (1987) *Forest Soils and Nutrient Cycles*. Melbourne University Press, Melbourne.

Attiwill, P.M., Polglase, P.J., Weston, C.J. & Adams, M.A. (1996) Nutrient cycling in forests of south-eastern Australia. In: Attiwill, P.M. & Adams, M.A., eds. *Nutrition of Eucalypts*, pp. 191–227. CSIRO, Melbourne.

Axelsson, B. (1983) Ultimate forest productivity: what is possible?. In: Ballard, R. & Gessel, S.P., eds. *IUFRO Symposium on Forest Site and Continuous Productivity*, pp. 61–9. US Department of Agriculture, Forest Service, Pacific Northwest Forest and Range Experiment Station, Portland, Oregon.

Ayers, G.P. & Granek, H. (1995) Sulfur in the Australian atmosphere. In: *Managing SO₂ in Australia. National Conference: Issues and Strategies*. Clean Air Society of Australia and New Zealand, Melbourne.

Baker, F.S. (1934) *Theory and Practice of Silviculture*. McGraw-Hill, New York.

Baule, H. (1973) World-wide forest fertilization: its present state, and prospects for the near future. *Potash Review* **6**.

Bengtson, G.W. (1979) Forest fertilization in the United States: progress and outlook. *Journal of Forestry* **77**, 222–9.

Berg, B. (1986) Nutrient release from litter and humus in coniferous forest soils: a mini review. *Scandinavian Journal of Forest Research* **1**, 359–69.

Berg, B., McClaugherty, C., Virzo de Santo, A., Johansson, M.B. & Ekbohm, G. (1995) Decomposition of litter and soil organic matter: can we distinguish a mechanism for soil organic matter buildup? *Scandinavian Journal of Forest Research* **10**, 108–19.

Berg, B., Johansson, M.B., Meentemeyer, V. & Kratz, W. (1998) Decomposition of tree root litter in a climatic transect of coniferous forests in northern Europe: a synthesis. *Scandinavian Journal of Forest Research* **13**, 402–12.

Binkley, C.S., Apps, M.J., Dixon, R.K., Kauppi, P.E. & Nilsson, L.O. (1997) Sequestering carbon in natural forests. *Critical Reviews in Environmental Science and Technology* **27**, 23–45.

Binkley, D. (1986) *Forest Nutrition Management.* John Wiley & Sons, New York.

Binkley, D. & Högberg, P. (1997) Does atmospheric deposition of nitrogen threaten Swedish forests? *Forest Ecology and Management* **92**, 119–52.

Binkley, D., Driscoll, C.T., Allen, H.L., Schoeneberger, P. & McAvoy, D. (1989) *Acidic Deposition and Forest Soils: Context and Case Studies of the Southeastern United States.* Springer-Verlag, New York.

Bird, M.I., Chivas, A.R. & Head, J. (1996) A latitudinal gradient in carbon turnover times in forest soils. *Nature* **381**, 143–5.

Bolan, N.S., Naidu, R., Mahimairaja, S. & Baskaran, S. (1994) Influence of low-molecular-weight organic acids on the solubilization of phosphates. *Biology and Fertility of Soils* **18**, 311–19.

Bolker, B.M., Pacala, S.W. & Parton, W.J. (1998) Linear analysis of soil decomposition: insights from the Century model. *Ecological Applications* **8**, 425–39.

Boomsma, D.B. & Hunter, I.R. (1990) Effects of water, nutrients and their interactions on tree growth, and plantation forest management practices in Australasia: a review. *Forest Ecology and Management* **30**, 455–76.

Boone, R.D. (1994) Light fraction soil organic matter: origin and contribution to net nitrogen mineralization. *Soil Biology and Biochemistry* **26**, 1459–68.

Brady, N.C. (1996) *The Nature and Properties of Soils*, 11th cdn. Prentice Hall, Upper Saddle River, New Jersey.

Brandes, B., Godbold, D.L., Kuhn, A.J. & Jentschke, G. (1998) Nitrogen and phosphorus acquisition by the mycelium of the ectomycorrhizal fungus *Paxillus involutus* and its effect on host nutrition. *New Phytologist* **140**, 735–43.

Brundrett, M.C. (1991) Mycorrhizas in natural ecosystems. *Advances in Ecological Research* **21**, 171–313.

Cadisch, G. & Giller, K.E. (eds) (1997) *Driven by Nature: Plant Litter Quality and Decomposition.* CAB International, Wallingford, UK.

Cannell, M.G.R., Dewar, R.C. & Pyatt, D.G. (1993) Conifer plantations on drained peatlands in Britain: a net gain or loss of carbon. *Forestry* **66**, 353–69.

Carter, M.R., Gergorich, E.G., Angers, D.A., Donald, R.G. & Bolinder, M.A. (1998) Organic C and N storage, and organic C fractions, in adjacent cultivated and forested soils of eastern Canada. *Soil Tillage and Research* **47**, 253–61.

Chambers, D.P. & Attiwill, P.M. (1994) The ash-bed effect in *Eucalyptus regnans* forest: chemical, physical and microbiological changes in soil after heating and partial sterilization. *Australian Journal of Botany* **42**, 739–49.

Chanway, C.P. (1997) Inoculation of tree roots with plant growth promoting soil bacteria: an emerging technology for reforestation. *Forest Science* **43**, 99–112.

Christensen, B.T. (1992) Physical fractionation of soil organic matter in primary particle size and density separates. *Advances in Soil Science* **20**, 1–90.

Clegg, S. & Gobran, G.R. (1997) Rhizospheric P and K in forest soil manipulated with ammonium sulfate and water. *Canadian Journal of Soil Science* **77**, 525–33.

Clemensson-Lindell, A. & Persson, H. (1993) Long-term effects of liming on the fine-root standing crop of *Picea abies* and *Pinus sylvestris* in relation to chemical changes in the soil. *Scandanavian Journal of Forest Research* **8**, 384–94.

Cole, D.W., Ford, E.D. & Turner, J. (1990) Nutrients, moisture and productivity of established forests. *Forest Ecology and Management* **30**, 283–99.

Coleman, K. & Jenkinson, D.S. (1996) RothC-26.3: a model for turnover of carbon in soil. In: Powlson, D.S., Smith, P. & Smith, J.U., eds. *Evaluation of Soil Organic Matter Models*, pp. 237–46. Springer-Verlag, Berlin.

Colinas, C., Molina, R., Trappe, J. & Perry, D. (1994a) Ectomycorrhizas and rhizosphere microorganisms of seedlings of *Pseudotsuga menziesii* (Mirb.) Franco. planted on a degraded site and inoculated with forest soils pretreated with selective biocides. *New Phytologist* **127**, 529–37.

Colinas, C., Perry, D., Molina, R. & Amaranthus, M. (1994b) Survival and growth of *Pseudotsuga menziesii* seedlings inoculated with biocide-treated soils at planting in a degraded clearcut. *Canadian Journal of Forest Research* **24**, 1741–9.

Currie, W.S. & Aber, J.D. (1997) Modeling leaching as a decomposition process in humid montane forests. *Ecology* **78**, 1844–60.

Dargavel, J., Hamilton, C. & O'Shaughnessy, P. (1995) *Logging and Water.* Discussion Paper 5, The Australia Institute, Canberra.

Dewar, R.C. (1991) Analytical model of carbon storage in trees, soils, and wood products of managed forests. *Tree Physiology* **8**, 239–58.

Dick, W.A., Blevins, R.L., Frye, W.W. *et al.* (1998) Impacts of agricultural management practices on C sequestration in forest-derived soils of the eastern corn belt. *Soil Tillage and Research* **47**, 235–44.

Dickerson, B.P. (1976) Soil compaction after tree-length skidding in northern Mississippi. *Soil Science Society of America Journal* **40**, 965–6.

Dixon, R.K. (1994) Carbon pools and flux of global forest ecosystems. *Science* **265**, 171.

Dixon, R.K. (1995) Agroforestry systems: sources or sinks of greenhouse gases. *Agroforestry Systems* **31**, 99–116.

Dixon, R.K., Brown, S., Houghton, R.A., Solomon, A.M., Trexler, M.C. & Wisniewski, J. (1994) Carbon pools and flux of global forest ecosystems. *Science* **263**, 185–90.

Dixon, R.K., Sathaye, J.A., Meyers, S.P. *et al.* (1996) Greenhouse gas mitigation strategies: preliminary results from the US country studies program. *Ambio* **25**, 26–32.

Ebermayer, E. (1876) *Die Gesammte Lehre der Wald-streu mit Rücksicht auf die Chemische Statik Des Waldbaues.* Springer, Berlin.

Entry, J.A. & Emmingham, W.H. (1998) Influence of forest age on forms of carbon in Douglas-fir soils in the Oregon coast range. *Canadian Journal of Forest Research* **28**, 390–5.

Evans, J. (1999) *Sustainability of Forest Plantations: the Evidence.* Issues Paper. Department for International Development (DFID), London.

Fahey, T.J. & Hughes, J.W. (1994) Fine root dynamics in a northern hardwood forest ecosystem, Hubbard Brook experimental forest, NH. *Journal of Ecology* **82**, 533–48.

FAO–UNESCO (1988) *Soil Map of the World 1 : 5 000 000, Volume 1, Revised Legend.* World Soil Resources Report 60, FAO, Rome.

Federer, C.A., Hornbeck, J.W., Tritton, L.M., Martin, C.W. & Pierce, R.S. (1989) Long-term depletion of calcium and other nutrients in eastern US forests. *Environmental Management* **13**, 593–601.

Fox, T.R. & Comerford, N.B. (1990) Low-molecular-weight organic acids in selected forest soils of the southeastern USA. *Soil Science Society of America Journal* **54**, 1139–44.

Fox, T.R., Comerford, N.B. & McFee, W.W. (1990) Phosphorus and aluminium release from a spodic horizon mediated by organic acids. *Soil Science Society of America Journal* **54**, 1763–7.

Gallardo, A. & Merino, J. (1999) Control of leaf litter decomposition rate in a Mediterranean shrubland as indicated by N, P and lignin concentrations. *Pedobiologia* **43**, 64–72.

Garbaye, J. (1994) Helper bacteria: a new dimension to the mycorrhizal symbiosis. *New Phytologist* **128**, 197–210.

Garbaye, J. & Duponnis, R. (1991) Moyens pour améliorer la croissance des plantes. French patent no. 267281 published on 31 December 1992, France.

Gessel, S.P. & Atkinson, W.A. (1984) Use of fertilizers in sustained productivity of Douglas-fir forests. In: Stone, E.L., ed. *Forest Soils and Treatment Impacts. Proceedings of the Sixth North American Forest Soils Conference*, pp. 67–87. Department of Forestry, Wildlife and Fisheries, University of Tennessee, Knoxville, Tennessee.

Gholz, H.L. & Fisher, R.F. (1984) The limits to productivity: fertilization and nutrient cycling in coastal plain slash pine forests. In: Stone, E.L., ed. *Forest Soils and Treatment Impacts. Proceedings of the Sixth North American Forest Soils Conference*, pp. 105–20. Department of Forestry, Wildlife and Fisheries, University of Tennessee, Knoxville, Tennessee.

Gobran, G.R., Clegg, S. & Courchesne, F. (1998) Rhizospheric processes influencing the biogeochemistry of forest ecosystems. *Biogeochemistry* **42**, 107–20.

Golchin, A., Oades, J.M., Skjemstad, J.O. & Clarke, P. (1994) Soil structure and carbon cycling. *Australian Journal of Soil Research* **32**, 1043–68.

Golchin, A., Oades, J.M., Skjemstad, J.O. & Clarke, P. (1995) Structural and dynamic properties of soil organic matter as reflected by C-13 natural abundance, pyrolysis mass spectrometry and solid-state C-13 NMR spectroscopy in density fractions of an oxisol under forest and pasture. *Australian Journal of Soil Research* **33**, 59–76.

Gonzalez, J.M.P., Manero, F.J.G., Probanza, A., Acero, N. & Decastro, F.B. (1995) Effect of alder (*Alnus glutinosa* L. Gaertn) roots on distribution of proteolytic, ammonifying, and nitrifying bacteria in soil. *Geomicrobiology Journal* **13**, 129–38.

Gorissen, A., Joosten, N.N. & Burgers, S.L.G.E. (1994) Ammonium deposition and the mycoflora in the rhizosphere of Douglas-fir. *Soil Biology and Biochemistry* **26**, 1011–22.

Grant, J.C., Laffan, M.D., Hill, R.B. & Neilsen, W.A. (1995) *Forest Soils of Tasmania.* Forestry Tasmania, Hobart.

Grayson, R.B., Haydon, S.R., Jayasuriya, M.D.A. & Finlayson, B.L. (1993) Water quality in mountain ash forests: separating the impacts of roads from those of logging operations. *Journal of Hydrology* **159**, 459–80.

Greacen, E.L. & Sands, R. (1980) Compaction of forest soils. A review. *Australian Journal of Soil Research* **18**, 163–89.

Grierson, P.F. (1992) Organic acids in the rhizosphere of *Banksia integrifolia* L. *Plant and Soil* **144**, 259–65.

Griffiths, R.P., Baham, J.E. & Caldwell, B.A. (1994) Soil

solution chemistry of ectomycorrhizal mats in forest soil. *Soil Biology and Biochemistry* **26**, 331–7.

Harmon, M.E., Baker, G.A., Spycher, G. & Greene, S.E. (1990) Leaf-litter decomposition in the *Picea/Tsuga* forests of Olympic National Park, Washington, U.S.A. *Forest Ecology and Management* **31**, 55–66.

Harrison, A.F., Howard, P.J.A., Howard, D.M., Howard, D.C. & Hornung, M. (1995) Carbon storage in forest soils. *Forestry* **68**, 335–48.

Hart, S.C. & Perry, D.A. (1999) Transferring soils from high- to low-elevation forests increases nitrogen cycling rates: climate change implications. *Global Change Biology* **5**, 23–32.

Herbauts, J., El Bayad, J. & Gruber, W. (1996) Influence of logging traffic on the hydromorphic degradation of acid forest soils developed on loessic loam in middle Belgium. *Forest Ecology and Management* **87**, 193–207.

Hill, R.B., Laffan, M.D. & Grant, J.C. (1995) *Soils of Tasmanian State Forests. 3. Fourth Sheet, North-West Tasmania*. Forestry Tasmania, Hobart.

Holland, E.A., Braswell, B.H., Lamarque, J.F. *et al.* (1997) Variations in the predicted spatial distribution of atmospheric nitrogen deposition and their impact on carbon uptake by terrestrial ecosystems. *Journal of Geophysical Research: Atmospheres* **102**, 15849–66.

Hopmans, P., Stewart, H.T.L. & Flinn, D.W. (1993) Impacts of harvesting on nutrients in a eucalypt ecosystem in southeastern Australia. *Forest Ecology and Management* **59**, 29–57.

Houghton, J.T., Callander, B.A. & Varney, S.K. (1992) *Climate Change 1992: the Supplementary Report to the IPCC Scientific Asssessment*. Cambridge University Press, Cambridge.

Houghton, R.A. (1996) Terrestrial sources and sinks of carbon inferred from terrestrial data. *Tellus Series B: Chemical and Physical Meteorology* **48**, 420–32.

Houghton, R.A. & Hackler, J.L. (1995) *Continental Scale Estimates of the Biotic Carbon Flux from Land Cover Change: 1850–1980*. Oak Ridge National Laboratory, Oak Ridge, Tennessee.

Houghton, R.A., Davidson, E.A. & Woodwell, G.M. (1998) Missing sinks, feedbacks, and understanding the role of terrestrial ecosystems in the global carbon balance. *Global Biogeochemical Cycles* **12**, 25–34.

Huang, J., Lacey, S.T. & Ryan, P.J. (1996) Impact of forest harvesting on the hydraulic properties of surface soil. *Soil Science* **161**, 79–86.

Hüttl, R.F. & Schneider, B.U. (1998) Forest ecosystem degradation and rehabilitation. *Ecological Engineering* **10**, 19–31.

Illmer, P., Barbato, A. & Schinner, F. (1995) Solubilization of hardly-soluble $AlPO_4$ with P-solubilizing microorganisms. *Soil Biology and Biochemistry* **27**, 265–70.

Isbell, R.F. (1996) *The Australian Soil Classification*. CSIRO, Melbourne.

Jackson, R.B., Canadell, J., Ehleringer, J.R., Mooney, H.A., Sala, O.E. & Schulze, E.D. (1996) A global analysis of root distributions for terrestrial biomes. *Oecologia* **108**, 389–411.

Jackson, R.B., Mooney, H.A. & Schulze, E.D. (1997) A global budget for fine root biomass, surface area, and nutrient contents. *Proceedings of the National Academy of Sciences USA* **94**, 7362–6.

Jansson, S.L. (1958) Tracer studies on nitrogen transformations in soil with special attention to mineralisation–immobilisation relationships. *Lantbrukshögskolans Annaler* **24**, 101–361.

Jarvis, P.G. (1981) Production efficiency of coniferous forest in the United Kingdom. In: Johnson, C.B., ed. *Physiological Processes Limiting Plant Productivity*, pp. 81–107. Butterworths, London.

Jenkinson, D.S. & Rayner, J.H. (1977) The turnover of soil organic matter in some of the Rothamsted classical experiments. *Soil Science* **123**, 298–305.

Jenny, H. (1941) *Factors of Soil Formation*. McGraw-Hill, New York.

Jenny, H. (1980) *The Soil Resource. Origin and Behavior*. Springer-Verlag, New York.

Johnson, M.G., Levine, E.R. & Kern, J.S. (1995) Soil organic matter: distribution, genesis, and management to reduce greenhouse gas emissions. *Water, Air and Soil Pollution* **82**, 593–615.

Jones, D.L. (1998) Organic acids in the rhizosphere: a critical review. *Plant and Soil* **205**, 25–44.

Judd, T.S. (1996) Simulated nutrient losses due to timber harvesting in highly productive eucalypt forests and plantations. In: Attiwill, P.M. & Adams, M.A., eds. *Nutrition of Eucalypts*, pp. 249–58. CSIRO, Melbourne.

Judd, T.S., Attiwill, M.A. & Adams, M.A. (1996) Nutrient concentrations in *Eucalyptus*: a synthesis in relation to differences between taxa, sites and components. In: Attiwill, P.M. & Adams, M.A., eds. *Nutrition of Eucalypts*, pp. 123–53. CSIRO, Melbourne.

Kauppi, P.E. & Mielikäinen, K.K. (1992) Biomass and carbon budget of European forests. 1971–90. *Science* **256**, 70–4.

Kauppi, P.E., Tomppo, E. & Ferm, A. (1995) C and N storage in living trees within Finland since 1950s. *Plant and Soil* **168/169**, 633–8.

Keeves, A. (1966) Some evidence of loss of productivity with successive rotations of *Pinus radiata* in the south-east of South Australia. *Australian Forestry* **30**, 51–63.

Kelly, R.H., Parton, W.J., Crocker, G.J. *et al.* (1997) Simulating trends in soil organic carbon in long-term experiments using the Century model. *Geoderma* **81**, 75–90.

Köstler, J. (1956) *Silviculture* (translated by M.L. Anderson). Oliver and Boyd, Edinburgh.

Kreutzer, K. & Weiss, T. (1998) The Höglwald field experiments: aims, concept and basic data. *Plant and Soil* **199**, 1–10.

Kuperman, R.G. (1999) Litter decomposition and nutrient dynamics in oak–hickory forests along a historic gradient of nitrogen and sulfur deposition. *Soil Biology and Biochemistry* **31**, 237–44.

Lamb, D. (1986) Forestry. In: Russell, J.S. & Isbell, R.F., eds. *Australian Soils: the Human Impact*, pp. 417–43. University of Queensland Press, St Lucia.

Lambert, M.J. (1981) *Inorganic Constituents in Wood and Bark of New South Wales Forest Tree Species.* Research Note 45, Forestry Commission of New South Wales, Sydney, NSW.

Landsberg, J.J., Kaufman, M.R., Binkley, D., Isebrands, J. & Jarvis, P.G. (1991) Evaluating progress toward closed forest models based on fluxes of carbon, water and nutrients. *Tree Physiology* **9**, 1–15.

Likens, G.E., Bormann, F.H., Pierce, R.S., Eaton, J.S. & Johnson, N.M. (1977) *Biogeochemistry of a Forested Ecosystem*. Springer-Verlag, New York.

Likens, G.E., Driscoll, C.T. & Buso, D.C. (1996) Long-term effects of acid rain: response and recovery of a forest ecosystem. *Science* **272**, 244–6.

Lindberg, S.E., Lovett, G.M., Richter, D.D. & Johnson, D.W. (1986) Atmospheric deposition and canopy interactions of major ions in a forest. *Science* **231**, 141–5.

Liski, J., Ilvesniemi, H., Makela, A. & Starr, M. (1998) Model analysis of the effects of soil age, fires and harvesting on the carbon storage of boreal forest soils. *European Journal of Soil Science* **49**, 407–16.

Loveday, J. (1955) *Reconnaissance Soil Map of Tasmania, Sheets 22 and 28: Table Cape and Burnie*. Divisional Report 10/55, CSIRO, Division of Soils, Adelaide.

Lutz, H.J. & Chandler, R.F. (1946) *Forest Soils*. John Wiley & Sons, New York.

McHale, P.J., Mitchell, M.J. & Bowles, F.P. (1998) Soil warming in a northern hardwood forest: trace gas fluxes and leaf litter decomposition. *Canadian Journal of Forest Research* **28**, 1365–72.

McKane, R.B., Rastetter, E.B., Melillo, J.M. *et al.* (1995) Effects of global change on carbon storage in tropical forests of South America. *Global Biogeochemical Cycles* **9**, 329–50.

McNulty, S.G., Aber, J.D. & Newman, S.D. (1996) Nitrogen saturation in a high elevation New England spruce–fir stand. *Forest Ecology and Management* **84**, 109–21.

Magill, A.H. & Aber, J.D. (1998) Long-term effects of experimental nitrogen additions on foliar litter decay and humus formation in forest ecosystems. *Plant and Soil* **203**, 301–11.

Majdi, H. & Bergholm, J. (1995) Effects of enhanced supplies of nitrogen and sulphur on rhizosphere and soil chemistry in a Norway Spruce stand in SW Sweden. *Water, Air and Soil Pollution* **85**, 1777–82.

Majdi, H. & Persson, H. (1995) Effects of ammonium sulphate application on the chemistry of bulk soil, rhizosphere, fine roots and fine root distribution in a *Picea abies* (L) Karst stand. *Plant and Soil* **169**, 151–60.

Majdi, H. & Rosengrenbrinck, U. (1994) Effects of ammonium sulphate application on the rhizosphere, fine-root and needle chemistry in a *Picea abies* (L) Karst stand. *Plant and Soil* **162**, 71–80.

Makipaa, R. (1995) Effect of nitrogen input on carbon accumulation of boreal forest soils and ground vegetation. *Forest Ecology and Management* **79**, 217–26.

Markewitz, D., Richter, D.D., Allen, H.L. & Urrego, J.B. (1998) Three decades of observed soil acidification in the Calhoun Experimental Forest: has acid rain made a difference? *Soil Science Society of America Journal* **62**, 1428–39.

Mathews, E. (1997) Global litter production, pools, and turnover times: estimates from measurement data and regression models. *Journal of Geophysical Research: Atmospheres* **102**, 18771–800.

Meentemeyer, V. (1978) Macroclimate and lignin control of litter decomposition rates. *Ecology* **59**, 465–72.

Meentemeyer, V., Box, E.O. & Thomson, R. (1982) World patterns and amounts of terrestrial plant litter production. *Bioscience* **32**, 125–8.

Merino, A. & Edeso, J.M. (1999) Soil fertility rehabilitation in young *Pinus radiata* D.Don. plantations from northern Spain after intensive site preparation. *Forest Ecology and Management* **116**, 83–91.

Miller, H.G. (1981) Forest fertilization: some guiding concepts. *Forestry* **54**, 157–67.

Miller, H.G. (1990) Management of water and nutrient relations in European forests. *Forest Ecology and Management* **30**, 425–36.

Mitchell, M.J., Iwatsubo, G., Ohrui, K. & Nakagawa, Y. (1997) Nitrogen saturation in Japanese forests: an evaluation. *Forest Ecology and Management* **97**, 39–51.

Moore, T.R., Trofymow, J.A., Taylor, B. *et al.* (1999) Litter decomposition rates in Canadian forests. *Global Change Biology* **5**, 75–82.

Nepstad, D.C., Decarvalho, C.R., Davidson, E.A. *et al.* (1994) The role of deep roots in the hydrological and carbon cycles of Amazonian forests and pastures. *Nature* **372**, 666–9.

Niklinska, M., Maryanski, M. & Laskowski, R. (1999) Effect of temperature on humus respiration rate and nitrogen mineralization: implications for global climate change. *Biogeochemistry* **44**, 239–57.

Nilsson, L.-O. (1995) Forest biogeochemistry interactions among greenhouse gases and N deposition. *Water, Air and Soil Pollution* **85**, 1557–62.

Northcote, K.H. (1971) *A Factual Key for the Recognition of Australian Soils*, 3rd edn. Rellim Technical Publications, Glenside, South Australia.

Norton, J.M. & Firestone, M.K. (1996) N dynamics in the rhizosphere of *Pinus ponderosa* seedlings. *Soil Biology and Biochemistry* **28**, 351–62.

Nussbaum, R., Anderson, J. & Spencer, T. (1995) Factors limiting the growth of indigenous tree seedlings planted on degraded rainforest soils in Sabah, Malaysia. *Forest Ecology and Management* **74**, 149–59.

Oades, J.M. (1988) The retention of organic matter in soils. *Biogeochemistry* **5**, 35–70.

Olsson, P.A. & Wallander, H. (1998) Interactions between ectomycorrhizal fungi and the bacterial community in soils amended with various primary minerals. *FEMS Microbiology Ecology* **27**, 195–205.

Ough, K. & Murphy, A. (1997) The effect of clearfell logging on tree-ferns in Victorian wet forest. *Australian Forestry* **59**, 178–88.

Ough, K. & Ross, J. (1992) *Floristics, Fire and Clearfelling in Wet Forests of the Central Highlands, Victoria*. VSP Technical Report no. 11, Department of Conservation and Environment, East Melbourne.

Page-Dumroese, D.S., Harvey, A.E., Jurgensen, M.F. & Amaranthus, M.P. (1998) Impacts of soil compaction and tree stump removal on soil properties and outplanted seedlings in northern Idaho, USA. *Canadian Journal of Soil Science* **78**, 29–34.

Parton, W.J., Stewart, J.W.B. & Cole, C.V. (1988) Dynamics of C, N, S and P in grassland soils: a model. *Biogeochemistry* **2**, 109–32.

Parton, W.J., Ojima, D.S., Cole, C.V. & Schimel, D.S. (1994) A general model for soil organic matter dynamics: sensitivity to litter chemistry, texture, and management. In: Bryant, R.B. & Arnold, R.W., eds. *Quantitative Modeling of Soil Forming Processes*, pp. 147–67. Soil Science Society of America Special Publication, Madison, Wisconsin.

Peng, C.H. & Apps, M.J. (1998) Simulating carbon dynamics along the boreal forest transect case study (BFTCS) in central Canada. 2. Sensitivity to climate change. *Global Biogeochemical Cycles* **12**, 393–402.

Peng, C.H., Apps, M.J., Price, D.T., Nalder, I.A. & Halliwell, D.H. (1998) Simulating carbon dynamics along the boreal forest transect case study (BFTCS) in central Canada. 1. Model testing. *Global Biogeochemical Cycles* **12**, 381–92.

Peterjohn, W.T., Melillo, J.M., Steudler, P.A., Newkirk, K.A., Bowles, F.P. & Aber, J.D. (1994) Response of trace gas fluxes and N availability to experimentally elevated soil temperatures. *Ecological Applications* **4**, 617–25.

Post, W.M. & Pastor, J. (1996) Linkages: an individual-based forest ecosystem model. *Climatic Change* **34**, 253–61.

Prescott, C.E. (1995) Does nitrogen availability control rates of litter decomposition in forests?. In: Nilsson, L.O., Hüttl, R.F. & Johansson, U.T., eds. *Nutrient Uptake and Cycling in Forest Ecosystems*, pp. 83–8. Kluwer Academic Publishers, Dordrecht.

Pritchett, W.L. (1979) *Properties and Management of Forest Soils*. John Wiley & Sons, New York.

Rab, M.A. (1994) Changes in physical properties of a soil associated with logging of *Eucalyptus regnans* forest in southeastern Australia. *Forest Ecology and Management* **70**, 215–29.

Rab, M.A. (1996) Soil physical and hydrological properties following logging and slash burning in the *Eucalyptus regnans* forests of southeastern Australia. *Forest Ecology and Management* **84**, 159–76.

Rab, M.A. (1999) Measures and operating standards for assessing Montreal process soil sustainability indicators with reference to Victorian Central Highlands forest, southeastern Australia. *Forest Ecology and Management* **117**, 53–73.

Raich, J.W. & Nadelhoffer, K.J. (1989) Below-ground carbon allocation in forest ecosystems: global trends. *Ecology* **70**, 1346–54.

Raison, R.J., Connell, M.J. & Khanna, P.K. (1987) Methodology for studying fluxes of soil mineral-N *in situ*. *Soil Biology and Biochemistry* **19**, 521–30.

Rapalee, G., Trumbore, S.E., Davidson, E.A., Harden, J.W. & Veldhuis, H. (1998) Soil carbon stocks and their rates of accumulation and loss in a boreal forest landscape. *Global Biogeochemical Cycles* **12**, 687–701.

Raven, J.A., Smith, S.E. & Smith, F.A. (1978) Ammonium assimilation and the role of mycorrhizae in climax communities in Scotland. *Transactions of the Botanical Society of Edinburgh* **43**, 27–35.

Remezov, N.P. & Pogrebnyak, P.S. (1969) *Forest Soil Science* (translated by A. Gourevitch). Israel Program for Scientific Translation, Jerusalem.

Rennie, P.J. (1955) The uptake of nutrients by mature forest growth. *Plant and Soil* **7**, 49–95.

Rosengrenbrinck, U., Majdi, H., Asp, H. & Widell, S. (1995) Enzyme activities in isolated root plasma membranes from a stand of Norway Spruce in relation to nutrient status and ammonium sulphate application. *New Phytologist* **129**, 537–46.

Sanchez, P.A., Palm, C.A., Szott, L.T., Cuevas, E. & Lal, R. (1989) Organic input management in tropical agroecosystems. In: Coleman, D.C., Oades, J.M. & Uehara, G., eds. *Dynamics of Soil Organic Matter in Tropical Ecosystems*, pp. 125–52. NifTAL Project and University of Hawaii Press, Hawaii.

Schaefer, D., Steinberger, Y. & Whitford, W.G. (1985) The failure of nitrogen and lignin decomposition in a North American desert. *Oecologia* **65**, 383–6.

Schimel, D.S. (1995) Terrestrial ecosystems and the carbon cycle. *Global Change Biology* **1**, 77–91.

Schimel, D.S., Parton, W.J., Kittel, T.G.F., Ojima, D.S. & Cole, C.V. (1990) Grassland biogeochemistry: links to atmospheric processes. *Climate Change* **17**, 13–25.

Schimel, D.S., Braswell, B.H., Holland, E.A. *et al.* (1994) Climatic, edaphic, and biotic controls over storage and turnover of carbon in soils. *Global Biogeochemical Cycles* **8**, 279–93.

Schimel, D.S., Enting, I.G., Heimann, M. *et al.* (1995) CO_2 and the carbon cycle. In: Houghton, J.T., Meiro Filho, L.G., Bruce, J. *et al.*, eds. *Climate Change 1994*, pp. 35–71. Cambridge University Press, Cambridge.

Schlesinger, W.H. (1977) Carbon balance in terrestrial detritus. *Annual Review of Ecology and Systematics* **8**, 51–81.

Schnitzer, M. & Schulten, H.-R. (1995) Analysis of organic matter in soil extracts and whole soils by pyrolysis–mass spectroscopy. *Advances in Agronomy* **55**, 168–217.

Schulze, E.-D. (1989) Air pollution and forest decline in a spruce (*Picea abies*) forest. *Science* **224**, 776–83.

Schulze, E.-D. & Ulrich, B. (1991) Acid rain: a large-scale, unwanted experiment in forest ecosystems. In: Mooney, H.A., Medina, E., Schindler, D.W., Schulze, E.-D. & Walker, B.H., eds. *Ecosystem Experiments*, pp. 89–106. John Wiley & Sons, New York.

Sedjo, R.A. (1992) Temperate forest ecosystems in the global carbon cycle. *Ambio* **21**, 1390–3.

Silver, W.L. (1998) The potential effects of elevated CO_2 and climate change on tropical forest soils and biogeochemical cycling. *Climate Change* **39**, 337–61.

Skeffington, R.A. (1990) Accelerated nitrogen inputs: a new problem or a new perspective? *Plant and Soil* **128**, 1–11.

Skelly, J.M. & Innes, J.L. (1994) Waldsterben in the forests of Central Europe and Eastern North America: fantasy or reality? *Plant Disease* **78**, 1021–32.

Smith, T.M., Cramer, W.P., Dixon, R.K., Leemans, R., Neilson, R.P. & Solomon, A.M. (1993) The global terrestrial carbon cycle. *Water, Air and Soil Pollution* **70**, 19–37.

Smith, W.H. (1976) Character and significance of forest tree root exudates. *Ecology* **57**, 324–31.

Soil Survey Staff (1996) *Keys to Soil Taxonomy*, 7th edn. United States Department of Agriculture, Natural Resources Conservation Service, Washington, DC.

Squire, R.O., Farrell, P.W., Flinn, D.W. & Aeberli, D.C. (1985) Productivity of first and second rotation stands of radiata pine on sandy soils. II. Height and volume growth at five years. *Australian Forestry* **48**, 127–37.

Stace, H.C.T., Hubble, G.D., Brewer, R. *et al.* (1968) *A Handbook of Australian Soils*. Rellim Technical Publications, Glenside, South Australia.

Stephens, C.G. (1941) *The Soils of Tasmania*. Bulletin 139, CSIRO, Melbourne.

Stohlgren, T.J. (1988) Litter dynamics in two Sierran mixed conifer forests. II. Nutrient release in decomposing leaf litter. *Canadian Journal of Forest Research* **18**, 1136–44.

Stone, D.M. & Elioff, J.D. (1998) Soil properties and aspen development five years after compaction and forest floor removal. *Canadian Journal of Soil Science* **78**, 51–8.

Stone, E.L. (1979) Nutrient removals by intensive harvest: some research gaps and opportunities. In: *Impact of Intensive Harvesting on Forest Nutrient Cycling*, pp. 366–86. State University of New York, College of Environmental Science and Forestry, Syracuse, New York.

Stone, E.L. & Kalisz, P.J. (1991) On the maximum extent of tree roots. *Forest Ecology and Management* **46**, 59–102.

Strickland, T. & Sollins, P. (1987) Improved method for separating light and heavy fraction organic matter from soil. *Soil Science Society of America Journal* **51**, 1390–3.

Tamm, C.O. (1964) Determination of nutrient requirements of forest stands. *International Review of Forest Research* **1**, 115–70.

Tamm, C.O. (1968) The evolution of forest fertilization in European silviculture. In: Bengston, G.W., Brendemuehl, R.H., Pritchett, W.L. & Smith, W.H., eds. *Forest Fertilization: Theory and Practice. Papers Presented at the Symposium on Forest Fertilization, Gainesville, Florida*, pp. 242–7. Tennessee Valley Authority, Muscle Shoals, Alabama.

Teng, Y. & Timmer, V.R. (1995) Rhizosphere phosphorus depletion induced by heavy nitrogen fertilization in forest nursery soils. *Soil Science Society of America Journal* **59**, 227–33.

Theodorou, C. (1990) Nitrogen transformations in particle size fractions from second rotation pine forest soil. *Communications in Soil Science and Plant Analysis* **21**, 407–13.

Tian, H.Q., Melillo, J.M., Kicklighter, D.W. *et al.* (1998) Effect of interannual climate variability on carbon storage in Amazonian ecosystems. *Nature* **396**, 664–7.

Torn, M.S., Trumbore, S.E., Chadwick, O.A., Vitousek, P.M. & Hendricks, D.M. (1997) Mineral control of soil organic carbon storage and turnover. *Nature* **389**, 170–3.

Turner, J. & Lambert, M.J. (1996) Nutrient cycling and forest management. In: Attiwill, P.M. & Adams, M.A., eds. *Nutrition of Eucalypts*, pp. 229–48. CSIRO, Melbourne.

Turvey, N.D., Attiwill, P.M., Cameron, J.N. & Smethurst, P.J. (1984) Growth of planted pine trees in response to variation in densities of naturally regenerated acacias. *Forest Ecology and Management* **7**, 103–17.

Ulrich, B. (1983) Effects of acid deposition. In: Beilke, S. & Elshout, A.J., eds. *Acid Deposition*, pp. 31–41. Reidel, Dordrecht.

Van Breemen, N. & Feijtel, T.C.J. (1990) Soil processes and properties involved in the production of greenhouse gases, with special reference to taxonomic systems. In: Bouwman, A.F., ed. *Soils and the Greenhouse Effect*, pp. 195–220. John Wiley & Sons, Chichester.

Van Cleve, K. & Powers, R.F. (1995) Soil carbon, soil formation, and ecosystem development. In: McFee, W.F. & Kelley, J.M., eds. *Carbon Forms and Functions in Soils*, pp. 155–200. Soil Science Society of America, Madison, Wisconsin.

Van Der Meer, P.J., Dignan, P. & Saveneh, A.G. (1999) Effect of gap size on seedling establishment, growth and survival at three years in mountain ash (*Eucalyptus regnans* F. Muell.) forest in Victoria, Australia. *Forest Ecology and Management* **117**, 33–42.

Veldkamp, E. (1994) Organic carbon turnover in 3 tropical soils under pasture after deforestation. *Soil Science Society of America Journal* **58**, 175–80.

Vitousek, P.M., Turner, D.R., Parton, W.J. & Sanford, R.L. (1994) Litter decomposition on the Mauna Loa environmental matrix, Hawaii: patterns, mechanisms, and models. *Ecology* **75**, 418–29.

Vogt, K.A., Grier, C.C. & Vogt, D.J. (1986) Production, turnover, and nutrient dynamics of above and belowground detritus of world forests. *Advances in Ecological Research* **15**, 303–77.

Vogt, K.A., Publicover, D.A. & Vogt, D.J. (1991) A review of the role of mycorrhizas in forest ecosystems. *Agricultural Ecosystems and Environment* **35**, 171–90.

Vogt, K.A., Vogt, D.J., Palmiotto, P.A., Boon, P., O'Hara, J. & Asbjornsen, H. (1996) Review of root dynamics in forest ecosystems grouped by climate, climatic forest types and species. *Plant and Soil* **187**, 159–219.

Weston, C.J. & Attiwill, P.M. (1990) Effects of fire and harvesting on nitrogen transformations and ionic mobility in soils of *Eucalyptus regnans* forests of south-eastern Australia. *Oecologia* **83**, 20–6.

Weston, C.J. & Attiwill, P.M. (1996) Clearfelling and burning effects on nitrogen mineralization and leaching in soils of old-age *Eucalyptus regnans* forests. *Forest Ecology and Management* **89**, 13–24.

White, D.P. & Leaf, A.L. (1956) *Forest Fertilization. A Bibliography, with Abstracts, on the Use of Fertilizers and Soil Amendments in Forestry*. Technical Publication 81, State University College of Forestry at Syracuse University, Syracuse, New York.

Whitman, A.A., Brokaw, N.V.L. & Hagan, J.M. (1997) Forest damage caused by selection logging of mahogany (*Swietenia macrophylla*) in northern Belize. *Forest Ecology and Management* **92**, 87–96.

Wild, A. (ed.) (1988) *Russell's Soil Conditions and Plant Growth*. Longman, Harlow, Essex.

Wild, A. (1993) *Soils and the Environment*. Cambridge University Press, Cambridge.

Wilde, S.A. (1946) *Forest Soils and Forest Growth*. Chronica Botanica Company, Waltham, Massachusetts.

Wingate-Hill, R. & Jakobsen, B.F. (1982) Increased mechanisation and soil damage in forests. *New Zealand Journal of Forest Science* **12**, 380–93.

Wollum, A.G. & Davey, C.B. (1975) Nitrogen accumulation, transformation, and transport in forest soils. In: Bernier, B. & Winget, C.H., eds. *Forest Soils and Land Management: Proceedings of the Fourth North American Forest Soils Conference*, pp. 67–106. Les Presses de l'Université Laval, Quebec.

Woods, R.V. (1976) *Early Silviculture for Upgrading Productivity on Marginal Pinus radiata Sites in the South-eastern Region of South Australia*. Bulletin 24, South Australia Woods and Forest Department, Adelaide.

Woodward, C.L. (1996) Soil compaction and topsoil removal effects on soil properties and seedling growth in Amazonian Ecuador. *Forest Ecology and Management* **82**, 197–209.

Working Group (1995) *Criteria and Indicators for the Conservation and Sustainable Management of Temperate and Boreal Forests (the Montreal Process)*. Canadian Forest Service, Natural Resources Canada, Hull, Quebec.

Worrell, R. & Hampson, A. (1997) The influence of some forest operations on the sustainable management of forest soils: a review. *Forestry* **70**, 61–85.

8: Ecophysiology of Forests

RICHARD H. WARING AND ANTHONY LUDLOW

8.1 INTRODUCTION

The global significance of forests has grown with our awareness that they contain a disproportionate amount of the earth's biodiversity and its terrestrial carbon stores. In addition, forests are recognized as major buffers to climate change because they can absorb energy efficiently and dissipate a large fraction of heat in the evaporation of water. Recently, trees have also been recognized as a major source of compounds that interact with ozone and create haze (Lerdau & Keller 1997). To understand the ramifications of these kinds of ecological interactions requires an appreciation of the underlying processes. If such understanding can be extended to large areas, the implications of climatic change and other alterations in the environment can be assessed more realistically.

Over the last half-century, the study of ecophysiology has slowly emerged from the laboratories of physiologists into the field where ecologists labour. Improvements in technology allowed this extension by the introduction of instruments suitable for making *in situ* measurements on individual trees. Portable gas-exchange cuvettes for measuring CO_2 consumption and O_2 outflow, branch bags, open-top chambers that allow experimental control of atmosphere for large trees, weighing lysimeters and heat pulse meters are a few examples of these technological improvements.

This emergence of ecophysiology has required that the forest be viewed as whole assemblages rather than the sum of individual components. Moreover, this expanded view of forests has led to collaboration with other types of scientists, such as micrometeorologists, who developed instruments and mathematical models to estimate energy and gas exchange from extensive areas. Starting in the 1960s, ecologists also contributed by constructing quantitative models to describe the movement of radioisotopes through food chains. During the International Biological Program in the 1970s, these food-chain models were extended to describe the capture, storage and release of carbon, water and minerals cycling through forest ecosystems (Reichle 1970, 1981).

In this chapter, we confine our discussion to living vegetation, which generally represents less than 50% of the total organic content in an ecosystem. Even with this restriction we illustrate that it is possible to predict the probability of disturbances that could result in tree death and lead to alterations in the pools of resources available to regenerating forests. The presentation is organized around a selected set of ecological concerns for which ecophysiological research has provided critical insights. These concerns include how alterations in climate, atmospheric chemistry and management practices affect the growth, composition and susceptibility of forests to insects and pathogens.

8.2 RESPONSES OF FORESTS TO VARIATION IN CLIMATE

The absolute limits on tree distribution are generally associated with climatic limitations. There has therefore been an effort to define these limits, as well as the optimum conditions of temperature and moisture for broadly representative types of forests. In this section we review the contributions of ecophysiology in defining functional limitations of temperature and moisture conditions (Mason & Langenheim 1957).

8.2.1 Determination of temperature optimum and limits

Ecophysiological studies have established the lower temperature limits for tropical, subtropical, temperate and boreal forests. Many subtropical species are able to concentrate solute and thus lower the freezing point in cells below that tolerated by tropical trees. Some temperate trees extend the freezing threshold down to –45°C via the mechanism of supercooling, whereas boreal species are able to transfer water out of the nucleus into cell walls and thereby withstand exposure to liquid nitrogen at –196°C. These adaptations to low temperatures serve to define the potential geographical limits of various types of forests (Table 8.1).

Defining the optimum temperature for growth or photosynthesis has traditionally been made in the laboratory on seedlings or from excised branches or buds of mature trees. By measuring the production rates of metabolic heat and CO_2 efflux of excised buds at varying temperatures, Anekonda *et al.* (1993, 1994) described relationships with growth and geographical origins of *Sequoia sempervirens*. However, the temperature optimum for photosynthesis often exhibits seasonal adjustment (Neilson *et al.* 1972; Strain *et al.* 1976; Slatyer & Morrow 1977). Intuitively, we would think that optimal growth for a species would occur at temperatures typical of those that

occur in the species' native range. In fact, this is not true. Studies by Hawkins and Sweet (1989) demonstrated that optimum temperatures for a broad array of tree species native to New Zealand averaged nearly 8°C above typical midsummer temperatures, whereas faster-growing species introduced from North America were closely adapted to the present New Zealand climate (Table 8.2).

Further refinements in growth-room experiments identified the importance of root temperature (Babalola *et al.* 1968; Bowen 1970; Donovan & McLeod 1985). Root temperature affects the rate that water may be taken up through root membranes, which helps explain the distribution patterns of closely related genera across broad climatic zones (Waring & Winner 1996).

8.2.2 Physiological definition of soil drought

Inadequate precipitation can lead to a soil water deficit. Ralph Slatyer (1967) set forth the basic principles of how plants respond to soil drought but until Per Scholander and colleagues at Scripps Institution of Oceanography constructed a portable pressure chamber, ecophysiologists were unable to measure water potential gradients in tall trees (Scholander *et al.* 1965). With the pressure chamber, relationships between relative water content, solute potential and incipient plasmolysis could be assessed and seasonal comparisons made to evaluate the physiological impact of drought.

In drought-prone areas, seedlings often experi-

Table 8.1 Lethal minimum temperatures limiting the geographical distribution of various types of forests. (From Woodward 1987.)

Type of forest	Lethal minimum temperature (°C)
Broadleaved tropical deciduous (rain-sensitive)	0–10
Broadleaved evergreen (frost-sensitive)	0
Broadleaved evergreen (frost-resistant)	15
Broadleaved temperate deciduous	–40
Broadleaved boreal deciduous	Not temperature limited
Needle-leaved evergreen (subtropical)	–15
Needle-leaved evergreen (temperate)	–45
Needle-leaved evergreen (boreal)	–60
Needle-leaved boreal deciduous	Not temperature limited

Table 8.2 Optimum temperature for photosynthesis compared with actual midsummer temperatures for five genera of native New Zealand and introduced European and North American tree species. (From Hawkins & Sweet 1989.)

	Optimum temperature (°C)	Midsummer temperature (°C)	Difference from midsummer (°C)
New Zealand (native spp.)			
Agathis australis	27.0	22.2	4.8
Dacrycarpus dacrydiodies	27.0	17.9	9.1
Dacrydium cupressinum	27.0	16.8	10.2
Nothofagus solandri	27.0	17.0	10.0
Podocarpus totara	27.0	21.5	5.5
Mean	27.0	19.1	7.9
Species introduced from Europe and North America			
Larix decidua	17.0	19.0	−2.0
Pinus radiata	23.0	21.0	2.0
Pseudotsuga menziesii	21.0	20.2	0.8
Sequoia sempervirens	19.0	17.0	2.0
Tsuga heterophylla	18.0	20.2	−2.2
Mean	19.6	19.4	0.2

ence near-lethal water potentials because their roots do not extend as deeply as those of mature trees. The pressure chamber provided the first direct measure of water availability to deeply rooted trees when measurements were taken at night under conditions when transpiration approaches zero. The fraction of water extracted from different depths within the soil profile has more recently been determined through analysis of the isotopic composition ($\delta^{18}O$ and δ^2H) of xylem sap (Fig. 8.1).

8.2.3 Sapwood as a water reservoir in trees

Trees contain a large volume of sapwood, which is significant because this tissue serves during the day and through periods of drought as a temporary reservoir of water that extends the opportunities for trees to photosynthesize (Waring & Running 1978; Holbrook & Sinclair 1992; Borchert 1994; Tognetti & Borghetti 1994; Goldstein *et al.* 1998). The extraction of large amounts of water from stems and branches involves some cavitation in the water-conducting elements of the xylem. Rapid recharge of the sapwood reservoir occurs through uptake of water by roots and through

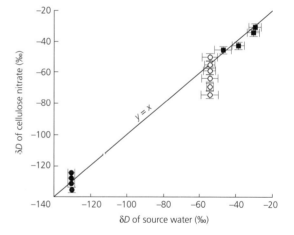

Fig. 8.1 Hydrogen isotopic composition, δD, of cellulose nitrate extracted from tree annual rings compares well with that of source water for white pine (*Pinus strobus*) growing in New York State (■) and for shallow-rooted (○) and deep-rooted (●) boxelder (*Acer negundo*) growing near a stream in Utah. (After Dawson 1993.)

foliage and twigs (Katz *et al.* 1989; Goldstein *et al.* 1998). The amount of water released from sapwood for a specific increase in the water potential gradient defines the 'capacitance', an essential term in modelling the water balance of trees (Edwards *et al.* 1986) and forests (Waring & Running 1978). As more water is extracted from the sapwood, the water transport pathway between the soil and evaporating surfaces of the foliage becomes less efficient, with implications for both transpiration and photosynthesis.

8.2.4 Transpiration by forests

Transpiration was the first physiological process directly measured for an entire forest by employing an energy budget analysis (Monteith & Unsworth 1990). A forest, unlike vegetation of shorter more uniform stature, represents an aerodynamically rough surface, a property that creates turbulence in the flow of air, even under calm conditions. As a consequence of this aerodynamic roughness, forests can absorb large amounts of solar radiation without experiencing temperatures that would be lethal to shorter vegetation, even when transpiration is restricted.

Stomatal response is closely linked to the photosynthetic process. Consequently, accurate transpiration models separate the photosynthetically active component of solar radiation (400–700 nm) from the total and take into account hydraulic constraints and variation in ambient concentrations of CO_2 (Williams *et al.* 1997). Most models that predict the mean canopy conductance of whole forests are still empirical, developed from responses recorded under a range of environmental conditions using gas-exchange cuvettes or inferred from flux measurements determined when the forest canopy is dry (Fig. 8.2).

From the extensive literature published on leaf and canopy stomatal conductance, some important generalizations have emerged. First, the maximum canopy conductance for a wide variety of forests averages two to four times the maximum value measured on individual leaves (Table 8.3). The leaf area of woodlands and forests is expressed in terms of leaf area index (m² of projected leaf surface per m² of ground) and ranges

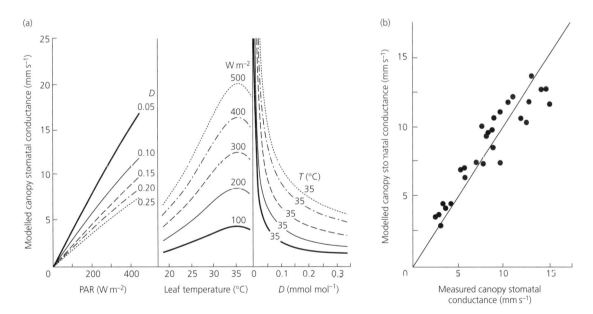

Fig. 8.2 (a) Empirical data collected above an Amazonian rain forest illustrate that canopy stomatal conductance (G_s) varies with meteorological conditions. In general, G_s increases with photosynthetically active radiation (PAR) and air temperature (T) and decreases with vapour pressure gradient between foliage and the air (D). (b) Combined into a model, predicted canopy conductance closely matched that measured. (After Lloyd *et al.* 1995.)

Table 8.3 Global average maximum leaf conductance (g_{smax}) and maximum canopy conductance (G_{smax}) to water. (From Kelliher *et al.* 1995.)

Type of forest	g_{smax} (mm s^{-1})	G_{smax} (mm s^{-1})
Coniferous forest	5.7 ± 2.4	21.2 ± 7.1
Eucalyptus forest	5.3 ± 3.0	17.0
Temperate deciduous forest	4.6 ± 1.7	20.7 ± 6.5
Tropical rain forest	6.1 ± 3.2	13.0

from less than $1 m^2 m^{-2}$ to $12 m^2 m^{-2}$. The maximum canopy conductance for forests is lower than the maximum that would be expected from the maximum canopy conductance of individual leaves because the illumination of individual leaves decreases as they shade each other when the total canopy leaf area increases. This reduces the stomatal conductance of individual leaves below that which would occur in full sunlight. As a result of self-shading, the maximum canopy conductance approaches a constant at leaf area indices greater than $3.0 m^2 m^{-2}$ (Kelliher *et al.* 1995). These two generalities greatly simplify the estimation of maximum stomatal conductance and reduce the need for accurate assessment of leaf area indices above $3.0 m^2 m^{-2}$ for purposes of estimating transpiration.

Although the *maximum* canopy conductance may not differ significantly among forests, this does not mean that reliable estimates of *actual* canopy stomatal conductance can be extrapolated from measurements made on seedlings or saplings to tall trees. As noted above, trees have deeper root systems and a larger volume of sapwood than seedlings or saplings, yet even with these benefits mature trees offer an increasingly long and tortuous path for water movement from roots to leaves (Mencuccini & Grace 1996). As a result, comparative studies between young and older trees often show that when water is readily available in the soil, older trees may be more sensitive to atmospheric vapour pressure deficits (Ryan & Yoder 1997). The differences are not simply in response to height and the hydrostatic gradient of $0.01 MPa m^{-1}$ (Goldstein *et al.* 1998). The spacing between individual trees also plays a role. As trees increase in size, branches elongate

as a function of the space between trees. Longer branches offer a less efficient pathway for water transport than the main stem, and can reduce transpiration and photosynthesis per unit of leaf area by hastening stomatal closure during the day (Walcroft *et al.* 1996; Hubbard *et al.* 1999).

8.3 RESPONSES OF FORESTS TO VARIATION IN NUTRIENT AVAILABILITY

One of the more challenging problems for eco-physiologists is to assess the supply of nutrients available to sustain forest growth. As noted in Chapter 7 (this volume), little insight is gained by simply analysing the elemental content in the soil because many nutrients are present in unavailable forms, and trees differ in their abilities to extract essential nutrients from throughout the soil profile. In addition, a range of nutrients can be acquired from the air via aerosol deposition on foliage (Tagliavini *et al.* 1998) and additional nitrogen can be acquired through root associations with symbiotic nitrogen-fixing bacteria. Physiologists and chemists, starting with Liebig in 1840, have established those elements essential for growth, defined specific deficiency symptoms and related concentrations in foliage to minimum requirements. In this section, we emphasize two additional components of nutrition that have significant impact on the growth of forests: balance and internal recycling. In a later section, we discuss how nutrient imbalances increase the susceptibility of forests to insects and pathogens.

8.3.1 Optimum nutrition

Torsten Ingestad (1979) developed a hydroponic system where the balance and total amount of nutrients could be varied daily to maintain relative growth rates constant over short periods. In a series of experiments he defined 'optimum nutritional balance' as that state when the ratios of nutrients in solution and leaf tissue maintain similar values. Under controlled laboratory conditions, the optimum nutrient balance, relative to the concentration of nitrogen, is similar for a wide range of tree species (Table 8.4).

Table 8.4 Optimum macronutrient weight proportions relative to N in leaves of seedlings grown in solution culture. (From Ingestad 1979; Ericsson 1994.)

Species	N	P	K	Ca	Mg
Alnus incana	100	16	41	10	14
Betula pendula	100	14	55	6	10
Eucalyptus globulus	100	10	37	10	9
Picea abies	100	16	50	5	5
Picea sitchensis	100	16	55	4	4
Pinus sylvestris	100	14	45	6	6
Populus simonii	100	11	48	7	7
Tsuga heterophylla	100	16	70	8	5
Mean	100	14	50	7	8

Further refinement in understanding and assessment of nutritional status is gained by defining 'metabolically active forms' in plant tissue (Attiwill & Adams 1993). For phosphorus, the inorganic form best reflects its availability (Chapin & Kedrowski 1983; Polglase *et al.* 1992). For nitrogen, most reserves are in the form of proteins. Above a certain concentration, excess nitrogen accumulates as free amino acids. This may be symptomatic of a generally unfavourable physiological status (Näsholm & Ericsson 1990; Ericsson *et al.* 1993) or it may be evidence that the rate of photosynthesis does not depend exclusively on substrate concentration.

Sune Linder, a compatriot of Ingestad, extended these optimum nutrient addition experiments into the field, providing convincing evidence that the growth of boreal forests of pine and spruce are more limited by slow release of nutrients through decomposition of organic matter than by climatic restrictions on photosynthesis and growth (Axelsson & Axelsson 1986; Linder & Flower-Ellis 1992).

8.3.2 Internal recycling of nutrients

It has been long recognized that evergreen forests can accumulate large amounts of foliage in excess of that required for efficient interception of solar radiation (Waring & Franklin 1979; Wang *et al.* 1991). It is also known that a linear correlation exists between the total nitrogen content in foliage and wood production (Ågren 1983; Hunter *et al.* 1987; Waring 1989). By comparing the nutri-

ent content in leaf litterfall with that in new and older foliage, it has been estimated that initially well-fertilized plantations can maintain favourable growth rates over decades by recycling stored nutrient capital acquired during development of a dense canopy (Millard & Proe 1992; Miller *et al.* 1992, 1996; Miller 1995).

Ecophysiologists have contributed to our understanding of nutrient uptake and internal recyling by defining seasonal patterns more precisely and by identifying the dynamic exchange from one internal pool to another. Insights gained from such analyses include recognition that:
• fast growth of evergreens results in a greater proportion of shoot nitrogen reserves going into new growth at the expense of older foliage (Müller 1949; Hawkins *et al.* 1998);
• when low growth rates prevail, elements in short supply are preferentially translocated to new growth (Müller 1949; Hawkins *et al.* 1998);
• deciduous trees may obtain more than 50% of their nitrogen from internal sources (Weinbaum & van Kessel 1998);
• during periods of rapid leaf expansion, uptake of nutrients from the soil plays a minor role in meeting demands for deciduous as well as evergreen trees (Mead & Preston 1994; Weinbaum & van Kessel 1998);
• understorey vegetation may contribute disproportionately to the acquisition and recycling of nutrients within forest ecosystems (Raison *et al.* 1993; Buchmann *et al.* 1996).

These insights have important implications in the management of vegetation and in the assessment of forest ecosystem response to atmospheric deposition of nitrogen and sulphur-rich compounds (Aber *et al.* 1989; Steudler *et al.* 1989; Johnson 1992). For example, significant changes in the ratio of calcium to aluminium and the concentrations of essential trace elements in annual growth rings of trees bear witness to the impact of air pollution (Baes & McLaughlin 1984; Bondietti *et al.* 1989).

8.4 CARBON BALANCE ANALYSES

With increased concerns about how forests may respond to changes in atmospheric chemistry and climate, there has been a major effort to gain a better understanding of the processes controlling tree growth. This effort has been accentuated by observations that historical measures of production no longer apply, possibly as a result of changes in genetic composition of trees, application of fertilizers, control of competing vegetation and alterations in atmospheric chemistry (CO_2, ozone, acid rain, etc.). Developing a more integrated picture of how processes such as photosynthesis, respiration and allocation combine to affect the structure and function of the autotrophic component of forest ecosystems is critical to predicting forest growth under changing global conditions. Although net primary production represents all carbon products produced annually from photosynthesis after subtracting losses from respiration, the components of respiration and growth vary significantly depending on the environment, composition of tissue, and allocation of resources above and below ground.

8.4.1 Canopy photosynthesis

Photosynthesis is among the best understood of all physiological processes. Historically, most research on photosynthesis has been done under laboratory conditions on individual trees. A series of basic equations have been developed for individual leaves that describe the capture of light energy, the diffusion of CO_2 and the enzymatic transformations involved in the photosynthetic process (Farquhar *et al.* 1980; Farquhar & Sharkey 1982). To apply these basic equations to whole canopies involves an analysis of how photosynthetic capacity and environment vary spatially. Sophisticated models have been constructed to predict the quantity of direct and diffuse radiation absorbed within a forest canopy (Norman & Jarvis 1975; Leuning *et al.* 1995; Le Roux *et al.* 1999).

Although such models are useful for predicting the maximum rates of photosynthesis, they are not able to predict rates under varying environments. To extend estimates of photosynthesis to whole canopies two important simplifications are generally made: (i) the canopy is considered to represent a single large leaf with varying ability to absorb radiation; and (ii) the relation between photosynthesis and light absorption is extended in time to achieve a more linear response than occurs instantaneously (Wang *et al.* 1991). Further simplifications are possible by combining the effects of subfreezing temperatures, drought, vapour pressure deficits and nutritional limitations on the conversion of absorbed radiation into photosynthetic products into a general equation as suggested by Landsberg (1986).

$$\text{Gross photosynthesis} = \alpha \sum \phi \, \epsilon f(H_2O)f(D)f(T)$$
[8.1]

where α is the light conversion efficiency established by nutrition, ϕ is the photosynthetically active radiation absorbed by the canopy, and modifying factors $f(i)$, which range from 1 to 0, describe fractional reductions in relation to soil water deficits (H_2O), unfavourable temperature (T) and vapour pressure deficits of the air (D). At weekly or monthly time steps, models based on these simplifying assumptions show good agreement with integrated estimates of gross photosynthesis obtained using eddy-flux covariance techniques (Dang *et al.* 1998) (Fig. 8.3).

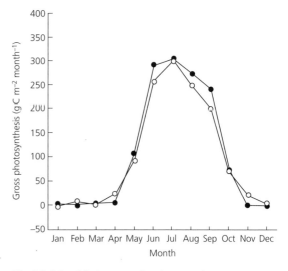

Fig. 8.3 Monthly integrated estimates of gross photosynthesis made with a stomatal constrained quantum-efficiency model (●) agreed well with integrated monthly values acquired through continuous flux measurements (○) made throughout an entire year ($r^2 = 0.97$). (After Waring *et al.* 1995.)

8.4.2 Autotrophic respiration

A large fraction of gross photosynthesis is expended through autotrophic respiration in the construction and maintenance of plant tissue. Physiologists generally separate respiration into three major components: photorespiration (R_p), construction respiration (R_c) and maintenance respiration (R_m).

Photorespiration

In light, photorespiration occurs in the process of generating the substrate for the enzyme ribulose bisphosphate carboxylase–oxygenase (Rubisco), essential for the primary fixation of CO_2. Under conditions where the CO_2 concentration within the leaf is depleted relative to the concentration of O_2, the enzyme releases CO_2 rather than fixing it into carbohydrates. In general, higher temperatures stimulate photorespiration to the extent that simple estimates of gross photosynthesis may be overestimated by as much as 10%. Conversely, rising CO_2 concentrations in the atmosphere tend to moderate the effects of rising temperature on photorespiration and shift the optimum temperature for photosynthesis upward. At temperatures above 40°C, gross photosynthesis decreases abruptly because of changes in chloroplast and enzyme activity (Berry & Downton 1982).

Construction respiration

The synthesis of plant tissue requires the metabolism of more resources than found in the final product. Rates of construction respiration for various tissues differ, depending on their composition. From an analysis of biochemical pathways Penning de Vries (1975) estimated that the production of 1 g of lipid would require 3.02 g of glucose, whereas 1 g of lignin, protein or sugar polymer might require 1.90, 2.35 and 1.18 g of glucose respectively. More recently, empirical relations have been derived to estimate the total cost of construction based on a correlation with the heat of combustion, ash, and organic nitrogen content of tissue and its biochemical composition (McDermitt & Loomis 1981;

Vertregt & Penning de Vries 1987; Williams *et al.* 1987; Griffin 1994). In general, 1.25–1.35 g of carbon are required to produce tissue containing 1 g of carbon. Although the amount of tissue produced may be a function of temperature, the construction cost per gram is fixed. This fact makes it possible to assess construction respiration at annual time steps from forest inventory data using allometric relations between measured increases in stem diameter and the production of foliage, branches, stemwood and large-diameter roots (Ryan 1991).

Maintenance respiration

Maintenance respiration (R_m), the basal rate of metabolism, includes the energy expended on ion uptake and transfer within plants. Repair of injured tissue may greatly increase respiration above the basal rate. Because trees accumulate a large amount of conducting and storage tissue as they age, the observed decrease in relative growth rates associated with age has often been assumed to reflect increasing maintenance costs (Whittaker 1975; Waring & Schlesinger 1985). However, most of the tissue in trees is wood and, of this, only the sapwood contains any living cells (5–30%). It is unlikely that the metabolism of this extra sapwood is enough to limit growth. Enzymatic activity associated with protein turnover can be predicted as a function of nitrogen content, the fraction of living cells in foliage, branches, stems and roots, and temperature of the tissue (Stockfer & Linder 1998). Maintenance respiration generally increases exponentially with temperature within the range of biological activity, as described by the formula

$$R_m(T) = R_0 Q_{10}^{[(T-T_0)/10]} \qquad [8.2]$$

where R_0 is the basal respiration rate at $T = 0$°C (or other reference temperature) and Q_{10} is the respiration quotient, which represents the change in the rate of respiration for a 10°C change in temperature (T). The Q_{10} is usually about 2.0–2.3. Under field conditions, where daily temperature variation is large, the non-linearity in the response of maintenance respiration can be important because the respiration rate will depend on the pattern of temperature variation even in sites with the same mean temperature. This has been modelled by fitting a sine function to minimum–maximum temperature differences and integrating the response not only daily but annually (Hagihara & Hozumi 1991; Ryan 1991). From analyses made in pine forests growing at mean annual temperatures from 5 to >20°C, Ryan *et al.* (1995) concluded that the annual above-ground woody tissue maintenance respiration represents 5–15% of gross photosynthesis and that other factors must therefore be responsible for commonly observed reductions of more than 50% in above-ground production as forests age.

8.4.3 Net primary production

Annual net primary production (NPP) is conceptually easy to define as the residual carbon products remaining after autotrophic respiration $(R_p + R_c + R_m)$ has been subtracted from gross photosynthesis. In practice, it is difficult to obtain accurate estimates of seasonal changes in foliage and fine-root mass, variation in metabolic reserves, and precise measurements of tissue tem-

peratures throughout a forest, thus making estimates of NPP difficult. None the less, carbon balance estimates that take into account construction and maintenance respiration along with estimates of annual growth come fairly close to independent estimates of gross primary production (GPP), modelled or derived from micrometeorological measurements (Fig. 8.4).

If one assumes that half the carbon allocated below ground goes into the production of roots while the rest is expended in construction and maintenance respiration, an estimate of total NPP can be made from datasets similar to those presented in Fig. 8.4. When this was done for 12 temperate forests, the ratio NPP:GPP approached a constant, 0.47 ± 0.04 (Waring *et al.* 1998). Similar analyses in boreal forests indicate that NPP : GPP $\cong 0.3$ (Ryan *et al.* 1997). This lower ratio in the boreal forest zone is attributed to delayed recovery in photosynthetic capacity of evergreen canopies in spring when solar radiation peaks (Jarvis *et al.* 1997; Ryan *et al.* 1997). Similar constant relationships between the ratio of autotrophic respiration and gross photosynthesis have been reported in growth-room experiments

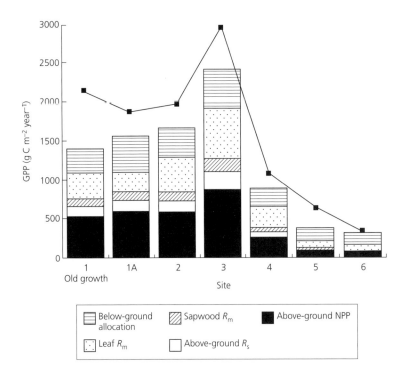

Fig. 8.4 For seven sites across a steep environmental gradient in western Oregon, USA, gross primary production (GPP) (■) estimated with a daily resolution process model is compared with a component carbon balance analysis that sums to GPP (stacked bars). (After Williams *et al.* 1997.)

where pine and eucalypt seedlings have been raised under a range of temperatures from 15 to 25°C (Gifford 1994). It is important to emphasize that the fraction of carbon allocated to wood production can vary by more than threefold as a result of differences in the way that carbon resources are allocated to growth.

8.4.4 Allocation of carbon resources

Although the overall NPP : GPP ratio derived from carbon balance analyses may approach a constant, the proportion of carbon allocated to roots increases with climatic limitations on gross photosynthesis (Runyon *et al.* 1994). The availability of nutrients alone can alter above- and below-ground allocation as much as climatic limitations, as demonstrated by deriving carbon budgets for plantations of *Pinus radiata* provided with different levels of available nitrogen (Fig. 8.5).

For ecologists, the seasonal dynamics of carbon uptake and allocation are critical for they define the growing season, the time of flowering and seed production and, often, the susceptibility of forests to damage from frost, herbivores and pathogens. To meet this challenge, ecophysiologists have applied insights developed from growth-room studies where photoperiod, nutrition, temperature and available water have been controlled (Partanen *et al.* 1998). A number of empirical models have been developed to predict budbreak, foliage and stem elongation, and bud set in temperate and boreal forests from accumulated daily temperature values (heat sums) above a threshold after photoperiodic and chilling requirements have been met (Cleary & Waring 1969; Hari & Hakkinen 1991; Dougherty *et al.* 1994; Whitehead *et al.* 1994). In tropical and subtropical forests, the onset of a wet season (or the end of a dry season) may initiate conditions that can be predicted to favour the initiation or cessation of growth (Borchert 1973, 1994; Reich 1995).

Although the theoretical basis for developing allocation models is still weak (Farrar 1996), some progress has been made. Various schemes have been proposed to model carbon allocation, as summarized by Cannell and Dewar (1994). These schemes are based on different assumptions about the controls on resource allocation as defined by: (i) sink strengths (Ford & Kiester 1990; Luxmoore 1991; Thornley 1991); (ii) resource deficiency (Ewel & Gholz 1991); (iii) functional balance (Davidson 1969); (iv) distance from essential resources (Weinstein *et al.* 1991); and (v) growth optimization (Ågren & Ingestad 1987; Johnson & Thornley 1987). Although the various schemes on allocation are interrelated, none deals with underlying processes, and some are in direct conflict with one another.

Models that relate root uptake of nitrogen and shoot uptake of carbon have been improved by incorporating more details. Initially, growth was assumed to be a simple function of the amount of carbon and nitrogen available, with the product of carbohydrate and nitrogen resources in foliage and roots determining the relative allocation. Then models began to include resistance to transport and demands for resources by intermediate structures between roots and shoots. More recently, models consider sugar and amino acid transport separately through phloem and sapwood (Cannell & Dewar 1994). Root-to-shoot gradients in water potential also influence rates at which resources are transported through the two vascular systems (Dewar 1993).

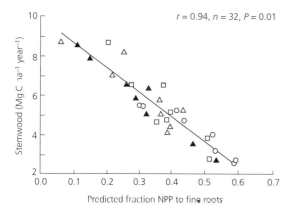

Fig. 8.5 Increasing the availability of nitrogen to *Pinus radiata* plantations in New Zealand did not change the concentration of nitrogen in foliage but did cause a proportional shift in carbon allocation of net primary production (NPP) away from fine roots into stemwood. Root production was estimated from a component analysis as a residual. (After Beets & Whitehead 1996.)

The majority of these models assume that growth at any instant is determined by the availability of substrate (C, N, etc.); however, availability is difficult to define. When plants are dormant, hormonal controls limit growth regardless of the concentration of carbohyrates and other resources in tissues. When shoot extension is rapid, root growth may be inhibited, even when starch concentrations are high (Ford & Deans 1977). Hormonal control is again involved in root development (evidence in McWilliam 1972 discussed by Dewar *et al.* 1994) and there is also evidence that current photosynthate rather than starch reserves is required to produce new roots (Marshall & Waring 1985) or to support the process of symbiotic nitrogen fixation (McNabb & Cromack 1983). With the present state of understanding, it is reasonable to focus on developing accurate phenological models to define seasonal timing in growth activity as a function of daylength, temperature, drought and chilling requirements. Such phenological models would predict the fluctuating strength of competing sinks, while substrate availability would determine how much growth is possible.

With knowledge of phenology and environmental conditions, the pools of available carbon and nitrogen can be allocated into biomass, storage reserves and losses through respiration. Weinstein *et al.* (1991) generated seasonal estimates of carbon allocation for red spruce (*Picea rubens*) by assuming priorities based on proximity to the resources (scheme iv in the list above). In this model, leaf growth has first priority for carbon after maintenance requirements of all living tissue are met. Storage of carbon in leaves is followed by growth and storage in branches, stem, coarse roots and, last, fine roots. Carbon in excess of that needed to meet the maximum growth rate of an organ was passed down the priority chain. The priority ranking for allocation of water and nutrients would, according to this proximity logic, be opposite, with fine roots ranked first and new foliage ranked last. Similar proximity logic was discussed by Thornley (1976). Because critical resources come from opposite directions, no given organ is likely to grow at its full potential unless phenology limits all growth elsewhere. Predictions of photosynthate allocation seasonally in *Picea rubens* based on phenology and proximity logic showed reasonable growth patterns and matched measured storage reserves within 10% and total carbon content within 2% at the end of the year (Fig. 8.6).

Most seasonal models of carbon allocation require specific knowledge of tree phenology, definition of the limits to growth of various organs, and specification of the size of storage reserves in all major organs. In reality, storage reserves should encompass not only starch but also amino acids, as well as a host of other compounds that protect against injury from temperature extremes

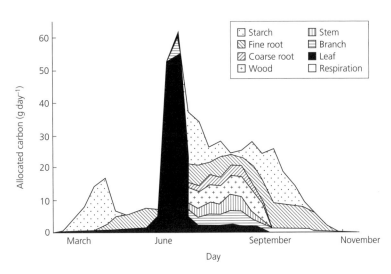

Fig. 8.6 Seasonal carbon allocation patterns generated with a simulation model for red spruce (*Picea rubens*) depend on phenology and proximity to resources. Starch reserves accumulate when growth is at a minimum compared to photosynthesis. Leaf growth attains priority in June while the production of other structural components peaks thereafter. (After Weinstein *et al.* 1991.)

and attacks from insects and diseases. Seasonally, trees are most susceptible to attack from insects and pathogens when storage reserves are low, a condition that occurs during rapid leaf expansion or when growth is prematurely halted during extended drought.

8.5 SUSCEPTIBILITY OF FORESTS TO HERBIVORES AND PATHOGENS

As emphasized in earlier sections, knowledge of the biochemical composition of tissue provides more insight into nutrient availability and the regulation of photosynthesis than a simple analysis of the total elemental content. In this section, we continue this emphasis in a review of the major types of biochemical defences that trees have evolved in response to herbivores and pathogens (see also Chapter 6, this volume). Although plant defence systems are highly variable, some general measures of stress have been developed to assess the susceptibility of forests to attack from herbivores and pathogens.

8.5.1 Biochemical and structural defences

Although each species has a unique suite of protective defences, some general symptoms of stress have emerged from comparative studies. Defensive chemicals present in plants are broadly classified into compounds that contain nitrogen and those that do not contain nitrogen. Compounds that contain nitrogen include cyanogenic glucosides, alkaloids and non-protein amino acids; compounds without nitrogen include tannins, terpenes, phytoalexins, steroids and phenolic acids. Each kind of compound may serve to protect the plant in a variety of ways.

Some trees produce fungistatic and bacteriostatic compounds that prevent colonization by pathogens. Other compounds act as physical barriers, such as waxes on the leaf surface or resins or lignin in cell walls. Increasing fibre and lowering water content decrease digestibility of plant tissue and reduce herbivore growth rates and survival (Scriber & Slansky 1981). Tannins precipitate protein, inhibiting most enzyme reactions and making protein present in plant tissue nutritionally unavailable to most animals and microbes (Zucker 1983). Phytoalexins are lipid-soluble compounds activated following an attack by pathogens and exhibit antibiotic properties (Harborne 1982). The alkaloids found in many angiosperms are particularly toxic to a variety of mammals (Swain 1977).

Changes in host biochemistry may also affect the colonization of organisms helpful to the host tree. These include protective ant colonies, macroorganisms that graze on bacteria, and symbiotic associations of nitrogen-fixing bacteria and mycorrhizal root fungi. These beneficial organisms may directly infect or prey on attacking organisms, release antibiotics or provide essential nutrients. Many of the compounds released as exudates, which include a variety of polysaccharides, organic acids and amino acids, are essential to beneficial associates, but when these organisms are absent they can also be assimilated by herbivores and pathogens.

In general, defensive compounds that lack nitrogen reach high concentrations in cells, often 10–15% by weight, whereas nitrogen-containing defensive compounds are usually at concentrations below 1%. Plants expend less total energy in the synthesis of small amounts of nitrogen-containing defensive compounds than when producing large amounts of carbon-rich compounds, although variations in the turnover rates of defensive compounds and differences in the relative growth rates must be considered in the assessment of synthesis and maintenance costs (Bryant *et al.* 1991). Nitrogen-containing compounds are most frequently found in deciduous fast-growing vegetation, whereas defensive compounds without nitrogen are more characteristic of slow-growing plants, particularly evergreens with long leaf lifespans (Bryant *et al.* 1986; Coley 1988).

Plants that depend on defensive compounds rich in nitrogen are at a competitive disadvantage where nitrogen is in short supply. On the other hand, plants producing carbon-rich defensive compounds are at a disadvantage when growing in shade with an abundant supply of nitrogen. Those plants adapted to more fertile soils may be expected to build a variety of defensive compounds from nitrogen. Thus, alkaloids predominate in the foliage of trees in many lowland tropical forests where nitrogen is relatively abundant (McKey *et al.* 1978). Plants growing in areas where nitrogen is scarce generally produce tannins and related carbon-based defensive compounds. Thus, in some tropical forests growing on sterile sands, primates must survive mainly by eating fruits (Gartlan *et al.* 1980). A similar pattern in distribution of vegetation with nitrogen-based or carbon-based defensive compounds to that observed in tropical forests has been reported in boreal and temperate forests (Rhoades & Cates 1976; Bryant *et al.* 1991).

To meet the challenge of insects and pathogens, many plants are able to produce toxic compounds quickly and construct barriers that consist of dead or resin/gum-filled tissue almost immediately after attack (Schultz & Baldwin 1982; Raffa & Berryman 1983). In response to localized insect activity, foliage throughout an entire tree may become less palatable (Haukioja & Niemelä 1979; Karban & Myers 1989). Morphological responses, such as stiffer thorns on *Acacia* trees, may also be induced by grazing (Seif el Din & Obeid 1971). Bark sloughing is a response to attack by woolly aphid (*Adeleges piceae*) and the aphid only reaches epidemic populations when balsam fir (*Abies balsamea*), which lacks the ability to shed bark, replaces native forest species (Kloft 1957).

The extent to which local environmental conditions can alter a plant's biochemical and structural properties, and affect their selection by insects and pathogens, has been demonstrated under precisely controlled laboratory conditions (Waring *et al.* 1985; Larsson *et al.* 1986; Coleman & Jones 1988a,b; Coleman *et al.* 1988). Field experiments have also demonstrated that plant susceptibility can be altered by varying the availability of light and nutrients (Bryant *et al.* 1985; Coley *et al.* 1985). Because many defensive compounds have high rates of turnover, it is essential that their assessment be made at critical times when the plants are being attacked (Lorio 1986; Reichardt *et al.* 1991).

The linking of ecology and physiology has explained unexpected responses to management practices designed to favour the abundance of certain game species. In the Lake States of the USA, protection of large areas of aspen (*Populus tremuloides*) forests beyond a certain age would be detrimental to the winter survival of ruffed grouse, which depends on a favourable biochemical composition of aspen buds that is found only with younger stands (Jakubas *et al.* 1989). Similarly, in the Alaskan coastal forest of western hemlock (*Tsuga heterophylla*), cutting old-growth forests increased the biomass of a prime browse species of blueberry (*Vaccinium ovalifolium*) by nearly fourfold but the population of Sitka black-tailed deer (*Odocoileus hemionus sitkensis*) decreased. Field experiments, combined with feeding trials, demonstrated that blueberry plants become less palatable to deer when grown under high light conditions (Fig. 8.7). An increase in tannins and non-tannin phenolics reduced the digestibility of protein nitrogen in plant tissue below that required to sustain female deer with one or more fawns through the winter months (McArthur *et al.* 1993). Structural properties, such as specific leaf weight, also change in parallel with leaf biochemistry (Fig. 8.7) (Rose 1990).

8.5.2 Biochemical and allocation ratios

Ecophysiologists have searched for simple indices of stress that might aid managers in quantifying the risk of attack to forests from native insects and pathogens. Various ratios were considered, following success in reinterpreting nutrient status in terms of elemental ratios. Ratios that contrast the relative concentrations of defensive compounds to those compounds that promote the growth of insects and pathogens have obvious merit. For example, the incidence of infection by pure cultures of the root pathogen *Armillaria ostoyae* shows marked sensitivity to the ratio of phenolics/sugars and lignin/sugars in the root cortex of Douglas fir (Entry *et al.* 1991) (Fig. 8.8).

Structural ratios also have value in the assessment of risk, as noted in the correlation between specific leaf weight and digestible nitrogen (see Fig. 8.7). The two most widely used indices are based on allocation differences: (i) light-use efficiency, i.e. annual above-ground growth per unit of absorbed radiation (Monteith 1972); and (ii) growth efficiency, i.e. wood growth per unit of

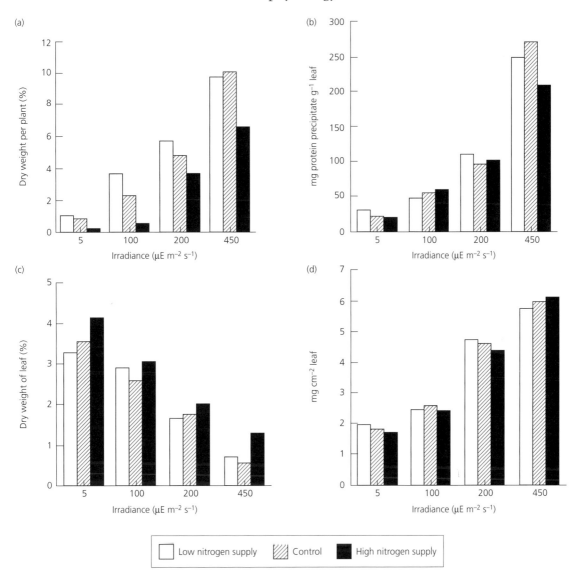

Fig. 8.7 Field-grown blueberry plants placed under zero to three layers of netting that received increasing amounts of nitrogen and other nutrients showed significant changes in (a) annual shoot production, (b) tannin concentrations in leaves, (c) digestible nitrogen and (d) specific leaf weight. These responses explain why open-grown plants were less palatable to deer than those found under more shaded conditions in gaps of an old-growth hemlock (*Tsuga heterophylla*) forest in south-western Alaska. (From Rose 1990 in Waring & Running 1998.)

foliage (Waring 1983). Light-use efficiency has the advantage that satellite technology can be used to predict solar radiation, the fraction absorbed by the canopy and above-ground growth (Goward *et al.* 1987; Dye & Goward 1993; Prince & Goward 1995; Lefsky *et al.* 1999; Means *et al.* 1999).

Growth efficiency has the advantage that it can be applied to assess the risk of insect or disease attack to individual trees (Oren *et al.* 1985; Christiansen *et al.* 1987). For some species, leaf area is correlated with sapwood basal area (Waring *et al.* 1982). In cases where sapwood converts to

(a)

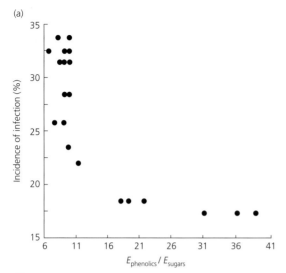

(b)

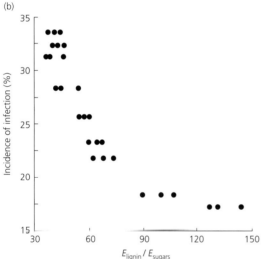

Fig. 8.8 (a) Incidence of infection by a root pathogen, *Armillaria*, on Douglas fir (*Pseudotsuga menziesii*) increased significantly once the ratio of energy required for phenolic degradation/energy available from sugars ($E_{phenolics}/E_{sugars}$) fell below about 15×10^{-4} kJ mol^{-1} g^{-1} root bark. (b) Similarly, when the ratio of energy required to degrade lignin/energy available from sugars (E_{lignin}/E_{sugars}) decreased below about 60×10^{-4} kJ mol^{-1} g^{-1} root bark, the incidence of infection increased significantly. (After Entry *et al.* 1991.)

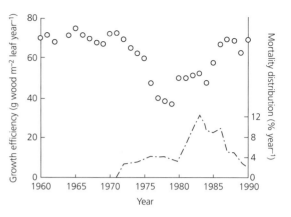

Fig. 8.9 Dynamics of a spruce budworm outbreak in a balsam fir (*Abies balsamea*) forest in Quebec, Canada, was reconstructed from extracted wood cores. The growth efficiency for the majority of trees decreased before the insect outbreak occurred in the 1970s. As trees were selectively killed between 1971 and 1990, surviving trees regained efficiency in wood production per unit leaf area. (After Coyea & Margolis 1994.)

heartwood after a fixed number of years, historical analyses of growth efficiency can be related to outbreaks and decline in insect populations (Fig. 8.9).

8.6 ECOPHYSIOLOGICAL MODELS OF FOREST GROWTH

There are two classes of forest growth models that are well developed: (i) those that focus on the whole ecosystem and (ii) those that concentrate on the life-cycle dynamics of trees in the forest (see also Chapter 9, this volume). Both classes are process models: ecosystem models compute growth from the seasonal dynamics of canopy carbon balances and allocation rules based on limiting resources; and life-cycle models emphasize disturbance, recruitment and mortality processes that affect the establishment, growth and death of individual trees. In this section we refer to only the first, ecosystem models. Examples of widely applied ecosystem models are reviewed and contrasted with life-cycle models by Ågren *et al.* (1991), Ryan *et al.* (1996), Landsberg and Gower (1997) and Waring and Running (1998). Most ecosystem models have a structure similar to that shown in Fig. 8.10.

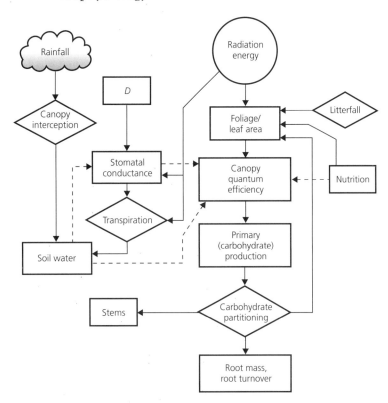

Fig. 8.10 A simplified process model to predict forest growth. (After Landsberg & Coops 1999.)

These models are driven within environmental inputs: rainfall, solar radiation, temperature and atmospheric vapour pressure deficit (*D*). Some models require daily weather summaries while others operate with weekly or monthly averages (see Aber & Federer 1992; Landsberg & Waring 1997). The canopy and forest floor intercept radiation, precipitation and aerosols. Some water recharges the soil and enters streams, while the rest is converted to water vapour and evaporated or transpired into the atmosphere. The upper limit on photosynthesis is set by the amount of visible light intercepted by the canopy and the maximum quantum efficiency or maximum photosynthetic capacity, the latter two being a function of nutrition. Reductions in stomatal conductance reduce the rates of water vapour loss via transpiration and CO_2 uptake via photosynthesis. Stomatal conductance is affected by all the climatic variables and by the availability of soil water. The products of photosynthesis are partitioned into growth and plant respiratory losses of CO_2 back to the atmosphere. Carbon is partitioned into foliage, stems and roots differentially, depending on the extent that different resources limit photosynthesis and growth. Eventually plant components are shed and converted into detritus that is incorporated into soil organic matter, leached or converted into CO_2 via heterotrophic respiration (not shown in the schematic diagram).

This type of process-linked model demands reasonable agreement in estimates of carbon, water and nutrient cycling. By extending the analyses to decades one can quickly identify unreasonable assumptions because they lead to unrealistic leaf areas, wood production and litter accumulation. The models generally predict the total light absorbed by the canopy as well as leaf area index so that calculations of light-use efficiency and growth efficiency are easily acquired. In addition, the content of nitrogen (if not other elements) can be predicted by the models. Some models keep track of diameter distributions of trees as well as stocking (Korol *et al.* 1991; Landsberg & Waring 1997), and all make separate estimates of above- and below-ground production.

One of the main advantages of process-linked models is that they can serve to diagnose the extent to which forest growth is now limited or may become limited in the future by climate. They can also be used to compare the influence of climate with that of site variables that may be manipulated, such as the floristic composition of the forest (changes in allocation) and the fertility of the soil (changes in allocation and photosynthetic capacity).

8.7 SUMMARY

By combining the traditional empiricism of ecology with the process orientation of physiology, ecophysiology has helped expand the reliability and generality of models designed to forecast the status of individual stands, forests and regional assemblages of vegetation. This merger of fields has contributed to a better understanding of the processes controlling forest growth and the susceptibility of forests to biotic disturbances. As a result, forest managers have acquired diagnostic tools that allow them to judge better the implications of various options today and in the future.

ACKNOWLEDGEMENTS

To provide this synthesis, we have relied heavily on material presented in a recently published textbook by Waring and Running (1998). All but Fig. 8.10 come directly from that source.

REFERENCES

Aber, J.D. & Federer, C.A. (1992) A generalized, lumped-parameter model of photosynthesis, evapotranspiration and net primary production in temperate and boreal forest ecosystems. *Oecologia* **92**, 463–74.

Aber, J.D., Nadelhoffer, K.J., Steudler, P. & Melillo, J. (1989) Nitrogen saturation in northern forest ecosystems. *Bioscience* **39**, 378–86.

Ågren, G.I. (1983) Nitrogen productivity of some conifers. *Canadian Journal of Forest Research* **13**, 494–500.

Ågren, G.I. & Ingestad, T. (1987) Root:shoot ratio as a balance between nitrogen productivity and photosynthesis. *Plant, Cell and Environment* **10**, 579–86.

Ågren, G.I., McMurtrie, R.E., Parton, W.J., Pastor, J. & Shugart, H.H. (1991) State-of-the-art models of production–decomposition linkages in conifer and grassland ecosystems. *Ecological Applications* **1**, 118–38.

Anekonda, T.S., Criddle, R.S., Libby, W.J. & Hansen, L.D. (1993) Spatial and temporal relationships between growth traits and metabolic heat rates in coast redwood. *Canadian Journal of Forest Research* **23**, 1793–8.

Anekonda, T.S., Criddle, R.S., Libby, W.J., Reidenbach, R.W. & Hansen, L.D. (1994) Respiration rates predict differences in growth of coast redwood. *Plant, Cell and Environment* **17**, 197–203.

Attiwill, P.M. & Adams, M.A. (1993) Nutrient cycling in forests. *New Phytologist* **124**, 561–82.

Axelsson, E. & Axelsson, B. (1986) Changes in carbon allocation patterns in spruce and pine trees following irrigation and fertilization. *Tree Physiology* **2**, 189–204.

Babalola, O., Boersma, L. & Youngberg, C.T. (1968) Photosynthesis and transpiration of Monterey pine seedlings as a function of soil water suction and soil temperature. *Plant Physiology* **43**, 515–21.

Baes, C.F. III & McLaughlin, S.B. (1984) Trace elements in tree rings: evidence of recent historical air pollution. *Science* **224**, 494–7.

Beets, P.N. & Whitehead, D. (1996) Carbon partitioning in *Pinus radiata* in relation to foliage nitrogen status. *Tree Physiology* **16**, 131–8.

Berry, J.A. & Downton, W.J.S. (1982) Environmental regulation of photosynthesis. In: Govindjee, R., ed. *Photosynthesis, Development, Carbon Metabolism, and Plant Productivity*, pp. 263–343. Academic Press, New York.

Bondietti, E.A., Baes, C.F. III & McLaughlin, S.B. (1989) Radial trends in cation ratios in tree rings as indicators of the impact of atmospheric deposition on forests. *Canadian Journal of Forest Research* **19**, 586–94.

Borchert, R. (1973) Simulation of rhythmic tree growth under constant conditions. *Physiologia Plantarum* **29**, 173–80.

Borchert, R. (1994) Water storage in soil or tree stems determines phenology and distribution of tropical dry forest trees. *Ecology* **75**, 1437–49.

Bowen, G.D. (1970) Effects of soil temperature on root growth and on phosphate uptake along *Pinus radiata* roots. *Australian Journal of Soil Research* **8**, 31–42.

Bryant, J.P., Chapin, F.S. III, Reihardt, P. & Clausen, T. (1985) Adaptation to resource availability as a determinant of chemical defense strategies in woody plants. *Recent Advances in Phytochemistry* **19**, 219–37.

Bryant, J.P., Chapin, F.S. III, Reichardt, P. & Clausen, T.

(1986) Chemical mediated interactions between plants and other organisms. *Recent Advances in Phytochemistry* **19**, 219–37.

Bryant, J.P., Provenza, F.D., Pastor, J., Reichardt, P.B., Clausen, T.P. & du Toit, J.T. (1991) Interactions between woody plants and browsing mammals mediated by secondary metabolites. *Annual Review of Ecology and Systematics* **22**, 431–46.

Buchmann, N., Gebauer, G. & Schulze, E.-D. (1996) Partitioning of ^{15}N-labeled ammonium and nitrate among soil, litter, below- and above-ground biomass of trees and understory in a 15-year-old *Picea abies* plantation. *Biogeochemistry* **33**, 1–23.

Cannell, M.G.R. & Dewar, R.C. (1994) Carbon allocation in trees: a review of concepts for modelling. *Advances in Ecological Research* **25**, 60–140.

Chapin, F.S. & Kedrowski, R.A. (1983) Seasonal changes in nitrogen and phosphorus fractions and autumn retranslocation in evergreen and deciduous taiga trees. *Ecology* **64**, 376–91.

Christiansen, E., Waring, R.H. & Berryman, A.A. (1987) Resistance of conifers to bark beetle attack: searching for general relationships. *Forest Ecology and Management* **22**, 89–106.

Cleary, B.D. & Waring, R.H. (1969) Temperature: collection of data and its analysis for the interpretation of plant growth and distribution. *Canadian Journal of Botany* **47**, 167–73.

Coleman, J.S. & Jones, C.G. (1988a) Plant stress and insect performance: cottonwood, ozone and leaf beetle. *Oecologia* **76**, 57–61.

Coleman, J.S. & Jones, C.G. (1988b) Acute ozone stress on eastern cottonwood (*Populus deltoides* Bartr.) and the pest potential of the aphid, *Chaitophorus populicola* Thomas (Homoptera: Aphididae). *Environmental Entomology* **17**, 207–12.

Coleman, J.S., Jones, C.G. & Smith, W.H. (1988) Interactions between an acute ozone dose, eastern cottonwood, and *Marssonina* leaf spot: implications for pathogen community dynamics. *Canadian Journal of Botany* **66**, 863–8.

Coley, P.D. (1988) Effects of plant growth rate and leaf lifetime on the amount and type of anti-herbivore defense. *Oecologia* **74**, 531–6.

Coley, P.D., Bryant, J.P. & Chapin, F.S. (1985) Resource availability and plant antiherbivore defense. *Science* **230**, 895–9.

Coyea, M.R. & Margolis, H.A. (1994) The historical reconstruction of growth efficiency and its relationship to tree mortality in balsam fir ecosystems affected by spruce budworm. *Canadian Journal of Forest Research* **24**, 2208–21.

Dang, Q.L., Margolis, H.A. & Collatz, G.J. (1998) Parameterization and testing of a coupled photosynthesis–stomatal conductance model for boreal trees. *Tree Physiology* **18**, 141–54.

Davidson, R.L. (1969) Effect of root–leaf temperature differentials on root–shoot ratios in some pasture grasses and clover. *Annals of Botany* **33**, 561–9.

Dawson, T.E. (1993) Water sources of plants as determined from xylem-water isotopic composition: perspectives on plant competition, distribution, and water relations. In: Ehleringer, J.R., Hall, A.E. & Farquhar, G.D., eds. *Stable Isotopes and Plant Carbon–Water Relations*, pp. 465–96. Academic Press, San Diego.

Dewar, R.C. (1993) A root–shoot partitioning model based on carbon–nitrogen–water interactions and Munch phloem flow. *Functional Ecology* **7**, 356–68.

Dewar, R.C., Ludlow, A.R. & Dougherty, P.M. (1994) Environmental influences on carbon allocation in pines. *Ecological Bulletins* **43**, 92–101.

Donovan, L.A. & McLeod, K.W. (1985) Morphological and root carbohydrate responses of bald cypress to water level and water temperature regimes. *Journal of Thermal Biology* **10**, 227–32.

Dougherty, P.M., Whitehead, D. & Vose, J.M. (1994) Environmental influences on the phenology of pine. *Ecological Bulletins* **43**, 64–75.

Dye, D.G. & Goward, S.N. (1993) Photosynthetically active radiation absorbed by global land vegetation in August 1984. *International Journal of Remote Sensing* **14**, 3361–4.

Edwards, W.R.N., Jarvis, P.G., Landsberg, J.J. & Talbot, H. (1986) A dynamic model for studying flow of water in single trees. *Tree Physiology* **1**, 309–24.

Entry, J.A., Cromack, K. Jr, Kelsey, R.G. & Martin, N.E. (1991) Response of Douglas-fir to infection by *Armillaria ostoyae* after thinning or thinning plus fertilization. *Phytopathology* **81**, 682–9.

Ericsson, A., Nordén, L.-G., Näsholm, T. & Walheim, M. (1993) Mineral nutrient imbalances and arginine concentrations in needles of *Picea abies* (L.) from two areas with different levels of airborne deposition. *Trees* **8**, 67–74.

Ericsson, T. (1994) Nutrient dynamics and requirements of forest crops. *New Zealand Journal of Forestry Science* **24**, 133–68.

Ewel, K.C. & Gholz, H.L. (1991) A simulation model of the role of below-ground dynamics in a slash pine plantation. *Forest Science* **37**, 397–438.

Farquhar, G.D. & Sharkey, T.D. (1982) Stomatal conductance and photosynthesis. *Annual Review of Plant Physiology* **33**, 317–45.

Farquhar, G.D., von Caemmerer, S. & Berry, J.A. (1980) A biochemical model of photosynthetic CO_2 assimilation in leaves of C_3 species. *Planta* **149**, 78–90.

Farrar, J.F. (1996) Sinks: integral parts of a whole plant. *Journal of Experimental Botany* **47**, 1273–9.

Ford, E.D. & Deans, J.D. (1977) Growth of a Sitka spruce plantation: spatial distribution and seasonal fluctuations of lengths, weights and carbohydrate concentrations of fine-roots. *Plant and Soil* **47**, 463–85.

Ford, E.D. & Kiester, R. (1990) Modeling the effects of pollutants on the process of tree growth. In: Dixon, R.K., Meldahl, R.S., Ruark, G.A. & Warren, W.G., eds. *Process Modeling of Forest Growth Responses to Environmental Stress*, pp. 324–37. Timber Press, Portland, Oregon.

Gartlan, J.S., McKey, D.B., Waterman, P.G., Struhsaker, C.N. & Struhsaker, T.T. (1980) A comparative study of the phytochemistry of two African rain forests. *Biochemical and Systematic Ecology* **8**, 401–22.

Gifford, R.M. (1994) The global carbon cycle: a viewpoint on the missing sink. *Australian Journal of Plant Physiology* **21**, 1–15.

Goldstein, G., Andrade, J.L., Meinzer, F.C. *et al.* (1998) Stem water storage and diurnal patterns of water use in tropical forest canopy trees. *Plant, Cell and Environment* **21**, 397–406.

Goward, S.N., Dye, D., Kerber, A. & Kalb, V. (1987) Comparison of North and South American biomes from AVHRR observations. *Geocarto International* **1**, 27–39.

Griffin, K.L. (1994) Calorimetric estimates of construction cost and their use in ecological studies. *Functional Ecology* **8**, 551–62.

Hagihara, A. & Hozumi, K. (1991) Respiration. In: Raghavendra, A.S., ed. *Physiology of Trees*, pp. 87–110. John Wiley & Sons, New York.

Harborne, J.B. (1982) *Introduction to Ecological Biochemistry*. Academic Press, New York.

Hari, P. & Hakkinen, H. (1991) The utilization of old phenological time series of budburst to compare models describing annual cycles of plants. *Tree Physiology* **8**, 281–7.

Haukioja, E. & Niemelä, P. (1979) Birch leaves as a resource for herbivores: seasonal occurrence of increased resistance in foliage after mechanical damage of adjacent leaves. *Oecologia* **39**, 151–9.

Hawkins, B.J. & Sweet, G.B. (1989) Photosynthesis and growth of present New Zealand forest trees relate to ancient climates. *Annales Des Sciences Forestières* **46**, 512–14.

Hawkins, B.J., Henry, G. & Kiiskila, B.R. (1998) Biomass and nutrient allocation in Douglas-fir and amabilis fir seedlings: influence of growth rates and nutrition. *Tree Physiology* **18**, 803–10.

Holbrook, N.M. & Sinclair, T.R. (1992) Water balance in the arborescent palm, *Sabal palmetto*. I. Stem structure, tissue water release properties and leaf epidermal conductance. *Plant, Cell and Environment* **15**, 393–9.

Hubbard, R.M., Bond, B.J. & Ryan, M.G. (1999) Evidence that hydraulic conductance limits photosynthesis in old *Pinus ponderosa* trees. *Tree Physiology* **19**, 165–72.

Hunter, I.R., Hunter, J.A.C. & Graham, J.D. (1987) *Pinus radiata* stem volume increment and its relationship to needle mass, foliar and soil nutrients, and fertiliser inputs. *New Zealand Journal of Forest Science* **17**, 67–75.

Ingestad, T. (1979) Nitrogen stress in birch seedlings. II. N, K, P, Ca and Mg nutrition. *Physiologia Plantarum* **45**, 149–57.

Jakubas, W.J., Gullion, G.W. & Clausen, T.P. (1989) Ruffed grouse feeding behavior and its relationship to secondary metabolites of quaking aspen flower buds. *Journal of Chemical Ecology* **15**, 1899–917.

Jarvis, P.G., Massheder, J.M., Hale, S.E., Moncrieff, J.B., Rayment, M. & Scott, S.L. (1997) Seasonal variation of carbon dioxide, water vapor, and energy exchanges of a boreal black spruce forest. *Journal of Geophysical Research* **102**, 28953–66.

Johnson, D.W. (1992) Nitrogen retention in forest soils. *Journal of Environmental Quality* **21**, 1–12.

Johnson, I.R. & Thornley, J.H.M. (1987) A model of shoot:root partitioning with optimal growth. *Annals of Botany* **60**, 133–42.

Karban, R. & Myers, J.H. (1989) Induced plant responses to herbivory. *Annual Review of Ecology and Systematics* **20**, 331–48.

Katz, C., Oren, R., Schulze, E.-D. & Milburn, J.A. (1989) Uptake of water and solutes through twigs of *Picea abies* (L.) Karst. *Trees* **3**, 33–7.

Kelliher, F.M., Leuning, R., Raupach, M.R. & Schulze, E.-D. (1995) Maximum conductances for evaporation from global vegetation types. *Agricultural and Forest Meteorology* **73**, 1–16.

Kloft, W. (1957) Further investigations concerning the interrelationship between bark conditions of *Abies alba* and infestation by *Adeleges piceae typica* and *A. nusslini schneideri*. *Zeitschrift für Angewante Entomologie* **41**, 438–42.

Korol, R.L., Running, S.W., Milner, K.S. & Hunt, E.R. Jr (1991) Testing a mechanistic carbon balance model against observed tree growth. *Canadian Journal of Forest Research* **21**, 1098–105.

Landsberg, J.J. (1986) *Physiological Ecology of Forest Production*. Academic Press, London.

Landsberg, J.J. & Coops, N. (1999) Modelling forest productivity across large areas and long periods. *Natural Resource Modelling* **12**, 383–411.

Landsberg, J.J. & Gower, S.T. (1997) *Applications of Physiological Ecology to Forest Management*. Academic Press, San Diego.

Landsberg, J.J. & Waring, R.H. (1997) A generalised model of forest productivity using simplified concepts of radiation-use efficiency, carbon balance and partitioning. *Forest Ecology and Management* **95**, 209–28.

Larsson, S., Wiren, A., Lundgren, L. & Ericsson, T. (1986) Effects of light and nutrient stress on leaf phenolic chemistry in *Salix dasyclados* and susceptibility to *Galerucella lineola* (Coleoptera). *Oikos* **47**, 205–10.

Lefsky, M.A., Harding, D., Cohen, W.B., Parker, G. & Shugart, H.H. (1999) Surface lidar remote sensing of basal area and biomass in deciduous forests of eastern Maryland, USA. *Remote Sensing of Environment* **67**, 83–98.

Lerdau, M. & Keller, M. (1997) Control on isoprene emission from trees in a subtropical dry forest. *Plant, Cell and Environment* **20**, 569–78.

Le Roux, X., Singoquet, H. & Vandame, M. (1999) Spatial distribution of leaf dry weight per area and leaf nitrogen concentration in relation to local radiation regime within an isolated tree crown. *Tree Physiology* **19**, 181–8.

Leuning, R., Kelliher, F.M., DePury, D.G.G. & Schulze, E.-D. (1995) Leaf nitrogen, photosynthesis, conductance and transpiration: scaling from leaves to canopies. *Plant, Cell and Environment* **18**, 1183–200.

Linder, S. & Flower-Ellis, J.G.K. (1992) Environmental and physiological constraints to forest yield. In: Teler, A., Mathy, T, & Jeffers, J.N.R., eds. *Responses of Forest Ecosystems to Environmental Changes*, pp. 149–64. Elsevier, The Hague.

Lloyd, J., Grace, J., Miranda, A.C. *et al.* (1995) A simple calibrated model of Amazon rainforest productivity based on leaf biochemical properties. *Plant, Cell and Environment* **18**, 1129–45.

Lorio, P.L. Jr (1986) Growth–differentiation balance: a basis for understanding southern pine beetle–tree interaction. *Forest Ecology and Management* **14**, 259–73.

Luxmoore, R.J. (1991) A source–sink framework for coupling water, carbon and nutrient dynamics of vegetation. *Tree Physiology* **9**, 267–80.

McArthur, C., Robbins, C.T., Hagerman, A.E. & Hanley, T.A. (1993) Diet selection by a ruminant generalist browser in relation to plant chemistry. *Canadian Journal of Zoology* **71**, 2236–43.

McDermitt, D.K. & Loomis, R.S. (1981) Elemental composition of biomass and its relation to energy content, growth efficiency, and growth yield. *Annals of Botany* **48**, 275–90.

McKey, D., Waterman, P.G., Mbi, C.N., Gartlan, J.S. & Struhsaker, T.T. (1978) Phenolic content of vegetation in two African rain forests: ecological implications. *Science* **202**, 61–4.

McNabb, D.H. & Cromack, K. Jr (1983) Dinitrogen fixation by a mature *Ceanothus velutinus* (Dougl.) stand in Western Oregon Cascades. *Canadian Journal of Microbiology* **29**, 1014–21.

McWilliam, A.A. (1972) *Some effects of the environment on the growth and development of* Picea sitchensis. PhD thesis, University of Aberdeen.

Marshall, J.D. & Waring, R.H. (1985) Predicting fine root production and turnover by monitoring root starch and soil temperature. *Canadian Journal of Forest Research* **15**, 791–800.

Mason, H.L. & Langenheim, J.H. (1957) Language analysis and the concept of environment. *Ecology* **65**, 1517–24.

Mead, D.J. & Preston, C.M. (1994) Distribution and retranslocation of ^{15}N in lodgepole pine over eight growing seasons. *Tree Physiology* **14**, 389–402.

Means, J.E., Acker, S.A., Harding, D.J. *et al.* (1999) Use of large-footprint scanning airborne Lidar to estimate forest stand characteristics in the western Cascades of Oregon. *Remote Sensing of Environment* **67**, 298–308.

Mencuccini, M. & Grace, J. (1996) Hydraulic conductance, light interception, and needle nutrient concentration in Scots pine stands (Thetford, U.K.) and their relation with net primary production. *Tree Physiology* **16**, 459–69.

Millard, P. & Proe, M.F. (1992) Storage and internal cycling of nitrogen in relation to seasonal growth of Sitka spruce. *Tree Physiology* **10**, L33–L43.

Miller, H.G. (1995) The influence of stand development on nutrient demand, growth and allocation. *Plant and Soil* **168**, 225–32.

Miller, J.D., Cooper, J.M. & Miller, H.G. (1992) Response of pole-stage Sitka spruce to applications of fertilizer nitrogen, phosphorus and potassium in upland Britain. *Forestry* **65**, 15–33.

Miller, J.D., Cooper, J.M. & Miller, H.G. (1996) Amounts and nutrient weights in litterfall, and their annual cycles, from a series of fertilizer experiments on pole-stage Sitka spruce. *Journal of Forestry* **69**, 289–302.

Monteith, J.L. (1972) Solar radiation and productivity in tropical ecosystems. *Journal of Applied Ecology* **9**, 747–66.

Monteith, J.L. & Unsworth, M.H. (1990) *Principles of Environmental Physics*, 2nd edn. Edward Arnold, London.

Müller, D. (1949) The physiological basis for the deficiency symptoms of plants. *Physiologia Plantarum* **2**, 11–23.

Näsholm, T. & Ericsson, A. (1990) Seasonal changes in amino acids, protein, and total nitrogen in needles of fertilized Scots pine. *Tree Physiology* **6**, 267–82.

Neilson, R.E., Ludlow, M.M. & Jarvis, P.G. (1972) Photosynthesis in Sitka spruce (*Picea sitchensis* (Bong.) Carr.). II. Response to temperature. *Journal of Applied Ecology* **9**, 721–45.

Norman, J.M. & Jarvis, P.G. (1975) Photosynthesis in Sitka spruce (*Picea sitchensis* (Bong.) Carr.). V. Radiation penetration theory and a test case. *Journal of Applied Ecology* **12**, 839–78.

Oren, R., Thies, W.G. & Waring, R.H. (1985) Tree vigor and stand growth of Douglas-fir as influenced by laminated root rot. *Canadian Journal of Forest Research* **15**, 985–8.

Partanen, J., Koski, V. & Hanninen, H. (1998) Effects of photoperiod and temperature on the timing of bud burst in Norway spruce (*Picea abies*). *Tree Physiology* **18**, 811–16.

Penning De Vries, F.W.T. (1975) The cost of maintenance processes in plant cells. *Annals of Botany* **39**, 77–92.

Polglase, P.J., Attiwill, P.M. & Adams, M.A. (1992) Nitrogen and phosphorus cycling in relation to stand age of *Eucalyptus regans* F. Muell. III. Labile inorganic and organic P, phosphatase activity and P availability. *Plant and Soil* **142**, 177–85.

Prince, S.D. & Goward, S.N. (1995) Global primary production: a remote sensing approach. *Journal of Biogeography* **22**, 815–35.

Raffa, K.F. & Berryman, A.A. (1983) Physiological aspects of lodgepole pine wound responses to a fungal symbiont of the mountain pine beetle, *Dendroctonus ponderosae* (Coleoptera: scolytidae). *Canadian Entomologist* **115**, 723–34.

Raison, R.J., O'Connell, A.M., Khanna, P.K. & Keith, H. (1993) Effects of repeated fires on nitrogen and phosphorus budgets and cycling processes in forest ecosystems. In: Trabaud, L. & Prodon, R., eds. *Fire in Mediterranean Ecosystems*, pp. 347–63. Commission of the European Community, Brussels.

Reich, P.B. (1995) Phenology of tropical forests: patterns, causes, and consequences. *Canadian Journal of Botany* **73**, 164–74.

Reichardt, P.B., Chapin, F.S. III, Bryant, J.P., Mattes, B.R. & Clausen, T.P. (1991) Carbon/nutrient balance as a predictor of plant defense in Alaskan balsam poplar: potential importance of metabolite turnover. *Oecologia* **88**, 401–6.

Reichle, D.E. (ed.) (1970) *Analysis of Temperate Forest Ecosystems*. Springer-Verlag, Berlin.

Reichle, D.F. (ed.) (1981) *Dynamic Properties of Forest Ecosystems*. Cambridge University Press, Cambridge.

Rhoades, D.F. & Cates, R.G. (1976) Toward a general theory of plant antiherbivore chemistry. *Recent Advances in Phytochemistry* **10**, 168–213.

Rose, C.L. (1990) *Application of the carbon/nutrient balance concept to predicting the nutritional qualities of blueberry foliage to deer in southeastern Alaska.* PhD thesis, Oregon State University, Corvallis, Oregon.

Runyon, J., Waring, R.H., Goward, S.N. & Welles, J.M. (1994) Environmental limits on net primary production and light-use efficiency across the Oregon transect. *Ecological Applications* **4**, 226–37.

Ryan, M.G. (1991) A simple method for estimating gross carbon budgets for vegetation in forest ecosystems. *Tree Physiology* **9**, 255–66.

Ryan, M.G. & Yoder, B.J. (1997) Hydraulic limits to tree height and growth. *Bioscience* **47**, 235–42.

Ryan, M.G., Gower, S.T., Hubbard, R.M. *et al.* (1995)

Stem maintenance respiration of four conifers in contrasting climates. *Oecologia* **101**, 133–40.

Ryan, M.G., Hunt, E.R. Jr, McMurtrie, R.E. *et al.* (1996) Comparing models of ecosystem function for temperate conifer forests. I. Model description and validation. In: Breymeyer, A.I., Hall, D.O., Melillo, J.M. & Ågren, G.I., eds. *Global Change: Effects on Coniferous Forests and Grasslands*, pp. 313–62. John Wiley & Sons, New York.

Ryan, M.G., Lavigne, M.B. & Gower, S.T. (1997) Annual carbon cost of autotrophic respiration in boreal forest ecosystems in relation to species and climate. *Journal of Geophysical Research* **102**, 28871–83.

Scholander, P.F., Hammel, H.T., Bradstreet, E.D. & Hemmingsen, E.A. (1965) Sap pressure in vascular plants. *Science* **148**, 339–46.

Schultz, J.C. & Baldwin, I.T. (1982) Oak leaf quality declines in response to defoliation by gypsy moth larvae. *Science* **217**, 149–51.

Scriber, R.A. & Slansky, F. (1981) The nutritional ecology of immature insects. *Annual Review of Entomology* **76**, 183–211.

Seif el Din, A. & Obeid, M. (1971) Ecological studies of the vegetation of the Sudan. IV. The effect of simulated grazing on the growth of *Acacia senegal* (L.) Willd. Seedlings. *Journal of Applied Ecology* **8**, 211–16.

Slatyer, R.O. (1967) *Plant–Water Relationships*. Academic Press, London.

Slatyer, R.O. & Morrow, P.A. (1977) Altitudinal variation in photosynthetic characteristics of snow gum, *Eucalyptus pauciflora* Sieb. Ex Spreng. 1. Seasonal changes under field conditions in the Snowy Mountains of south-eastern Australia. *Australian Journal of Botany* **24**, 1–20.

Steudler, P.A., Bowden, R.D., Melillo, J.M. & Aber, J.A. (1989) Influence of nitrogen fertilizer on methane uptake in temperate forest soils. *Nature* **341**, 314–16.

Stockfors, J. & Linder, S. (1998) Effect of nitrogen on the seasonal course of growth and maintenance respiration in stems of Norway spruce trees. *Tree Physiology* **18**, 155–66.

Strain, B.R., Higginbotham, K.O. & Mulroy, J.C. (1976) Temperature preconditioning and photosynthetic capacity of *Pinus taeda* L. *Photosynthetica* **10**, 47–53.

Swain, T. (1977) Secondary compounds as protective agents. *Annual Review of Plant Physiology* **28**, 479–501.

Tagliavini, M., Millard, P. & Quartier, M. (1998) Storage of foliar-absorbed nitrogen and remobilization for spring growth in young nectarine (*Prunus persica* var. *nectarina*) trees. *Tree Physiology* **18**, 203–7.

Thornley, J.H.M. (1976) *Mathematical Models in Plant Physiology*. Academic Press, London.

Thornley, J.H.M. (1991) A transport-resistance model of

forest growth and partitioning. *Annals of Botany* **68**, 211–26.

Tognetti, R. & Borghetti, M. (1994) Formation and seasonal occurrence of xylem embolism in *Alnus cordata*. *Tree Physiology* **14**, 241–50.

Vertregt, N. & Penning de Vries, F.W.T. (1987) A rapid method for determining the efficiency of biosynthesis of plant biomass. *Journal of Theoretical Biology* **128**, 109–19.

Walcroft, A.S., Silvester, W.B., Grace, J.C., Carson, S.D. & Waring, R.H. (1996) Effects of branch length on carbon isotope discrimination in *Pinus radiata*. *Tree Physiology* **16**, 281–6.

Wang, Y.P., Jarvis, P.G. & Taylor, C.M.A. (1991) PAR absorption and its relation to above-ground dry matter production of Sitka spruce. *Journal of Applied Ecology* **28**, 547–60.

Waring, R.H. (1983) Estimating forest growth and efficiency in relation to canopy leaf area. *Advances in Ecological Research* **13**, 327–54.

Waring, R.H. (1989) Ecosystems: fluxes of matter and energy. In: Cherrett, J.M., ed. *Ecological Concepts*, pp. 17–41. Blackwell Scientific Publications, Oxford.

Waring, R.H. & Franklin, J.F. (1979) Evergreen coniferous forests of the Pacific Northwest. *Science* **204**, 1380–6.

Waring, R.H. & Running, S.W. (1978) Sapwood water storage: its contribution to transpiration and effect upon water conductance through the stems of old-growth Douglas-fir. *Plant, Cell and Environment* **1**, 131–40.

Waring, R.H. & Running, S.W. (1998) *Forest Ecosystems: Analysis at Multiple Scales*. Academic Press, San Diego.

Waring, R.H. & Schlesinger, W.H. (1985) *Forest Ecosystems: Concepts and Management*. Academic Press, Orlando, Florida.

Waring, R.H. & Winner, W.E. (1996) Constraints on terrestrial primary production in temperate forests along the Pacific Coast of North and South America. In: Lawford, R.G., Alaback, P. & Fuentes, E.R., eds. *High Latitude Rain Forests and Associated Ecosystems of the West Coast of the Americas: Climate, Hydrology, Ecology and Conservation*, pp. 89–102. Springer-Verlag, New York.

Waring, R.H., Schroeder, P.E. & Oren, R. (1982) Application of the pipe model theory to predict canopy leaf area. *Canadian Journal of Forest Research* **12**, 556–60.

Waring, R.H., McDonald, A.J.S., Larsson, S., Ericsson, T., Wiren, A. & Arwidsson, E. (1985) Differences in chemical composition of plants grown at constant relative growth rates with stable mineral nutrition. *Oecologia* **66**, 157–60.

Waring, R.H., Law, B., Goulden, M. *et al.* (1995) Scaling gross ecosystem production at Harvard Forest with remote sensing: a comparison of estimates from a constrained quantum-use efficiency model and eddy correlation. *Plant, Cell and Environment* **18**, 1201–13.

Waring, R.H., Landsberg, J.J. & Williams, M. (1998) Net primary production of forests: a constant fraction of gross primary production? *Tree Physiology* **18**, 129–34.

Weinbaum, S. & Van Kessel, C. (1998) Quantitative estimates of uptake and internal cycling of ^{14}N-labeled fertilizer in mature walnut trees. *Tree Physiology* **18**, 795–801.

Weinstein, D.A., Beloin, R.M. & Yanai, R.D. (1991) Modeling changes in red spruce carbon balance and allocation in response to interacting ozone and nutrient stresses. *Tree Physiology* **9**, 127–46.

Whitehead, D., Kelliher, F.M., Frampton, C.M. & Godfrey, M.J.S. (1994) Seasonal development of leaf area in a young, widely spaced *Pinus radiata* D. Don stand. *Tree Physiology* **14**, 1019–38.

Whittaker, R.H. (1975) *Communities and Ecosystems*, 2nd edn. Macmillan, New York.

Williams, K., Percival, F., Merino, J. & Mooney, H.A. (1987) Estimation of tissue construction cost from heat of combustion and organic nitrogen content. *Plant, Cell and Environment* **10**, 725–34.

Williams, M., Rastetter, E.R., Rernandes, D.N., Goulden, M.L. & Shaver, G.R. (1997) Predicting gross primary productivity in terrestrial ecosystems. *Ecological Applications* **4**, 882–94.

Woodward, F.I. (1987) *Climate and Plant Distribution*. Cambridge University Press, New York.

Zucker, W.V. (1983) Tannins: does structure determine function? An ecological perspective. *American Naturalist* **121**, 335–65.

9: Models for Pure and Mixed Forests

HANS PRETZSCH

9.1 INTRODUCTION

The previous chapter showed how forest growth is the sum of their species interactions with environment as mediated through physiological processes, and the dynamic nature of the responses of trees and stands. Long before forest scientists understood even some of the more basic processes governing the way trees grow, considerable empirical knowledge had accumulated through observation to quantify tree growth. While based on observation and not first causes, it is no less a contribution to understanding how trees grow and what affects them by looking at the way they respond in the 'field' or more properly the 'forest'!

This chapter reviews our understanding of tree and forest stand behaviour gained from empiricism (what has been observed), through quantifying systematically such observations (the yield table) to process-based modelling of both pure and mixed forest stands. For a deeper understanding of the introduced model types, the bibliography offers the most important references. A model's objective and existing knowledge about the observed system determine how complex the model approach has to be. Single-tree models, ecophysiological-based gap models and hybrid models are of particular interest for forest management as they are suitable across many stand types and forest conditions.

Forest growth models aggregate knowledge of individual processes of forest growth to predict stand or whole system functioning. Forest ecosystems may be modelled with varying degrees of temporal and spatial resolution. The time-scale may range from seconds to millenia, while the spatial scale may encompass anything from cells and mineral surfaces to continents (Fig. 9.1). The slow processes on large spatial scales fix the boundary for quicker processes on smaller scales. Conversely, the rapid and spatially bounded processes determine the processes on higher levels. Model approaches that take into consideration these feedback loops between the different system levels can provide important contributions to system understanding as well as to management decision support. At our present state of system knowledge, single-tree and stand models, which model processes on temporal scales from a year to a century and on spatial scales from tree to stand level, best fullfil the demands of forest management. They model the stand dynamic on the basis of the classical growth and yield variables, like diameter, height, crown length, etc. Process models have a higher spatial and temporal resolution and approximate to the classical variables by scaling-up. However, the behaviour of the whole system can be more than the sum of the underlying processes. Forest sucession and biome shift models become an important tool for global change research in forest ecosystems.

9.2 PATTERNS AND DYNAMICS OF GROWTH: EMPIRICAL OBSERVATIONS

9.2.1 Periodicity and pattern in individual tree growth

Tree growth is periodic over short time-spans and follows a definite pattern in the long term. In temperate countries growth is overwhelmingly determined by the season, with growth confined to a period of a few weeks to perhaps several months

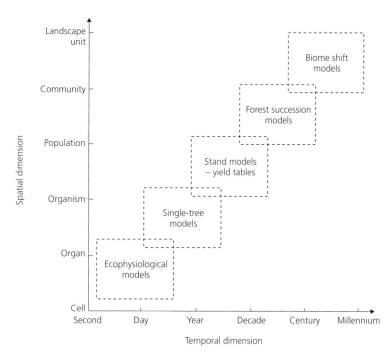

Fig. 9.1 Spatial and temporal dimensions of processes in forest ecosystems and models with increasing aggregation from ecophysiological models to biome shift models.

in any one year while warmth and moisture are adequate. Even in the moist tropics, with year-round favourable growing conditions, trees show periodicity under a measure of genetic control as recorded, for example, from measurements of multinodal tropical pines such as *Pinus caribaea* (Chudnoff & Geary 1973) and fast-growing hardwoods such as *Cordia alliodora* (Blake *et al.* 1976).

As well as periodicity in one year, trees exhibit a strong pattern of growth during their life. In relation to age the pattern is typically sigmoid, whether recording total height, diameter or volume over time.

9.2.2 Dynamics of stand growth

The dynamic nature of stand growth is a little less obvious than for a single tree, since superficially a stand seems to be a collection of individual trees the relations of which appear constant. That this is not so is seen by considering a small tree in a mature even-aged stand. At regeneration (or planting) this small tree will have been about the same size as its neighbours but subsequently will have competed less successfully for light, nutrients or moisture to become inferior to its neighbours. As a stand develops any apparent uniformity disappears, with some trees developing vigorously and neighbours becoming suppressed, moribund and even dying. Thus the stand is continually changing – it is dynamic. It is not only a collection of individual trees, each with its own genetic potential for using the site, but also a collection of trees that interact and compete with one another.

Seedlings and regeneration

From the outset trees appear to grow at different rates. Even in new plantations of well-spaced trees of one species, and long before onset of between-tree competition, the growth of individual trees is not identical. Moreover, there is often no correlation between initial postgermination vigour and subsequent growth rate. Small seedlings, or plants from the forest nursery, do not necessarily lead to small slow-growing trees (see illustration of this for young *Pinus caribaea* in Evans 1992, p. 221).

Between-tree competition

As trees grow they eventually begin to compete with their neighbours. This has both an above-ground component, mainly competition for light, and a below-ground element in terms of root competition for nutrients and moisture. Such competition is most readily seen in suppression of side branches on the lower crown and a measurable impact on diameter increment. The timing of the onset of between-tree competition depends on distance between trees, but is usually first measurable in forest stands from about the time of canopy closure. The principal exception is in arid climates where competition between root systems for moisture greatly exceeds that for light, and trees and shrubs remain widely dispersed and rarely in contact above ground.

Differentiation into crown classes

As competition begins, growth of some trees slows more than others. These tend to be the smaller trees at the time of canopy closure and the competition reinforces their inferior status. Once dominated, few trees can recover unless a gap develops owing to death of a neighbour. Thus as a stand develops, a range of tree sizes emerges and, traditionally, these are classified into different crown classes according to the tree's relative position in the canopy.

9.2.3 Interventions and manipulations

These empirical relationships of how trees are observed to grow and how they develop and interact in a stand provide the basis for manipulating their behaviour. Adjusting spacing between trees, both when planted or through thinnings, or influencing the balance of a mix of species in a stand all impact on how a stand develops, on how the increment is distributed and hence on the composition of tree types, trees sizes and the total of woody growth that will result.

In summary, densely stocked forest leads to high volumes per unit area but small mean tree size compared with less well-stocked forest on a similar site and of the same age. In the latter case there will be fewer but larger diameter trees, although total volume of timber may be somewhat reduced. This allows forests to be managed in different ways to yield different assortments of products. How to model these relationships and outcomes in detail, beyond the purely empirical, forms the bulk of this chapter.

9.3 GROWTH MODELS

The history of forest growth models is not simply characterized by the development of continuously improved models replacing former inferior ones. Instead, different model types with diverse objectives and concepts were developed simultaneously. The objectives and structure of a model reflect the state of the respective research area at its time and document the contemporary approach to forest growth prediction. The history of growth modelling thus also documents the advancement of knowledge in the science of forest growth.

Beginning with yield tables for large regions as a basis for taxation and planning (such as those by Schwappach 1893 and Wiedemann 1932, 1939a,b, 1942), model development led to regional yield tables and site-specific yield tables and culminated in the construction of growth simulators for the evaluation of stand development under different management schemes. Vanclay (1994) provided an overview about growth and yield management models and their application to mixed tropical forests. The 1980s brought a new trend with the development of ecophysiological models, which give insight into the complex causal relationships in forest growth and predict growth processes under various ecological conditions. The emphasis in model research has shifted towards ecophysiological models and away from models aimed only at providing growth and yield information for forest management. These models attempt to simulate forest growth on the basis of fundamental ecophysiological processes. The scientific value of ecophysiological models cannot be overrated; however, they will not be applied in forest management for the next few years as they are in many ways not yet sufficiently validated. Also, input and output variables do not yet meet the demand of forest management practice.

A major change has taken place in model conception, i.e. the understanding of forest growth on which the model is based. The tables by Weise (1880), Schwappach and Wiedemann resulted from a purely descriptive analysis of sample area data in the form of total and mean values of observed processes of stand development. These descriptions were later combined with theoretical model concepts that also considered natural growth relationships and causal relations as far as they were known at the time. For example, yield tables for mixed stands of pine and beech created by Bonnemann (1939) characterize growth of beech in the middle and lower storey by mean values. The FOREST model of Ek and Monserud (1974) controls increment behaviour of lower-storey trees by geometrical competition indices, and the ecophysiological growth models of Bossel (1994), Mäkelä and Hari (1986) and Mohren (1987) derive increment behaviour of lower-storey trees from light availability and performance in terms of photosynthesis.

The change in model objectives and concepts is closely related to a change in quality of the information generated. Pure management models aim at reliable prediction of forest yield values that are crucial for planning and control in forest management, e.g. height and diameter increment and associated economic value. Ecophysiological models aim at biomass development, nutrient input and loss, etc.; variables relevant to forest management are only of secondary importance in these models. For future planning in modern forestry, models meeting the information demands of ecology as well as of economy will gain in importance.

Ecophysiological models and stand management models can give specific decision support (Fig. 9.2). Ecological and socioeconomic conditions define the framework and thus the 'foundations' for management decisions. Ecophysiological models can support the ecological elements of the framework, for example the effect of site conditions, species mixture and thinning variants on critical loads, water quality or acidification. Stand treatments of interest can be judged in this way as ecologically acceptable or unacceptable. Management models help to optimize the path from starting point to objective via the given framework, for example they support the decision between different thinning and pruning strategies.

With the shift from tree and stand management models with low resolution to more complex ecophysiological models, different source data are needed for model construction and for the determination of model parameters. Standard datasets derived from research sample plots (diameter, height, etc.) were used for the development of stand growth models for applied forestry. For the construction of single-tree models, additional data are required (crown dimension, tree position, etc.). The transition to ecophysiological models requires an additional database that can only be provided by broadening experimental concepts and cooperation with neighbouring disciplines.

Models are always an abstraction of reality and

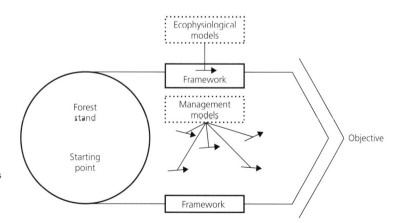

Fig. 9.2 Stand management models and ecophysiological models for decision support in forest ecosystem management.

are greatly influenced by the modeller's knowledge and perception of nature. This applies to the construction of yield tables as well as for eco-physiological models.

9.3.1 Stand growth models based on mean stand variables

With a history of over 250 years, yield tables for pure stands may be considered the oldest growth models in forestry science and forest management. They are representations of stand growth within defined rotation periods and are based on a series of measurements of diameter, height, biomass, etc. reaching far back into the past. From the late eighteenth to the middle of the nineteenth century, German scientists such as Paulsen (1795), von Cotta (1821), R. Hartig (1868), Th. Hartig (1847), G.L. Hartig (1795), Heyer (1852), Hundeshagen (1825) and Judeich (1871)

created the first generation of yield tables based on a restricted dataset. These original yield tables soon revealed great gaps in scientific knowledge. A series of long-term data-collection campaigns on experimental areas was therefore started. This was the birth of a unique network of long-term experimental plots in Europe that is still under survey. The old British yield table for Scots pine is shown as Fig. 9.3.

The second generation of yield tables, initiated towards the end of the nineteenth century and continued into the 1950s, follows uniform construction principles proposed by the Association of Forestry Research Stations (the predecessor organization of the International Union of Forest Research Organizations (IUFRO)), in 1874 and 1888 and has a solid empirical basis. The list of protagonists involved in this work includes Weise (1880), von Guttenberg (1915), Zimmerle (1952), Vanselow (1951), Krenn (1946), Grundner (1913)

Scots Pine
Normal yield table: yield class 160

Age (years)	Main crop after thinning							Yield from thinnings						Total production		Increment	
	Number of trees	Top height	Mean BHQG	Basal area	Vol. (h. ft.) to top diameter o.b. of			Number of trees	Mean BHQG	Av. vol. per tree	Vol. (h. ft.) to top diameter o.b. of			Basal area	Vol. to 3 in.	CAI		MAI
				sq. ft q. g.	3 in.	7 in.	9 in.		(in.)	h. ft.	3 in.	7 in.	9 in.	sq. ft q. g.	h. ft.	Basal area	Vol. to 3 in.	Vol. to 3 in.
		(feet)	(in.)															
15	1650	27.5	2.75	86	750	–	–	–	–	–	–	–	–	86	750	7.3	130	50
20	765	36.0	3.5	65	1020	–	–	885	3.0	0.80	480	–	–	122	1500	7.2	168	75
25	478	44.0	4.75	71	1380	120	–	287	4.0	1.95	560	10	–	158	2420	7.0	194	97
30	333	51.0	6.0	80	1830	610	95	145	5.0	3.86	560	80	–	192	3430	6.7	208	114
35	250	57.5	7.25	90	2330	1500	580	83	6.25	6.74	560	240	50	224	4490	6.3	213	128
40	199	63.5	8.5	100	2840	2350	1420	51	7.5	11.00	560	385	170	254	5560	5.8	214	139
45	166	69.0	9.75	110	3350	3015	2270	33	8.75	16.7	560	470	290	282	6630	5.3	210	147
50	142	74.5	110.0	119	3820	3590	3020	24	10.0	23.0	560	510	400	308	7660	4.9	201	153
55	125	79.0	12.25	128	4255	4095	3650	17	11.25	30.6	535	500	440	331	8630	4.5	189	157
60	112	83.5	13.5	135	4685	4540	4290	13	12.25	38.0	490	470	430	352	9550	4.0	177	159
65	102	87.0	14.25	142	5085	4950	4650	10	13.5	46.2	450	435	410	371	10400	3.6	163	160
70	94	90.5	15.0	147	5455	5310	5050	8	14.5	54.3	410	395	375	388	11180	3.2	149	160
75	88	93.5	15.75	152	5790	5650	5390	6	15.25	62.4	370	360	340	403	11885	2.8	134	159
80	83	96.0	16.25	156	6095	5970	5700	5	16.0	70.0	330	320	310	416	12520	2.4	120	157
85	79	98.0	17.0	159	6365	6240	5980	4	16.5	77.2	295	275	275	427	13085	2.1	106	154
90	76	100.0	17.5	162	6600	6480	6220	3	17.0	83.6	260	240	240	437	13580	1.8	92	151
95	73	101.5	18.0	165	6805	6680	6410	3	17.25	88.6	225	210	210	446	14010	1.5	79	147
100	71	103.0	18.5	167	6970	6850	6580	2	18.0	94.5	195	180	180	453	14370	1.3	68	144

Fig. 9.3 Facsimile of normal yield table of Scots pine used in the 1960s. Normal yield tables model the development of fully stocked stands on the basis of mean stand variables, e.g. number of trees per acre, mean diameter at breast height or basal area per acre. (From Bradley *et al.* 1966.) (Uses traditional (Imperial) measures.)

and, in particular, Schwappach (1893), Wiedemann (1932) and Schober (1967), who designed yield tables that were conceptually related and is still being used to this day. A brilliant example of their work is the yield tables for European beech. In the 1930s and 1940s the first models of mixed stands were contructed under the direction of Wiedemann. Data from some 200 experimental areas established by the Prussian Research Station led to the yield tables for even-aged mixed stands of pine and beech (Bonnemann 1939), spruce and beech (Wiedemann 1942), pine and spruce (Christmann 1949), and oak and beech (Wiedemann 1939a). The Second World War prevented Wiedemann from bringing the development of yield tables for uneven-aged pure and mixed stands to an end, but his studies initiated systematic research on mixed stands. Yield tables for mixed stands of this generation were never consistently used in forestry practice as they were restricted to specific site conditions, intermingling patterns and age structures.

Yield tables developed by Gehrhardt (1909, 1923) in the 1920s effected a transition from purely empirical models to models based on theoretical principles and biometric formulae and led to a third generation of yield tables. These models were designed by, among others, Assmann and Franz (1963), Hamilton and Christie (1973, 1974), Vuokila (1966), Schmidt (1971) and Lembcke *et al.* (1975), and at their core is a flexible system of functional equations. These functional equations are based as far as possible on natural growth relationships and are generally parameterized by means of statistical methods. The biometric models are usually transferred into computer programs and predict expected stand development for different spectra of yield and site classes. A wealth of data were available for the construction of these models and processed with modern statistical methods.

Since the 1960s a fourth generation of yield table models has been created, i.e. the stand growth simulations of Franz (1968), Hoyer (1975), Hradetzky (1972), Bruce *et al.* (1977) and Curtis *et al.* (1981, 1982), which simulate expected stand development under given growth conditions for different stem numbers at stand establishment and for different tending regimes. They describe stand development at different sites and for varying treatments and varying numbers of trees at the time of establishment. Expected stand development under given growth conditions is simulated by means of computer programs and controlled by systems of suitable functions forming the core of the growth simulator. All information available on forest growth is synthesized into a complex biometric model that simulates stand development for a wide range of possible management alternatives and summarizes the results in tabular form similar to yield tables. These yield tables reflect the stand dynamic for a wide range of imaginable management scenarios. While table and model were identical for the yield tables of earlier generations, simulator-created yield tables now describe just one of many potentially computable stand development courses.

Despite a number of drawbacks, yield tables still form the backbone of sustainable forest management planning. When computing capacities and available data for model construction increased and with the rising demand for information in forestry, mean-value and sum-orientated growth models and yield tables were increasingly replaced by stand-orientated growth models, predicting stem number frequencies, and by single-tree growth models. Prodan (1965, p. 605) commented on the significance of yield tables in the context of silviculture and forest sciences: 'Undoubtedly, yield tables are still the most colossal positive advance achieved in forest science research. The realization that yield tables may no longer be used in the future except for more or less comparative purposes in no way detracts from this achievement'.

9.3.2 Stand-orientated management models predicting stem number frequency

With the transition towards new intensive treatment concepts, the demand for information in forestry has changed the emphasis from mean stand values towards single-tree dimensions of selected parts of a stand. This changed demand for information resulted in the 1960s in the creation of the first growth models, which enabled prediction of mean stand values as well as frequencies of single-tree dimensions. Until then, a stand served as the usual information unit on which all predictions were based; these predictions were now strengthened by statements about stem number frequencies in diameter classes (Fig. 9.4), which are needed for precise prediction of assortment yield and value of a stand. Depending on their concept and construction, stand-orientated growth models predicting stem number frequency are classified into differential equation models, distribution prediction models and stochastic evolution models.

Many natural processes in various disciplines of the natural sciences can be described by differential equations. Examples are the differential equations formulating change of yield descriptors for diameter classes of a stand, i.e. change of stem number, basal area and growing stock, depending on current yield state values. Stand development then results from the numerical solution of the differential equations. In the 1960s and 1970s, Buckman (1962), Clutter (1963), Leary (1970), Moser (1972, 1974) and Pienaar and Turnbull (1973) developed stand-orientated growth models based on differential equations.

In the mid 1960s, Clutter and Bennett (1965) proposed a completely new approach to stand growth modelling. They characterized the condi-

tion of a tree population by its diameter and height distribution and described stand development by extrapolation of these frequency distributions. The precision of such models is decisively determined by the flexibility of the distribution type on which it is based. The suitability of different distribution types, for example beta, gamma, lognormal, Weibull or Johnson, has to be assessed individually. Compared with those reviewed earlier, in these models stand development is not controlled by the age function of the individual yield descriptors but by the parameters of the underlying frequency distribution. Models of this type were initially constructed by Clutter and Bennett for North American spruce stands and further developed by McGee and Della-Bianca (1967), Burkhart and Strub (1974), Bailey (1973) and Feduccia *et al.* (1979).

The term 'evolution models' for stochastic growth models is derived from the fact that in these models stand development evolves from an initial frequency distribution, for example from a diameter distribution known from forest inventory. Thus these models, like distribution prediction models, predict frequencies of single-stem dimensions (see Fig. 9.4). However, the mechanism accounting for the extrapolation is based on a Markov process, giving the transition probabil-

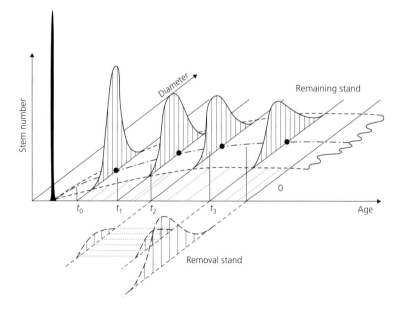

Fig. 9.4 Principle of management models predicting the shift of the diameter or height distribution along the time-axis. (After Sloboda 1976.)

ity for the shift between the diameter classes. Stochastic growth models were introduced to forestry science with the pioneering investigations by Suzuki (1971, 1983), and they continue to be linked to his name today. His growth models, for Japanese *Chamaecyparis* pure stands for example, have been consistently developed by Sloboda (1976) and his team since the mid 1970s; they are mainly interested in adapting the models, which are orientated to Japanese conditions, to the issues of German forestry and in model validation based on permanent test plot data. Stand-orientated growth models based on stochastic processes have been developed by Bruner and Moser (1973) and Stephens and Waggoner (1970) also for mixed stands.

9.3.3 Single-tree orientated management models

Single-tree models describe the stand as a mosaic of single trees and model individual growth and interactions with or without consideration of tree position. This has paved the way for the design of models of pure and mixed stands of all age structures and intermingling patterns. An equation system that controls growth behaviour of single trees depending on their constellation within the stand is the central module of all single-tree models. Position-independent or position-dependent competition indices are used to quantify the spatial growth constellation of each tree and to predict its increment of height, diameter, etc. in the following period. Compared with stand-orientated growth models based on mean stand descriptors and those predicting stem number frequencies, single-tree models work on higher resolution. The information unit in single-tree models is the individual tree. However, results of lower-resolution models, for example mean tree development or diameter frequency distributions, can also be derived from single-tree model results by integration. Information about stand growth then results from summarizing and aggregating each individual single-tree development for a given growth period. Recent single-tree models are programmed to enable the user to influence a simulation run interactively. This allows stand development to be followed step by step during the simulation and permits the user to

specify other factors (e.g. thinning or influence of disturbance) at any time during the simulation process, thus influencing or diverting the current course of stand development.

After parameters for the control of the single-tree model have been set, tree characteristics at the beginning of the prediction phase for the test area to be investigated are fed into the computer as initial values for the simulation (Fig. 9.5). This tree list can contain data on tree species, stem dimensions, crown morphology, stem position and other data about the stand individuals. These data usually originate from single-tree-based inventories of indicator plots. Starting with these initial values, change (e.g. mortality or development of diameter, height or crowns) for all stand members depending on individual growth conditions is predicted using an appropriate control function; this is done for a first growth period, for example 5 years. Once the tree list has been processed, change of growth conditions (e.g. due to thinning or disturbance) can be specified prior to continuing to the next increment period. This will now influence single-tree growth in the following period. The modified state values of all trees resulting at the end of the first growth period also represent the initial values for the second growth period. These values are repeatedly extrapolated in every simulation cycle and interim results are given. The simulation continues until the envisaged prediction period has been completed step by step. In most models, time steps are 5 years, sometimes only 1 or 2 years. By removing single trees during a simulation run, the growth constellation and growth behaviour of the remaining individuals change in the next growth period. Growth reaction of the stand is thus explained by the reactions of all single trees to this intervention. By relating stand development back to growth behaviour of single trees and by modelling single-tree dynamics depending on growth constellation within the stand, single-tree models, after being initialized accordingly, enable evaluation of a wide range of treatment programs.

The first single-tree model was developed for pure Douglas fir stands by Newnham (1964). It was followed by the development of models for pure stands by Arney (1972), Bella (1970)

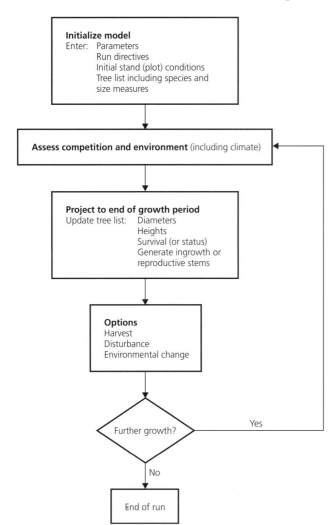

Fig. 9.5 Flow chart of single tree models. (After Ek & Dudek 1980.)

and Mitchell (1969, 1975) and colleagues. In the mid 1970s, Ek and Monserud applied the construction principles for single-tree orientated growth models for pure stands to uneven-aged pure and mixed stands (Ek & Monserud 1974; Monserud 1975). Munro (1974) distinguished distance-dependent and distance-independent single-tree models, the former being able to refer to data about stem position and stem distance for the control of single-tree growth. The worldwide bibliography of single-tree growth models compiled by Ek and Dudek (1980) lists

more than 40 different single-tree models, which are grouped into 20 distance-dependent and 20 distance-independent models. Single-tree models developed since the 1980s (Wykoff *et al.* 1982; van Deusen & Biging 1985; Wensel & Koehler 1985; Pretzsch 1992, 1998, 1999; Sterba *et al.* 1995; Nagel 1996) in many ways go back to the methodological bases of their predecessors; however, owing to the rapidly improving technology of modern computers they are far more user-friendly than older single-tree models.

9.3.4 Ecophysiological growth models

All the models mentioned above rely on growth and yield data from long-term observation plots and hence have the advantage of being validated empirically. However, there is a drawback to historically deduced data in as much as growth conditions undergo changes, and reaction patterns from the past cannot simply be projected into the future. In the 1970s, model research was pointed in a new direction with the creation of high-resolution ecophysiological process models, which account for metabolism, organ formation, assimilation and respiration as well as biochemical and soil chemistry reactions. Pioneers of the ecophysiological process model for forest stands are Bossel (1994), Mäkelä & Hari (1986) and Mohren (1987). The term 'process model' is slightly misleading in the sense that all forest growth models describe processes. Only the temporal and spatial scales of modelled processes become more detailed and accurate in the transition from yield table models via single-tree management models and succession models to growth models based on ecophysiological data (see Fig. 9.1).

The development of modern process models begins with a systems analysis and the selection of characteristic system components. A system to be analysed and modelled is first described using methods of systems analysis. Results of this description can be transferred into a system diagram (Fig. 9.6). The description breaks the system down into system components characteristic for all biological systems and identified by different symbols in the system diagram. By system parameters we mean those that remain constant during the lifetime of the system. Exogenous parameters are variables that control the system but which cannot be influenced by the system, e.g. stress caused by air pollutants. State variables are the actual output value of the model; their current values reflect the system's state. Important state variables in stand models are accumulated carbon quantities in needles, branches, stem and roots. The initial values of the state variables give the starting values of a system and thus crucially influence its further development. In a growth model for example, stem

number and initial stand structure have to be specified as initial values. The rate of change of the state variables controls change, i.e. input and output of state variables. Examples are mortality rates or respiration rates, which control the change of the carbon quantities accumulated in the different components. Intermediary variables change simultaneously with the state variables and feed back into the system. The system components are indicated in the system diagram with different symbols and their interrelations are identified by arrows.

The model thus outlined is transferred into a mathematical model and subsequently into a computer program. For this, the system components and links are described by mathematical or logical relationships. Once the complete model is constructed, the causal relations implemented are parameterized. The system behaviour can be simulated with the developed computer programs. All suitable information known about the system is therefore consolidated in the system components and the system structure. The process of system analysis and model development concludes in the validation of the final model. For validation, i.e. testing if the causal relations assumed in the model realistically reflect growth of stands or single trees, empirical yield data can be used. If necessary, individual model assumptions are corrected or model parts revised.

A vastly improved understanding of ecophysiological processes in forest ecosystems paved the way for this model approach and it was the actual modelling of these processes that provided an idea of the functioning of the overall system. A further impetus to process model development was the need to understand and predict the reactions of forest ecosystems to an increasing number of adverse effects, such as industrial emissions, rise in atmospheric CO_2 and climate change. In the context of environmental instability, high-resolution and accurately detailed process models are certainly the ideal approach for understanding and predicting forest ecosystem behaviour. However, there are particular constraints in developing and applying process models due to considerable gaps in our knowledge of part processes in assimilation organs and in the soil.

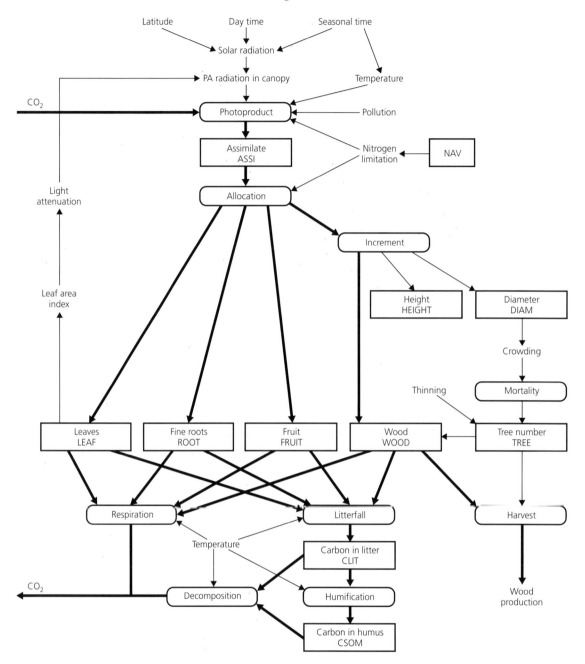

Fig. 9.6 Carbon flow (heavy arrows) in the TREEDYN3 forest simulation model, with the most important state variables, processes and flows indicated by boxes, ovals and arrows respectively. (After Bossel 1994.)

Also, the scaling-up of part processes to the behaviour of the overall system is still largely unresolved. Moreover, the introduction of process models still requires intensive research and extremely high-powered computers that are only rarely available in practice. To date, process models are therefore primarily research instruments rather than forest management planning tools.

9.3.5 Gap models and biome shift models

In the view of modern theoretical ecology, a spatially extensive system is composed of mosaic-like subunits and can be studied by analysing these subunits. Watt (1925, 1947), Bormann and Likens (1979) and others transferred this view of extensive ecosystems to the study and model representation of the growth dynamics of pure and mixed stands. This laid the foundations for the concept of gap models suitable to predict succession. According to this concept, a forest stand is an aggregation of gaps. The size of these gaps corresponds to the extent of a potential crown area of a dominant tree or tree group (areas of 0.04–0.08 ha). The actual information unit is the tree group in the gap; stand development results as the sum of the total spectrum of contributing gaps. Gap models imply that forest development in a gap occurs in a fixed cycle: A gap results from exploitation or death of a dominant tree, and thus the growth conditions of under-storey trees improve and natural regeneration occurs. Growing trees successively close the gap and a new overstorey develops. The cycle is repeated with further losses of dominant trees (Fig. 9.7). Growth models using this approach were predominantly employed for investigations of competition and succession in semi-natural stands.

Gap models, such as those designed by Shugart (1984), Pastor and Post (1985), Aber and Melillo (1982) and Leemans and Prentice (1989), are primarily aimed at mixed stands. While in the models described above increment-determining factors have effects on stands or individuals respectively, gap models describe tree growth that depends on growth conditions in the individual gap. Gap models simulate growth dynamics for

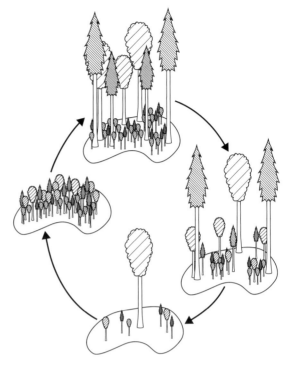

Fig. 9.7 Gap models imply that the stand dynamic occurs in a characteristic cycle: a gap results from exploitation or death of a dominant tree; growth conditions in the gap improve; young trees close the gap and form a new upper layer. (After Shugart 1984.)

single trees or tree classes in a gap; it is therefore possible to generate information about the development of diameter, height and volume of single trees as well as stands. However, regarding input and output variables they are less dependent on information available from, or required by, forestry practice; rather, they aim at predicting long-term succession in natural forest stands and the effects of altered growth conditions. The FORMIX2 model (Bossel & Krieger 1994) for virgin and logged Malaysian lowland dipterocarp forests is an example of an ecophysiological-based gap model with output variables that is useful as decision support in forest management.

Biome shift models, such as those of Box and Meentemeyer (1991) and Prentice *et al.* (1992), establish statistical relationships between regional climate and vegetation type. Based on relevant climatic conditions, the nature of potential biomes, i.e. communities, may be predicted on a regional and even global scale. Of all the models under discussion, these are the ones that provide the highest aggregation of data on vegetation development and forest growth. They have therefore gained increasing importance in research on global change.

9.3.6 Hybrid models for forest management

The transfer of specific components of ecophysiological models (based on solid process knowledge) into stand or single-tree management models (based on long-term experimental plots and increment series) leads to what Kimmins called 'hybrid growth models'. Models of this type were constructed by, among others, Botkin *et al.* (1972) and Kimmins (1993). Their objective is to make the best possible use of the newly acquired knowledge of ecophysiological processes combined with historical increment observations to assist in forest planning and management. On account of the implemented relationship between site conditions and species-specific growth, they can be used for pure and mixed stands. In the past 100 years mixed stands have gradually become the focus of forest research, particularly on account of studies by Gayer (1886), Wiedemann (1939b) and Assmann (1961), but to this day growth models for mixed stands are scarcely used as quantitative planning tools.

Only very recently have models created by Kolström (1993), Nagel (1996), Pretzsch (1992), Pukkala (1987) and Sterba *et al.* (1995) found use in forestry practice for planning work in pure and mixed stands. These are in effect site-sensitive single-tree models constructed from a broad base of ecophysiological and growth and yield data. Version 2.2 of the SILVA model, developed in Germany for pure and mixed stands, belongs to the category of hybrid models (Pretzsch 1992; Pretzsch & Kahn 1996; Kahn & Pretzsch 1997) and may be used as an example to explain the functional principles underlying this approach.

9.3.7 Management model SILVA 2.2 for pure and mixed stands

SILVA reflects the spatial and dynamic character of mixed-stand systems in as much as it models spatial stand structures at 5-year intervals. This permits the recording of the individual growth constellation of every tree and the control of tree increment in relation to growth constellation and the original dimensions of the tree (Fig. 9.8). The external variables determining tree increment and stand structure are treatment, risk and site factors. The model simulates the effects that tending, thinning, regeneration and natural hazards such as storms and wind have on the stand dynamic. The feedback loop, stand structure → tree growth → state of tree → stand structure, forms the backbone of the model. The step-by-step modelling of the growth of all individual trees via differential equation systems provides information about the development of assortment yield, financial yield, stand structure, stability and diversity of the stand over and above the data, required in yield calculations, on height, diameter at breast height, number of stems, etc.

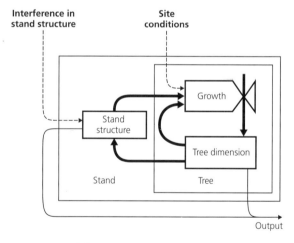

Fig. 9.8 Simplified system diagram of the growth model SILVA 2.2 showing the levels 'stand' and 'tree', the external variables 'interference in stand structure' and 'site conditions' and the feedback loop 'stand structure → growth → tree dimension → stand structure'.

Input and output data used in the model correspond to the data available from, or required in, forestry practice, for example only site variables available on a large scale are considered. With models of this type a weighting between yield-related, socioeconomic and ecological aspects of stand development in pure and mixed stands becomes possible. Parametrization relies on yield and site characteristics of pure and mixed stands that have been under observation for over 100 years.

The position-dependent individual tree model SILVA 2.2 breaks down forest stands into a mosaic of individual trees and reproduces their interactions as a space–time system (Fig. 9.9). It can therefore be used for pure and mixed stands of all age combinations. Primarily it is designed to assist in the decision-making processes in forest management. Based on scenario calculations SILVA 2.2 is able to predict the effects of site conditions, silvicultural treatment and stand structure on stand development, and therefore also serves as a research instrument.

A first model element reflects the relationship between site conditions and growth potential and aims at adapting the increment functions in the model to actual observed site conditions (Fig. 9.10). With the aid of nine site factors reflecting nutritional, water and temperature conditions, the parameters of the growth functions are determined in a two-stage process (Kahn 1994). The stand structure generator STRUGEN facilitates the large-scale use for position-dependent individual tree growth models. The generator converts verbal characterizations as commonly used in forestry practice (e.g. mixture in small clusters, single tree mixture, row mixture) into a concrete initial stand structure with which the growth model can subsequently commence its forecasting run (Pretzsch 1997). The three-dimen-

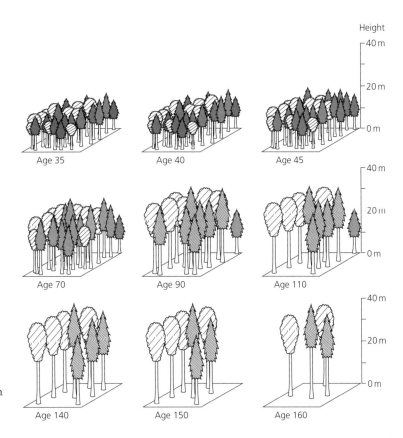

Fig. 9.9 SILVA 2.2 breaks down forest stands into a mosaic of individual trees and reproduces their interactions as a space–time system. Extract of a simulation run for a mixed stand with two species (slight thinning from below).

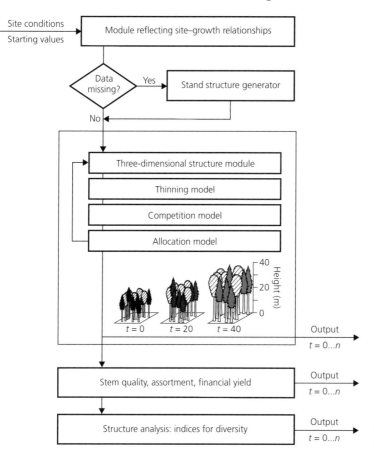

Fig. 9.10 Essential elements of the SILVA 2.2 simulator for pure and mixed stands.

sional structure module uses tree attributes such as stem position, tree height, diameter, crown length, crown diameter and species-related crown form to construct a spatial model of the stand in question. The thinning model is also based on individual trees and can model a wide spectrum of treatment programs (Kahn 1995). The core of the thinning model is a fuzzy logic controller. In the simulation studies described below the thinning model simulates various thinning methods (thinning from below and selective thinning) and thinning intensities (slight, moderate and heavy). The competition model employs the light-cone method (Pretzsch 1992) and calculates a competition index for every tree on the basis of the three-dimensional stand model. The allocation model controls the development of individual stand elements. Tree diameter at height 1.3 m, tree height, crown diameter, height

of crown base, crown shape and survival status are controlled, at 5-year intervals, in relation to site conditions and interspecific and intraspecific competition. Finally classical yield information on stand and single-tree level for the prognosis period are compiled in listings and graphs. Additional information on stem quality, assortment and financial yield complete the growth and yield characteristic. At every stage of the simulation run, a program routine for structural analysis calculates a vector of structural indices that serve as indicators for habitat and species diversity and form a link to the ecological assessment of forest stands.

The algorithmic sequence for predicting forest development comprises the following steps (Fig. 9.10). Step 1 is the input of data on the initial structure and site conditions of the monitored stand. In step 2, the parameters of the growth functions are adapted to actual site conditions. Once the starting values for the prognostic run are complete, monitoring can begin. If there are no initial values, for example stem positions are unknown, the missing data can be realistically complemented with the help of the stand structure generator (step 3). Once the spatial model has been constructed (step 4) the silvicultural treatment program is specified in step 5. The competition index calculated for each tree through the three-dimensional model in step 6 is used, in step 7, to control individual tree development. Steps 4–7 are repeated until the entire prognostication period has been run through in 5-year steps.

To date, model research has had little success in substituting the yield tables for pure stands by an improved information system for pure and mixed stands. This can in no way be attributed to a deficit in methodological principles, data or technical equipment. Rather, the causes lie in the fact that new models are not properly adapted to practical requirements. The recent introduction of the growth model SILVA 2.2 for forest management use led to a range of operational requirements and outputs demanded from the management models that will be used in decision-making processes at stand and forest enterprise levels.

1 The natural management of forests is currently making great headway. In the long run only those growth models capable of simulating the growth of pure and mixed stands of all age compositions and structural patterns will find approval.

2 Models need to be operable at stand and forest enterprise levels and able to simulate growth behaviour under different thinning regimes and different processes of artificial and natural regeneration.

3 Flexibility of the model is essential so as to permit simulation of growth reactions to site alterations and interference factors on a large regional scale.

4 Apart from tree and stand characteristics such as volume production, assortment yield, wood quality and financial yield should also include structural parameters determining the recreational and protective functions of forests as well as indicators showing the impact of hazards or ecological instability.

5 Forestry practice is interested, first and foremost, in calculating scenarios at stand and forest enterprise levels. This can only be achieved if input and output data of the model consider what information is available and which data are needed in forestry practice. Furthermore, achieving this goal also depends on whether the model forms part of a comprehensive forestry information system and, lastly, whether hardware specifications are acceptable in practice.

For decades forestry practice has been hoping for improved growth models to assist with research, planning, operations and control in forest management. The general acceptance of new models by practitioners calls for close cooperation between forest science and forest practice, from the design and development of the model to its actual introduction in forest management.

REFERENCES

Aber, J.D. & Melillo (1982) *FORTNITE: A Computer Model of Organic Matter and Nitrogen Dynamics in Forest Ecosystems*. Research Bulletin R3130, College of Agriculture and Life Sciences, University of Wisconsin, Madison, Wisconsin.

Arney, J.D. (1972) *Computer simulation of Douglas-fir tree and stand growth*. PhD thesis, Oregon State University.

Assmann, E. (1961) *Waldertragskunde*. BLV Verlagsgesellschaft, München.

Assmann, E. & Franz, F. (1963) *Vorläufige Fichten-Ertragstafel für Bayern*. Institut für Ertragskunde der Forstlichen Forschungsanstalt, München.

Bailey, R.L. (1973) *Development of unthinned stands of Pinus radiata in New Zealand*. PhD thesis, University of Georgia, Athens, Georgia.

Bella, I.E. (1970) *Simulation of growth, yield and management of aspen*. PhD thesis, Faculty of Forestry, University of British Columbia, Vancouver.

Blake, J., Rosero, P. & Lojan, L. (1976) Interaction between phenology and rainfall in the growth of *Cordia alliodora*. *Commonwealth Forestry Review* **55**, 37–40.

Bonnemann, A. (1939) Der gleichaltrige Mischbestand von Kiefer und Buche. *Mitt. a. Forstwirtschaft und Forstwissenschaft* **10**(4), 45 pp.

Bormann, F.H. & Likens, G.E. (1979) *Pattern and Process in Forested Ecosystems*. Springer-Verlag, New York.

Bossel, H. (1994) Treedyn 3 Forest simulation model. *Berichte des Forschungszentrums Waldökosysteme, Reihe B* **35**, 118 pp.

Bossel, H. & Krieger, H. (1994) Simulation of multispecies tropical forest dynamics using a vertically and horizontally structured model. *Forest Ecology and Management* **69**, 123–44.

Botkin, D.B., Janak, J.F. & Wallis, J.R. (1972) Some ecological consequences of a computer model of forest growth. *Journal of Ecology* **60**, 849–72.

Box, E.O. & Meentemeyer, V. (1991) Geografic modeling and modern ecology. In: Esser, G. & Overdieck, D., eds. *Modern Ecology: Basic and Applied Aspects*, pp. 773–804. Elsevier, Amsterdam.

Bradley, R.T., Christie, J.M. & Johnston, D.R. (1966) *Forest Management Tables*. Forestry Commission Booklet no. 16, HMSO, London.

Bruce, D., De Mars, D.J. & Reukema, D.C. (1977) *Douglas-fir Managed Yield Simulator: DFIT User's Guide*. USDA Forest Service General Technical Report PNW-57, PNW Forest and Range Experimental Station, Portland, Oregon.

Bruner, H.D. & Moser, J.W. (1973) A Markov chain approach to the prediction of diameter distributions in uneven-aged forest stands. *Canadian Journal of Forest Research* **3**, 409–17.

Buckman, R.E. (1962) *Growth and Yield of Red Pine in Minesota*. USDA Technical Bulletin no. 1272, US Forest Service, Lake States Forest Experiment Station, St. Paul, Minnesota.

Burkhart, H.E. & Strub, M.R. (1974) A model for simulation of planted loblolly pine stands. In: Fries, J., ed. *Growth Models for Tree and Stand Simulation*, pp. 128–35. Royal College of Forestry, Research Notes No. 30, Stockholm, Sweden.

Christmann (1949) Ertragstafel für den Kiefern-Fichten-Mischbestand. In: Wiedemann, E., ed. *Ertragstafeln der Wichtigen Holzarten bei Verschiedener Durchforstung Sowie Einiger Mischbestandsformen*. Schaper, Hannover, 1949.

Chudnoff, M. & Geary, T.F. (1973) Terminal shoot elongations and cambial growth rhythms in *Pinus caribaea*. *Commonwealth Forestry Review* **52**, 317–24.

Clutter, J.L. (1963) Compatible growth and yield models for loblolly pine. *Forest Science* **9**, 354–71.

Clutter, J.L. & Bennett, F.A. (1965) *Diameter Distributions in Old-field Slash Pine Plantations*. Georgia Forestry Research Council Report, no. 13. US Forest Service SE Experiment Station, Ashville, Carolina.

Curtis, R.O., Clendenen, G.W., De Mars, D. & J. (1981) *A New Stand Simulator for Coast Douglas-fir: User's Guide*. USDA Forest Service General Technical Report PNW-128, PNW Forest and Range Experimental Station, Portland, Oregon.

Curtis, R.O., Clendenen, G.W., Reukema, D.C., De Mars, D. & J. (1982) *Yield Tables for Managed Stands of Coast Douglas-fir*. USDA Forest Service General Technical Report PNW-135, PNW Forest and Range Experimental Station, Portland, Oregon.

Ek, A.R. & Dudek, A. (1980) *Development of Individual Tree Based Stand Growth Simulators: Progress and Applications*. Department of Resources Staff Paper no. 20, College of Forestry, University of Minnesota, St. Paul.

Ek, A.R. & Monserud, R.A. (1974) Trials with program FOREST: growth and reproduction simulation for mixed species even- or uneven-aged forest stands. In: Fries, J., ed. *Growth Models for Tree and Stand Simulation*, pp. 56–73. Royal College of Forestry, Research Notes No. 30, Stockholm, Sweden.

Evans, J. (1992) *Plantation Forestry in the Tropics*, 2nd edn. Oxford University Press, Oxford.

Feduccia, D.P., Dell, T.R., Mann, W.F. & Polmer, B.H. (1979) *Yields of Unthinned Loblolly Pine Plantations on Cutover Sites in the West Gulf Region*. USDA Research Paper SO-148, Southern Forest Experiment Station, New Orleans, Louisiana.

Franz, F. (1968) *Das EDV-Programm STAOET: zur Herleitung mehrgliedriger Standort-Leistungstafeln*. Manuskriptdruck, München.

Gayer, K. (1886) *Der gemischte Wald*. Paul Parey, Berlin.

Gehrhardt, E. (1909) Über Bestandes-Wachstumsgesetze und ihre Anwendung zur Aufstellung von Ertragstafeln. *Allgemeine Forst-und Jagdzeitung* **85**, 117–28.

Gehrhardt, E. (1923) *Ertragstafeln für Eiche. Buche, Tanne Fichte und Kiefer*. Verlag Julius Springer, Berlin.

Grundner, F. (1913) *Normalertragstafeln für Fichtenbestände*. Springer-Verlag, Berlin.

Hamilton, G.J. & Christie, J.M. (1973) *Construction and Application of Stand Yield Tables*, British Forestry Commission Research and Development Paper no. 96, Forestry Commission, London.

Hamilton, G.J. & Christie, J.M. (1974) Construction and application of stand yield models. In: Fries, J., ed. *Growth Models for Tree and Stand Simulation*, pp. 222–39. Royal College of Forestry, Research Notes No. 30, Stockholm, Sweden.

Hartig, G.L. (1795) *Anweisung Zu Taxation der Forsten Oder Zur Bestimmung Des Holzertrages der Wälder*. Heyer-Verlag, Gießen.

Hartig, R. (1868) *Die Rentabilität der Fichtennutzholz-Und Buchenbrennholzwirtschaft Im Harze und Im Wesergebirge*. Cotta-Verlag, Stuttgart.

Hartig, Th. (1847) *Vergleichende Untersuchungen Über Den Ertrag der Rotbuche*. Förstner-Verlag, Berlin.

Heyer, G. (1852) *Über die Ermittlung der Masse, des Alters und des Zuwachses der Holzbestände*. Verlag Katz, Dessau.

Hoyer, G.E. (1975) *Measuring and Interpreting Douglas-fir Management Practices*. Washington State Deartment of Natural Resources Report no. 26, Olympia, Washington.

Hradetzky, J. (1972) *Modell eines integrierten Ertragstafel-Systems in modularer Form.* PhD thesis, University of Freiburg.

Hundeshagen, J.C. (1825) *Beiträge zur gesamten Forstwirtschaft.* Verlag Laupp, Tübingen.

Judeich, F. (1871) *Die Forsteinrichtung.* Schönfeld-Verlag, Dresden.

Kahn, M. (1994) Modellierung der Höhenentwicklung ausgewählter Baumarten in Abhängigkeit vom Standort. *Forstliche Forschungsberichte. München* **141**.

Kahn, M. (1995) Die Fuzzy Logik basierte Modellierung von Durchforstungseingriffen. *Allgemeine Forst-und Jagdzeitung* **166**, 169–76.

Kahn, M. & Pretzsch, H. (1997) Das Wuchsmodell SILVA: Parametrisierung für Rein- und Mischbestände aus Fichte und Buche. *Allgemeine Forst-und Jagdzeitung* **168**, 115–23.

Kimmins, J.P. (1993) *Scientific Foundations for the Simulation of Ecosystem Function and Management in FORCYTE-11.* Information Report NOR-X-328, Forestry Canada, Northwest Region Northern Forestry Centre, Vancouver, Canada.

Kolström, T. (1993) Modelling the development of unevenaged stand of *Picea abies. Scandinavian Journal of Forest Research* **8**, 373–83.

Krenn, K. (1946) Ertragstafeln für Fichte. *Schriftenreihe der Badischen Forstlichen Versuchsanstalt,* Freiburg. **3**.

Leary, R.A. (1970) *System Identification Principles in Studies of Forest Dynamics.* USDA North Central Forest Experimental Station Research Paper NC-45, St. Paul, Minnesota.

Leemans, R. & Prentice, I.C. (1989) *FORSKA: a General Forest Succession Model* Vol. 2, pp. 1–45. Institute of Ecology and Botany, University of Uppsala, Meddelanden fran Växtbiologiska Institutionen.

Lembcke, G., Knapp, E. & Dittmar, O. (1975) Die neue DDR-Kiefernertragstafel 1975. *Beiträge für die Forstwirtschaft* **15**, 55–64.

McGee, C.E. & Della-Bianca, L. (1967) *Diameter Distributions in Natural Yellow-poplar Stands.* USDA South East Forest Experimental Station Research Paper SE-25, US Forest Service, Ashville, Carolina.

Mäkelä, A. & Hari, P. (1986) Stand growth model based on carbon uptake and allocation in individual trees. *Ecological Modelling* **33**, 205–29.

Mitchell, K.J. (1969) *Simulation of the Growth of Evenaged Stands of White Spruce.* Yale University School of Forestry Bulletin no. 75, New Haven, Connecticut.

Mitchell, K.J. (1975) Dynamics and simulated yield of douglas-fir. *Forest Science Monograph* **17**.

Mohren, G.M.J. (1987) *Simulation of forest growth, applied to douglas fir stands in the Netherlands.* PhD thesis, Agricultural University Wageningen.

Monserud, R.A. (1975) *Methodology for simulating Wisconsin northern hardwood stand dynamics.* PhD thesis, University of Wisconsin-Madison, *Dissertation Abstracts* **36**(11).

Moser, J.W. (1972) Dynamics of an uneven-aged forest stand. *Forest Science* **18**, 184–91.

Moser, J.W. (1974) A system of equations for the components of forest growth. In: Fries, J., ed. *Growth Models for Tree and Stand Simulation,* pp. 260–87. Royal College of Forestry, Research Notes No. 30, Stockholm, Sweden.

Nagel, J. (1996) Anwendungsprogramm zur Bestandesbewertung und zur Prognose der Bestandesentwicklung. *Forst und Holz* **51**, 76–8.

Newnham, R.M. (1964) The *development of a stand model for Douglas-fir.* PhD thesis, Faculty of Forestry, University of British Columbia, Vancouver.

Pastor, J. & Post, W.M. (1985) *Development of a Linked Forest Productivity–Soil Process Model.* Oak Ridge National Laboratory, Environmental Science Division, Oak Ridge, Tennessee.

Paulsen, J.C. (1795) *Kurze praktische Anleitung zum Forstwesen. Verfaßt von einem Forstmanne.* Hrsg. von Kammerrat G. F. Führer, Detmold.

Pienaar, L.V. & Turnbull, K.J. (1973) The Chapman–Richards generalization of von Bertalanff's growth model for basal area growth and yield in even-aged stands. *Forest Science* **19**, 2–22.

Prentice, I.C., Cramer, W., Harrison, S.P., Leemans, R., Monserud, R.A. & Solomon, A.M. (1992) A global biome model based on plant physiology and dominance, soil properties and climate. *Journal of Biogeography* **19**, 117–34.

Pretzsch, H. (1992) Konzeption und Konstruktion von Wuchsmodellen für Rein- und Mischbestände. *Forstliche Forschungsberichte München* **115**.

Pretzsch, H. (1997) Analysis and modeling of spatial stand structures. Methological considerations based on mixed beech larch stands in Lower Saxony. *Forest Ecology and Management* **97**, 237–53.

Pretzsch, H. (1998) Structural diversity as a result of silvicultural operations, *Lesnictví—Forestry* **44**(10), 429–39.

Pretzsch, H. (1999) Modelling growth in pure and mixed stands: a historical overview. In: Olsthoorn, A.F.M., Bartelink, H.H., Gardiner, J.J., Pretzsch, H., Hekhuis, H.J. & Franc, A., eds. *Management of Mixed-species Forest: Silviculture and Economics,* pp. 102–7. Wageningen.

Pretzsch, H. & Kahn, M. (1996) Wuchsmodelle für die Unterstützung der Wirtschaftsplanung im Forstbetrieb, Anwendungsbeispiel: Variantenstudie Fichtenreinbestand versus Fichten/Buchen-Mischbestan. *Allgemeine Forstzeitschrift* **51**, 1414–19.

Prodan, M. (1965) *Holzmesslehre.* J. D. Sauerländer's Verlag, Frankfurt am Main.

Pukkala, T. (1987) Simulation model for narural regener-ation of *Pinus sylvestris, Picea abies, Betula pendula* and *Betula pubescens.Silva Fennica* **21**, 37–53.

Schmidt, A. (1971) Wachstum und Ertrag der Kiefer auf wirtschaftlich wichtigen Standorteinheiten der Oberpfalz. *Forstliche Forschungsberichte. München* **1**.

Schober, R. (1967) Buchen-Ertragstafel für mäßige und starke Durchforstung. In: Schober, R., ed. *Die Rotbuche 1971*. J.D. Sauerländer's Verlag, Frankfurt am Main.

Schwappach, A. (1893) *Wachstum und Ertrag Normaler Rotbuchenbestände*. Verlag Julius Springer, Berlin.

Shugart, H.H. (1984) *A Theory of Forest Dynamics. The Ecological Implications of Forest Succession Models*. Springer-Verlag, New York.

Sloboda, B. (1976) *Mathematische und stochastische Modelle zur Beschreibung der Statik und Dynamik von Bäumen und Beständen: insbesondere das bestandesspezifische Wachstum als stochastischer Prozess*. Habil.-Schrift, University of Freiburg.

Stephens, G.R. & Waggoner, P.E. (1970) *The Forests Anticipated from 40 years of Natural Transitions in Mixed Hardwoods*. Connecticut Agricultural Experimental Station Bulletin no. 707, New Haven.

Sterba, H., Moser, M. & Monserud, R. (1995) Prognaus-Ein Waldwachstumssimulator für Rein- und Mischbestände. *Österreichische Forstzeitung* **5**, 1–2.

Suzuki, T. (1971) Forest transition as a stochastic process. *Mitteilungen der Forstlichen Bundesversuchsanstalt Wien* **91**, 137–50.

Suzuki, T. (1983) Übergang des Waldbestandes als ein stochastischer Prozess. *Beiträge zur biometrischen Modellbildung in der Forstwirtschaft* **76**, 23–58.

Vanclay, J.K. (1994) *Modelling Forest Growth and Yield: Applications to Mixed Tropical Forests*. CAB International, Wallingford, UK.

van Deusen, P.C. & and Biging (1985) *STAG: A Stand Generator for Mixed Species Stands, Version 2.0*. Research Note 11, Northern California Forest Yield Cooperative, Department of Forestry and Research Management, University of California, Berkeley, USA.

Vanselow, K. (1951) Fichtenertragstafel für Südbayern. *Forstwissenschaftliches Centralblatt* **70**, 409–45.

von Cotta, H. (1821) *Hülfstafeln für Forstwirte unf Forsttaxatoren*. Arnoldische Buchhandlung, Dresden.

von Guttenberg, A. (1915) *Wachstum und Ertrag der Fichte im Hochgebirge*. Wien, Leipzig.

Vuokila, Y. (1966) Functions for variable density yield tables of pine based on temporary sample plots. *Communicationes Instituti Forestalis Fenniae* **60**.

Watt, A. & S. (1947) Pattern and process in the plant community. *Journal of Ecology* **35**, 1–22.

Weise, W. (1880) *Ertragstafeln für die Kiefer*. Verlag Springer, Berlin.

Wensel, L.C. & Koehler, J.R. (1985) *A Tree Growth Projection System for Northern California Coniferous Forests*. Research Note no. 12, Northern California Forest Yield Cooperative, Department of Forestry and Research Management, University of California, Berkeley.

Wiedemann, E. (1932) Die Rotbuche 1931. *Mitteilungen aus Forstwirtschaft und Forstwissenschaft* **3**.

Wiedemann, E. (1939a) Untersuchungen der Preußischen Versuchsanstalt über Ertragstafelfragen. *Mitteilungen aus. Forstwirtschaft und Forstwissenschaft* **10**.

Wiedemann, E. (1939b) Ertragskundliche Fragen des gleichaltrigen Mischbestandes aus der Preußischen Versuchsanstalt. *Der deutsche Forstwirt* **51**.

Wiedemann, E. (1942) Der gleichaltrige Fichten-Buchen-Mischbestand. *Mitteilungen aus Forstwirtschaft und Forstwissenschaft* **13**.

Wykoff, W.R., Crookston, N.L.U. & Stage, A.R. (1982) *User's Guide to the Stand Prognosis Model*. US Forestry Service General Technical Report INT-133, Ogden, Utah.

Zimmerle, H. (1952) *Ertragszahlen für Grüne Douglasie. Japaner Lärche und Roteiche in Baden-Württemberg*. Ulmer-Verlag, Stuttgart.

10: Forests and the Atmosphere

PAUL G. JARVIS AND DAVID FOWLER

10.1 INTRODUCTION

Forests function as the major terrestrial interface between the atmosphere and the earth: they interact with the atmosphere to condition life on the earth. On the one hand, the atmosphere provides driving variables for processes in forests; on the other, forests feed back to modify the atmosphere, very efficiently removing gases and particles from the atmosphere. This chapter is concerned with the processes of exchange between forests and the atmosphere and with the consequences of these exchanges for the growth and production of forests and for the composition of the atmosphere. For this purpose, we may think of forests in a number of different ways. For example, we can divide them on a phytogeographical basis into boreal, temperate and tropical forests (Malhi *et al.* 1999), on a phenological basis into deciduous and evergreen forests, or on a morphological basis into broadleaved and needle-leaved (largely coniferous) forests. However, at the present time, we have scarcely sufficient data to classify them on the basis of exchange processes into a range of functional types. A succession of recent international programmes has very substantially increased the extent of quantitative information available about forest processes, particularly in the boreal and temperate regions, compared with the situation 25 years ago (Jarvis *et al.* 1976) but our knowledge base is still to a large extent anecdotal. In this chapter, we attempt to pull together the strands of new information relating to exchange processes resulting from the development of new technology and the explosion of measurement programmes over the last 10 years and to derive some generalizations, based on functional attributes.

10.1.1 Spatial and temporal scales

Spatial scales

On a large spatial scale, say 100 km, a natural forest is a haphazard mosaic of individual trees and patches of different species and ages, occasioned by microenvironment, topography and disturbance, such as windthrow and fire events. At the other extreme, intensively managed forest generally comprises uniform compartments (woodlots or coupes) each of a single species of tree of similar age and size, comprising a range of species and ages, on a spatial scale of up to 1 km. A sustainably managed forest, whether maintained by natural regeneration or planting, should comprise compartments covering a range of ages from regeneration through to harvest.

The unit for intensive study of exchange processes may be several compartments or part of a compartment, commonly referred to as a stand, the size of which depends on the processes investigated and the methodology used. For measurements of net ecosystem exchange (NEE) of scalars* such as CO_2, water vapour, CH_4 and other trace gases by eddy covariance, the scale of the stand is necessarily similar to that of the compartment (< 100 ha), whereas for measurements of net primary productivity (NPP) the

---

* In this context, a scalar is any constituent of the atmosphere that is transported by turbulence without a defined directional component such as is possessed by a vector.

sample plot may be as small as 0.1 ha. While from an economic perspective a stand is generally defined in terms of the keystone species, i.e. the trees, from a functional ecosystem perspective the understorey, ground vegetation and, particularly, the soil components can be just as important.

While the stand is an appropriate scale at which to assess forest–atmosphere interactions, it is necessary to move to the smaller scales of tree, branch and leaf, coarse and fine roots and soil microorganisms to explain the magnitude and to understand the constraints on NEE (Fig. 10.1). Conversely, to appreciate interactions between the atmosphere and a forest comprising compartments of different species and ages on an overall scale of about 10–100 km, it is necessary to adopt different strategies to scale up or to obtain spatially integrated measurements.

Temporal scales

Exchanges between the atmosphere and a young forest stand regenerating or regrowing after dis-

turbance such as harvesting or fire is critically dependent on the phase of the successional cycle. During the first 5 or 10 years after disturbance in temperate forests, the trees are small, the understorey vegetation damaged and the soil may be churned up and exposed. Because of the changing height, structure and inherent physiological activity of trees and stands with age and development, processes within the stand and coupling of the stand to the atmosphere change during the course of a rotation. For example, a recently disturbed area is likely to be losing carbon to the atmosphere as CO_2; subsequently, as the trees come to occupy the site fully, the area will accumulate carbon strongly during a phase of rapid growth that may last for decades, depending on the species of trees and site conditions, before old-growth forest may become a net source of carbon to the atmosphere in some circumstances.

However our capability to measure exchanges of H_2O, CO_2, CH_4, N_2O and other trace gases is sufficiently new that only specimen, usually mature, stands during a rotation have so far been studied, over periods as short as a few months up

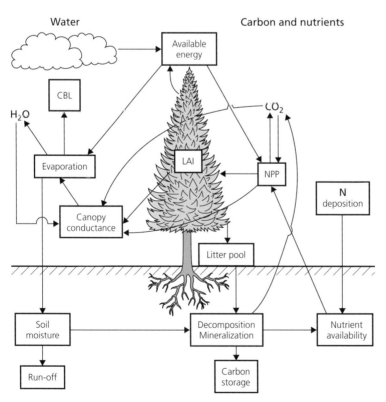

Fig. 10.1 The main flows of energy, water, carbon and nitrogen gases between a forest canopy and the atmosphere and how they relate to canopy and soil properties and processes. CBL, convective boundary layer; LAI, leaf area index; NPP, net primary productivity. (From Baldocchi & Meyers 1998.)

to 5 years. Stages during the natural or managerial succession have not so far been systematically studied. We now recognize that investigations of chronosequences are needed to obtain temporally integrated estimates of the exchanges of water vapour and trace gases for a range of forest types to be able fully to understand the interactions and to construct lifetime budgets. Furthermore, management operations (e.g. planting, thinning and harvesting) and natural disturbances (e.g. windthrow and fire) occur at irregular intervals with major consequences for pools and fluxes, particularly at the larger spatial scales, and thus also impose a requirement for long sequences of data or, at the least, for campaigns to sample different stages in a chronosequence, for example substituting spatial variability for temporal variability.

Interannual variability

In addition, at any time during growth of a new plantation or natural forest succession, there may be appreciable variability from year to year in the exchanges of trace gases as a result of variability in the weather. For example, dry tropical forests may become a source of CO_2 in years of relatively low solar radiation inputs and high temperature, such as were caused by the eruption of Mt Pinatubo (Grace *et al.* 1995a,b), and may occur particularly during El Niño years (Tian *et al.* 1998). Water vapour and CO_2 exchanges of temperate deciduous forest are particularly sensitive to the length of the growing season as defined by the times of budburst and leaf senescence, which may vary by 2–3 weeks from year to year (Goulden *et al.* 1996); temperate coniferous forests are similarly affected by the length of the photosynthetic period, particularly in northern maritime latitudes (Valentini *et al.* 2000). In the boreal forest, exchanges are even more sensitive to the timing of the onset of the spring thaw, which can vary from year to year by 2–3 weeks and may result in a change in the annual sequestration of carbon of 1 Mg ha^{-1} so that a stand may be a sink for CO_2 one year and a source the next (Goulden *et al.* 1998; Lindroth *et al.* 1998; Bergh & Linder 1999). Thus, measurements made in a single year at some arbitrary time in the succession, or even over a few years such as the initial

5-year 'commitment period' of the Kyoto Protocol, may give a misleading picture of the long-term exchange capacity of the forest. For this reason, measurements should extend over periods of at least 5 years, in order to define the major drivers and constraints on the fluxes. However, such lengthy periods of observation have so far been made on only a handful of stands.

10.1.2 Historical development

Although the basic concept of measurement of vegetation–atmosphere exchanges by eddy covariance was enunciated in the 1930s (see Sutton 1953), the technique was essentially confined to the measurement of sensible heat flux because of the difficulty in obtaining sufficiently fast sensors for water vapour, CO_2 and other trace gases (Swinbank 1951). In the 1960s, water vapour fluxes were successfully measured continuously over short periods using fine-wire thermojunctions wetted with single strands of cotton (the Evapotron and Fluxatron: Dyer & Maher 1965; Dyer & Hicks 1967) but generally flux measurements were made by the Bowen ratio or flux-gradient method, sometimes automated into an energy balance system. This approach, while practical over crops and grassland (e.g. Biscoe *et al.* 1975), was particularly difficult (and expensive) to apply over forest because of the extremely small size of the gradients as a result of canopy height and surface roughness (Thom *et al.* 1975; Jarvis *et al.* 1976). Consequently between the mid 1970s and the early 1990s there was a major gap in measurement of trace gas fluxes while researchers awaited improvements in the eddy covariance and eddy accumulation approaches. Developments in sensor technology since 1990 (Baldocchi *et al.* 1996; Moncrieff *et al.* 1996, 1997; Aubinet *et al.* 2000) have led to near-continuous measurements by eddy covariance of canopy–atmosphere exchanges of momentum, water vapour, CO_2 and other trace gases, first in campaigns lasting months, then over a year and now over periods of up to 5 years.

Exchanges of radiation, momentum, sensible heat and water vapour by forests have been relatively well understood for some time (e.g. Stewart & Thom 1973; Biscoe *et al.* 1975; Rutter 1975; Thom 1975; McNaughton & Jarvis 1983; Raupach

1988; Shuttleworth 1989) and stand-scale fluxes have been progressively documented over the past 30 years (e.g. Jarvis *et al.* 1976; Ruimy *et al.* 1995; Baldocchi & Meyers 1998). Consequently, these exchanges are treated relatively briefly, with the emphasis on new concepts and data. A broader, more comprehensive treatment is given to fluxes of the several trace gases of importance as pollutants and to CO_2, of particular importance in relation to 'global warming' and the reduction of net CO_2 emissions of anthropogenic origin. Understanding fluxes of trace gases has increased substantially in recent years in direct relation to the development and application of the new measurement technologies. Considerable impetus has been provided by major national and international experiments on land–atmosphere interactions that have included forests as a major land surface cover (e.g. HAPEX-MOBILHY, Andre *et al.* 1986; HAPEX-SAHEL, Moncrieff *et al.* 1997; ABRACOS, Gash *et al.* 1996; EUROFLUX, Valentini *et al.* 1999; MEDEFLU, FLUXNET, Running *et al.* 1999; BOREAS, Sellers *et al.* 1995, 1997; LBA, Kabat *et al.* 1999) as well as by more restricted and local campaigns (e.g. Hollinger *et al.* 1994; Greco & Baldocchi 1996; Baldocchi 1997; Kelliher *et al.* 1997; Schulze *et al.* 1999).

10.1.3 The stand system

For simplicity we consider forest stands as being extensive and relatively uniform in the horizontal and treat exchanges with the atmosphere as largely one-dimensional processes in the vertical, acknowledging none the less that advection may at times exercise a major influence. We define the upper boundary as a horizontal plane at the height of the eddy covariance flux station, approximately twice the height of the trees, and the lower boundary as a horizontal plane either at the soil surface or at some depth below the rooting zone.

At the outset, we consider the environmental variables driving exchange processes as independent variables at the scale of the stand, subsequently recognizing the increasing importance of feedbacks with increase in the spatial scale, so that the driving variables become progres-

sively dependent variables (McNaughton & Jarvis 1991).

10.2 STAND ENERGY BALANCE

Conservation of energy requires that the influxes and effluxes of energy should equate, i.e.

$$R_n = H + \lambda E + G + J + S + \mu F_c \qquad [10.1]$$

where R_n is the net absorbed all-wave radiation flux, H the sensible heat flux, λE the latent heat flux (i.e. the energy equivalent of the evaporation and transpiration of water), G the flux of sensible heat into storage in the soil, J the flux of sensible and latent heat into storage in the air column, S the flux of sensible heat into storage in the trees and other vegetation, and μF_c the energy equivalent of CO_2 exchange (Fig. 10.2). Each of the quantities in Eqn 10.1 can be both an influx and an efflux. Conventionally, effluxes are treated as positive and influxes as negative, with the exception of R_n for which a net influx is treated as positive and a net efflux as negative.

Conceptually this is a one-dimensional model and supposes an area of forest sufficiently extensive and uniform for horizontal gradients and

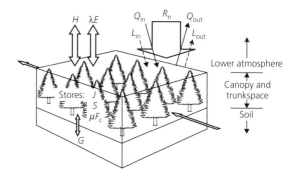

Fig. 10.2 Representation of the energy and radiation budget components of a forest stand. R_n, net radiation; H, sensible heat flux; λE, latent heat flux; G, soil heat flux; μF_c, biochemical energy stored in photosynthesis and released in respiration; S, heat storage within the stand biomass; J, sensible and latent heat storage in the canopy air column; Q_{in}, short-wave radiation to the surface; Q_{out}, reflected short-wave radiation from the surface; L_{in}, long-wave radiation to the surface; L_{out}, long-wave radiation from the surface to the atmosphere. (Adapted from Oke 1987.)

fluxes to be negligible. However, with a few exceptions in the tropical and boreal regions, most forests, and more particularly stands within forests, are too small to meet this requirement. Advection terms can be added to this equation but the practical difficulties in evaluating small horizontal gradients are such that this is rarely done and horizontal transport is neglected. None the less, 'footprint analysis' of flux source areas around a flux measurement tower in a stand (Schuepp *et al.* 1990) indicates that about 90% of the measured flux originates within a radius of about 50–700 m from the tower for flux-measuring instrumentation placed about 10 m above the canopy surface; various empirical estimates suggest that horizontal gradients become negligible within 200–300 m of a stand edge, so that any horizontal flux component neglected is likely to be very small.

The first two terms on the right-hand side of Eqn 10.1 $(H + \lambda E)$ are together generally much larger than the storage and photosynthetic terms, accounting for around 90% of the energy exchange, except in very open, sparse forest when the exchange with the soil may at times be large ($\approx 100\,W\,m^{-2}$) (Jarvis *et al.* 1976). The ratio of sensible heat transfer to latent heat transfer, the Bowen ratio ($\beta = H/\lambda E$), is a functional descriptor of energy use by forest stands. Originally conceived to represent partitioning of absorbed net radiation, the Bowen ratio is particularly useful now that we are able to measure the sensible and latent heat fluxes independently and routinely by

eddy covariance (Fig. 10.3) and to appreciate that advection and entrainment of warm or dry air are the norm rather than the exception where forests are concerned.

Similar stands in different climates partition the available energy in different ways, as do dissimilar stands in the same environment. In a maritime environment on the Atlantic fringe of Europe, for example, high humidities suppress evaporation so that transpiration rates of spruce are low, with midday summer values of β of 2 or more; in the more continental climate of central Europe, with higher temperatures and larger water vapour saturation deficits, values of β for spruce are 1 or less. Similarly, stands of different species on different soils in the same climatic region have very different ranges of β (Fig. 10.4).

Because forest areas are on the whole small or fragmented, the variables driving fluxes in a stand have very largely been determined elsewhere, frequently outwith the forested area. Air both advected and entrained into the forested area is likely to be very different in temperature, humidity and trace gas composition to the air in the surface layer, particularly in the early morning, so that these variables can be regarded initially as independent driving variables.

Energy balance closure provides a useful test of the competence of the measurement techniques. Rearranging Eqn 10.1, net radiation, less all the storage terms, should equate to the sum of the sensible and latent heat fluxes for complete closure, i.e.

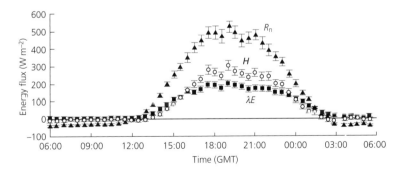

Fig. 10.3 Ensemble daily courses of net radiation R_n (▲), sensible heat flux H (○) and latent heat flux λE (■) densities of a boreal black spruce stand during a 22-day period in the summer of 1994. Each point is the value of the half-hourly average of the flux before the time shown on every day in the period. The error bars show ± 1 SE. Six hours should be subtracted from the time to give local time. (From Jarvis *et al.* 1997.)

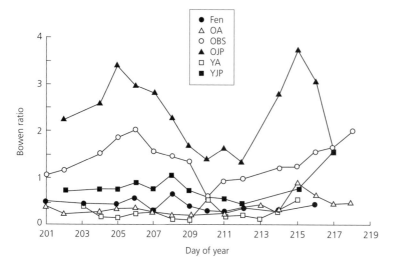

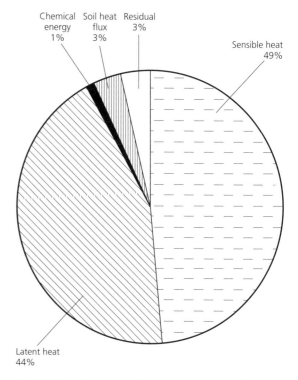

Fig. 10.4 Examples of the Bowen ratio of four forest sites and fen at midday during days 201–218 in Canadian boreal forest during 1994. OJP, stand of old jack pine on sandy soil; OBS, stand of old black spruce on peaty soil; YJP, stand of young jack pine on sandy soil; OA, stand of mature aspen with an understorey of hazel on brown earth soil; YA, stand of young aspen on a brown forest soil; and fen. All stands are located in the BOREAS Southern Study Area (see Sellers *et al.* 1995, 1997.)

$$R_n - (G + J + S + \mu F_c) = A = H + \lambda E \qquad [10.2]$$

where A is the so-called available energy. In practice, integrating over 24 h, excellent agreement is obtained, i.e. closure within 2–5% (Fig. 10.5). However, considering the daytime alone, a problem is evident. The discrepancy then can amount to 10–15%, depending on site. There are several possible explanations for this. For example, there are known errors in the measurement of net radiation, which in any case is measured over a fixed area at the tower whereas the flux footprint covers a variable area some distance away. However, the consistency of this underestimation across a number of sites indicates that a part of the scalar fluxes transported in large eddies is possibly not being measured and efforts are currently being made to resolve this.

10.3 RADIATION EXCHANGE

10.3.1 Solar radiation albedo

Daily average canopy reflectance (albedo) of short-wave solar radiation (Q) ranges between about 8 and 13% for conifers and about 12 and 16% for broadleaved species (Jarvis *et al.* 1976; Baldocchi *et al.* 2000). The range of values depends strongly on zenith angle, canopy structure, phenology and snow cover, and there are interactions among these factors. In general, the

Fig. 10.5 Partitioning of the total net radiation flux density of boreal black spruce over the entire growing season of 120 days from 23 May (day 143) to 21 September (day 264) day and night at the BOREAS Southern Study Area Site. (From Jarvis *et al.* 1997.)

larger the zenith angle, the higher the reflectance, so that on sunny days as the day progresses, albedos decline to a minimum at midday rising again in the afternoon, the minimum becoming progressively lower as the season advances to midsummer. The lower albedos for coniferous canopies result from the spire-like form of the tree crowns, with multiple reflections and scattering among them, so that when seen from above, the spaces between the tree crowns appear as black holes. The more planar surfaces of broadleaved canopies result in higher albedos that depend more strongly on larger leaf area densities in the upper canopy and leaf inclination angles closer to horizontal than in coniferous canopies. In the tropics albedos of evergreen species tend to be lower than in the boreal region on sunny days because of the smaller zenith angles (Pinker *et al.* 1980; Pinker 1982); in the boreal region they are up to 5% higher in winter when there is some snow retained on the canopy (Betts & Ball 1997; Baldocchi *et al.* 2000). The albedos of deciduous broadleaved trees are up to 5% lower in the winter when they are leafless.

Midday or mean daily canopy reflectances for photosynthetically active radiation (PAR) or photosynthetic photon flux density (PPFD) from closed forest canopies are no more than 3–5% because of strong absorption in the waveband 350–700 nm but may reach 7% in winter, and in open stands vary depending on canopy leaf area density and the type of understorey (e.g. Hassika & Berbigier 1998). Above canopies the relation between incident PPFD and solar radiation is relatively conservative ($2.0 \pm 0.1\,\mu\text{mol J}^{-1}$), although within canopies selective absorption of PPFD, particularly in the red waveband, reduces this ratio substantially (Sattin *et al.* 1997). The red:far-red ratio, important for phytochrome photoequilibrium and for neighbour detection in canopies (Gilbert *et al.* 1995), also changes with depth into a canopy as a result of preferential absorption of red radiation by chlorophyll. From a typical value of 1.3 in sunlight, this ratio falls to an order of magnitude less below canopies of leaf area index larger than about 4.

10.3.2 Long-wave radiation exchange

There have been few regular or continuous measurements of incident, reflected or emitted thermal radiation above forest canopies until very recently because of the lack of appropriate network-standard instruments. Thus estimates in the past have been based on the relationship between measured solar radiation exchange and net radiation balance (e.g. Monteith & Szeicz 1961) or on calculation based on simple assumptions regarding canopy and sky temperatures. Forest canopies are in general well coupled to the atmosphere (see section 10.5) so that leaf temperatures, particularly of small narrow leaves in unstable conditions, are close to air temperature, and the emissivity of leaves in the waveband range 10–14 μm is 0.97 ± 0.02 (Gates 1980), i.e. very close to that of a black body. Energy balance models of forest canopies have also been solved iteratively to yield leaf temperatures to provide more accurate estimates of thermal radiation emission (e.g. Baldocchi 1997). Both the incoming and outgoing long-wave fluxes are large, although the difference between the net long-wave flux is small relative to the net solar radiation in daytime but is crucial to canopy temperature at night.

10.3.3 Net all-wave radiation balance

The net all-wave radiation absorbed by a canopy (i.e. the measured sum of net short-wave and net long-wave radiation, R_n) defines the energy available to drive energy-requiring processes, in the absence of advected and entrained convective energy. As such, it is particularly important for driving transpiration from the understorey and the ground vegetation but less so for driving transpiration from the overstorey, which is usually more strongly driven by the atmospheric water vapour saturation deficit, because of the strong coupling of forest canopies to the atmosphere for much of the time (see section 10.6). Because the net long-wave exchange is comparatively small there is a close relation between net all-wave radiation and net solar radiation. The net all-wave radiation measured over a broadleaved canopy is about $0.7Q$ to $0.75Q$, whereas over conifers it may exceed $0.85Q$ because of the small albedo. This is one reason why coniferous canopies are particularly effective in warming the atmosphere.

10.4 MOMENTUM EXCHANGE

The aerodynamically rough surfaces of forests enhance rates of exchange of momentum and scalars between forests and the atmosphere relative to surfaces of other shorter, terrestrial vegetation. The frictional and form drag of forests on the boundary layer airflow very effectively captures the momentum from the airflow. Expressed in energy terms, the flux of momentum over typical forest surfaces exceeds that over shorter vegetation by a factor of five or more for the same wind speed (Table 10.1). The exchange of momentum between forests and the atmosphere provides the primary driving energy for the local exchange of scalars entrained in the atmospheric surface layer. A detailed mathematical description of the turbulent structure within the vegetation boundary layer is given by Thom (1975) and lies beyond the scope of this chapter, though see also Chapter 12. More recent descriptions of the subject are provided by Garratt (1992), Kaimal and Finnigan (1994) and Raupach *et al.* (1996).

In considering a simple, dimensional Fickian approach to mass and momentum transfer across the boundary layer of the forest canopy, the turbulence generated by the frictional drag of terrestrial surfaces is the primary regulator of scalar transport. The enhanced momentum exchange is reflected in the potential for a greatly enhanced exchange of scalar quantities, such as water vapour, trace gases, CO_2 and particles, and is manifest in large values of the leaf boundary layer and canopy aerodynamic conductances and small

values of the corresponding transfer resistances (see section 10.5).

None the less, the degree to which the potential rates of exchange are realized is also determined by processes at the leaf, branch and stem surfaces which regulate the source or sink strength. Examples of the rates of exchange of a range of trace gases are included in Table 10.2 to illustrate typical upper limits for rates of exchange of these gases.

10.5 EXCHANGE OF SCALARS

Exchanges of scalars (water vapour, trace gases, particles) between forests and the atmosphere can be expressed as a flux (F) driven by a difference in 'concentration' (ΔX) and constrained by a conductance (g) or resistance (r), i.e.

$$F = g \cdot \Delta X = \Delta X / r \qquad [10.3]$$

Depending on the scalar and whether the net flux is towards or away from the canopy, the flux is constrained by a catena of conductances or resistances along the pathway between sources or sinks on the surfaces of leaves, twigs and branches, or within leaves, and the source or sink in the atmosphere, notably the conductances of the stomata and stomatal antechambers, the conductance of the leaf cuticle and other tree surfaces, the conductances across the leaf boundary layers, and the aerodynamic conductances through the canopy and across the canopy boundary layer to a reference location above (Fig. 10.6).

Table 10.1 Momentum exchange properties of a Sitka spruce canopy for six ensemble wind speed profiles above the canopy. (From Jarvis *et al.* 1976.)

	Wind speed $u(h)$ (m s^{-1})					
	1	2	3	4	5	6
r_{aM} (s m^{-1})	12.0	6.00	3.96	2.98	2.38	1.98
g_{aM} (m s^{-1})	0.08	0.17	0.25	0.33	0.42	0.50
$K_M(h)$ $(\text{m}^2\,\text{s}^{-1})$	0.16	0.33	0.45	0.60	0.75	0.91
$\tau(h)$ (N m^{-2})	0.10	0.40	0.91	1.61	2.52	3.63

$u(h)$, wind speed at canopy top; r_{aM}, aerodynamic resistance for transfer of momentum; g_{aM}, corresponding aerodynamic conductance; K_M, momentum exchange coefficient; τ, momentum flux density or shear stress.

Table 10.2 Potential effects of forests on deposition fluxes. (From Fowler *et al.* 1999.)

		Moorland	Forest	Percentage increase forest/moor
Canopy height	H (m)	0.15	10	–
Zero plane displacement	D (m)	0.1	7	–
Roughness length	z_0 (m)	0.01	1.0	–
Friction velocity	$u^\star$ (m s^{-1})	0.32	0.82	156%
Momentum flux	τ (N m^{-2})	131	840	541%
Maximum deposition velocity for SO_2	$V_{max} SO_2$ (mm s^{-1})	18.6	35.1	89%
Maximum deposition velocity for NO_2	$V_{max} NO_2$ (mm s^{-1})	20.0	43.5	118%
Maximum deposition velocity for NH_3	$V_{max} NH_3$ (mm s^{-1})	21.4	55.5	160%

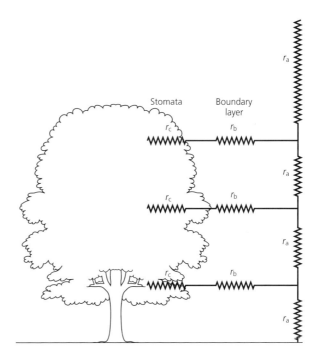

Fig. 10.6 A resistance/conductance model for a tree in a stand. The surface resistance (r_c) is made up of the sum of the individual leaf stomatal resistances, usually including the leaf boundary layer resistance (r_b). The aerodynamic resistance (r_a) comprises the resistances within the canopy and from the upper limit of the canopy to the reference level above. (Adapted from Monteith & Unsworth 1990.)

While the application of conductance and resistance approaches are the norm for current applications, it is necessary in important cases to recognize that fluxes are bidirectional and that multiple sources and sinks within forest canopies increasingly require more complex treatments to describe the interaction of biological production, uptake and chemical processing with the turbulent exchange processes above and within canopies.

At stand scale, canopy stomatal and boundary layer conductances are derived as the product of the leaf conductances and the leaf area index of leaves, appropriately weighted according to the scalar (e.g. McNaughton 1994). The scaled-up leaf stomatal conductance is commonly referred to as the canopy surface conductance, or simply the canopy conductance.

Historically, resistances (the inverse of conductances) were first used but recently conductances have become more widespread. The conductance terminology is now commonly used for transfer of CO_2 and, to a lesser extent, water vapour, whereas the preferred convention for quantifying the affinity of vegetation surfaces for uptake of *reactive gases* remains the canopy or surface resistance (Fowler & Unsworth 1979), because the full range of surface affinities must be included and this would result in infinite surface conductances. The resistance approach is therefore not only conventional but also preferable on theoretical grounds. In the cases of both HNO_3 and HCl, for example, the surface resistance is negligible and tree surfaces generally are considered to be 'perfect sinks' for these gases. There are few measurements of HNO_3 (and especially HCl) canopy fluxes, although the work so far indicates zero canopy resistance for these gases (Huebert 1983; Muller *et al.* 1993). The combination of large atmospheric conductance (g_a) (or small atmospheric resistance, r_a) and zero canopy resistance (r_c) leads to significant inputs of atmospheric nitrogen as HNO_3 to forests even when ambient concentrations are very small and, conversely, to large losses of intercepted precipitation from wet canopies although atmospheric saturation deficits are then small.

More problematical, at least two sets of units have been, and continue to be, widely used by different sectors of the community, depending largely on whether fluxes are required in units of mass (kg), volume (m^3) or amount (mol) and whether 'concentration' differences are expressed as mixing ratios, volume fractions, mole fractions, partial pressures, etc. Conveniently, where concentrations are dimensionless, as in the case of mole fraction for example, the conductance has the same dimensions as the flux and the dimensions of the resistance are the inverse. Where the 'flux' is expressed as a volume flux density, i.e. $m^3 m^{-2} s^{-1}$, the conductance has the units of $m s^{-1}$ and the resistance the somewhat curious units of $s m^{-1}$!

10.5.1 Aerodynamic conductance/resistance

The aerodynamic conductance/resistance should be formulated to give an accurate representation of the transfer of the appropriate scalar of interest (water vapour, CO_2 and other trace gases) from the source or sink in the canopy, across the atmospheric surface layer to the sink or source in the mixed layer. Aerodynamic conductances of canopies (g_a) have most frequently been estimated using the transfer of momentum to canopies or the transfer of water vapour away from wet canopies as tracers. The transfer of momentum is driven by pressure forces and all canopy elements provide the sink, whereas the transfer of water vapour and trace gases is by molecular or turbulent diffusion and the sources and sinks in the canopy are usually only the leaves, depending on the scalar. The difference between the two can be accounted for (Chamberlain 1975; Thom 1975) but since the aerodynamic resistance of forest canopies is generally at least one order of magnitude smaller than the surface resistance because of the strong turbulence in and near forest canopies, little is gained by this and either may be used.

Canopy aerodynamic conductance increases in proportion to wind speed. Analysis of wind profiles showed g_a for momentum to increase from $0.08\,m\,s^{-1}$ ($3.2\,mol\,m^{-2}\,s^{-1}$) for wind speed of $1\,m\,s^{-1}$ at canopy height to $0.5\,m\,s^{-1}$ ($20\,mol\,m^{-2}\,s^{-1}$) for wind speed of $6\,m\,s^{-1}$ above a Sitka spruce stand. From a review of wind profile data, the following general relation was deduced for coniferous forest stands (Jarvis *et al.* 1976):

$$g_a = 0.1\,u(h)\,m\,s^{-1}, \text{ i.e. } g_a = 4\,u(h)\,mol\,m^{-2}\,s^{-1}$$

$$[10.4]$$

where $u(h)$ is wind speed in $m\,s^{-1}$ at canopy height.

In a Sitka spruce spacing experiment g_a for water vapour was found to increase asymptotically in relation to number of trees, and hence leaf area, per hectare, whereas the conductance on a per tree crown basis increased linearly with distance between the trees (Teklehaimanot & Jarvis 1991; Teklehaimanot *et al.* 1991). This can be attributed to enhanced ventilation of the trees and increases in a number of turbulence parameters (Green 1990). Typical values of g_a for water vapour were in the range 3–9 mol m^{-2} s^{-1} (Jarvis 1993). In a comparison between selectively logged and unlogged plots in tropical forest, Asdak *et al.* (1998a,b) found a comparable situation: the boundary layer conductance, on a per tree basis, of the trees remaining on the logged plot was higher than that of the trees on the comparable unlogged plot. It seems likely that g_a will also be influenced by vertical distribution of leaf area density in canopies: a high leaf area density in the upper layers of a multistoried tropical forest canopy inhibits penetration of downdrafts and effectively decouples the lower layers from the air above (Shuttleworth 1989; Kruijt *et al.* 1999). However, effects of crown and canopy structure on g_a have not yet been systematically investigated.

10.5.2 Surface conductance/resistance

Surface conductances of canopies (g_c) have frequently been estimated using water vapour transfer from dry canopies as a tracer and scaled for other trace gases such as CO_2 by the ratio of their molecular diffusivities. For a range of deposition fluxes, using a resistance analogy allows the aerodynamic resistance to be subtracted from the total transfer resistance to give the surface resistance. The absence of a residual term implies very efficient exchange and large deposition rates (velocities) as illustrated in Table 10.2. However, more commonly a residual or surface resistance is present which, in the case of reactive trace gases, may be considered as a measure of the chemical affinity of the absorbing surface for the trace gas in question. In general, highly reactive gases such as HCl or HNO_3 are characterized by very small or zero surface resistances.

Surface conductances of dry canopies for water vapour (g_c) lie in the range 0.02–2 mol m^{-2} s^{-1}. Despite high leaf area index in some species, very low maximum values of g_c occur in conifers, partly because the stomatal antechambers are filled with wax rods and tubules. Higher values of g_c occur in broadleaved trees, particularly in multistorey tropical forest. In addition to the area of leaves in a canopy, g_c is sensitive to the several environmental and physiological variables that influence stomatal aperture (Schulze *et al.* 1994). Stomatal conductance increases with PPFD, has a temperature optimum, and decreases with atmospheric water vapour saturation deficit and CO_2 concentration, and with shortage of soil water (Jarvis & Morison 1981). Canopy surface conductances are usually at least an order of magnitude smaller than canopy aerodynamic conductances so that fluxes from forest canopies are generally effectively constrained by g_c. In coniferous forests in particular, atmospheric conditions at the leaf and canopy surfaces are very similar to the conditions overhead so that feedbacks affecting the local atmosphere play little or no role in constraining transpiration and exchanges of non-reactive trace gases, which are highly sensitive to changes in g_c (McNaughton & Jarvis 1991).

There are several alternative models of stomatal action in relation to environmental variables and leaf properties (e.g. Jarvis 1976) and these have been incorporated into large-scale soil–vegetation atmosphere transfer schemes (SVATS). In due course dynamic models incorporating surface chemistry will be able to simulate surface reactive processes within SVATS, and the first steps in this process have been made (Flechard *et al.* 1999). However, for most forests and trace gases, current understanding limits the scope for such modelling to a very few specific cases.

10.6 EVAPORATION AND TRANSPIRATION

Evaporation and transpiration of water vapour from forests have major influences on forest production, local water supplies and regional climates (refer also to Chapters 8 and 12). Depending on the precipitation climate, significant amounts of water are returned to the atmosphere by evaporation after interception by the overstorey and understorey canopies without ever reaching the soil, the so-called interception loss (Rutter 1975). A proportion of the precipitation reaches the ground by through-fall and stemflow and infiltrates the soil. Some of this water is taken up by roots of the trees and understorey vegetation, and having traversed from roots to leaves is returned to the atmosphere in transpiration, defined as evaporation of water from within the leaves and its subsequent diffusion and turbulent transfer from the sites of evaporation to the air above (Jarvis 1976; Whitehead & Jarvis 1981). Total evaporation and transpiration rates from forests may reach 7 mm daily, although annual totals mostly average 1–3 mm daily (e.g. Gash & Stewart 1977).

10.6.1 Driving variables and constraints

The driving variables for both the evaporation of intercepted water and for transpiration are the net radiation and sensible heat absorbed by the leaves, both of which raise leaf temperature and power the phase change in evaporation from liquid water on leaf surfaces or within leaves to water vapour, together with the water vapour mole fraction or vapour pressure deficit at the sink for water vapour in the atmosphere outwith the canopy. The process is constrained by the catena of conductances or resistances along the pathway from the sources of water vapour on the surfaces of wet leaves or from within dry leaves to the sink in the atmosphere. This catena of conductances, together with the difference in water vapour mole fraction between the sources of water vapour on the surfaces of, or within, the canopy leaves and the sink in the atmosphere, determines the tendency for molecules of water vapour to escape from the canopy to the atmosphere. Surface conductances (or resistances) strongly determine the exchange of water and other scalars between vegetation and the atmosphere at leaf, stand, forest and regional scales (e.g. Lindroth 1985; Kelliher *et al.* 1994; Raupach 1995).

10.6.2 Evaporation of intercepted water

Although models using a statistical description of raindrop size and frequency may give a somewhat better predictive representation of the interception process (Calder 1996), the evaporation of water from wet canopies (E_i) has generally been represented using the Penman equation (Monteith & Unsworth 1990) in the Rutter model (e.g. Rutter *et al.* 1975, Rutter & Morton 1977, Teklehaimanot *et al.* 1991; Whitehead & Kelliher 1991; Valente *et al.* 1997; Asdak *et al.* 1998b), i.e.

$$E_i = [\varepsilon/(\varepsilon+1)](A/\lambda) + [1/(\varepsilon+1)](g_a D/p) \qquad [10.5]$$

where ε is a coefficient for change in the sensible and latent heat contents of air with respect to temperature, λ the molar latent heat of vaporization of water, and D and P the ambient water vapour pressure saturation deficit and atmospheric pressure of the air at the reference location above the canopy. Both the available energy and saturation deficit of the atmosphere are usually small ($< 100\,W\,m^{-2}$ and $< 0.1\,kPa$, respectively) during and immediately after rainfall events, when evaporation of intercepted water occurs. None the less, evaporation rates are high, up to

0.6 mm h^{-1} (400 W m^{-2}, 10 mmol m^{-2} s^{-1}) for short periods as measured by eddy covariance or inversion of the Penman equation (Mizutani *et al.* 1997). This is a result of the high aerodynamic conductance to sensible heat transfer towards the forest surface, and to water vapour transfer away from the source to the sink in the atmosphere above, and because the saturation deficit, although small, is maintained by advection of drier air from the landscape surrounding the area of the rainstorm (Stewart 1977; Asdak *et al.* 1998b). The water vapour pressure at the canopy surface may fall to the saturation vapour pressure at the wet-bulb temperature (Teklehaimanot & Jarvis 1991) and then the wet canopy behaves like a large psychrometer: the rate of evaporation is balanced by the input of sensible heat and thus given solely by the second term on the right-hand side of Eqn 10.5 (Jarvis 1993), i.e. the net radiation term can be neglected, a useful convenience when deriving g_a from inversion of Eqn 10.5.

The proportion of gross precipitation that is intercepted and directly evaporated back to the atmosphere depends strongly on the frequency and intensity of the rainfall (Rutter 1975). In the tropics, where much of the rainfall comes in relatively short intense storms, interception loss may amount to no more than 11% of the gross precipitation (Lloyd *et al.* 1988; Asdak *et al.* 1998a), although some higher values have been reported (e.g. Sinun *et al.* 1992). At the other extreme, in maritime climates where the rainfall comes in gentle showers over long periods, so that the canopy is wet for much of the time, interception loss can amount to 40% of the measured precipitation (e.g. Jarvis & Stewart 1979; Gash *et al.* 1980). Furthermore, much of the fog and, on hills and mountains, the cloud water intercepted by the foliage of trees is not measured by conventional rain gauges and is readily evaporated back to the atmosphere, frequently leaving a nasty deposit on leaf surfaces (see section 10.8). In 'cloud forest' however, fog days may add significantly to the precipitation, although not measured as such, and consequently is known as 'occult' precipitation (see further comment in Chapter 12).

The evaporation of intercepted water is influenced by tree spacing and logging. One might expect the amount of water evaporated to decline in proportion to the reduction in number of trees because of the reduction in surface area of wet foliage. However, in a Sitka spruce spacing experiment this was found to be compensated to a significant degree by the increase in g_a on a per tree basis associated with the increase in distance between the trees (Teklehaimanot & Jarvis 1991) and similar results were obtained in a comparison between logged and unlogged areas of tropical forest (Asdak *et al.* 1998b). Valente *et al.* (1997) added an additional coefficient to the Rutter model to take account of the wide open spaces in the sparse canopies of Mediterranean forests and Gash *et al.* (1999) have shown, by comparison with eddy covariance data, that evaporation is extremely well represented by their sparse-canopy Rutter model in combination with the *momentum* aerodynamic conductance.

10.6.3 Evaporation of transpired water

A forest comprises a system of parallel conduits through which water is passively transferred from soil to atmosphere as the result of evaporation of water from the leaves and the consequent development of a gradient in water potential (Jarvis 1975). Recently, attention has become strongly focused on the capacity of the liquid water conducting system and its vulnerability to cavitation and entry of air into the vessels and tracheids. However, it is relevant to re-emphasize the point made long ago (Rawlins 1963; Philip 1966; Weatherley 1970) that by far and away the largest drop in water potential (or equivalent vapour pressure) in the soil–plant–atmosphere continuum (SPAC) occurs across the stomatal pores, and consequently this is the one location in the SPAC where the flow of water from soil to atmosphere can effectively be regulated. Thus the conductances of the stomata in a dry canopy provide the fine-scale, variable control of the transpiration and latent heat fluxes. While the mechanism of stomatal action is still somewhat of a 'black box', one hypothesis that has been incorporated into some models (e.g. Williams *et al.* 1996, 1998) has it that stomatal conductance adjusts to minimize the risk that the conducting system is impaired through cavitation and air entry.

The driving forces and constraints on transpiration are combined in the now very familiar Penman–Monteith equation (Monteith 1965; Monteith & Unsworth 1990), which provides the basis for the representation of transpiration in most canopy models. While based on sound physics at the scales of the pore and the leaf, use of this equation at canopy scale does involve some assumptions and approximations because of the different spatial distribution of the sources and sinks for heat and water vapour in a canopy compared with a single leaf (Jarvis & McNaughton 1986; Finnigan & Raupach 1987; McNaughton 1994).

Coupling to the atmosphere

While the temperature and humidity of the air may in some circumstances be regarded as independent driving variables, vegetation influences its own local environment to varying degrees, depending particularly on its structure (McNaughton & Jarvis 1983, 1991). At one extreme, feedback from the exchanges of heat and water vapour in extensive areas of crops, for example, determines the temperature and humidity within the vegetation and in the adjacent surface layer. Equilibrium profiles of temperature and humidity develop in relation to the fluxes of heat and water vapour that are driven by the absorbed net radiation. Consider the consequences of a change in conditions that lead to an increase in transpiration, such as an increase in stomatal conductance or saturation deficit. There will be a transient increase in transpiration, but if the additional molecules of water vapour are not able to escape from the vicinity of the foliage, transpiration will subsequently be damped down and will return to its original rate, as determined by the net radiation. From an analysis of transpiration datasets obtained from extensive areas of field crops, Priestley and Taylor (1972) arrived at the following model for equilibrium transpiration:

$$E_{eq} = [\varepsilon/(\varepsilon+1)]\alpha(A/\lambda) \qquad [10.6]$$

where α is a coefficient, in principle unity but found by Priestley and Taylor to have an empirical value of about 1.3 for field crops, that essentially takes into account effects of entrainment and advection when this equation is used alone. More recently, this model for transpiration has been shown to represent transpiration by a range of field crops reasonably well (McNaughton & Jarvis 1983) but was found to be completely inadequate for plantations of trees, for which wholly improbable values of α of about 11 were required (Shuttleworth & Calder 1979).

At the other extreme, consider a stand of tall trees with an irregular, 'rough' canopy surface. Partly because the foliage of tall trees is in a higher wind-speed regime but also because of the 'rough' surface, air movement and turbulence are much greater among the foliage, with the result that the ambient environmental conditions are imposed close to the leaf surfaces. Consequently, when an initial increase in transpiration occurs because of an increase in stomatal conductance or ambient saturation deficit, the additional molecules of water vapour have little difficulty in escaping so that the increase in transpiration is maintained at a new higher rate that is largely independent of the net radiation input. At this limit, leaf and air temperatures are equal and the ambient saturation deficit is imposed right up to the leaf surfaces so that transpiration is given by

$$\lambda E_{imp} = g_c D/(P\lambda) \qquad [10.7]$$

McNaughton and Jarvis (1983) and Jarvis and McNaughton (1986) redeveloped the Penman–Monteith equation to give expression to these two limit situations and in the process derived a decoupling coefficient, Ω, that defines the extent to which actual transpiration approaches one or other of the two limits, i.e.

$$E = \Omega E_{eq} + (1-\Omega)E_{imp} \qquad [10.8]$$

where $\Omega = (\varepsilon + 1)/(\varepsilon + 1 + g_a/g_c)$ $\qquad [10.9]$

Ω has a range of zero to 1.0, approaching zero as g_a increases relative to g_c and approaching 1.0 as g_c increases relative to g_a. Since g_a is a function of wind speed and turbulence and g_c a function of water stress, the ratio g_a/g_c is large for tall stands of trees in windy environments, particularly when saturation deficits are large and soil water stress

intervenes, and is small for well-watered short vegetation in less windy conditions. The ratio also varies somewhat with stand and canopy structural features such as tree spacing and the spatial distribution of leaf area density, both of which affect g_a, and with leaf area, structure, anatomy and physiology, all of which affect g_c.

Estimates of Ω are in the range 0.1–0.3 for stands of conifers and 0.3–0.6 for stands of broadleaved trees, the lower values of each range being in windier maritime climates, the higher values in less windy, more continental climates. These low values of Ω show that E_{imp} is the much more important component of transpiration in forests than is E_{eq} and thus that saturation deficit is a more important driving variable than net radiation, especially in conifers, and indeed that stomatal conductance is a more important constraint than is the availability of absorbed net radiation. Furthermore, E_{eq} is necessarily limited by the availability of net radiation whereas E_{imp} is open-ended in response to saturation deficit, unless or until stomatal closure or leaf fall intervenes.

Dynamics of transpiration

The daily dynamics of forest transpiration are driven in particular by the saturation deficit, and to a lesser extent by the net radiation, and follow very closely changes in surface conductance, which are moderated especially by the fluctuating incident PPFD and saturation deficit of the ambient atmosphere and by the availability of soil water (Fig. 10.7). The seasonal dynamics are also strongly constrained by the surface conductance, but for rather different reasons in different localities. In the boreal forests, the stomata remain closed during the winter while the soil is frozen so that it is the timing of the thaw in the spring and the freeze-up in the autumn that determines the period over which the surface conductance can respond to the environmental driving variables. In deciduous temperate forests, the phenology of leafing and senescence of the foliage determines the period of transpiration, and in Mediterranean climates the seasonal dynamics are determined by lack of water. In humid tropical forests, the seasonality is much less evident, although

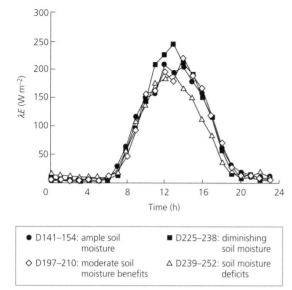

● D141–154: ample soil moisture
■ D225–238: diminishing soil moisture
◇ D197–210: moderate soil moisture benefits
△ D239–252: soil moisture deficits

Fig. 10.7 Ensemble diurnal courses of latent heat fluxes (λE) of a mixed oak, hickory and maple temperate broadleaved forest in Tennessee. The data were sorted by hour and averaged for 2-week periods and represent varying conditions of soil moisture, temperature and photosynthetic capacity. (From Baldocchi 1997.)

even here periods of water stress occur and the canopy leaf area also fluctuates somewhat for other reasons, thereby affecting the surface conductance.

Role of the understorey

In widely spaced open stands the evaporating surface area, and hence the canopy conductance, of the trees is less than in denser stands and it might be thought that the amounts of water returned to the atmosphere in transpiration would be correspondingly less (Whitehead *et al.* 1994). However, Roberts (1983) compiled data on annual transpiration from a number of investigations in stands of different kinds and stocking across Europe, including some of his own, and found that the annual amounts of transpiration were very similar among the stands, with an overall mean of 333 ± 35 (s.d.) mm ($n = 12$). He drew the conclusion that strong development of the understorey in widely spaced stands compensated for the reduction in the overstorey as far as transpiration is con-

cerned. Is this likely? A priori one might assume that there would be little transpiration from the understorey, firstly because it is shaded from above and consequently has less net absorbed radiation, and secondly because it would be much less well coupled to the atmosphere above the stand (Baldocchi & Vogel 1996). In the latter case, transpiration would tend more towards the equilibrium rate and would thus be small because of the small input of net radiation. However, measurements of vertical profiles of air temperature, water vapour pressure, CO_2 and other trace gas concentrations, particularly in coniferous stands, show only very small differences with height at any one time and measurements of these variables in the trunk space differ little on average from those above the canopy. The explanation for this lies in the frequent penetrating downdrafts of air from above every few minutes (Denmead & Bradley 1987). In other words, the understories of coniferous stands are surprisingly well coupled to the atmosphere above. The combination of good coupling with the comparatively high stomatal conductances in understorey foliage seems to be sufficient to account for the compensation noted by Roberts (1983). The degree of coupling of the understorey to the atmosphere above is somewhat less in stands of broadleaved trees, possibly because of less deep canopies with higher leaf area density, and is very much less in multistoried, humid, tropical forest stands where there are bands of high leaf area density in the upper canopy (e.g. Shuttleworth *et al.* 1984; Shuttleworth 1989; Kruijt *et al.* 1999).

Bowen ratio

As discussed earlier, β is a functional descriptor of the partitioning of energy between warming or cooling the forest and evaporating water, and varies widely depending on forest type and climate. Analysis of β in terms of the likely variables (Jarvis *et al.* 1976) shows that these observations are explicable in terms of the so-called 'isothermal' or 'climatological' resistance of Monteith (1965) and Stewart and Thom (1973). Combining this resistance with the Penman–Monteith equation (Jarvis *et al.* 1976) leads to the following expression for the Bowen ratio

$$\beta + 1 = (\varepsilon + 1 + g_a/g_c)/(\varepsilon + g_a/g_i) \qquad [10.10]$$

where the climatological conductance, g_i, is given by:

$$g_i = (A/D) \cdot (P/\lambda) \qquad [10.11]$$

It is apparent that β is approximately proportional to the climatological conductance, g_i, essentially the ratio A/D, and inversely proportional to the stomatal conductance, becoming larger in a predictable manner as D declines relative to A, or as g_c changes with vegetation type and structure and as stomata close. In the case of different climates cited earlier, it is the differences in g_i that are important; in the case of different forest types (see Fig. 10.4), it is the differences in g_c among the stands that account for the different values of β.

Equation 10.10 can be rearranged to provide a simple direct means of deriving the canopy conductance from a knowledge of β, g_a and g_i as follows (Thom 1975; Jarvis *et al.* 1976):

$$g_c = 1/[(\varepsilon\beta - 1)/g_a + (\beta + 1)/g_i] \qquad [10.12]$$

This is not only a useful means of derivation of g_c (e.g. Gash & Stewart 1975) but also provides a practical description of the interrelationships among energy partitioning, the pathway properties and the driving variables of available energy and saturation deficit. In essence this expression encapsulates the most significant functional attributes of forest canopies with respect to the exchanges of heat and water vapour with the atmosphere.

10.6.4 Some conclusions

There are some paradoxes evident here with respect to water use by forests. Canopy aerodynamic conductances are generally high, especially so in comparison with crops, grassland and heathland, decreasing somewhat from coniferous boreal forest, through mixed broadleaved temperate forest to multistorey, humid, tropical forest. Conversely, canopy surface conductances are generally low in coniferous boreal forest, close to zero during the period of frozen soils in winter; somewhat higher in broadleaved temperate forest, at least during the foliated period; and possibly somewhat higher still in humid tropical

forest, the year round. In general, forest canopies, particularly coniferous forest canopies with large g_a/g_c, are extremely well coupled to the atmosphere.

High g_a leads to high rates of evaporation of intercepted water from canopy surfaces; low g_c leads to low rates of transfer of transpired water vapour from canopies to the atmosphere. Thus in climates where a large proportion of the precipitation occurs during long periods of low-intensity rainfall, such as in temperate maritime regions, evaporation of intercepted rainfall, particularly from coniferous forest, can amount to a very large direct return of water to the atmosphere. On the other hand, the low g_c of temperate and boreal coniferous forest results in low transpiration rates from dry canopies. Thus the local forest water budget depends on the balance between the periods of time that the canopy is wet or dry and thus on the precipitation regime in a particular area (Jarvis & Stewart 1979).

Because of low g_c, forest canopies, even when not superficially wet, present a particularly dry surface to the atmosphere. High g_a/g_c also leads to relatively high rates of sensible heat transfer and to large Bowen ratios which drive regional climate. However, the magnitude of this effect is moderated by another aspect of local climate, the relative size of g_i, the climatological conductance. At identical g_c, β increases as g_i increases and transpiration is inhibited by small saturation deficits.

We conclude this section on evaporation and transpiration, with its emphasis on Ω and β, by suggesting that these two coefficients are useful summarizing descriptors of the functional properties of forest stands and could well form the basis for a classification of forest stands into functional types.

10.7 CARBON DIOXIDE EXCHANGE

10.7.1 The policy imperative

The exchange of CO_2 between forests and the atmosphere has become a topic of intense interest over the past 3 years as the role of forests in the global atmospheric CO_2 budget has become more widely appreciated. The Kyoto Protocol, arising from the Third Conference of the Parties (COP3) of the United Nations Framework Convention on Climate Change (UNFCCC) held in Kyoto in December 1997, explicitly implicated forests as agents in the net removal of CO_2 of anthropogenic origin from the atmosphere. The developed countries of the world were assigned legally binding targets for the reduction of their industrial emissions of CO_2 to the atmosphere, a part of which could be offset against the net uptake of CO_2 by new, young forests. The Intergovernmental Panel on Climate Change (IPCC) was commissioned to produce a special report on 'Land Use, Land Use Change and Forestry' for submission to the UNFCCC at COP6 in November 2000 and there is an expectation that 'full accounting' of all forests may then be included. The overall result of these developments has been a major upsurge of interest and concern regarding the exchanges of CO_2 between forests and the atmosphere on stand, forest, national, regional and global scales, as is happening now and is likely to eventuate over the next 100 years.

10.7.2 Forests as a carbon pump

Forests are the major terrestrial interface for C transfer between the atmosphere and the soil. Like the net exchange of short- and long-wave radiation and the balance between precipitation and evaporation, the NEE of CO_2 between forests and the atmosphere is a balance between two largely simultaneous, co-located, opposed sets of processes. In all three cases it is the balance between the inflows and outflows that is of major consequence. We are concerned with the small shift in the radiation balance responsible for global warming, with changes in the spatial and temporal balance between precipitation and evaporation that may lead to water shortage and, in forest stands, with the size and permanence of the balance between the inflows of C in photosynthesis by the trees, understorey and ground vegetation and outflows resulting very largely from the heterotrophic respiration in the soil and from the corresponding autotrophic respiration of the trees (Fig. 10.8).

We may regard the trees in forests as 'pumps' that effect a net transfer of C from the atmosphere to the soil where it is sequestered. The large

amounts of C that have accumulated in forest soils in the Northern Hemisphere since the retreat of the ice sheets after the last glaciation are evidence of this. While the trees in a forest are juvenile and growing rapidly, they may also accumulate significant amounts of C in the wood; however, even when the trees 'mature' and are growing very slowly, if at all, they none the less continue to transfer C from the atmosphere to the soil where it accumulates. This transfer is effected through leaf and branch fall, internal translocation of assimilate to the root system and fine-root turnover. For example, trees in large areas of the boreal forest today are small in diameter (e.g. stands of black spruce with a median diameter of 10 cm at 120 years old) and are growing on deep peat-rich soils, the C of which has all accumulated over the past 8000 years or so.

On the basis of the 'climax concept' (Clements 1916), it was long believed that pristine forests, particularly tropical forests, were in a state of equilibrium such that over a period of several years they were C neutral, with neither a net gain nor loss of C. This view has been challenged in recent years with increasing evidence from eddy covariance measurements of NEE of a net gain in the exchange of CO_2 (Malhi *et al.* 1999) and from sample plot studies of NPP which show that tree

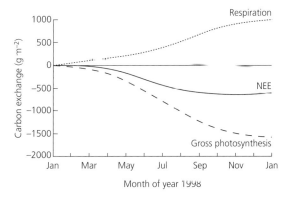

Fig. 10.8 Cumulative carbon budget of a young (18-year-old) stand of Sitka spruce in the maritime environment of central Scotland during 1998, showing net ecosystem exchange (NEE) as the difference between gross primary production and forest respiration. (R. Clement, J.B. Moncrieff & P.G. Jarvis, unpublished results.)

growth in undisturbed tropical moist forests continues to sequester C (e.g. Lugo & Brown 1992; Phillips *et al.* 1998).

Regarding forests as a pump transferring C from the atmosphere to the soil in this way is closely analogous to the 'biotic pump' in the photic zone of the oceans, which transfers C from the atmosphere to long-term sequestration in the deep ocean.

10.7.3 Dynamics of carbon dioxide exchange

Daily dynamics

Over the course of a day (Fig. 10.9), C sequestration essentially follows solar radiation (provided there is no major constraint such as frozen or dry soil), with C being accumulated during the daylight hours and lost at night. For example, large daytime gains by tropical forest stands are offset by large night-time losses (Malhi *et al.* 1999). Depending on the balance between these hourly gains and losses, for any stand there may be a net gain or loss of C over any 24 h in the year. Typically, through a growing season, there may be a net gain of C on about three-quarters of the days but a loss on days of particularly low solar radiation input (i.e. with cloud and rain, or smoke in the atmosphere) and high temperatures (e.g. Jarvis *et al.* 1997).

Seasonal dynamics

Over a year, the length of the photosynthetic season has a major influence on C transfer (e.g. Berg *et al.* 1998). In boreal coniferous forest, for example, one-third to a half of the C gained in the five summer months is lost by respiration over the seven winter months, during which a small efflux of C continues although the ground is frozen (Fig. 10.10). In broadleaved temperate forest the length of the photosynthetic season is determined by budburst in the spring and leaf senescence in the autumn (Fig. 10.11), whereas in temperate coniferous forest the season of net photosynthetic gain is determined by daylength and daily total radiation input in the winter months and may be 10 months long for coniferous forest in maritime climates (Fig. 10.12). In boreal, temperate and Mediterranean climates the vegetation is strongly seasonal and NEE may be strongly constrained by water availability for a large part of the time (Fig. 10.13); only in tropical moist forests is net C gain near continuous throughout the year, being reduced by occasional short periods of low temperature and moderate seasonal water shortage (Fig. 10.14). Examples of the net seasonal dynamics of boreal, temperate and moist tropical forest stands are shown in Fig. 10.15.

Annual dynamics

Transfer of CO_2 to young forest stands regenerating after disturbance such as fire or harvesting is critically dependent on the phase of the successional cycle. During the first 5 or 10 years the disturbed area of young trees is likely be losing C to the atmosphere; subsequently, as the trees come

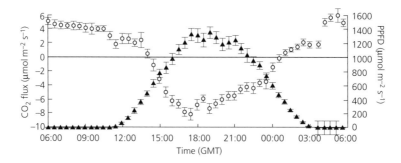

Fig. 10.9 Ensemble daily courses of PPFD (photosynthetic photon flux density) (▲) and CO_2 flux density (○) of a boreal black spruce stand during a 22-day period in the summer of 1994. Each point is the value of the half-hourly average of the flux before the time shown on every day in the period. The error bars show ± 1 SE. Six hours should be subtracted from the time to give local time. (From Jarvis *et al.* 1997.)

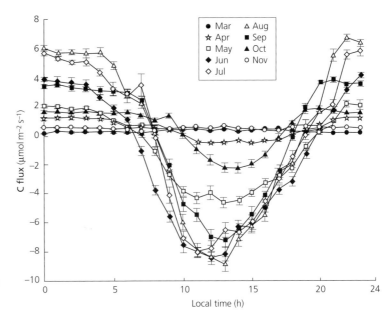

Fig. 10.10 Ensembles of the daily course of net carbon flux throughout the year of old black spruce at the BOREAS Southern Study Area Site in 1996. (P.G. Jarvis, J. Massheder & J.B. Moncrieff, unpublished results.)

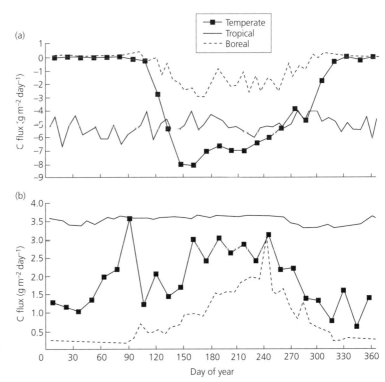

Fig. 10.11 Annual cycles of (a) daytime uptake and (b) night-time release of carbon of a boreal black spruce stand (---) at the BOREAS Southern Study Area, a temperate forest stand (——■——) in Tennessee and a moist tropical forest stand (—) in Amazonia. (From Malhi *et al.* 1999.)

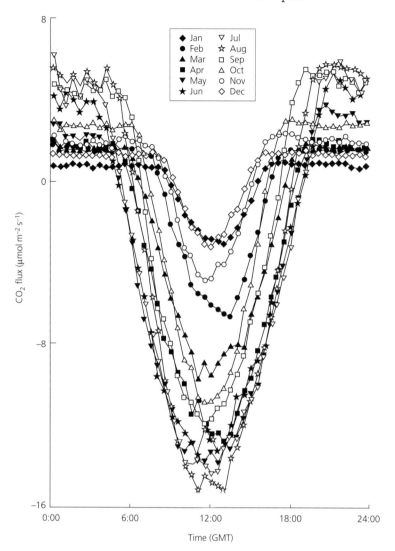

Fig. 10.12 Ensembles of the daily course of net carbon flux throughout the year of Sitka spruce in the maritime environment of central Scotland during 1998. (R. Clement, J.B. Moncrieff & P.G. Jarvis, unpublished results.)

to occupy the site fully, the area will accumulate C strongly during a phase of rapid growth that may last for decades, depending on the species of trees and site conditions. Mature forests take up C from the atmosphere at much slower rates, but even as the growth increment of the trees declines towards zero, C continues to be transferred from the atmosphere to the soil via the trees in the form of above- and below-ground detritus. None the less, old-growth forests do become net sources of C to the atmosphere under some circumstances, for example if the forest soil becomes

warmer or more aerobic, thus promoting oxidation of the soil organic matter (Lindroth *et al.* 1998).

10.7.4 Net transfer of carbon dioxide from the atmosphere to forests

Tropical forests

So far there are few C flux measurements over periods long enough to give annual estimates of NEE in tropical forests, although they are likely to

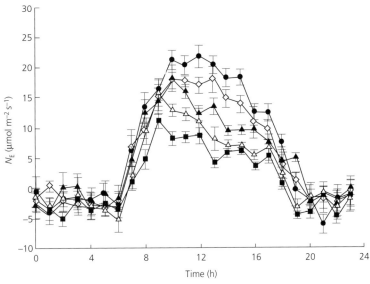

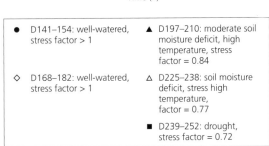

Fig. 10.13 Ensemble diurnal courses of net CO_2 uptake of a mixed oak, hickory and maple temperate, broadleaved forest in Tennessee. The data were sorted by hour and averaged for 2-week periods and represent varying conditions of soil moisture and temperature, as in Fig. 10.7. (From Baldocchi, 1997.)

● D141–154: well-watered, stress factor > 1

◇ D168–182: well-watered, stress factor > 1

▲ D197–210: moderate soil moisture deficit, high temperature, stress factor = 0.84

△ D225–238: soil moisture deficit, stress high temperature, factor = 0.77

■ D239–252: drought, stress factor = 0.72

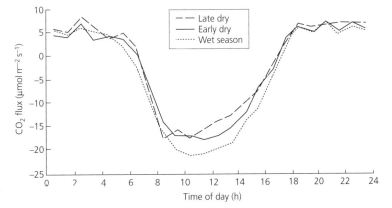

Fig. 10.14 Ensembles of net ecosystem exchange of CO_2 at a moist tropical forest site in Amazonia to show seasonality at three times of the year. (From Malhi *et al.* 1998.)

increase appreciably over the next 5 years with the onset of the LBA experiment. Eddy covariance measurements indicate an annual net C sink of about 1.0 Mg ha^{-1} in pristine, sparse, dry, rain forests in Amazonia (Fan *et al.* 1990; Grace *et al.* 1995a,b, 1996) and about 5.5 Mg ha^{-1} for dense, moist, rain forest (Malhi *et al.* 1998, 1999). Plot studies of biomass increase in the humid Neotropics also indicate annual net transfer of C from the atmosphere to the stands of 0.7 to

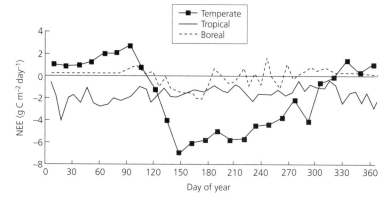

Fig. 10.15 Comparative net ecosystem exchanges (NEE) over a year for the temperate, tropical and boreal forest sites referred to in Fig. 10.11. (From Malhi *et al.* 1999.)

1.1 Mg ha^{-1} (Phillips *et al.* 1998; Grace & Malhi 1999).

Temperate forests

Forests in the temperate region are virtually all managed to a greater or lesser extent and there are but few patches that might be regarded as pristine. Recent measurements of NEE by eddy covariance over a year or more in Europe in the EUROFLUX experiment (Fig. 10.16) and in North America (e.g. Goulden *et al.* 1996; Greco & Baldocchi 1996; Valentini *et al.* 1996; Baldocchi 1997; Anthoni *et al.* 1999) indicate rates of C sequestration in the range 1.5 to 7 Mg ha^{-1} over about 20 sites differing in latitude, altitude, climate (continental vs. maritime), species, sparseness of cover and degree of management. The values at the lower end of the range derive from sparse sites where lack of water is a major factor or from northern sites where length of growing season, lack of nutrients and low temperatures are major factors. Estimates of NPP of sample plots indicate annual carbon sequestration of approximately 3 Mg ha^{-1} (range 1 to 8 Mg ha^{-1}) based on numerous yield tables (e.g. Spiecker *et al.* 1996). A preliminary scaling-up exercise of the EUROFLUX measurements suggests that the largely temperate forests of the European Union are a current annual C sink of 0.17 to 0.35 Gt (Martin *et al.* 1998).

Boreal forests

The extensive boreal forests across Siberia and Canada are fire-driven climax forests, the near

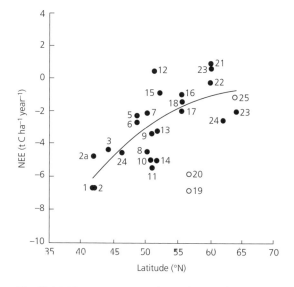

Fig. 10.16 Net ecosystem exchange (NEE) of the EUROFLUX sites plotted against latitude: (●), forests of natural origin and planted stands with traditional forest management; (○), intensively managed plantations at maritime sites (numbers 19 and 20 refer to the Sitka spruce site in central Scotland over the years 1997 and 1998 respectively). (From Valentini *et al.* 2000.)

end-point of succession following retreat of the ice after the last glaciation about 10 000 years ago. These forests have been subject to increasing exploitation over the past 150 years but much of the area standing can be regarded as largely pristine. Measurements of C sequestration by eddy covariance on a campaign basis in Siberia (Hollinger *et al.* 1994, 1995, 1998; Kelliher *et al.* 1997; Schulze *et al.* 1999), over periods of several months up to 5 years in northern Canada, in the BOREAS experiment (Sellers *et al.* 1997) and in northern Europe (Valentini *et al.* 2000) have demonstrated old-growth spruce stands that are C neutral in Canada (Goulden *et al.* 1998) and a C source in Sweden (Lindroth *et al.* 1998), losing C at an annual rate of up to 1 Mg ha^{-1}. However, the majority of boreal forest stands investigated, some 12 or more, are C sinks sequestering C at annual rates of up to 2.5 Mg ha^{-1} (Black *et al.* 1996; Blanken *et al.* 1997, 1998; Jarvis *et al.* 1997; McCaughey *et al.* 1997; Chen *et al.* 1999; Schulze *et al.* 1999). Estimates of NPP from sample plots also indicate that boreal forests are weak C sinks of about 2.5 Mg ha^{-1} (range 1 to 6 Mg ha^{-1}) (Gower *et al.* 1997; Jarvis *et al.* 2000).

10.7.5 Carbon balance components

Carbon stocks and flows in forest biomes

To understand the magnitude of the net transfer of CO_2 between stands and the atmosphere at a particular stage in a succession and to appreciate its sensitivity to current conditions, disturbance and climate change, it is helpful to consider the principal component processes within the stand that add up to give the overall net exchange. The feedbacks among these processes, their speed of response and their sensitivity to environmental change determine the present and future flows of CO_2 and the C sequestration capacity of the stand.

Figure 10.17 shows the measured annual fluxes among the components of NEE in a tropical moist, a temperate and a boreal forest stand compiled from a number of sources (see Malhi *et al.* 1999). It is evident that the annual NEE is the difference between a very large gain term (i.e. photosynthesis of the foliage) and a very large loss term dominated by respiration of heterotrophs in the soil (see Fig. 10.8). Thus the net exchange is particularly sensitive to relatively small proportional changes in the opposed gain and loss terms, particularly if they are out of phase.

A mass balance of the component fluxes above ground (photosynthesis; foliage, branch and stem respiration; leaf, branch and stem litter of all kinds, i.e. detritus; and above-ground NPP) enables an estimate to be made of the amount of C internally translocated below ground. A mass balance of the component fluxes below ground, comprising the inputs from above (i.e. the detritus and translocate), root and heterotrophic respiration, fine-root turnover, and mycorrhizal and root system NPP, enables an approximate estimate to be made of net changes in the pool of soil C.

In principle, summation of the component fluxes over a year should add up to the independently measured annual NEE. However, there are appreciable errors in measuring the below-ground components in particular, so that close agreement between measurements or estimates of the components of the budget is not as yet to be expected.

Exchanges of volatile organic compounds and methane

The net exchange of C between forests and the atmosphere also includes the emission of biogenic volatile organic C compounds (VOC) such as isoprene and terpenes from the tree canopy (Baldocchi *et al.* 1995). Simple algorithms have been developed to calculate emissions of individual compounds from individual tree species based mainly on PAR and leaf temperature (Guenther *et al.* 1991). The emissions of VOC increase with temperature and PPFD and it has been argued that the emissions of certain VOC compounds, notably isoprene, reduce the physiological effects of water stress (Sharkey & Loreto 1993). The absolute magnitude of C exchange represents a small fraction of the total. Emissions of terpenes from Norway spruce in Denmark (Christensen *et al.* 1997) amount to no more than 0.3% of the annual C assimilated, but short-term losses from the Mediterranean species *Quercus ilex* may exceed 6% at high temperatures (Staudt & Bertin 1998) Globally emissions are small, around 0.1% of annual carbon emissions from vegetation (Guenther *et al.* 2000).

Undrained peatlands in high latitudes have accumulated appreciable amounts of C from the atmosphere since the retreat of the ice and are currently significant annual CO_2 sinks (0.2–0.5 Mg ha^{-1} of C) but are also small annual sources of methane (CH_4) (0.03–0.3 Mg ha^{-1} of C). By contrast, drained peatlands, whether intended for agriculture or afforestation, release C as CO_2 because of accelerated decomposition of the aerobic peat (Cannell *et al.* 1993) but no longer

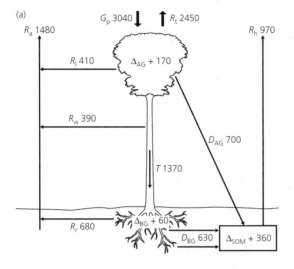

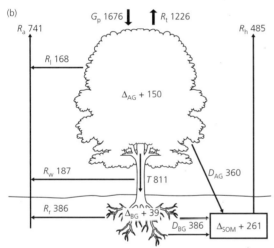

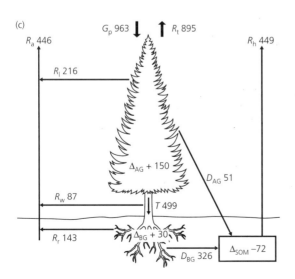

release CH_4 in significant amounts. Peatlands drained for agriculture continue to be a sustained C source while the peat remains. Depending on species of tree and time in the rotation, peatlands drained for afforestation may continue to be a source of C in spite of forest biomass growth (Zoltai & Martikainen 1996; Cannell *et al.* 1999).

Thus drainage and afforestation of peatlands reduces CH_4 emissions and provides a larger CO_2 sink than the peatland plant community. However, the net exchange of C by peatlands is an upward flux for some years, until the C fixed by the canopy exceeds the losses of C from the oxidizing peat with the lower water table. The primary interest in forest–atmosphere C fluxes is the contribution to the global atmosphere C budgets and the consequent radiative forcing of climate. As CH_4 on a per molecule basis in radiative equivalents has a global warming potential 30 times that of CO_2, it is not simply the mass fraction that should be used to identify priorities. The change in the net direction of CH_4 flux from emissions to deposition needs to be included in C budgets and the overall net radiative forcing calculated.

10.7.6 Future carbon sequestration potential

Across Europe forests are now growing faster than previously (Spiecker *et al.* 1996) and similarly in the tropics (Phillips *et al.* 1998) We currently seem to be in a period of especially strong C uptake by forests, although it has been suggested that this will not continue indefinitely, and

Fig. 10.17 Estimated annual total carbon flows for (a) a moist tropical forest near Manaus, Amazonas, Brazil; (b) temperate deciduous oak–hickory forest, Tennessee, USA; (c) boreal evergreen old black spruce forest, Saskatchewan, Canada. G_p, gross primary production; R_t, total respiration; R_a, autotrophic respiration; R_h, heterotrophic respiration; R_l, leaf respiration; R_w, above-ground wood respiration; R_r, root respiration; D_{AG} above-ground detritus (litterfall and mortality); D_{BG} below-ground detritus (root mortality, fine-root turnover and exudation); T, below-ground carbon translocation; Δ_{AG}, above-ground net biomass carbon increment; Δ_{BG} below-ground net biomass carbon increment; Δ_{SOM}, net increment in soil organic carbon. (see Malhi *et al.* 1999 for details)

indeed there is concern that the rate will diminish significantly as the 21st century proceeds (Walker *et al.* 1999).

It has frequently been suggested that the present rates of CO_2 uptake and growth are the combined result of the recent global increases in atmospheric CO_2 concentration (e.g. Jarvis 1998), N deposition (e.g. Kauppi *et al.* 1992; Hudson *et al.* 1994) and temperature. Quantitative evaluation of the relative impacts of all three of these factors is a focus of current research.

CO_2-stimulated carbon sequestration

It is difficult to measure the impact of the slow progressive rise in atmospheric CO_2 concentration on the C sequestration capacity of stands of trees but some relevant indications can be obtained from recent experimental programmes to evaluate the impacts of elevated atmospheric CO_2 concentration on the growth, development and physiology of young trees.

Doubling atmospheric CO_2 concentration speeds up development, photosynthesis and growth by up to 60% in a wide range of species of young trees grown rooted in the ground in the field over periods of up to 5 years in both North America and Europe (Curtis & Wang 1998; Medlyn *et al.* 1999; Peterson *et al.* 1999). Moderate lack of nutrients and of water reduce growth but have little adverse effect on the *relative* impact of the increase in CO_2 concentration. To all intents and purposes the impact of the increase in CO_2 concentration is that trees get bigger quicker but otherwise remain similar in most respects to trees of the same size in the current conditions. Meta-analysis of the results from many experiments shows that the key physiological parameters defining capacity for photosynthetic CO_2 uptake by trees, V_{cmax} and J_{max}, are downregulated on average by no more than 12% and stomatal conductance by 15% as a result of growth in elevated CO_2 concentration (Curtis & Wang 1998; Medlyn *et al.* 1999).

A key question, as yet unresolved, is to what extent will the responses of key parameters to increased CO_2 concentration change when a stand of young trees reaches canopy closure. This is not readily solvable other than by long-term, free air elevated CO_2 concentration exposure experiments (FACE), starting with seedlings or clonal propagules. So far, the few FACE experiments in progress in forests have been imposed on existing stands (e.g. Ellsworth 1999), which are unlikely to acclimate fully over, say, a 5-year period of exposure. Thus major uncertainties remain as to the likely long-term effects of the rise in atmospheric CO_2 concentration on C sequestration and the contribution of CO_2 fertilization to the global terrestrial C sink (Friedlingstein *et al.* 1995).

Temperature-stimulated carbon return to the atmosphere

Photosynthesis is particularly dependent on solar radiation, CO_2 concentration and N availability and thus is likely to be reduced by increase in cloudiness but increased by elevations of the global atmospheric CO_2 concentration and by N deposition. All respiratory processes are very sensitive to temperature, as is the multiplication of the populations of respiring organisms, particularly the fine roots and heterotrophs in the soil. Thus soil respiration increases strongly in boreal and temperate forests in relation to the rise in soil temperature as the spring progresses (Boone *et al.* 1998; Rayment & Jarvis 1999) and consequently is likely to be very sensitive to the increase in global temperature. Increase in soil temperature leads to enhanced mineralization of soil organic matter and release of N, which feed back to stimulate photosynthesis, tree growth and transfer of CO_2 from atmosphere to forest (Bergh & Linder 1999; Jarvis & Linder 2000). The balance between these conflicting effects of rising temperature on the net exchange of CO_2 (enhanced respiration and enhanced photosynthesis) is only now being resolved experimentally in the short term, although models indicate a potentially complex time-course in the long term (Dewar *et al.* 2000).

Carbon sink saturation

It has been proposed that on time-scales of decades to centuries saturation of C sinks will occur and that existing C sinks will become significant C sources (Scholes 1999; Scholes *et al.* 1999). This effect, it is suggested, arises from a difference in phase and functional relationship between the responses to global climate change of the principal gain and loss terms in NEE. The CO_2 stimulation of photosynthesis increases with increasing atmospheric CO_2 concentration (the CO_2-fertilization effect) but does so at a diminishing rate. On the other hand, heterotrophic respiration (decomposition of litter and soil C, resulting in emission of CO_2 to the atmosphere) increases somewhat later in time as detritus production builds up and increases exponentially with increasing temperature (Boone *et al.* 1998; Rayment & Jarvis 1999). Because so much C is stored in forest soils, and because this C is potentially vulnerable to climate change, it is essential to take full account of the C stocks in soil. It has been postulated on the basis of models (e.g. Cao & Woodward 1998; Scholes *et al.* 1999) that changes in climate, particularly the increase in temperature, will lead to higher rates of heterotrophic respiration and oxidation of the C stock in the soils, it is presumed without acclimation. Thus, as both atmospheric CO_2 concentration and temperature rise, the rate of heterotrophic respiration may be expected to increase relative to the rate of photosynthesis, and the overall capacity of the terrestrial biosphere to take up C from the atmosphere to diminish. However there are a number of possible feedbacks not included in these model analyses. These include the effects of rising CO_2 and temperature on:

1 litter quantity, which may increase as a result of any short-term CO_2-fertilization effect and may lead to enhanced nutrient availability and C storage in soils;

2 litter quality (i.e. the C:N ratio), which may decrease and retard decomposition and nutrient release;

3 allocation of C below ground, which may well increase and lead to increased production of roots, mycorrhizae and exudates;

4 increased nutrient availability and temperature leading to higher photosynthetic rates and NPP; and

5 longer photosynthetic and growing seasons.

These feedbacks raise major uncertainties, currently being addressed by soil-warming ex-periments and modelling. The results of recent modelling analyses that have taken these feedbacks into account (McMurtrie & Dewar 1999; McMurtrie *et al.* 2000a,b; Medlyn *et al.* 2000) suggest that the dominant *short-term* constraint on the CO_2-fertilization effect is an increase in litter quantity. On the other hand, *long-term* increases in C sequestration are possible as atmospheric CO_2 concentration increases, if the net N input also increases (either through fixation or deposition) or if the N:C composition of newly formed litter substantially decreases. If below-ground C allocation increases, as observed in many experiments, and leads to increased production of roots, mycorrhizae and exudates, with consequent increase in N fixation, a large long-term increase in NPP and C sequestration is likely (Medlyn *et al.* 2000).

The simulated *long-term* effect of rising temperature is a large increase in NPP, and hence litter production, as a result of enhanced N mineralization and consequent increased photosynthesis, whereas *short-term* effects of temperature on photosynthesis and respiration are relatively unimportant (Dewar *et al.* 2000).

These simulations highlight the importance of litter production on the one hand and turnover processes in the soil on the other, the former being strongly constrained by N availability and air temperature, the latter by soil microbiological processes and soil temperature. One cannot escape the conclusion that more emphasis needs to be placed on understanding better the processes in the soil, and the link between soil and atmosphere through the medium of the trees.

Indications from soil-warming experiments are that after an early initial flush of CO_2, the CO_2 efflux is similar in the heated and unheated control plots; furthermore, the feedbacks are actually increasing the efficiency of the 'forest pump' through the release of N (Lückwille & Wright 1997; Ineson *et al.* 1998) and may override the initial losses of C (Jarvis & Linder 2000). The effect of climate change on the efficiency of the pump in transferring C from the atmosphere to the soil via enhanced nutrient availability, photosynthesis and detritus production needs to be evaluated in addition to the effect of climate change on the return of C from the soil to the atmosphere, in order to obtain a complete picture of the net change in C sequestration by forests in a warmer climate.

A further key question is whether the current annual rate of N deposition ($\approx 20\,kg\,ha^{-1}$) adequately balances the current annual rise in atmospheric CO_2 concentration ($\approx 1.5\,\mu mol\,mol^{-1}$) to sustain the NEE over the next 50 years. This is addressed in section 10.9. As things are at the present, the possibility of C sinks becoming sources in the mid twenty-first century remains highly speculative.

10.8 TRACE GASES, AEROSOLS AND CLOUD DROPLETS

The interactions between forests and the atmosphere outlined so far identify the key role of forests in carbon exchange between the global atmosphere and the biosphere, in regulating regional climates and in the exchange of water vapour. These interactions have drawn attention to forests as a central issue within the global warming debate. The loss of forest through biomass burning, especially in tropical regions, is closely associated with global carbon emissions but is also a major contributor to regional air pollution through the production of photochemical smog, with important consequences for human health.

Forests have been at the centre of earlier environmental issues, for example the acid rain debate, a major issue of the 1970s and 1980s throughout Europe and North America (Last & Watling 1991). Forests were regarded as one of the ecosystems most sensitive to deposited acidity and 'forest decline' was widely associated with air pollutants derived from anthropogenic emissions of oxides of sulphur and nitrogen (Ulrich *et al.* 1980; Schulze & Freer-Smith 1991). Multinational research programmes investigated the underlying processes in field and laboratory studies and many important links between the pollutants and effects were established. It is notable that these links often proved subtle and complex, and some of the more obvious connections that were anticipated were never established. For example, deposited acidity has been shown to be associated with the decline of red

spruce in the Appalachians of North America (De Hayes 1992); however, the mechanism involves susceptibility to winter injury and frost damage (Sheppard 1991) and is restricted mainly to stands at moderate or high elevations that are subject to orographic cloud containing large concentrations of NO_3^-, SO_4^{2-} and H^+ (Fowler *et al.* 1989). There is also evidence that the same high-elevation stands of red spruce have been influenced by photochemical oxidant pollutants (McLaughlin & Kohut 1992). Other examples include induced magnesium deficiency in European silver fir stands on acidified soils (Roberts *et al.* 1989).

Forests are efficient 'scavengers' of pollutants from the atmosphere as a result of their low canopy aerodynamic resistance. The term 'filtering effect' has been extensively used to emphasize the relative efficiency of capture of pollutants by trees relative to shorter vegetation (Johnson *et al.* 1991). It has also been suggested that trees provide an effective means of reducing urban air pollution, largely as a consequence of the assumed 'filtering effect' (Freer-Smith *et al.* 1997), although field data in support of such claims are scarce. There are, therefore, effects of reactive atmospheric pollutants on forests and effects of forests on the atmosphere.

In this section we examine the underlying mechanisms that control the interactions between the major groups of trace gases and aerosols and forests. Table 10.3 lists the trace gases and aerosols studied in forests during the last two decades. Particles are present in sizes ranging from a few nanometres to $10\,\mu m$, and the aerosols contain many further minor and trace plant nutrients. This list is far from exhaustive, and within the scope of the chapter a complete and detailed treatment of each pollutant is not possible. In the following sections the pollutants are classified largely by environmental issue, which in the majority of cases also groups them by their general chemical properties. Deposition of all the compounds on forest canopies therefore contributes to biogeochemical cycling at regional scales.

10.8.1 Acid gases: HCl, HNO_3, HF

For the very reactive gases, including HNO_3, HCl and HF, forests represent an important sink, even when ambient concentrations are small. The highly reactive nature of these gases causes them to be removed rapidly at terrestrial surfaces.

The canopy or surface resistance can be derived as the residual when the aerodynamic turbulent transfer and boundary layer resistances have been subtracted from the total resistance to transfer from a reference height in the atmosphere to the effective surface within the forest canopy (Fig. 10.18). For both HNO_3 and HCl, the surface resistance is negligible and terrestrial surfaces

Table 10.3 Trace gases and aerosols studied in forests.

___

Gases
Acid gases: HCl, HNO_3, HF
Ammonia (NH_3)
Nitrogen oxides: NO, NO_2, HONO
Radiatively active gases: N_2O, CH_4, CO
Photochemical oxidants: O_3, PAN (peroxyacetyl nitrate), PPN, H_2O_2
Sulphur compounds: SO_2, H_2S, COS, CS_2
Biogenic volatile organic compounds (VOCs): isoprene, α-pinene, β-pinene, β-myrcene, β-thajene
Anthropogenic VOCs: i-butene, propene, ethene, benzene (and 100 other hydrocarbons)

Aerosols
Major ions in atmospheric aerosols: SO_4^{2-}, NO_3^-, NH_4^+, H^+
Base cations: Ca^{2+}, Mg^{2+}, Na^+
Heavy metals: Pb, Zn, Mn, Cu

___

are generally considered to be 'perfect sinks' for these gases, as discussed in section 10.5.

The presence of atmospheric aerosols, clouds and rain provide additional atmospheric sinks for these gases and, as a consequence, their lifetime in the atmosphere and their ambient concentrations are small. In the case of HCl and HNO_3, there are both natural and anthropogenic sources, whereas HF is largely anthropogenic in origin and is of importance only at local scales.

Ambient HNO_3 concentrations in Europe vary from 0.1 μg m^{-3} in the maritime regions of northern and western Europe to 1–5 μg m^{-3} in polluted areas of central Europe (Erisman *et al.* 1998). For a forest with an atmospheric resistance of $10 s m^{-1}$ such concentrations would lead to N inputs ranging from 0.7 to $70 kg ha^{-1}$. Clearly, forests exposed to large ambient HNO_3 concentrations will receive appreciable atmospheric inputs through dry deposition. Such areas include substantial areas of southern Europe and the forests in southern California. The use of mean values, for example, is appropriate for rough approximations such as these but in practice ambient HNO_3 concentrations are closely coupled to the synoptic meteorological conditions, being largest during periods of persistent anticyclonic weather in which the sources of primary pollutants (NO_x and VOC) are processed through photochemical reactions to generate significant HNO_3 concentrations, and smallest in

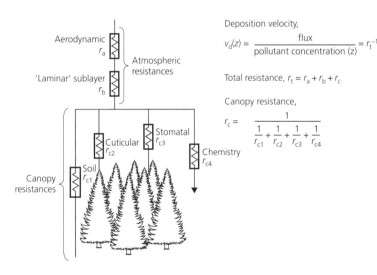

Deposition velocity,

$$v_d(z) = \frac{flux}{pollutant\ concentration\ (z)} = r_t^{-1}$$

Total resistance, $r_t = r_a + r_b + r_c$

Canopy resistance,

$$r_c = \frac{1}{\dfrac{1}{r_{c1}} + \dfrac{1}{r_{c2}} + \dfrac{1}{r_{c3}} + \dfrac{1}{r_{c4}}}$$

Fig. 10.18 Resistance analogy applied to reactive trace gas exchange between forests and the atmosphere. Canopy resistance components: r_{c1}, stomata; r_{c2}, leaf surfaces; r_{c3}, soil; r_{c4}, chemical reaction within the canopy.

cyclonic conditions with wind and rain (PORG 1997; Kley *et al.* 1999).

Concentrations of HNO_3 are monitored at few locations and we currently rely primarily on long-range transport models for regional concentration fields and estimates of deposition. The large atmospheric conductances of forests relative to shorter vegetation (or open water or soil surfaces) make the trees the primary sink for dry deposition in a mixed land-use landscape. It is of interest to calculate the fractional cover within the landscape necessary for the woodland to 'capture' say 50% of the dry deposition. As the relative deposition rates to forest and short vegetation differ typically by a factor of 5, then in a landscape containing 20% woodland the cumulative input to the woodland would be similar to the combined input to the remaining 80% of the land cover. In practice, the concentration fields are not spatially uniform, but the general point is clear, and the importance of forests as a sink for fixed N is considered further in section 10.9.

While the current uncertainty in estimating inputs of HNO_3 (or HCl) to forests lies primarily in knowledge of the ambient concentrations of the two gases, there are aspects of the underlying deposition process that remain poorly understood. The exchange of momentum between forests and the atmosphere is both well described and easily measured, and forms the basis of our understanding of forest–atmosphere exchanges. However, the exchange processes at the surface differ in an important way as there is no equivalent in mass transfer to the transfer of momentum by form drag (Thom 1975). The conventional practical solution used to overcome this problem is to introduce a further resistance (r_b) in series with the aerodynamic resistance (r_a). Many different formulations for r_b may be found in the literature, and are discussed by Brutsaert (1991) and, in the case of HNO_3 deposition, have been considered by Muller *et al.* (1993). However, the available field data over forest are too limited to be able to identify the most appropriate of the formulations. These uncertainties will remain until a substantial quantity of flux measurements with much smaller uncertainties in the measurement become available. Such developments await an instrument able to detect fluctuations

in ambient HNO_3 with similar accuracy to the instruments used to measure atmospheric CO_2 fluctuations.

This section has focused on HNO_3 rather than HCl or HF because HNO_3 represents a much more important component of the trace gas climate above forests and constitutes an important source of N deposition in the warmer polluted regions of the planet. However, both the two other gases are believed to behave in broadly similar ways.

10.8.2 Ammonia

Gaseous NH_3 is a trace gas that interacts with vegetation in similar ways to CO_2, even though the two gases differ greatly in their chemical properties and roles in plant biochemistry. Both gases are processed through all plants and animals, play a major role in plant biochemistry and physiology, exhibit bidirectional exchange between the atmosphere and terrestrial ecosystems, and exhibit a 'compensation point', i.e. a surface concentration at which no net surface–atmosphere exchange occurs. Both CO_2 and NH_3 are also important atmospheric pollutants, their emission as a consequence of anthropogenic activities representing an important driver of the change in the chemical composition of the atmosphere.

These similarities have largely been overlooked because knowledge of NH_3 is much more recent, the gas is more difficult to measure and, most important of all, NH_3 has a very short atmospheric lifetime relative to CO_2. This highly soluble reactive gas readily undergoes gas to particle formation and is a major component of aerosols over the mid-latitude continental regions (Dentener 1993). Ammonia also deposits rapidly on vegetation, soil and water surfaces (Sutton *et al.* 1993). The atmospheric lifetime and fate of NH_3 is illustrated in Fig. 10.19 and is generally considered to be of the order of 0.5–2 h, depending on the presence of acidic aerosols and wet surfaces (Erisman *et al.* 1998). Ambient concentrations of NH_3 are therefore small, generally in the range 0.1–3.0 $\mu g\,m^{-3}$. However, close to major sources, such as intensive livestock units, mean ambient NH_3 concentrations can reach 30 $\mu g\,m^{-3}$ (Pitcairn *et al.* 1998).

The exchange of NH_3 between forests and the

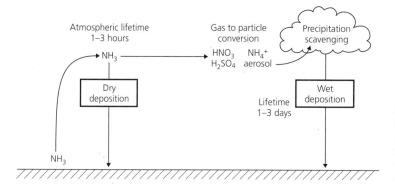

Fig. 10.19 Schematic representation of the atmospheric cycle of NH_3 illustrating the relative lifetimes of gaseous and aerosol forms of reduced nitrogen and the removal pathways.

atmosphere is bidirectional. Within foliage, the presence of NH_4^+ in the apoplast (the liquid phase within cell walls) leads, via Henry's law, to a potential equilibrium gas-phase NH_3 concentration. The partial pressure of NH_3 within the substomatal cavity drives the exchange with the atmosphere within or above the forest canopy. The net transfer of NH_3 molecules through the stomata represents uptake or loss from the foliage and is regulated by the direction of the concentration gradient (Fig. 10.20). However, the uptake of NH_3 is not limited to stomatal exchange and the primary sinks in forests for NH_3 are the external surfaces of the foliage (Wyers & Erisman 1998). Unlike the very reactive HNO_3, the forest canopy is not a perfect sink for NH_3. The magnitude of the surface resistance is regulated by a combination of chemical processes on the external surfaces and the net exchange between apoplastic NH_4^+ and substomatal gaseous NH_3 (Flechard & Fowler 1998). The effect of a forest on net annual inputs of NH_3 may be quantified from recent studies of NH_3 concentration and deposition at three European sites: Auchencorth in southern Scotland, Speulde in the central region of The Netherlands, and Melpitz in south-east Germany close to Leipzig (Table 10.4). By applying a constant value ($37\,s\,m^{-1}$) for canopy resistance (r_c) at all three sites, based on continuous measurements of r_c over a year, the net sink strength of terrestrial surfaces may be quantified using data from the three sites. The enhancement in NH_3 deposition by a forest relative to short vegetation is 65%. The problem with presenting comparisons between different surfaces using measured variables is that there are no field data available

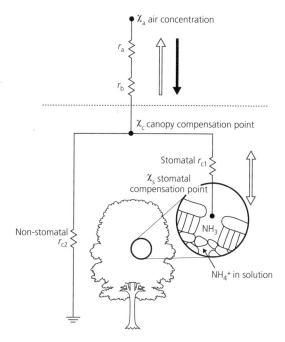

Fig. 10.20 Schematic representation of forest–atmosphere exchange of NH_3, including the exchange with apoplast NH_4^+. Resistance components denoted: r, r_a, aerodynamic resistance. r_b boundary resistance; $r_{c(1-2)}$ canopy resistance. χ, Atmospheric concentration of NH_3 as defined on the figure.

for both forest and short vegetation at the same location. Furthermore, it is clear from the studies to date that the concentrations of major pollutants present, in this case SO_2 and NH_3, interact with one another on the surfaces of vegetation (Flechard *et al.* 1999). The local meteorology at the site, especially the frequency of wet surfaces, and the N status of the canopy, and hence apoplastic NH_4^+ and H^+, also influence r_c and no rigorous comparison based on measurements has been attempted to date. The main point of this exercise is that NH_3 inputs to forest may be large, even at sites unaffected by large local sources of NH_3, and are larger than inputs to short vegetation (Erisman *et al.* 1998).

Canopy–atmosphere exchange of NH_3 and interactions with other pollutants

The preceding arguments may be formalized within a model that simulates the interaction of turbulent exchange processes with leaf surface chemistry and exchange between apoplastic NH_4^+ and the ambient NH_3 concentration with plant canopies.

The first dynamic canopy–atmosphere NH_3 exchange model, developed by Flechard (1998) and described in Flechard *et al.* (1999), is outlined in diagrammatic form in Fig. 10.21. The leaf surface solution chemistry is simulated using the Henry and solution equilibria for the trace gases NH_3, SO_2, CO_2, HONO and HCl. Oxidation rates for SO_2 in solution and ion exchange between the leaf surface and the interior of the leaves characterized by apoplasm solution are also simulated. As the detailed chemistry simulated is limited to bulk solution phase processes, the model is designed to describe the exchange processes for a wet (or partially wet) canopy, and activity coefficients are included within the modelling scheme for pH and solute concentrations up to ionic strengths of $0.3\,mol\,l^{-1}$. For higher concentrations

and for 'dry surfaces', which are assumed at relative humidities less than 70%, simple transfer resistances are applied. The site used to provide test data for the model (Auchencorth) is 'wet' within the definition applied by the model for almost 70% of the time.

The model is initialized using measured precipitation chemistry and ambient concentrations of the trace gases, and fluxes and surface concentrations simulated for periods of 2 to 11 days. The results of the comparison of 1629 h of flux measurements with the model show the net NH_3 deposition over the period (68 days) to be within 3% of the measured fluxes (i.e. $-6.3\,ng\,m^{-2}\,s^{-1}$ vs. $-6.1\,ng\,m^{-2}\,s^{-1}$). This represents a major improvement over application of a simple canopy resistance model, which overestimated the net deposition flux by 27%, largely because such models are unable to simulate emission fluxes. The limitation of this approach is the requirement for extensive chemical data as input variables. However, the successful simulations show that the major processes that regulate NH_3 fluxes into and out of forests are now understood and can be quantified for situations where the major gases and the precipitation chemistry are continuously monitored.

One important consequence of the recent long-term flux measurements and dynamic modelling is that interactions of pollutants on the external surfaces of vegetation have now been recognized as important. In particular, the interaction of NH_3 with SO_2 effectively regulates the canopy resistance for SO_2. In effect the presence of sufficient NH_3 prevents the pH of the leaf surface water

Table 10.4 A comparison of NH_3 deposition to forest and short seminatural vegetation at three sites in Europe along a transect from north-east Europe in Scotland, through The Netherlands to south-east Germany. (From Erisman *et al.* 1998.)

	Auchencorth, southern Scotland	Speulde, Netherlands	Melpitz, south-east Germany
Annual forest NH_3-N deposition $(kg\,ha^{-1})$	4.2	36.8	31.5
Deposition of NH_3-N to short seminatural vegetation $(ng\,m^{-2}\,s^{-1})$	2.7	23.6	20.3
Ambient NH_3 concentration $(\mu g\,m^{-3})$	0.4	3.5	3.0
r_c for NH_3 assumed constant for all sites and surfaces $(s\,m^{-1})$	37	37	37

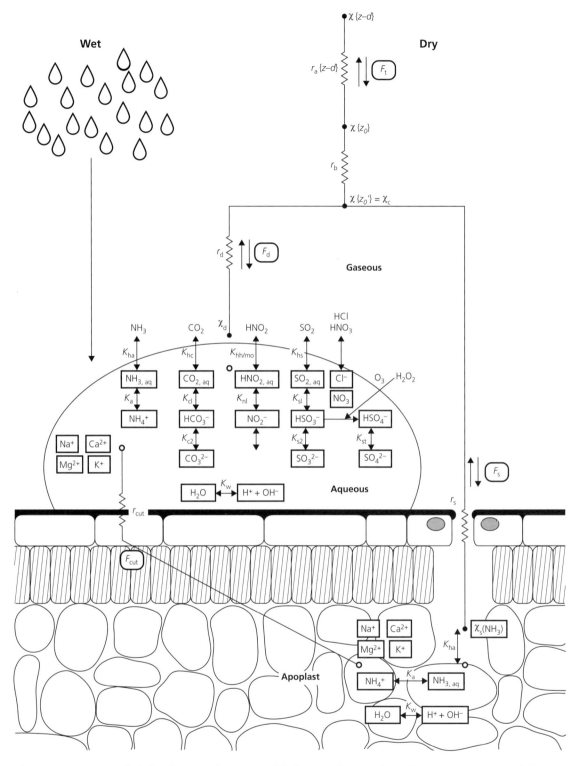

Fig. 10.21 Processes included within a mechanistic model of NH_3 and SO_2 exchange between vegetation and the atmosphere. χ, Concentrations in air at reference height (z) as defined; z, height; d, zero plane displacement; z_0, roughness length; χ_s, concentration of NH_3 (or SO_2) within substomatal cavity; r, resistances: r_a, aerodynamic; r_b, boundary layer; r_{cut}, cuticular; r_s, stomatal; r_d, air–water exchange resistance. K_h, Henry's law constant; $K_{a,cl,etc.}$, dissociation constant and equilibrium constants for the reactions defined on the figure. F, fluxes of NH_3 or SO_2; d, non-stomatal; s, stomatal; t, total. (From Flechard 1998.)

falling to values that would suppress further SO_2 uptake, just as NH_3 in the boundary layer maintains the pH of cloud water sufficiently high to promote heterogeneous SO_2 oxidation by O_3 in low-level stratiform cloud.

A consequence of the chemical control of forest–atmosphere exchange of pollutants is that control strategies to regulate effects of one pollutant may inadvertently influence the deposition footprint and effects of another pollutant, and this has already introduced non-linearity into the relationship between sulphur emission and deposition at a regional scale (RGAR 1997).

Local influence of forest on NH_3 deposition and dispersion

The rapid deposition of NH_3 on forest leads directly to large inputs of fixed N in forests in areas with large ambient concentrations. Such inputs may represent a valuable nutrient supply and contribute, at least in the short term, to net sequestration of atmospheric CO_2 (see section 10.7). The rapid 'sequestration' of NH_3 from the atmosphere may also be regarded as an opportunity to use forest or woodland cover as a means of reducing the impact of NH_3 from intensive livestock sources on the local environment. In principle, a shelterbelt plantation of woodland surrounding a ground-level source of NH_3 from an intensive livestock unit would capture gaseous NH_3 more rapidly than shorter seminatural vegetation and much more rapidly than heavily fertilized grassland or arable crops. The latter would simply introduce a surface with a much higher compensation concentration, and hence reduce net NH_3 deposition.

The preceding section shows that a rough estimate of local NH_3 deposition can be made from first principles using the dispersion equations on the assumption that forests have a constant canopy resistance (and compensation concentration) at all distances from the source. However, recent field data are now available to examine the NH_3 concentration profile downwind of an NH_3 source surrounded by woodland and also the influence of ambient NH_3 concentration on r_c for NH_3 (Pitcairn *et al.* 1998). The field data obtained

from transects of NH_3 concentration downwind of a poultry farm show a rapid decline in NH_3 concentration, from almost $30\,\mu g\,m^{-3}$ at a distance of 15 m from the source to $2\,\mu g\,m^{-3}$, effectively background, at 270 m from the source. The relationship between canopy resistance and ambient concentration from measurements at a site in the same region allow concentration-specific values of r_c for NH_3 to be applied in order to calculate the local dry deposition rates for surfaces that are wet approximately 60% of the time. The resulting local mass balance is illustrated in Fig. 10.22 (from Fowler *et al.* 1998a).

The local mass balance shows large annual N inputs within 20 m of the source (42 kg ha^{-1}) yet the total deposition within 270 m of the source represents only 3.2% of the annual source of NH_3. The quantity deposited is subject to considerable uncertainty but even taking quite generous estimates of the likely errors in the estimates, local removal is unlikely to capture more than 20% of the emissions from the source within this zone. The conclusion therefore is that the use of woodland to capture even rapidly deposited pollutants such as NH_3 must be regarded as only a partial solution, unless very large areas of woodland are used or methods of introducing the NH_3 within the trunk space are developed.

10.8.3 Sulphur dioxide

The deposition of SO_2 on forests has been one of the most intensively studied aspects of atmospheric trace gas deposition. The interest stems from two quite different requirements. Firstly, there was the need to parameterize the regional deposition fluxes in long-range transport models for both Europe and North America, where forests represent a substantial fraction of the land cover. Secondly, and equally important, was the interest in the potential effects of sulphur on forest health and productivity. The concern during the early years (1970–80) of the acid rain issue was centred in Scandinavia on the effects of deposited acidity on freshwater ecosystems and forests (SNSF 1981). This early work identified clear effects on fish and fresh water but effects on forests remained equivocal. However, within the last decade the limited understanding of SO_2 transfer

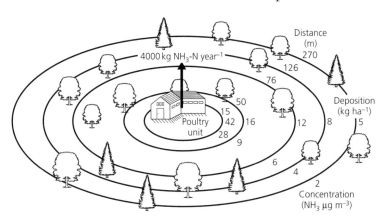

Fig. 10.22 Atmospheric mass balance of NH$_3$ in woodland within 0.25 km of a livestock source of NH$_3$. (From Fowler *et al.* 1998a.)

to forests has been enhanced by the development and application of continuous SO$_2$ flux measurement stations over forests (Erisman & Wyers 1993).

The deposition of SO$_2$ on forests is regulated by a combination of surface chemical processes within surface moisture on the external surfaces of foliage (see Fig. 10.21) and by stomatal uptake in parallel. However, for regions with significant NH$_3$ concentrations, the processes on leaf surfaces represent the primary sink and provide effective control over the overall rates of exchange. The chemistry regulating SO$_2$ canopy resistance is primarily the ratio of the NH$_3$ concentration and the sum of the acid gases present (SO$_2$ + HNO$_3$ + HCl); because SO$_2$ is the dominant acid gas present in most of the polluted regions of northern Europe, it is the NH$_3$/SO$_2$ ratio that regulates r_c for SO$_2$.

At Speulde Forest in The Netherlands, the r_c value for SO$_2$ obtained from continuous flux monitoring (32 s m^{-1}) for a site with an NH$_3$/SO$_2$ molar ratio of 3.7 is much smaller than the corresponding r_c value at Auchencorth Moss (240 s m^{-1}) where the NH$_3$/SO$_2$ ratio is 2.0. These observations show that the sink strength of forests for SO$_2$ depends on both the physical and chemical climatology (Erisman *et al.* 1998). Thus, as emissions and concentrations of SO$_2$ have declined, the canopy resistance for SO$_2$ has also changed as a consequence of changes in the NH$_3$/SO$_2$ concentration ratio, and the direction of the change has led to an increase in the deposition velocity (of canopy conductance) of SO$_2$ with

time. However, ambient concentrations of NH$_3$ have only recently been monitored continuously anywhere in Europe and both emissions and concentrations are very poorly known for periods prior to 1995. The practical consequences of the new understanding of SO$_2$ dry deposition have yet to find their way into the policy forum. However, the international protocols developed within Europe to regulate emissions of SO$_2$ will further increase the NH$_3$/SO$_2$ concentration ratio throughout the region and accelerate the dry deposition and near-field removal of SO$_2$. Such effects will increase non-linearity in the spatial patterns of the deposition footprint. Forest canopies remain the surfaces with the largest sink capacity given a constant chemical composition, and in the source regions especially will continue to receive large dry deposition inputs of SO$_2$ and NH$_3$.

The areas of greatest interest for SO$_2$ deposition globally are the rapidly expanding zones of China and South-East Asia, which experience large ambient SO$_2$ concentrations (Galloway & Rodhe 1991). It has been estimated that the areas of forest subject to sulphur inputs equivalent to 2 keq ha^{-1} will increase by a factor of 6 between 1985 and 2050, and that the largest increases will be in southern and eastern Asia (Fowler *et al.* 1999).

10.8.4 Nitric oxide and nitrogen dioxide

Exchanges of these gases between forests and the atmosphere are closely linked, at least in polluted regions. The respective sources and sinks are well defined and reasonably well understood but there are complexities that have made progress towards a complete understanding of overall rates of exchange of the individual gases and their interactions rather slow.

NO is not deposited at significant rates within forests but is generated within soils largely by nitrification processes (Davidson & Kingerlee 1997). The emission fluxes from forest soils range from 0.1 to 50 ng m^{-2} s^{-1}, increase with soil N and temperature, and are optimal at intermediate water-filled pore space (WFPS) values (Skiba *et al.* 1997).

NO$_2$ is readily taken up by leaves via the stomata and detailed laboratory studies show no evidence of internal resistances to uptake (Thoene *et al.* 1991). The rates of stomatal uptake of NO$_2$ are therefore readily quantified from a knowledge of ambient NO$_2$ concentrations within the canopy and the canopy conductances for NO$_2$. This relative simplicity in the NO$_2$ exchange process is complicated by gas-phase processes within the canopy in which emissions of NO from the soil are readily oxidized to NO$_2$ by O$_3$. These processes are illustrated in Fig. 10.23. A consequence of these interactions is that while NO$_2$ is possibly being absorbed by the tree canopy, the source of the NO$_2$ molecules is ambiguous: a significant fraction may result from cycling between forest soils and the foliage. Furthermore, in warm summer conditions, a forest may be a net source of NO$_x$ (NO + NO$_2$), while NO$_2$ is being absorbed by the canopy as a consequence of a substantial soil NO flux (Pilegaard *et al.* 1995). This may arise as a consequence of the rapid increase in soil emission of NO with soil temperature and relatively small rates of NO$_2$ uptake by the canopy and is therefore characteristic of locations with small ambient NO$_2$ concentrations.

At a global scale, emissions of NO from soil have been estimated to range from 6 to 20 Tg of N annually. The larger values (Davidson & Kingerlee 1997) are based on the assumption that a negligible proportion of the soil emission flux is captured as NO$_2$ by canopy uptake; similarly, the smaller values assume that the canopy captures a substantial proportion of the soil NO emission (Yienger & Levy 1995). The current uncertainty in the net forest–atmosphere exchange of NO$_x$ is therefore large and there are clearly important feedbacks between the NO soil flux and the overall quantity of N cycling within forests as NO and NO$_2$ (Skiba *et al.* 1997).

The direct measurement of NO$_2$ deposition using micrometeorological approaches is further complicated by photochemical processes, notably the photolysis of NO$_2$ above the canopy. Such effects have been simulated and incorporated within models of NO$_x$ exchange between plant canopies and the atmosphere (Duyzer *et al.* 1995). Overall, uptake by forest of NO$_2$ is similar to uptake by shorter vegetation with relatively small average fluxes. Current interest in NO$_x$ and forests centres on the interactions of soil sources of NO with canopy sinks of NO$_2$ and their influence on the net surface–atmosphere exchange of NO$_x$.

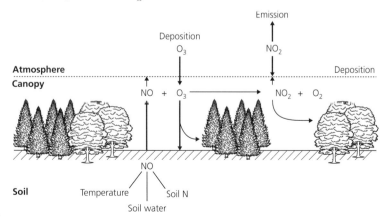

Fig. 10.23 Interaction of emission and atmospheric chemistry of NO, NO$_2$ and O$_3$ above and within a forest canopy.

10.8.5 Ozone

Deposition on to natural surfaces represents the major sink for O_3 within the convective boundary layer. Quantifying the magnitude of the sink and understanding the underlying mechanisms are essential for modelling regional and global concentrations and budgets (Stevenson *et al.* 1998). Ozone is also recognized as the major regional phytotoxic pollutant globally and has been identified as an important contributor to forest decline in Europe and North America (McLaughlin & Kohut 1992; Chappelka & Samuelson 1998; Skarby *et al.* 1998).

The O_3 flux from the atmosphere to forest canopies may be partitioned between stomatal uptake and an external surface flux. The available literature suggests that physiological damage and growth and yield reductions are caused almost entirely by stomatal uptake. However, it is also important to understand and quantify the overall flux in order to be able to define the fate and atmospheric residence time for O_3, because over seasonal time-scales the non-stomatal flux is the largest component of the total flux to semi-natural vegetation (Fowler *et al.* 1998b) (Fig. 10.24).

Ozone uptake and effects on forests

The available literature for both agricultural crops and forests has identified an ambient concentration above which accumulated exposure to O_3 leads to growth reductions (Skarby *et al.* 1998). The concentrations used in experiments on the effects of O_3 have been used to develop simple linear relationships between accumulated exposure to O_3 over a threshold concentration and reductions in dry matter yield. The concentration threshold used (40 ppb) is arbitrary and there may be physiologically important effects at concentrations less than 40 ppb. However, the concentration of 40 ppb is quite close to the free tropospheric background O_3 concentration in northern mid-latitudes during spring and summer (Roemer 1996).

Ozone concentrations in Europe have increased from about 10 ppb around 1900) to values in excess of 20 ppb, but early measurements are few (Voltz & Kley 1988). Application of three-dimensional chemistry-transport models to simulate global concentration fields of O_3 provide an alternative measurement. The United Kingdom Meteorological Office unified model, used at climate resolution with a full description

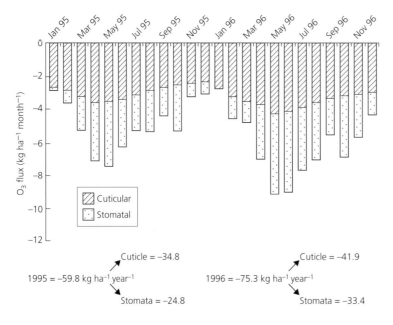

Fig. 10.24 Relative magnitudes of stomatal and non-stomatal ozone uptake over moorland. (From Fowler *et al.* 1998b.)

of the atmospheric chemical processing of the pollutants leading to O_3 formation (Collins *et al.* 1997), shows the increase in surface concentrations during the last 240 years and may be used to quantify the exposure of forests to O_3.

The magnitude of the effects of elevated O_3 concentrations on forest health or productivity remains uncertain but there are areas of forest that have clearly been damaged visibly by O_3, especially in the San Bernadino Mountains in southern California (Miller *et al.* 1963). There is also substantial evidence of regional effects of O_3 on tree growth and productivity in the southeastern states of the USA (especially *Pinus taeda*, *Pinus echinata*, *Pinus virginiana* and *Pinus strobus*) (Chappelka & Samuelson 1998). Data from similar studies are available in Europe, with beech, spruce and poplar species all showing reductions in the biomass of saplings (Skarby *et al.* 1998). The index of exposure currently used is the cumulated exposure above an arbitrary threshold, in this case 40 ppb, i.e. AOT_{40}, which provides a measure of the dose, even though dimensionally it is simply the product of concentration and time and is not a flux. The dry matter reduction found in many of the studies to date declines linearly with increasing AOT_{40} and has been used to define critical levels of O_3 in Europe (Fuhrer *et al.* 1997). An AOT_{40} of 10^4 ppb·h has been set as the critical level for forests.

While it is not possible to quantify the current magnitude of yield loss in global forests attributable to O_3 with current AOT_{40} procedures, it is possible to quantify the area of forest that is exposed to O_3 concentrations in excess of 60 ppb, where reduced productivity or physiological injury may occur. This is a somewhat larger ambient O_3 concentration than the threshold used for definition of the critical level but has the advantage that it is substantially larger than background (which has not been shown to be causing damage to forests) and is more likely to identify the key areas in which O_3 presents a potential problem for forest health. The global surface O_3 concentrations for 1990 were provided by the STOCHEM model (see Stevenson *et al.* 1998).

The 1990 area of forest exposed to O_3 concentrations in excess of 60 ppb represents 18% of the total global forest area (Table 10.5). This large area of exposure is a consequence of the extensive zones in both the tropics and mid-latitude regions over which elevated O_3 concentrations occur. Table 10.5 also illustrates the increase in exposure to these large concentrations since 1860 and provides very speculative estimates of possible exposure in the year 2100.

To quantify the magnitude of loss of forest production, the relationship between O_3 dose and forest productivity must be defined. As the available evidence shows that uptake through the stomata is the primary cause of physiological effects, a relationship between the stomatal dose above a threshold and the dry matter loss per unit uptake is required. Such an approach has been proposed by many. The concept of a threshold is necessary because O_3 is a natural component of surface air and there is no evidence that

Table 10.5 Areas of global forest exposed to elevated concentrations of surface ozone (>60 ppb). (From Fowler *et al.* 1999.)

	All forests	Tropical and subtropical	Temperate and subpolar
1860	No exceedance of 60 ppb		
1950	2.1×10^6 km^2 (6.3%)	0.5×10^6 km^2 (2.9%)	$1\ 7 \times 10^6$ km^2 (9.2%)
1970	6.3×10^6 km^2 (18.5%)	1.9×10^6 km^2 (12.1%)	4.4×10^6 km^2 (24.1%)
1990	8.3×10^6 km^2 (24.4%)	3.0×10^6 km^2 (19.1%)	5.3×10^6 km^2 (29.1%)
2100	17.0×10^6 km^2 (49.8%)	6.0×10^6 km^2 (37.9%)	11.0×10^6 km^2 (60.2%)

the preindustrial concentrations or fluxes were damaging to vegetation. A rough estimate of the likely threshold flux below which no damage is likely to occur may be obtained from the maximum concentration at which no effects have been observed and the canopy conductance. While very approximate, this approach indicates a likely threshold flux of O_3 of about $0.5\,\mu g\,m^{-2}\,s^{-1}$ (Fowler *et al.* 1998b). A quantitative basis for estimates of forest yield loss, therefore, seems a reasonable goal provided that there are properly replicated field studies to provide the necessary validation.

The conclusion is that O_3 appears to present a threat to forest productivity on a global scale and the current global models suggest that concentrations and exposure may progressively increase over the coming decades. The simplistic approaches adopted to date are not adequate to quantify yield loss. Many feedbacks have been ignored and some may play an important part. In particular, interactions between drought stress and the effects of O_3 may be expected (Mansfield 1998). In drought stress conditions the reduced stomatal conductance to water vapour reduces stomatal O_3 influx and the overall deposition flux, and this would be expected to feed back to generate larger O_3 concentrations in the convective boundary layer (CBL). There is evidence that peak O_3 concentrations in the UK occur in precisely these conditions, yet the response of vegetation to these peak concentrations is largely unknown.

Volatile organic compounds

Terrestrial vegetation represents a major global source of the primary reactants required for production of photochemical oxidants including O_3, H_2O_2 and peroxyacetyl nitrate (PAN). The compounds emitted by trees include isoprene, terpenes (including α-pinene, β-pinene and Δ^3-carene) and alcohols and organic acids (Steinbrecher *et al.* 1997a). While emissions of VOC from anthropogenic sources dominate the source inventory in small heavily industrialized countries such as the UK or The Netherlands, in remote areas biogenic sources dominate production of precursors of photochemical oxidants.

Globally, vegetation contributes more than two-thirds of total VOC emissions (Guenther 1997). The spectrum of organic compounds emitted is very specific to plant species and to environmental conditions (Guenther *et al.* 2000). The options for agencies wishing to regulate ambient O_3 concentrations must therefore be tailored to the region and the relative magnitudes of the anthropogenic and vegetation sources.

10.8.6 Aerosols and cloud droplets

Atmospheric aerosols contain a substantial fraction of the reaction products of the pollutant gases and therefore include a wide spectrum of ions from the acid gases, including SO_4^{2-}, NO_3^-, Cl^-, the base cations Ca^{2+}, Na^+ and Mg^{2+}, and H^+, as well as the ubiquitous NH_4^+. Aerosols also contain the majority of the heavy metals in the atmosphere (Pb, Cu, Zn, Cr) and a wide range of other trace metals. Organic compounds are also present in substantial quantities in aerosols. These elements and compounds are returned to the surface in rain and by dry deposition. The rates of dry deposition on forest are strongly influenced by particle size and the characteristic dimensions of the foliage, with the most effective sinks being conifer forests with large leaf area indices. Wind tunnel studies show a very striking relationship between particle size and deposition velocity (see dotted line in Fig. 10.25). There is a minimum deposition velocity for aerosols in the range $0.3\,\mu m$ to $0.5\,\mu m$ diameter, but at sizes both above and below this deposition velocity increases, for rather different reasons (Chamberlain 1975). However, recent field data obtained by a wide range of methods, including eddy covariance flux measurements using active scattering laser spectrometers, passive flux measurement techniques and leaf washing methods, all imply much larger aerosol deposition rates on forests than the wind tunnel studies indicated (Fig. 10.25) (Gallagher *et al.* 1997a). A physical model to provide a mechanistic interpretation of extensive field data is currently lacking but it is clear that forests capture aerosols much more effectively than short vegetation. For example, the inventory of aerosol-deposited ^{210}Pb beneath shelterbelt woodland is 40–50% larger than that under grass. The filtering effect is clearly, therefore, also a feature of forest canopies in the case of atmospheric aerosols, and aerosol, deposition inputs to catchments are greatly reduced following clearfelling.

Cloud droplet deposition

Unlike the bulk of atmospheric aerosols, in which the mass is contained primarily within a diameter range 0.5–1.5 µm, the cloud droplets present in orographic cloud and coastal and radiation fogs are much larger, typically 3–10 µm in diameter. Such droplets are too large to follow the stream-lines of airflow around leaves or branches and are very efficiently captured by a forest canopy. In fact the deposition of cloud water is similar to that of momentum, when expressed as conductivities (or, more commonly, as deposition veloci-ties) for the canopy (Beswick *et al.* 1991). The efficiency of cloud droplet capture by forests leads to two important ecological consequences (see also Chapter 12). Firstly, and very widely known, are the cloud forests throughout the world, which receive an important water supply directly through the capture of cloud water; such a water supply frequently maintains a unique ecosystem.

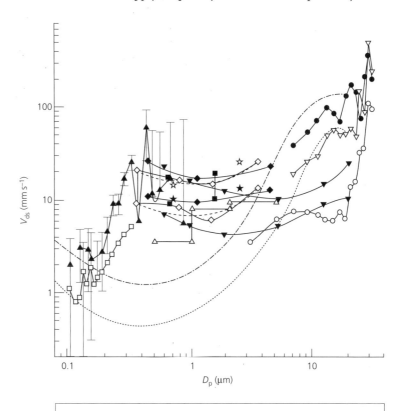

Fig. 10.25 Relationship between aerosol size and deposition velocity, including field data and model results. V_{ds}, deposition velocity (mm s^{-1}); D_p, diameter of particle (µm); u, wind speed (m s^{-1}); v^*, friction velocity (m s^{-1}); z_0, roughness length (m); h_c, canopy height (m).

◇ Höfken and Gravenhorst (1983)
◆ Grosch and Schmitt (1988)
■ Brückman (1988)
⊓ Joutsenoja (1992) $u^* = 0.08$ m s^{-1}, $u = 5.6$ m s^{-1}, $z_0 = 0.08$–0.13m
▼ Waragai and Gravenhorst (1989)
▽ Gallagher *et al.* (1992) $u^* = 0.71$ m s^{-1}, $z_0 = 15.7$cm
● Gallagher *et al.* (1992) $u^* = 0.75$ m s^{-1}, $z_0 = 18.6$cm
○ Beswick *et al.* (1991) $u^* = 0.37$ m s^{-1}, $z_0 = 30.1$cm

▲ Gallagher *et al.* (1997b) $u^* = 0.5$ m s^{-1}, $z_0 = 13.5$m
△ Lorenz and Murphy (1989) Gradient Loblolly pine, $h_c = 9$ m
★ Höfken & Gravenhorst (1983) TF beech
☆ Höfken & Gravenhorst (1983) TF spruce
······ Slinn (1982), $u^* = 0.65$ m s^{-1} Unmodfied model result
—·— Slinn (1982), $u^* = 1.30$ m s^{-1} Unmodified model result

Secondly, and potentially much less beneficial, is the capture of orographic cloud in polluted regions. In polluted air the oxidation product of SO_2 oxidation is ultimately aerosol SO_4^{2-}, which is generally accompanied by variable amounts of NH_4^+ (depending on available NH_3), NO_3^- and H^+ as the four dominant ions. These aerosols are readily activated into cloud droplets as moist air is advected up hillsides. The droplets formed grow rapidly with height to provide typically 200 mg m^{-3} of liquid water at heights 100 m above cloud base (Fig. 10.26). The effective capture of this cloud water by forest represents a major input to high-elevation forests in windy, polluted regions. Measurements, especially in Europe and North America, show inputs of N to high-elevation forests in the range 20–50 kg ha^{-1} annually and even larger inputs of sulphur in polluted regions (Unsworth & Fowler 1988; Crossley *et al.* 1992). Even in the relatively unpolluted areas of southern Scotland the N input to a Sitka spruce forest at an elevation of 600 m is approximately 50 kg ha^{-1} annually, of which 60% is NH_4^+-N and 40% NO_3^--N (Crossley *et al.* 1992). A further property of cloud droplets is that the concentrations of the major ions may be sufficiently large to cause foliar injury or reduction in frost hardiness (Leith *et al.* 1989; Cape *et al.* 1991; Sheppard *et al.* 1993).

10.8.7 Diurnal and seasonal cycles

The reactive gases that are primary pollutants exhibit marked diurnal cycles in ambient concentration. In part these cycles are generated by diurnal changes in the sources of the pollutants. For example, sources of the oxides of nitrogen are closely coupled to sources of combustion and anthropogenic activity. Sources of these gases within the boundary layer also lead to marked diurnal cycles through the daily growth and decay in boundary layer depth. The forest–atmosphere exchange of NH_3 would also be expected to exhibit a marked diurnal cycle because the partitioning between liquid-phase NH_4^+ concentrations and gaseous NH_3 is strongly coupled to temperature, with the result that vegetation may change systematically from a sink to a source as temperature increases during the day. This effect is further enhanced by leaf surface humidity and/or surface water. In this case the sink presented by high humidity at night is removed as the canopy dries and warms during the day. Thus marked diurnal cycles in the exchanges of NH_3 and NO_2 have been widely reported (Sutton *et al.* 1995; Hargreaves *et al.* 1992).

The reactive gases that are secondary pollutants, such as HNO_3 or PAN, also show marked diurnal cycles of concentration as a result of

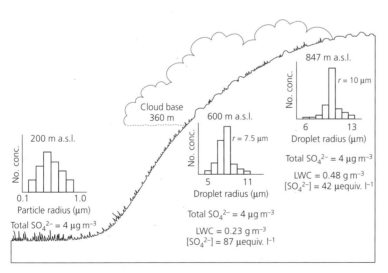

Fig. 10.26 Influence of altitude on cloud liquid water content (LWC) and concentration of major ions in solution. a.s.l., above sea level; *r*, mean radius. (From Fowler *et al.* 1991.)

depletion at the surface during the night, and reduced downward transport from the free troposphere. These two gases also show marked seasonal cycles, with pronounced maxima coinciding with a broad seasonal peak in photochemical activity during the summer months.

The seasonal cycles in the exchanges of SO_2 and NH_3 with forests are controlled more by the relative frequency of surface water and its effect on canopy resistance than by the ambient concentrations. As the frequency of wet surfaces and ambient concentrations is at a maximum in the winter months, the deposition rates of SO_2 and NH_3 show strong seasonal patterns with a winter maximum and summer minimum. While the close link between deposition rates and surface wetness in the control of SO_2 and NH_3 fluxes generates both diurnal and seasonal cycles in the net exchanges between forests and atmosphere, it also leads to different geographical patterns in the seasonality of trace gas exchange. For example, the pronounced wet and dry seasons of South-East Asia almost certainly influence the wet and dry deposition rates of SO_2 and its atmospheric lifetime, although these seasonal changes are quite different from those in Europe and North America. Generalizations at a global scale about the seasonality of these fluxes is, therefore, inappropriate.

Spatial variability in forest–atmosphere exchange of trace gases and aerosols

This discussion of trace gases according to their chemical properties identifies a range of chemical affinities, ranging from HNO_3 and HCl, which are captured by forest canopies as rapidly as atmospheric conductivity permits, to N_2O, which is entirely unreactive in the troposphere (Wayne 1991). Spatial variability in the deposition or emission of the reactive gases results from spatial variability in canopy conductance. For example forest edges, which are particularly well coupled to the atmosphere with very efficient exchange at all heights within the canopy, represent hotspots of deposition within the landscape. Likewise hilltops, where wind speeds are substantially enhanced and which are frequently in cloud, represent areas in which inputs are enhanced (Fowler

et al. 1989; Crossley *et al.* 1992). It is relatively straightforward to quantify the local enhancement in the deposition field for these reactive pollutants at a site, given appropriate measurements of the gas concentrations and meteorological variables. However, to quantify the cumulative effects of edges and exposed hilltops and ridges on the deposition field at a regional scale is a more complex task, and there is evidence that current regional deposition amounts are underestimated for this reason. The magnitude of edge effects has been examined in a range of studies of NH_3 deposition (see Fig. 10.22) and aerosol deposition of ^{210}Pb and has been found to range from an enhancement of 30 to 50% (Fowler *et al.* 1998c) to up to a factor of 4 for hilltops or woodland close to point sources of NH_3.

10.9 ECOLOGICAL EFFECTS OF NITROGEN DEPOSITION

In areas of the world in which HNO_3 concentrations are enhanced by anthropogenic emissions of NO_x, the atmospheric inputs of fixed N may represent an important contribution to the N economy of the forests. Similarly, the annual inputs of fixed N as NH_3 to forests in regions with elevated ambient NH_3 concentrations range from 10 to 50 kg ha^{-1} and the areas of Europe and North America with large inputs is extensive and increasing. The high-elevation forests enveloped in orographic cloud for a significant fraction of the time are also subject to large inputs of ions within the cloud water, including NH_4^+ and NO_3^-. These areas are also hot-spots of N deposition and occur widely in polluted regions of North America, Europe and South-East Asia.

Rather than considering broad regional N deposition values for forests, as has been the focus within global analyses, consideration should be given to the effect of the spatial variability in N deposition on forests at much smaller scales. In this way we may be able to show whether the effects of the deposited N are simply proportional to the total input or whether there are threshold separating effects that are beneficial (e.g. in C sequestration) from those we may regard as damaging. Indicators of negative effects may include NO_3^- leaching, stimulation of N_2O and

NO production, or a reduction in CH_4 oxidation. Such approaches have commonly been applied in the development of *critical load* concepts (Hornung *et al.* 1995), which have been very useful in the development of control strategies for pollutants in Europe. To put this argument into perspective, consider the atmospheric budget of fixed N over the UK. The annual emissions of oxidized N within the UK are 780 Gg, of which 150 Gg is deposited on terrestrial surfaces within the UK. For reduced N the annual emission is much smaller at 260 Gg and the annual deposition larger at 230 Gg. Thus the deposition budget is dominated by reduced N even though these emissions represent only 25% of the total fixed N emissions (Fig. 10.27). The 380 Gg of N deposited in the UK is equivalent to an average deposition of 16 kg ha^{-1}, although the input to forests (33 kg ha^{-1}) exceeds that to non-forest-land by a factor of 2 (Table 10.6). These averages also conceal a substantial range in annual N

deposition, which from the measurement networks may be roughly estimated to be 3 to 90 kg ha^{-1} at 5×5 km resolution (Fig. 10.28). There is substantial evidence that at the upper end of this distribution there are clear effects on the forest, including increased leaching loss of NO_3^-, increased emissions of both NO and N_2O, and loss of species diversity in the ground flora (Pitcairn *et al.* 1998).

This approach can be extended to other European countries on the basis of measured data, although insufficient measured data are available to provide more extensive regional or global analysis and we must rely entirely on models to estimate budgets of net deposited N. It is probable that the relative deposition inputs of forest vs. non-forest are similar in many parts of the world simply because the enhancement in deposition to forests includes HNO_3, NH_3, aerosol NO_3^- and NH_4^+ and cloud droplet inputs of NH_4^+ and NO_3^-, and is not limited to a single compound or

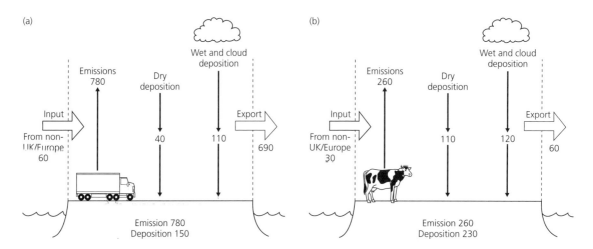

Fig. 10.27 Mass balance of (a) oxidized and (b) reduced nitrogen (Gg) over the UK for the period 1992–94.

Table 10.6 Partitioning the annual N deposition in the UK to different land cover.*

	Forest	Moorland	Grassland	Arable
Total N deposition (Gg)	68	124	98	124
Area (10^6 ha)	2	7.9	6.5	7.9
Mean N deposition (kg ha^{-1})	33	16	15	16
Percentage reduced N	78	65	56	65

* Total N deposition 380 Gg; reduced N, 230 Gg; oxidized N, 150 Gg.

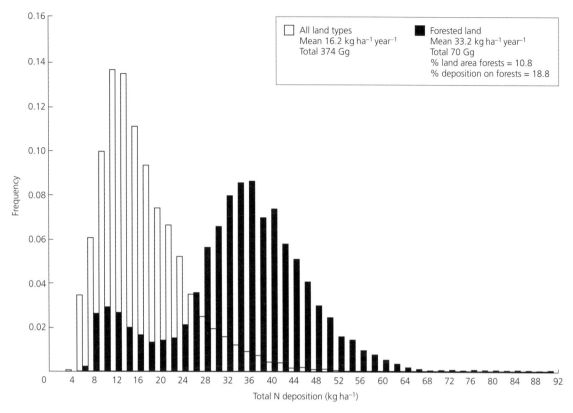

Fig. 10.28 Frequency distribution and average deposition of nitrogen on forests and other land uses in the UK for 1996 (see Fowler *et al.* 1998d).

geographical area. The exception is NO_2, which appears to be deposited on short vegetation and forest at broadly similar rates.

10.9.1 Nitrogen-stimulated carbon sequestration

In general, the forests of the world are chronically deficient in nitrogen (or in some cases phosphorus), particularly the extensive forests in the northern temperate and boreal regions from where the glacial ice cover has retreated relatively recently. In general, annual total wet and dry deposition of N (oxidized and reduced) to forests in rural areas is in the range 5–40 kg ha^{-1}. (Exceptions are forests close to industrial sources and in the near vicinity of intensive agricultural pig and poultry enterprises, where deposition may be substantially larger.) It is likely that such inputs are supporting the rapid growth and C sequestration of young temperate forests, of particular rele-

vance to the Kyoto Protocol (Cannell *et al.* 1999; Valentini *et al.* 2000), but the causal link remains to be firmly established.

A number of experiments over the past 60 years have demonstrated that the growth of forests in northern Europe is very responsive to application of fertilizers, particularly nitrogen (Tamm 1991). Recent long-term experiments in Sweden have shown a fivefold increase in growth of Norway spruce in response to annual fertilizer N applications of 75 kg ha^{-1} at 64°N over the past 12 years and a threefold increase at 57°N over the past 10 years (Bergh *et al.* 1999). In northern Britain, growth of commercial forest is also responsive to applications of N (Taylor & Tabbush 1990; Wang *et al.* 1991). Separate consideration of daytime and night-time C fluxes of an 18-year-old maritime forest of Sitka spruce, now at canopy closure (yield class 16 m^3 ha^{-1}) and initially fertilized with urea, indicates a large annual gross primary production of about $19\,Mg\,ha^{-1}$ and a large autotrophic plus heterotrophic respiration of about $13\,Mg\,ha^{-1}$, i.e. an annual NEE of approximately $6\,Mg\,ha^{-1}$ (Fig. 10.8).

The extent to which N deposition is currently maintaining high rates of C sequestration by forests requires more thorough study, including hypothesis testing with field experiments and modelling (Vitousek 1994). On balance, of the current hypotheses to explain the observed rapid growth of European forests (increases in N deposition and CO_2 concentration and changes in climate), N deposition appears to be more consistent with spatial patterns in the observed increases in growth. The forward projection of these effects will rely heavily on models of N deposition to forests over the coming decades. There are control strategies for oxidized N emissions in Europe and North America but controls on NH_3 emissions are very limited. The expansion of fossil fuel combustion elsewhere in the world more than compensates for the reductions in Europe and North America. Thus we can anticipate further increases in N deposition on to global forests through the coming decades. The net result of interactions between the effects of N deposition and changes in climate and CO_2 concentration on the net land–atmosphere exchange of C are unknown and present a major challenge. Current approaches to quantify the effects rely on models that are somewhat simplistic and lacking in relevant feedbacks. As the questions are very important, both scientifically and politically, it is necessary to develop experimental studies to test hypotheses developed from the models. Field experiments in this area have very largely been limited to the short term and to one or two of the variables. There is a clear requirement to develop large-scale field experiments in order to track the fate and effects of deposited N, its effect on ecosystem function and the interactions with ambient CO_2 concentration and temperature.

10.10 FORESTS AND REGIONAL CLIMATES

The potential migration of treelines in response to changes in global climate may also signifi-
cantly change the regional exchange of C as CO_2 and CH_4. In the high latitudes, the peatlands have steadily accumulated C since the last ice age but at modest rates (Clymo 1984) relative to net C sequestration by young forests. However, if forests migrate into peatlands, the effective lowering of the water table will suppress CH_4 emission and lead to oxidation of the surface organic layers to CO_2. While supporting forest growth, such changes may represent a large net source of CO_2 in the atmosphere over substantial areas, as currently observed when peat wetlands are drained and planted with trees. The extent to which these changes may occur at the northern boundary of the boreal forest during the coming decades will be controlled by the changes in climate at these high latitudes, and on the balance between precipitation, and evapotranspiration and the length of growing season. At present the possible role of changes in nutrient inputs from the atmosphere remain very uncertain at these remote locations and the major controls are likely to be driven by climate.

It came as a considerable surprise when Bowen ratios with values of 2 or more for coniferous forest, measured by flux-gradient methods, were first reported in the 1960s because virtually all previous measurements over agricultural crops had resulted in values between about –0.5 and +1.0, except in a few cases of extreme water stress. Suspicion was placed on the methods being used, particularly on the problems of measuring with sufficient accuracy over sustained periods the very small gradients of temperature and humidity that occur in the surface layer over forests and which are required for the so-called Bowen ratio, flux-gradient method (Stewart & Thom 1973; Thom *et al.* 1975; Jarvis *et al.* 1976). For those who had worked previously only with prairie grasslands and crops, it still came as a surprise 25 years later to discover that values of β measured by eddy covariance for stands of black spruce and jack pine, which comprise the most extensive areas of the Canadian boreal forest, were frequently larger than 2 (Goulden *et al.* 1997; Jarvis *et al.* 1997; Baldocchi *et al.* 2000), and this has revolutionized thinking with respect to climatic process at larger scales in the boreal region (Sellers *et al.* 1995, 1997). The significance of a large

Bowen ratio at regional scale lies not so much in the transpiration rate but in the large sensible heat flux which, in extensive continental areas drives the CBL up to heights of 3000 m during just an hour or two in the morning, with consequent entrainment of cold dry air rich in CO_2 from above the capping inversion which, as it warms up, strongly forces evaporation and transpiration and replenishes CO_2 over cropping areas as well as the forest within the region (Nobre *et al.* 1996; Barr & Betts 1997) (for further discussion see McNaughton 1989; Betts *et al.* 1996; Pielke *et al.* 1998; Raupach 1998).

10.11 CONCLUSIONS

Drawing brief conclusions from the diversity of forest interactions with the atmosphere considered in this chapter seems bound to oversimplify reality. However, there are several scientific and political issues that merit emphasis here. In considering a relatively short list of the trace gases and aerosols and their interactions with forest there are common threads and important exceptions.

These new developments in understanding result directly from recent large-scale field studies that, while readily incorporated into models of land–atmosphere exchange, were not expected. Such lessons are common in the development of understanding environmental issues and include the serendipitous route to the discovery of acid rain by Swedish scientists during the 1960s (Oden 1968) and, more recently, to the discovery of the ozone hole from a long series of direct measurements (Farman *et al.* 1985). The lesson from such examples is that the important new signals arise largely from measurement-based research rather than from models. The developments in measurement methods to study the net exchange of CO_2, heat and water vapour continuously over years has already revealed much in understanding the coupling of forest ecosystems to short-term variability in climate and has shown the need for long-term measurements (> 10 years) to separate real trends in the annual net forest–atmosphere exchange from the effects of seasonal variability in the weather.

REFERENCES

André, J.-C., Goutorbe, J.-P. & Perrier, A. (1986) HAPEX-MOBILHY: a hydrologic atmospheric experiment for the study of water budget and evaporation flux at the climatic scale. *Bulletin of the American Meteorological Society* **67**, 138–44.

Anthoni, P.M., Law, B.E. & Unsworth, M.H. (1999) Carbon and water vapor exchange of an open-canopied ponderosa pine ecosystem. *Agricultural and Forest Meteorology* **95**, 151–68.

Asdak, C., Jarvis, P.G., van Gardingen, P. & Fraser, A. (1998a) Rainfall interception loss in unlogged and logged forest areas of Central Kalimantan, Indonesia. *Journal of Hydrology* **206**, 237–44.

Asdak, C., Jarvis, P.G. & van Gardingen, P. (1998b) Evaporation of intercepted precipitation based on an energy balance in unlogged and logged forest areas of central Kalimantan, Indonesia. *Agricultural and Forest Meteorology* **92**, 173–80.

Aubinet, M., Grelle, A., Ibrom, A. *et al.* (2000) Estimates of the annual net carbon and water exchange of European forests: the EUROFLUX methodology. *Advances in Ecological Research*, **30**, 113–75.

Baldocchi, D.D. (1997) Measuring and modeling carbon dioxide and water vapor exchange over a temperate broad-leaved forest during the 1995 summer drought. *Plant, Cell and Environment* **20**, 1108–22.

Baldocchi, D.D. & Meyers, T.P. (1998) On using eco-physiological, micrometeorological and biogeochemical theory to evaluate carbon dioxide, water vapor and gaseous deposition fluxes over vegetation. *Agricultural and Forest Meteorology* **90**, 1–26.

Baldocchi, D.D. & Vogel, C.A. (1996) A comparative study of water vapor, energy and CO_2 flux densities above and below a temperate broadleaf and a boreal pine forest. *Tree Physiology* **16**, 5–16.

Baldocchi, D.D., Guenther, A., Harley, P. *et al.* (1995) The fluxes and air chemistry of isoprene above a deciduous hardwood forest *Philosophical Transactions of the Royal Society of London B* **350**, 279–96.

Baldocchi, D.D., Valentini, R., Running, S., Oechel, W. & Dahlmann, R. (1996) Strategies for measuring and modelling carbon dioxide and water vapour fluxes over terrestrial ecosystems. *Global Change Biology* **2**, 159–68.

Baldocchi, D.D., Kelliher, F.M., Black, T.A. & Jarvis, P.G. (2000) Climate and vegetation controls on boreal zone energy exchange. *Global Change Biology*, in press.

Barr, A.G. & Betts, A.K. (1997) Radiosonde boundary layer budgets above a boreal forest. *Journal of Geophysical Research* **102**, 29205–13.

Bergh, J. & Linder, S. (1999) Effects of soil warming during spring on photosynthetic recovery in boreal Norway spruce stands. *Global Change Biology* **5**, 245–53.

Bergh, J., McMurtrie, R. & Linder, S. (1998) Climatic factors controlling the productivity of Norway spruce: a model-based analysis. *Forest Ecology and Management* **110**, 127–39.

Bergh, J., Linder, S., Lundmark, T. & Elfving, B. (1999) The effect of water and nutrient availability on the productivity of Norway spruce in northern and southern Sweden. *Forest Ecology and Management* **119**, 51–62.

Beswick, K.M., Hargreaves, K.J., Gallagher, M.W., Choularton, T.W. & Fowler, D. (1991) Size resolved measurements of cloud droplet deposition velocity to a forest canopy using an eddy correlation technique. *Quarterly Journal of the Royal Meteorological Society* **117**, 623–45.

Betts, A.K. & Ball, J.H. (1997) Albedo over the boreal forest. *Journal of Geophysical Research* **102**, 28901–9.

Betts, A.K., Ball, J.H., Beljaars, A.C., Miller, M.J. & Viterbo, P.A. (1996) The land surface–atmosphere interaction: a review based on observational and global modelling perspectives. *Journal of Geophysical Research* **101**, 7209–25.

Biscoe, P.V., Clark, J.A., Gregson, K., McGowan, M., Monteith, J.L. & Scott, R.K. (1975) Barley and its environment. I. Theory and practice. *Journal of Applied Ecology* **12**, 227–57.

Black, T.A., den Hartog, G., Neumann, H. *et al.* (1996) Annual cycles of CO_2 and water vapor fluxes above and within a boreal aspen stand. *Global Change Biology* **2**, 219–30.

Blanken, P.D., Black, T.A., Yang, P.C. *et al.* (1997) Energy balance and canopy conductance of a boreal aspen forest: partitioning overstory and understory components. *Journal of Geophysical Research* **102**, 28915–28.

Blanken, P.D., Black, T.A., Neumann, H.H. *et al.* (1998) Turbulent flux measurements above and below the overstory of a boreal aspen forest. *Boundary Layer Meteorology* **89**, 109–40.

Boone, R.D., Nadelhoffer, K.J., Canary, J.D. & Kaye, J.P. (1998) Roots exert a strong influence on the temperature sensitivity of soil respiration. *Nature* **396**, 570–2.

Brückmann, A. (1988) *Radionuklidbilanz von 4 Waldoekosystemen nach dem Reaktorunfall in Tschernobyl und eine Bestimmung der trockenen Deposition.* Diplomarbeit Forstwissenschaftisher Fachbereich, Gottingen.

Brutsaert, W. (1991) *Evaporation Into the Atmosphere.* Kluwer Academic Publishers, Dordrecht.

Calder, I.R. (1996) Dependence of rainfall interception on drop size. I. Development of the two-layer stochastic model. *Journal of Hydrology* **185**, 363–78.

Cannell, M.G.R., Dewar, R.C. & Pyatt, D.G. (1993) Conifer plantations on drained peatlands in Britain: a net gain or loss of carbon? *Forestry* **66**, 353–69.

Cannell, M.G.R., Milne, R., Hargreaves, K.J. *et al.* (1999) National inventories of terrestrial carbon sources and sinks: the UK experience. *Climatic Change* **42**, 505–30.

Cao, M. & Woodward, F.I. (1998) Dynamic responses of terrestrial ecosystem carbon cycling to global climate change. *Nature* **393**, 249–52.

Cape, J.N., Leith, I.D., Fowler, D. *et al.* (1991) Sulphate and ammonium in mist impair the frost hardening of red spruce seedlings. *New Phytologist* **118**, 119–26.

Chamberlain, A.C. (1975) The movement of particles in plant communities. In: Monteith, J.L., ed. *Vegetation and the Atmosphere, Vol. 1 Principles*, pp. 155–203. Academic Press, London.

Chappelka, A.H. & Samuelson, J. (1998) Ambient ozone effects on forest trees of the eastern United States: a review. *New Phytologist* **139**, 91–108.

Chen, W.J., Black, T.A., Yang, P.C. *et al.* (1999) Effects of climatic variability on the annual carbon sequestration by a boreal aspen forest. *Global Change Biology* **5**, 41–53.

Christensen, C.S., Hummelshøj, P., Jensen, N.O., Lohse, C. & Skov, H. (1997) Biogenic emissions from a Norway spruce (*Picea abies*) forest at Ulfborg, Denmark measured by relaxed eddy accumulation (unpublished).

Clements, F.E. (1916) *Plant Succession: an Analysis of the Development of Vegetation.* Carnegie Institute of Washington Publication 242, facsimile reprint of Haffner.

Clymo, R.S. (1984) The limits to peat growth. *Philosophical Transactions of the Royal Society of London B* **303**, 605–54.

Collins, W., Stevenson, D.S., Johnson, C.E. & Derwent, R.G. (1997) Tropospheric ozone in a global-scale three-dimensional Lagrangian model and its response to NO_x emission controls. *Journal of Atmospheric Chemistry* **26**, 223–74.

Crossley, A., Wilson, D.B. & Milne, R. (1992) Pollution in the upland environment. *Environmental Pollution* **75**, 81–8.

Curtis, P.S. & Wang, X. (1998) A meta-analysis of elevated CO_2 effects on woody plant mass, form, and physiology. *Oecologia* **113**, 299–313.

Davidson, E.A. & Kingerlee, A. (1997) A global inventory of nitric oxide emissions from soils. *Nutrient Cycling Agroecosystems* **48**, 91–104.

De Hayes, D.H. (1992) Winter injury and developmental cold tolerance of red spruce. In: Billings, W.D., Golley, R.G.F., Lange, O.L., Olson, J.S. & Remmert, H., eds. *Ecology and Decline of Red Spruce in the Eastern United States*, pp. 295–337. Springer-Verlag, New York.

Denmead, O.T. & Bradley, E.F. (1987) On scalar transport in plant canopies. *Irrigation Science* **8**, 131–49.

Dentener, F.J. (1993) *Heterogeneous chemistry in the troposphere.* PhD thesis University of Utrecht, The Netherlands.

Dewar, R.C., Medlyn, B.E. & McMurtrie, R.E. (2000) Acclimation of the respiration/photosynthesis ratio to temperature: insights from a model. *Global Change Biology* (in press).

Duyzer, J., Deinum, G. & Baak, J. (1995) The interpretation of measurements of surface exchange of nitrogen oxides: correction for chemical reactions. *Philosophical Transactions of the Royal Society London B* **351**, 231–48.

Dyer, A.J. & Hicks, B.B. (1967) The fluxatron: a revised approach to the measurement of eddy fluxes in the lower atmosphere. *Journal of Applied Meteorology* **6**, 408–13.

Dyer, A.J. & Maher, F.J. (1965) Automatic eddy-flux measurement with the evapotron. *Journal of Applied Meteorology* **4**, 622–5.

Ellsworth, D.S. (1999) CO_2 enrichment in a maturing pine forest: are CO_2 exchange and water status in the canopy affected? *Plant, Cell and Environment* **22**, 461–72.

Erisman, J.W. & Wyers, G.P. (1993) Continuous measurements of surface exchange of SO_2 and NH_3: implications for their possible interaction in the deposition process. *Atmospheric Environment* **A27**, 1937–49.

Erisman, J.W., Mennen, M., Fowler, D. *et al.* (1998) Deposition monitoring in Europe. *Environmental Monitoring Assessment* **53**, 279–95.

Fan, S.-M., Wofsy, S.C., Bakwin, P.S. & Jacob, D.J. (1990) Atmosphere–biosphere exchange of CO_2 and O_3 in the Central Amazon Forest. *Journal of Geophysical Research* **95**, 16851–64.

Farman, J.C., Gardiner, B.G. & Shanklin, J.D. (1985) Large losses of total ozone in Antartica reveal seasonal ClO_x/NO_x interaction. *Nature* **315**, 207–10.

Finnigan, J.J. & Raupach, M.R. (1987) Transfer processes in plant canopies in relation to stomatal characteristics. In: Zeiger, E., Farquhar, G.D. & Cowan, I.R., eds. *Stomatal Function*, pp. 385–429. Stanford University Press, Stanford, California.

Flechard, C.R. (1998) *The turbulent exchange of ammonia above vegetation.* PhD thesis. University of Nottingham.

Flechard, C.R. & Fowler, D. (1998) Atmospheric ammonia at a moorland site. II. Long-term surface/atmosphere micrometeorological flux measurements. *Quarterly Journal of the Royal Meteorological Society* **124**, 733–57.

Flechard, C.R., Fowler, D., Sutton, M.A. & Cape, J.N. (1999) A dynamic chemical model of bi-directional ammonia exchange between semi-natural vegetation and the atmosphere. *Quarterly Journal of the Royal Meteorological Society* **125**, 1–33.

Fowler, D. & Unsworth, M.H. (1979) Turbulent transfer of sulphur dioxide to a wheat crop. *Quarterly Journal of the Royal Meteorological Society* **105**, 767–83.

Fowler, D., Cape, J.N. & Unsworth, M.H. (1989) Deposition of atmospheric pollutants on forests. *Philosophical Transactions of the Royal Society of London B* **324**, 247–65.

Fowler, D., Baldocchi, D. & Duyzer, J.H. (1991) Inputs of trace gases, particles and cloud droplets to terrestrial surfaces. In: Last, F.T. & Watling, R., eds. *Acidic Deposition: its Nature and Impacts*, pp. 35–59. Royal Society of Edinburgh, Edinburgh.

Fowler, D., Pitcairn, C.E.R., Sutton, M.A. (1998a) The mass budget of atmospheric ammonia in woodland within 1 km of livestock buildings. *Environmental Pollution* **102**, 343–8.

Fowler, D., Flechard, C., Skiba, U., Coyle, M. & Cape, J.N. (1998b) The atmospheric budget of oxidized nitrogen and its role in ozone formation and deposition. *New Phytologist* **139**, 11–23.

Fowler, D., Smith, R.I., Crossley, A. *et al.* (1998c) Quantifying fine scale variability in deposition of pollutants in complex terrain using ^{210}Pb inventories in soil. *Water, Air and Soil Pollution* **105**, 459–70.

Fowler, D., Cape, J.N., Coyle, M.L., Flechard, C., Kuylenstierna, J., Derwent, D., Johnson, C. & Stevenson, D. (1999) The global exposure of forests to air pollutants. *Water, Air and Soil Pollution* **116**, 5–32.

Freer-Smith, P.H., Holloway, S. & Goodman, A. (1997) The uptake of particulates by an urban woodland: site description and particulate composition. *Environmental Pollution* **95**, 27–35.

Friedlingstein, P., Fung, I., Holland, E. *et al.* (1995) On the contribution of CO_2 fertilization to the missing biospheric sink. *Global Biogeochemical Cycles* **9**, 541–56.

Fuhrer, J., Skarby, L. & Ashmore, M.R. (1997) Critical levels for ozone effects on vegetation in Europe. *Environmental Pollution* **97**, 91–106.

Gallagher, M.W., Beswick, K. & Choularton, T.W. (1992) Measurements and modelling of cloud water deposition to a snow covered forest. *Atmospheric Environment*, **26A**, 2893–904.

Gallagher, M., Fontan, J., Wyers, P. *et al.* (1997a) Atmospheric particles and their interactions with natural surfaces. In: Slanina, S., ed. *Biosphere–Atmosphere Exchange of Pollutants and Trace Substances*, pp. 45–92. Springer-Verlag, Berlin.

Gallagher, M.W., Beswick, K., Duyzer, J., Westrate, H., Choularton, T.W. & Hummelshøj, P. (1997b) Measurements of aerosol fluxes to Speulder forest using a micrometeorological technique. *Atmospheric Environment* **31**, 359–73.

Galloway, J.N. & Rodhe, H. (1991) Regional atmospheric budgets of S and N fluxes: how well can they be quantified?. In: Last, F.T. & Watling, R., eds. *Acidic Deposition: its Nature and Impacts*, pp. 61–80. Royal Society of Edinburgh, Edinburgh.

Garratt, J.R. (1992). *The Atmospheric Boundary Layer.* Cambridge University Press, Cambridge.

Gash, J.H.C. & Stewart, J.B. (1975) The average surface

resistance of a pine forest derived from Bowen ratio measurements. *Boundary-Layer Meteorology* **8**, 453–64.

Gash, J.H.C. & Stewart, J.B. (1977) The evaporation from Thetford Forest during 1975. *Journal of Hydrology* **35**, 385–96.

Gash, J.H.C., Wright, I.R. & Lloyd, C.R. (1980) Comparative estimates of interception loss from three coniferous forests in Great Britain. *Journal of Hydrology* **48**, 89–105.

Gash, J.H.C., Nobre, C.A., Roberts, J.M. & Victoria, R.L. (1996). *Amazonian Deforestation and Climate*. John Wiley & Sons, Chichester.

Gash, J.H.C., Valente, F. & David, J.S. (1999) Estimates and measurements of evaporation from wet, sparse pine forest in Portugal. *Agricultural and Forest Meteorology* **94**, 149–58.

Gates, D.M. (1980). *Biophysical Ecology*. Springer-Verlag, New York.

Gilbert, I.R., Seavers, G.P., Jarvis, P.G. & Smith, H. (1995) Photomorphogensis and canopy dynamics. Phytochrome-mediated proximity perception accounts for the growth dynamics of canopies of *Populus trichocarpa* x *deltoides* 'Beaupré'. *Plant, Cell and Environment* **18**, 475–97.

Goulden, M.L., Munger, J.W., Fan, S.-M., Daube, B.C. & Wofsy, S.C. (1996) Exchange of carbon dioxide by a deciduous forest: response to interannual climate variability. *Science* **271**, 1576–8.

Goulden, M.L., Daube, B.C., Fan, S.-M. *et al.* (1997) Physiological responses of a black spruce forest to weather. *Journal of Geophysical Research (BOREAS Special Issue)* **102** (D24), 28987–96.

Goulden, M.L., Wofsy, S.C., Harden, J.W. *et al.* (1998) Sensitivity of boreal forest carbon balance to soil thaw. *Science* **279**, 214–17.

Gower, S.T., Vogel, J.G., Norman, J.M., Kucharik, C.J., Steele, S.J. & Stow, T.K. (1997) Carbon distribution and aboveground net primary production in aspen, jack pine, and black spruce stands in Saskatchewan and Manitoba, Canada. *Journal of Geophysical Research* **102**, 29029–41.

Grace, J. & Malhi, Y. (1999) The role of rain forests in the global carbon cycle. *Progress in Environmental Science* **1**, 177–93.

Grace, J., Lloyd, J., McIntyre, J. *et al.* (1995a) Fluxes of carbon dioxide and water vapour over an undisturbed tropical rainforest in south-west Amazonia. *Global Change Biology* **1**, 1–12.

Grace, J., Lloyd, J., McIntyre, J. *et al.* (1995b) Carbon dioxide uptake by an undisturbed tropical rain forest in South-West Amazonia 1992–93. *Science* **270**, 778–80.

Grace, J., Malhi, Y., Lloyd, J. *et al.* (1996) The use of eddy covariance to infer the net carbon dioxide uptake of a Brazilian rain forest. *Global Change Biology* **2**, 209–17.

Greco, S. & Baldocchi, D.D. (1996) Seasonal variations of CO_2 and water vapour exchange rates over a temperate deciduous forest. *Global Change Biology* **2**, 183–97.

Green, S.R. (1990) *Air flow through and above a forest of widely spaced trees*. PhD thesis, University of Edinburgh.

Grosch, S. & Schmitt, G. (1988) Experimental investigations on the deposition of trace elements in forest area. In: Greefen, K. & Löbel, L., eds. *Environmental Meteorology*, pp. 201–16. Kluwer Academic Publishers, Dordrecht.

Guenther, A. (1997) Seasonal and spatial variations in the natural volatile organic compound emissions. *Ecological Applications* **7**, 34–45.

Guenther, A.B., Monson, R.K. & Fall, R. (1991) Isoprene and monoterepene rate variability: observations with *Eucalyptus* and emission rate algorithm development. *Journal of Geophysical Research* **96**, 799–808.

Guenther, A., Geron, C., Pierce, T., Lamb, B., Hanley, P. & Fall, R. (2000) Natural emissions of non-methane volatile organic compounds, carbon monoxide, and oxides of nitrogen from North America. *Atmospheric Environment* **34**, 2205–30.

Hargreaves, K.J., Fowler, D., Storeton-West, R.L. & Duyzer, J.H. (1992) The exchange of nitric oxide, nitrogen dioxide and ozone between pasture and the atmosphere. *Atmospheric Pollution* **75**, 53–9.

Hassika, P. & Berbigier, P. (1998) Annual cycle of photosynthetically active radiation in maritime pine forest. *Agricultural and Forest Meteorology* **90**, 157–71.

Höfken, K.D. & Gravenhorst, G. (1983) Untersuchung uber die Deposition Atmosphärischer Spurentoffe an Buchtel-und-Fichten Wald. In UBA—Berichte 6/83, Teil II, Schmidt Verlag, Berlin.

Hollinger, D.Y., Kelliher, F.M., Byers, J.N., Hunt, J.E., McSeveny, T.M. & Weir, P.L. (1994) Carbon dioxide exchange between an undisturbed old-growth temperate forest and the atmosphere. *Ecology* **75**, 134–50.

Hollinger, D.Y., Kelliher, F.M., Schulze, E.-D. *et al.* (1995) Initial assessment of multi-scale measures of CO_2 and H_2O flux in the Siberian taiga. *Journal of Biogeography* **22**, 425–31.

Hollinger, D.Y., Kelliher, F.M., Schulze, E.-D. *et al.* (1998) Forest–atmosphere carbon dioxide exchange in eastern Siberia. *Agricultural and Forest Meteorology* **90**, 291–306.

Hornung, M., Sutton, M.A. & Wilson, R.B. (1995) *Mapping and Modelling of Critical Loads for Nitrogen: a Workshop Report*. Proceedings of the Grange-over-Sands UNECE Workshop, Institute of Terrestrial Ecology, Edinburgh.

Hudson, R.J.M., Gherini, S.A. & Goldstein, R.A. (1994) Modeling the global carbon cycle–nitrogen fertilization of the terrestrial biosphere and the missing CO_2 sink. *Global Biogeochemical Cycles* **8**, 307–33.

Huebert, B.J. (1983) Measurements of the dry deposition flux of nitric acid vapour to grasslands and forests. In: Prupaccher, H.R., Semonin, R.G. & Slinn, W.G.N., eds. *Precipitation Scavenging, Dry Deposition and Resuspension*, pp. 785–94. Elsevier, New York.

Ineson, P., Benham, D.G., Poskitt, J., Harrison, A.F., Taylor, K. & Woods, C. (1998) Effects of climate change on nitrogen dynamics in upland soils. 2. A soil warming study. *Global Change Biology* **4**, 153–61.

Jarvis, P.G. (1975) Water in plants. In: De Vries, D.A. & Afgan, N.H., eds. *Heat and Mass Transfer in the Biosphere. I. Transfer Processes in the Plant Environment*, pp. 69–394. Scripta Book Company, Washington; John Wiley & Sons, New York.

Jarvis, P.G. (1976) Exchange properties of coniferous forest canopies. In: *Proceedings 16th IUFRO World Congress. Division II. Forest Plants and Forest Protection.* Norwegian IUFRO Congress Committee, pp. 90–8. International Union of Forest Research Organisations, Vienna.

Jarvis, P.G. (1993) Water losses of crowns, canopies and communities. In: Smith, J.A.C. & Griffiths, H., eds. *Water Deficits: Plant Responses from Cell to Community*, pp. 285–315. Bios Scientific Publishers, Oxford.

Jarvis, P.G. (ed.) (1998) *European Forests and Global Change: the Likely Impacts of Rising CO$_2$ and Temperature.* Cambridge University Press, Cambridge.

Jarvis, P.G. & Linder, S. (2000) Constraints to growth of boreal forest. *Nature* **405**, 904–5.

Jarvis, P.G. & McNaughton, K.G. (1986) Stomatal control of transpiration: scaling up from leaf to region. *Advances in Ecological Research* **15**, 1–49.

Jarvis, P.G. & Morison, J.I.L. (1981) The control of transpiration and photosynthesis by the stomata. In: Jarvis, P.G. & Mansfield, T.A., eds. *Stomatal Physiology*, pp. 247–79. Cambridge University Press, Cambridge.

Jarvis, P.G. & Stewart, J.B. (1979) Evaporation of water from plantation forest. In: Ford, E.D., Malcolm, D.C. & Atterson, J., eds. *The Ecology of Even-aged Plantations*, pp. 327–49. Institute of Terrestrial Ecology, Cambridge.

Jarvis, P.G., James, B.G. & Landsberg, J.J. (1976) Coniferous forest. In: Monteith, J.L., ed. *Vegetation and Atmosphere*, pp. 171–204. Academic Press, London.

Jarvis, P.G., Massheder, J.M., Hale, S.E., Moncrieff, J.B., Rayment, M. & Scott, S.L. (1997) Seasonal variation of carbon dioxide, water vapor, and energy exchange of a boreal black spruce forest. *Journal of Geophysical Research* **102**, 28953–66.

Jarvis, P.G., Saugier, B. & Schulze, E.-D. (1999) Productivity of boreal forest. In: Ray, J. & Mooney, H.A., eds. *Primary Productivity*. Springer-Verlag, Berlin, in press.

Johnson, D.W., Cresser, M.S., Nilsson, S.I. *et al.* (1991) Soil changes in forest ecosystems: evidence for and probable causes. In: Last, F.T. & Watling, R., eds. *Acidic Deposition: its Nature and Impacts*, pp. 81–116. Royal Society of Edinburgh, Edinburgh.

Joutsenoja, T. (1992) Measurements of aerosol deposition to a cereal crop. In: Choularton, T.W., ed. *Measurements and Modelling of Gases and Aerosols to Complex Terrain. NERC Report 1992, GR3/7259* A.P.I. Appendix XI.2.

Kabat, P., Hutjes, R.W.A., Dolman, A.J. *et al.* (eds) (1999) *The Effects of Changes in Land Use and Climate on the Sustainability of Natural and Man-made Ecosystems in Amazonia: the European Contribution to the Large-scale Biosphere Atmosphere Experiment in Amazonia (LBA). Part 1, Thematic LBA Implementation Plan; Part 2, Thematic LBA Implementation Plan: Executive Summary; Part 3, Report of the European First Open Science Conference on LBA, 23–25 June 1997, Wageningen, Netherlands.* DLO Winand Staring Centre, Wageningen, The Netherlands.

Kaimal, J.C. & Finnigan, J.J. (1994). *Atmospheric Boundary Layer Flows.* Oxford University Press, New York.

Kauppi, P.E., Mielkäinen, K. & Kuusela, K. (1992) Biomass and carbon budget of European forests, 1971 to 1990. *Science* **256**, 70–4.

Kelliher, F.M., Leuning, R., Raupach, M.R. & Schulze, E.-D. (1994) Maximum conductances for evaporation from global vegetation types. *Agricultural and Forest Meteorology* **73**, 1–16.

Kelliher, F.M., Hollinger, D.Y., Schulze, E.-D. *et al.* (1997) Evaporation from an eastern Siberian larch forest. *Agricultural and Forest Meteorology* **85**, 135–47.

Kley, D., Kleinman, M., Sandermann, H. & Krupa, S. (1999) Photochemical oxidants: state of the science. *Environmental Pollution* **100**, 19–42.

Kruijt, B., Malhi, Y., Lloyd, J. *et al.* (1999) Turbulence statistics above and within two Amazon rain forest canopies. *Boundary Layer Meteorology*, **94**, 297–331.

Last, F.T. & Watling, R. (eds) (1991) *Acidic Deposition: its Nature and Impacts.* Royal Society of Edinburgh, Edinburgh.

Leith, I.D., Murray, M.B., Sheppard, L.J. *et al.* (1989) Visible foliar injury of red spruce seedlings subjected to simulated acid mist. *New Phytologist* **113**, 313–20.

Lindroth, A. (1985) Seasonal and diurnal variation of energy budget components in coniferous forests. *Journal of Hydrology* **82**, 1–15.

Lindroth, A., Grelle, A. & Morén, A.S. (1998) Long-term measurements of boreal forest carbon balance reveal large temperature sensitivity. *Global Change Biology* **4**, 443–50.

Lloyd, C.R., Gash, J.H.C., Shuttleworth, W.J. & Marques, A.D.O. (1988) The measurement and modelling of rainfall interception by Amazonian rain forest. *Agricultural and Forest Meteorology* **43**, 277–94.

Lorenz, R. & Murphy Jr, C.E. (1989) Dry deposition of particles to a pine plantation. *Boundary-Layer Met.* **46**, 355–66.

Lugo, A.E. & Brown, S. (1992) Tropical forests as sinks of atmospheric carbon. *Forest Ecology and Management* **54**, 239–55.

Lükewille, A. & Wright, R.F. (1997) Experimentally increased soil temperature causes release of nitrogen at a boreal forest catchment in southern Norway. *Global Change Biology* **3**, 13–21.

McCaughey, J.H., Lafleur, P.M., Joiner, D.W. *et al.* (1997) Magnitudes and seasonal patterns of energy, water, and carbon exchanges at a boreal young jack pine forest in the BOREAS northern study area. *Journal of Geophysical Research* **102**, 28997–9009.

McLaughlin, S.B. & Kohut, R.J. (1992) The effects of atmospheric deposition and ozone on carbon allocation and associated physiological processes in red spruce. In: Eagar, C. & Adams, M.B., eds. *Ecology and Decline of Red Spruce in the Eastern United States.* Springer-Verlag, Berlin.

McMurtrie, R.E. & Dewar, R.C. (1999) Ecosystem modelling of the CO_2-response of forests on sites limited by nitrogen and water. In: Luo, Y. & Mooney, H.A., eds. *Carbon Dioxide and Environmental Stress*, pp. 347–69. Academic Press, San Diego.

McMurtrie, R.E., Medlyn, B.E., Dewar, R.C. & Jeffreys, M.P. (2000a) Effects of rising CO_2 on growth and carbon sequestration in forests: a modelling analysis of the consequences of altered litter quantity and quality. *Plant and Soil* (in press).

McMurtrie, R.E., Medlyn, B.E. & Dewar, R.C. (2000b) Increased understanding of nutrient immobilisation in soil organic matter is critical for predicting the carbon sink strength of forest ecosystems over the next 100 years. *Global Change Biology* (in press).

McNaughton, K.G. (1989) Regional interactions between canopies and the atmosphere. In: Russell, G., Marshall, B. & Jarvis, P.G., eds. *Plant Canopies: their Growth, Form and Function*, pp. 63–81. Cambridge University Press, Cambridge.

McNaughton, K.G. (1994) Effective stomatal and boundary-layer resistances of heterogeneous surfaces. *Plant, Cell and Environment* **17**, 1061–8.

McNaughton, K.G. & Jarvis, P.G. (1983) Predicting effects of vegetation changes on transpiration and evaporation. In: Kozlowski, T.T., ed. *Water Deficits and Plant Growth VII*, pp. 1–47. Academic Press, London.

McNaughton, K.G. & Jarvis, P.G. (1991) Effect of spatial scale on stomatal control of transpiration. *Agricultural and Forest Meteorology* **54**, 279–302.

Malhi, Y., Nobre, A.D., Grace, J. *et al.* (1998) Carbon dioxide transfer over a Central Amazonian rain forest. *Journal of Geophysical Research* **103**, 31593–612.

Malhi, Y., Baldocchi, D.D. & Jarvis, P.G. (1999) The carbon balance of tropical, temperate and boreal forests. *Plant, Cell and Environment* **22**, 715–40.

Mansfield, T.A. (1998) Stomata and plant water relations: does air pollution create problems? *Environmental Pollution* **101**, 1–11.

Martin, P.H., Valentini, R., Jacques, M. *et al.* (1998) New estimate of the carbon sink strength of EU forests integrating flux measurements, field surveys and space observations: 0.17–0.35 Gt (C). *Ambio* **27**, 582–4.

Medlyn, B.E., Badeck, F.-W., de Pury, D.G.G. *et al.* (1999) Effects of elevated [CO_2] on photosynthesis in European forest species: a meta-analysis of model parameters. *Plant, Cell and Environment* **22**, 1475–95.

Medlyn, B.E., McMurtrie, R.E., Dewar, R.C. & Jeffreys, M.P. (2000) Soil processes dominate long-term response of net primary productivity of forests to increased temperature and atmospheric CO_2 concentration. *Canadian Journal of Forest Research* (in press).

Miller, P.R., Parmeter, J.R. Jr, Taylor, O.C. & Cardiff, E.A. (1963) Ozone injury to the foliage of *Ponderosa pine. Phytopathology* **53**, 1072–6.

Mizutani, K., Yamanoi, K., Ikeda, T. & Watanabe, T. (1997) Applicability of the eddy correlation method to measure sensible heat transfer to forest under rainfall conditions. *Agricultural and Forest Meteorology* **86**, 193–203.

Moncrieff, J.B., Malhi, Y. & Leuning, R. (1996) The propagation of errors in long-term measurements of land–atmosphere fluxes of carbon and water. *Global Change Biology* **2**, 231–40.

Moncrieff, J.B., Massheder, J.M., deBruin, H. *et al.* (1997) A system to measure surface fluxes of momentum, sensible heat, water vapour and carbon dioxide *Journal of Hydrology* **189**, 589–611.

Monteith, J.L. (1965) Evaporation and environment. *Symposia of the Society for Experimental Biology* **19**, 205–34.

Monteith, J.L. & Szeicz, G. (1961) The radiation balance of bare soil and vegetation. *Quarterly Journal of the Royal Meteorological Society* **87**, 159–70.

Monteith, J.L. & Unsworth, M.H. (1990). *Principles of Environmental Physics*, 2nd edn. Edward Arnold, London.

Muller, H., Kramm, G., Meixner, F., Dollard, G.J., Fowler, D. & Possanzini, M.F. (1993) Determination of HNO_3 dry deposition by modified Bowen-ratio and aerodynamic profile techniques. *Tellus* **45B**, 346–67.

Nobre, C.A., Fisch, G., da Rocha, H.R.F. *et al.* (1996) Observations of the atmospheric boundary layer in Rondonia. In: Gash, J.H.C., Nobre, C.A., Roberts, J.M. & Victoria, R.L., eds. *Amazonian Deforestation and Climate*, pp. 413–24. John Wiley & Sons, Chichester.

Oden, S. (1968) *The Acidification of Air and Precipitation and its Consequences on the Natural Environ-*

ment (in Swedish). Ecology Committee, Bulletin 1, National Science Research Council of Sweden. (Translated by Translation Consultations Ltd, Virginia, no. TR-1172.)

Oke, T.R. (1987). *Bounday-Layer Climates*. Cambridge University Press, Cambridge.

Peterson, A.G., Ball, J.T., Luo, Y., *et al.* (1999) The photosynthesis–leaf nitrogen relationship at ambient and elevated atmospheric carbon dioxide: a meta-analysis. *Global Change Biology* **5**, 331–46.

Philip, J.R. (1966) Plant water relations: some physical aspects. *Annual Review of Plant Physiology* **17**, 245–68.

Phillips, O.L., Malhi, Y., Higuchi, N. *et al.* (1998) Changes in the carbon balance of tropical forests: evidence from long-term plots. *Science* **282**, 439–42.

Pielke, R.A., Avissar, R., Raupach, M., Dolman, A.J., Zeng, X. & Denning, A.S. (1998) Interactions between the atmosphere and terrestrial ecosystems: influence on weather and climate. *Global Change Biology* **4**, 461–75.

Pilegaard, K., Jensen, N.O. & Hummelshoj, P. (1995) Deposition of nitrogen oxides and ozone to Danish forest sites. In: Heij, G.J. & Erisman, J.W., eds. *Acid Rain Research: Do We Have Enough Answers?*, pp. 31–40. Elsevier Science, Amsterdam.

Pinker, R.T. (1982) The diurnal asymmetry in the albedo of tropical forest vegetation. *Forest Science* **28**, 297–304.

Pinker, R.T., Thompson, O.E. & Eck, T.F. (1980) The albedo of a tropical evergreen forest. *Quarterly Journal of the Royal Meteorological Society* **106**, 551–8.

Pitcairn, C.E.R., Leith, I.D., Sheppard, L.J. *et al.* (1998) The relationship between nitrogen deposition, species composition and foliar nitrogen concentrations in woodland flora in the vicinity of livestock farms. *Environmental Pollution* **102** (S1), 41–8.

PORG (1997). *Ozone in the UK*. Fourth report of the Photochemical Oxidants Review Group, Department of the Environment, Transport and the Regions, London.

Priestley, C.H.B. & Taylor, R.J. (1972) On the assessment of surface heat flux and evaporation using large-scale parameters. *Monthly Weather Review* **100**, 81–92.

Raupach, M.R. (1988) Canopy transport processes. In: Steffen, W.L. & Denmead, O.T., eds. *Flow and Transport in the Natural Environment: Advances and Applications*, pp. 95–127. Springer-Verlag, Berlin.

Raupach, M.R. (1995) Vegetation–atmosphere interaction and surface conductance at leaf, canopy and regional scales. *Agricultural and Forest Meteorology* **73**, 151–80.

Raupach, M.R. (1998) Influences of local feedbacks on land–air exchanges of energy and carbon. *Global Change Biology* **4**, 477–94.

Raupach, M.R., Finnigan, J.J. & Brunet, Y. (1996) Coherent eddies and turbulence in vegetation canopies: the mixing layer analogy. *Boundary-Layer Meteorology* **78**, 351–82.

Rawlins, S.L. (1963) Resistance to water flow in the transpiration stream. In: Zelitch, I., ed. *Stomata and Water Relations of Plants*, pp. 69–85. Connecticut Agricultural Experiment Station, New Haven.

Rayment, M.B. & Jarvis, P.G. (2000) Temporal and spatial variation of soil respiration in a Canadian boreal forest. *Soil Biology and Biochemistry*, **32**, 35–45.

RGAR (1997) Acid deposition in the United Kingdom 1992–1994. In: *The Fourth Report of the United Kingdom Review Group on Acid Rain*, pp. 129–30. Department of the Environment, Transport and the Regions, London.

Roberts, J. (1983) Forest transpiration: a conservative hydrological process? *Journal of Hydrology* **66**, 133–41.

Roberts, T.M., Skeffington, R.A. & Blank, L.W. (1989) Causes of Type I spruce decline in Europe. *Forestry* **62**, 179–222.

Roemer, M. (1996) *Trends of tropospheric ozone over Europe*. PhD thesis University of Utrecht, The Netherlands.

Ruimy, A., Jarvis, P.G., Baldocchi, D.D. & Saugier, B. (1995) CO_2 fluxes over plant canopies and solar radiation: a review. *Advances in Ecological Research* **26**, 1–68.

Running, S.W., Baldocchi, D., Cohen, W., Gower, S.T., Turner, D., Bakwin, P. & Hibbard, K. (1999) A global terrestrial monitoring network integrating tower fluxes, flask sampling, with ecosystem modelling and EOS satellite data. *Remote Sens. Environ.* **70**(1): 108–27.

Rutter, A.J. (1975) The hydrological cycle in vegetation. In: Monteith, J.L., ed. *Vegetation and the Atmosphere*, Vol. 1, pp. 111–54. Academic Press, New York.

Rutter, A.J. & Morton, A.J. (1977) A predictive model of rainfall interception in forests. III. Sensitivity of the model to stand parameters and meteorological variables. *Journal of Applied Ecology* **14**, 567–88.

Rutter, A.J., Morton, A.J. & Robins, P.C. (1975) A predictive model of rainfall interception in forests. II. Generalization of the model and comparison with observations in some coniferous and hardwood stands. *Journal of Applied Ecology* **12**, 367–80.

Sattin, M., Milne, R., Deans, J.D. & Jarvis, P.G. (1997) Radiation interception measurement in poplar: sample size and comparison between tube solarimeters and quantum sensors. *Agricultural and Forest Meteorology* **85**, 209–16.

Saugier, B., Granier, A., Pontailler, J.-Y. & Baldocchi, D.D. (1997) Transpiration of a boreal pine forest measured by branch bags, sapflow and micrometeorological methods. *Tree Physiology* **17**, 511–20.

Scholes, R.J. (1999) Will the terrestrial carbon sink saturate soon? *Global Change Newsletter* **37**, 2–3.

Scholes, R.J., Schulze, E.-D., Pitelka, L.F. & Hall, D.O. (1999) Biogeochemistry of terrestrial ecosystems. In: Walker, B.H., Steffen, W.L., Canadell, J. & Ingram, J.S.I., eds. *The Terrestrial Biosphere and Global Change. Implications for Natural and Managed Ecosystems*, pp. 271–303. Cambridge University Press, Cambridge.

Schuepp, P.H., Leclerc, M.Y., Macpherson, J.I. & Desjardins, R.L. (1990) Footprint predictions of scalar fluxes from analytical solutions of the diffusion equation. *Boundary-Layer Meteorology* **50**, 355–73.

Schulze, E.-D. & Freer-Smith, P.H. (1991) An evaluation of forest decline based on field observations focussed on Norway spruce, *Picea abies*. In: Last, F.T. & Watling, R., eds. *Acidic Deposition: its Nature and Impacts*, pp. 155–68. Royal Society of Edinburgh, Edinburgh.

Schulze, E.-D., Kelliher, F.M., Körner, C., Lloyd, J. & Leuning, R. (1994) Relationships among maximum stomatal conductance, ecosystem surface conductance, carbon assimilation rate, and plant nitrogen nutrition: a global ecology scaling exercise. *Annual Review of Ecological Systems* **25**, 629–60.

Schulze, E.-D., Lloyd, J., Kelliher, F.M. *et al.* (1999) Productivity of forests in the Eurosiberian boreal region and their potential to act as a carbon sink: a synthesis. *Global Change Biology* **5**, 703–22.

Sellers, P.J., Hall, F.G., Margolis, H. *et al.* (1995) Boreal Ecosystem–Atmosphere Study (BOREAS): an overview and early results from the 1994 field year. *Bulletin of the American Meteorological Society* **76**, 1549–77.

Sellers, P.J., Hall, F.G., Kelly, R.D. *et al.* (1997) Boreas in 1997: scientific results, experimental overview and future directions. *Journal of Geophysical Research* **102**, 28731–70.

Sharkey, T.D. & Loreto, F. (1993) Water-stress, temperature and light effects on the capacity for isoprene emission and photosynthesis of Kudzu leaves. *Oecologia* **95**, 328–33.

Sheppard, L.J. (1991) Causal mechanisms by which sulphate, nitrate and acidity influence forest hardiness in red spruce: review and hypothesis. *New Phytologist* **127**, 69–82.

Sheppard, L.J., Cape, J.N. & Leith, I.D. (1993) Influence of acidic mist on frost hardiness and nutrient concentrations in red spruce seedlings. *New Phytologist* **124**, 595–605.

Shuttleworth, W.J. (1989) Micrometeorology of temperate and tropical forest. *Philosophical Transactions of the Royal Society of London B* **324**, 299–334.

Shuttleworth, W.J. & Calder, I. (1979) Has the Priestley–Taylor equation any relevance to forest evaporation? *Journal of Applied Meteorology* **18**, 639–46.

Shuttleworth, W.J., Gash, J.H.C., Lloyd, C.R. *et al.* (1984) Daily variations in temperature and humidity within and above Amazonian forest. *Weather* **40**, 102–8.

Sinun, W., Meng, W.W., Douglas, I. & Spencer, T. (1992) Throughfall, stemflow, overland flow and throughflow in the Ulu Segama rainforest, Sabah. Malaysia. *Philosophical Transactions of the Royal Society of London B* **335**, 389–95.

Skarby, L., Ro-Poulsen, H., Wellburn, F.A.M. & Sheppard, L.J. (1998) Impacts of ozone on forests: a European perspective. *New Phytologist* **139**, 109–22.

Skiba, U., Fowler, D. & Smith, K.A. (1997) Nitric oxide emissions from agricultural soils in temperate and tropical climates: sources, controls and mitigation options. *Nutrient Cycling in Agroecosystems* **48**, 139–53.

Slinn, W.G.N. (1982) Predictions for particle deposition to vegetative canopies. *Atmospheric Environment* **16**, 1785–94.

SNSF (1981) *Acid Precipitation: Effects on Forest and Fish*. Final report of the Norway SNSF-project 1972–80, SNSF, Oslo.

Spiecker, H., Mielikäinen, K., Köhl, M. & Skovsgaard, J. (eds) (1996). *Growth Trends in European Forests*. Springer-Verlag, Berlin.

Staudt, M. & Bertin, N. (1998) Light and temperature dependence of the emission of cyclic and acyclic monoterpenes from holm oak (*Quercus ilex*) leaves. *Plant, Cell and Environment* **21**, 385–95.

Steinbrecher, R., Hahn, J., Stahl, K. *et al.* (1997a) Investigations on emissions of low molecular weight compounds (C_2-C_{10}) from vegetation. In: Slanina, S., ed. *Biosphere–Atmosphere Exchange of Pollutants and Trace Substances*, pp. 342–51. Springer-Verlag, Berlin.

Steinbrecher, R., Ziegler, H., Fichstädter, C. *et al.* (1997b) Monoterpene and isoprene emissions in Norway spruce forests. In: Slanina, S., ed. *Biosphere–Atmosphere Exchange of Pollutants and Trace Substances*, pp. 352–65. Springer-Verlag, Berlin.

Stevenson, D.S., Johnson, C.E., Collins, W.J., Derwent, R.G., Shine, K.P. & Edwards, J.M. (1998) Evolution of tropospheric ozone radiative forcing. *Geophysical Research Letters* **25**, 3819–22.

Stewart, J.B. (1977) Evaporation from the wet canopy of a pine forest. *Water Resources Research* **13**, 915–21.

Stewart, J.B. & Thom, A.S. (1973) Energy budgets in pine forest. *Quarterly Journal of the Royal Meteorological Society* **99**, 154–70.

Sutton, M.A., Pitcairn, C.E.R. & Fowler, D. (1993) The exchange of ammonia between the atmosphere and plant communities. *Advances in Ecological Research* **24**, 301–93.

Sutton, M.A., Schjørring, J.K. & Wyers, G.P. (1995) Plant–atmosphere exchange of ammonia. *Philosophical Transactions of the Royal Society of London A* **351**, 261–78.

Sutton, O.G. (1953). *Micrometeorology*. McGraw-Hill, New York.

Swinbank, W.C. (1951) The measurement of the vertical transfer of heat. *Journal of Meteorology* **8**, 135–45.

Tamm, C.O. (1991) *Nitrogen in Terrestrial Ecosystems: Questions of Productivity, Vegetational Changes and Ecosystem Stability*. Springer-Verlag, Berlin.

Taylor, C.M.A. & Tabbush, P.M. (1990). *Nitrogen Deficiency in Sitka Spruce Plantations*. Forestry Commission Bulletin 89, HMSO, London.

Teklehaimanot, Z. & Jarvis, P.G. (1991) Direct measurement of evaporation of intercepted water from forest canopies. *Journal of Applied Ecology* **28**, 603–18.

Teklehaimanot, Z., Jarvis, P.G. & Ledger, D.C. (1991) Rainfall interception and boundary layer conductance in relation to tree spacing. *Journal of Hydrology* **123**, 261–78.

Thoene, B., Schröder, P., Papen, H., Egger, A. & Rennenberg, H. (1991) Absorption of atmospheric NO_2 by spruce (*Picea abies* L. Karts.) trees. 1. NO_2 influx and its correlation with nitrate reduction. *New Phytologist* **117**, 575–85.

Thom, A.S. (1975) Momentum, mass and heat exchange of plant communities. In: Monteith, J.L., ed. *Vegetation and the Atmosphere*, Vol. 1, pp. 57–109. Academic Press, London.

Thom, A.S., Stewart, J.B., Oliver, H.R. & Gash, J.H.C. (1975) Comparison of aerodynamic and energy budget estimates of fluxes over a pine forest. *Quarterly Journal of the Royal Meteorological Society* **101**, 93–105.

Tian, H., Melillo, J.M., Kicklighter, D.W. *et al.* (1998) Effect of interannual climate variability on carbon storage in Amazonian ecosystems. *Nature* **396**, 664–7.

Ulrich, B., Mayer, R. & Khann, P.K. (1980) Chemical changes due to acid precipitation in a loess-derived soil in central Europe. *Soil Science* **130**, 193–9.

Unsworth, M.H. & Fowler, D. (eds) (1988). *Acid Deposition at High Elevation Sites*. Kluwer Academic, Dordrecht.

Valente, F., David, J.S. & Gash, J.H.C. (1997) Modelling interception loss for two sparse eucalypt and pine forests in central Portugal using reformulated Rutter and Gash analytical models. *Journal of Hydrology* **190**, 141–62.

Valentini, R., De Angelis, P., Matteucci, G., Monaco, R., Dore, S. & Scarascia Mugnozza, G.E. (1996) Seasonal net carbon dioxide exchange of a beech forest with the atmosphere. *Global Change Biology* **2**, 199–207.

Valentini, R., Matteucci, G., Dolman, A.J. *et al.* (2000) Respiration as the main determinant of carbon balance in European forests. *Nature* **404**, 861–5.

Vitousek, P.M. (1994) Beyond global warming: ecology and global change. *Ecology* **75**, 1861–76.

Voltz, A. & Kley, D. (1988) Evaluation of the Montsouris series of ozone measurements made in the nineteenth century. *Nature* **332**, 240–2.

Walker, B.H., Steffen, W. & Langridge, J. (1999) Interactive and integrated effects of global change on terrestrial ecosystems. In: Walker, B.H., Steffen, W.L., Canadell, J. & Ingram, J.S.I., eds. *The Terrestrial Biosphere and Global Change: Implications for Natural and Managed Ecosystems*, pp. 329–75. Cambridge University Press, Cambridge.

Wang, Y.P., Jarvis, P.G. & Taylor, C.M.A. (1991) PAR absorption and its relation to above-ground dry matter production of Sitka spruce. *Journal of Applied Ecology* **28**, 547–60.

Waraghai, A. & Gravenhorst, G. (1989) Dry deposition of atmopheric particles to an old spruce stand. In: Georgii, H.-W., ed. *Mechanisms and Effects of Pollutant Transfer into Forests*, pp. 77–96. Kluwer Academic Publishers, Dordrecht.

Wayne, R.P. (1991). *Chemistry of the Atmospheres*. Clarendon Press, Oxford.

Weatherley, P.E. (1970) Some aspects of water relations. *Advances in Botanical Research* **3**, 171–206.

Whitehead, D. & Jarvis, P.G. (1981) Coniferous forests and plantations. In: Kozlowski, T.T., ed. *Water Deficits and Plant Growth VI*, pp. 49–152. Academic Press, London.

Whitehead, D. & Kelliher, F.M. (1991) Modeling the water balance of a small *Pinus radiata* catchment. *Tree Physiology* **9**, 17–33.

Whitehead, D., Kelliher, F.M., Lane, P.M. & Pollock, D.S. (1994) Seasonal partitioning of evaporation between trees and understory in a widely spaced *Pinus radiata* stand. *Journal of Applied Ecology* **31**, 528–42.

Williams, M., Rastetter, E.B., Fernades, D.N. *et al.* (1996) Modelling the soil–plant–atmosphere continuum in a *Quercus Acer* stand at Harvard forest: the regulation of stomatal conductance by light, nitrogen and soil/plant hydraulic properties. *Plant, Cell and Environment* **19**, 911–27.

Williams, M., Malhi, Y., Nobre, A.D., Rastetter, E.B., Grace, J. & Pereira, M.G.P. (1998) Seasonal variation in net carbon exchange and evapotranspiration in a Brazilian rain forest: a modelling analysis. *Plant, Cell and Environment* **21**, 953–68.

Wyers, G.P. & Erisman, J.W. (1998) Ammonia exchange over coniferous forest. *Atmospheric Environment* **32**, 441–52.

Yienger, J.J. & Levy, I.I.H. (1996) Empirical model of the global soil-biogenic NO_x emissions. *Journal of Geophysical Research* **100**, 11447–64.

Zoltai, S.C. & Martikainen, P.J. (1996) Estimated extent of forest peatlands and their role in the global C cycle. In: Apps, M.J. & Price, D.T., eds. *Forest Ecosystems, Forest Management and the Global C Cycle*, Chapter 4. Springer-Verlag, Heidelberg.

11: Environmental Stresses to Forests

PETER FREER-SMITH

11.1 INTRODUCTION

The physical environment determines growth rates, the survival of individual trees and thus the type, structure and extent of forest ecosystems. For managed forest stands, we can manipulate some of the abiotic factors that control growth rate. In plantations, we also specifically control species composition, spacing and other factors. At its simplest, environmental stress can be considered to occur when one or more environmental factors are present at levels that lie outside the range tolerated by an individual species or forest stand. Where tree growth has been checked or dieback or decline are occurring we are seeing the effects of environmental stress, that is unless fungal pathogens or insect pests are a clear primary cause of damage. However, foresters and forest ecologists are also concerned when abiotic factors, particularly anthropogenic factors, are at values that, although not apparently detrimental, may be suboptimal for growth and biodiversity of the forest ecosystem. Worries over the slow imperceptible deterioration of forest ecosystems and of possible crises led to consideration of forest 'stability' even before 'sustainability' came to prominence as a concept after the United Nations Conference on Environment and Development in Rio de Janeiro in 1992. Furthermore, experience has shown that there are few forests where temporary or chronic stress does not occur due to limitations of water or nutrient availability and commonly both.

Forest managers require an understanding of the ecological processes that occur in order to be confident that their intervention will achieve the desired objectives. Stress occurs in forests but how detrimental is it to growth, sustainability, conservation, recreational value and other benefits? What actions can be taken to minimize adverse impacts, particularly those resulting from anthropogenic factors? Indeed, most forest ecosystems are influenced by human activities, in many cases severely, and remaining areas of 'pristine' forest are small. The largest areas of relatively undisturbed forest occur in the boreal regions, although even here logging and mining affect a significant area. Many tropical forests are generally severely disturbed and the effects of pollutants are clear in many temperate forests and also in the Pacific Rim. Finally elevated CO_2 concentrations and climate change will have a widespread effect. Thus human influence is a major factor.

Since we are concerned here with the response of forests to environmental stresses and these responses are not simple, it is necessary to give brief consideration to definitions. Levitt (1972, 1980) derived a definition of biological stress and strain from physical science. Physical stress is force applied to an object and strain is the change in that object caused by this stress. Levitt suggested that biological stress is the change in environmental conditions that reduce or adversely affect plant growth or development, biological strain being the reduction of, or change in, function. Levitt also distinguished elastic and plastic biological strain on the basis of whether plant function returned to its normal level when conditions returned to initial conditions (elastic) or did not (plastic). Interestingly, Levitt's definition applies the term 'stress' to factors that result in adverse effects alone. In forest ecosystems a change of environmental conditions may have adverse effects on one species but beneficial impacts for a second, i.e. species composition may alter. The effect or 'strain' may be succession or evolutionary adaptation. Biological systems are complex and the physical analogy has clear limitations. Perhaps more usefully, Levitt also distinguished 'stress avoidance' from 'stress tolerance', stress avoidance being typified by responses like exclusion of toxic metals or pollutants. An example might be closure of stomata to exclude the entry of gaseous air pollutant. In contrast, stress tolerance is exemplified by a plant that allows uptake into the substomatal cavity and responds by chemical detoxification mechanisms. It is worth noting at the outset that the detailed interactions between the plant and the environment will differ for trees and forests compared with herbaceous plants. This is because of the size, longevity and anatomy (i.e the presence of the tall woody stem) in trees.

In this chapter I have interpreted Levitt's 'environmental conditions' as meaning changes to the physical conditions experienced by plants. Biotic factors, such as pests, pathogens, genotypes (species and provenance) and grazing by animals, are clearly important to forest ecosystems and their effects are known to interact with those of the key abiotic factors; however, they have not been my primary concern here. Plant growth is influenced fundamentally by availability of nutrients, light and water and, at the practical level, by climatic and soil conditions, which determine the availability of resources (nutrients, light and water) and also influence forests directly. In plantation forestry, the intention is to select species and provenance suited to the particular conditions of a site and/or to manage conditions of the site so that they favour the selected species. Stress should not be present or at least can be considered to be present at low enough levels that it is 'tolerated' by the system, to the extent that there are not changes in the relative competitive ability of species leading to species decline or creation of new niches that enable new species to gain a foothold. In unmanaged forest ecosystems, plant competition and, in the long term, natural selection will control species composition. Disturbance by wind, fire and, in some instances such as *Metrosideros* forests of Hawaii (Hüttl & Mueller-Dembois 1992), nutrient limitations are a

normal part of succession in unmanaged forest ecosystems. However, after the climax high forest has been established, species composition and ecosystem structure should be such that they are suited to the prevailing environmental conditions. Undisturbed first-growth forests have evolved over millennia and are considered to be stable; they are well suited to the environmental conditions in which they grow except perhaps for species at the edge of their natural distribution, or where occasional infrequent extreme events occur which damage or even kill trees that have become established in benign intervening years. Ecosystem productivity is still limited by the availability of light, water and nutrients or temperature, but changes in these factors do not elicit ecosystem responses. Thus, Levitt's definition helps us in that it draws attention to *change* in environmental conditions. Of the abiotic factors that influence plant growth, light is perhaps the least likely to be changing at the ecosystem level. Competition for light is an important factor within forest ecosystems. However, except in the equatorial zone where daylength does not vary seasonally, daylength and light quality are important mechanisms by which plants respond to season; tree response to photoperiod is thus of major importance when introducing trees to new latitudes. However, in the sections that follow I have chosen to focus on climate, pollution, water deficit (drought), and soil conditions, nutrient deficiency/enrichment and soil degradation as factors which are now subject to change to the extent that forest ecosystems could experience stress on a worldwide basis. On the time-scales with which we are concerned, humans are the major cause of these changes.

11.2 CLIMATE

Climate is the dominant control factor in the distribution of the world's main vegetation types or biomes. Biomes are defined as a distinctive ecological system, characterized primarily by the nature of their vegetation (Houghton 1997). On a global scale and in a historical and geological time frame, response to climatic stress is of overriding importance. The relevance of understanding tree and forest ecosystem response to climatic stress has increased because we are currently experiencing an unprecedented change of global climate which is, at least in part, the result of anthropogenic factors.

Most contemporary texts on the influence of climate and vegetation date our understanding of the relationship between climate and the distribution of vegetation from the work of Von Humboldt at the beginning of the nineteenth century (Woodward 1987; Landsberg & Gower 1997). Holdridge (1947 and subsequent work) demonstrated a strong correlation between the main groups of life form or biomes (e.g. deserts, grasslands, deciduous or coniferous forests) and two major climatic characteristics, temperature and water availability. These climatic factors are the main influence at the biome scale, while soil type (nutrient availability) topography and anthropogenic factors are major influences on vegetation patterns within biomes on a smaller local scale.

It is of considerable interest and relevance that where temperature and moisture availability or moisture budget (defined as precipitation–evapotranspiration) are similar, the same tree species or similar species of one tree genera are present or prove to be satisfactory forestry species if introduced. This reflects the relationship between climate and suitable physiology and anatomy, and also the pattern of species' evolution and migration. The timing of the separation of the earth's present-day continents relative to the evolutionary history of tree species and taxa determines presence or absence of particular species within suitable climatic zones. Eventually the land bridges between the continents became impassable to tree species during the period of their rapid evolution in the Cretaceous Era (Daubenmire 1978). In the absence of human influence, the geographical extent (natural distribution) of the various forest biomes is thus determined mainly by climatic conditions and the natural species composition will be further influenced by evolutionary and geological factors.

There are several critically important temperature thresholds that affect physiological processes and thus define the environmental limits of the major forest biomes (Woodward 1987). The first threshold is a critical minimum temperature in the range +12 to 1°C, which results in chilling injury to plant membranes (Larcher & Bauer 1981). The next critical threshold is at about –15°C and roughly corresponds to the temperature at which frost damage occurs to leaves and buds (depending on the degree of frost acclimation). The leaves of broadleaved species with a damage threshold at about –15°C are less resistant to frost than coniferous needles (Woodward 1987). The buds of broadleaved and coniferous species have similar minimum temperature thresholds but the xylem of broadleaved trees (angiosperms) is more susceptible to ice formation than that of conifers. This is because coniferous xylem is made up of tracheids, which have a smaller lumen diameter than the vessels that make up the xylem in broadleaved trees. Similarly broadleaved species with diffuse-porous xylem, which is of smaller diameter than ring-porous xylem, are more resistant to ice-induced xylem damage. Broadleaved trees with ring-porous xylem are notably absent from boreal forests.

The next critical temperature threshold is at –39 to –40°C, below which only a few boreal tree species can survive (e.g. *Picea glauca*, *Larix laricina* and *Larix sibirica*) because this temperature will freeze the xylem and buds. At their extreme northern limits, these cold-tolerant species often have a parenchyma pith cavity beneath the crown of the primordial bud that prevents ice formation from spreading from the xylem to the bud (Richards & Bliss 1986).

It has been argued that in harsh climates, particularly low temperatures, the evergreen habit gives a more favourable carbon balance (Schulze *et al.* 1997) and thus we see the dominance of coniferous evergreens in cold montane and boreal forests. Climates with ample rainfall, well distributed throughout the year and with rare frosts, as experienced in temperate regions (30–55° of latitude), also favour evergreen rather than deciduous forests. This is considered to have led to the dominance of broadleaved evergreen forests in the temperate regions of the Southern Hemisphere, with the notable exception of southern Chile where deciduous broadleaved forests occur. Broadleaved deciduous tree species predominate in the northern temperate regions.

The distribution of the various forest biomes is thus determined by climatic conditions. However within the major biomes individual species gradually decrease in numbers near their altitudinal or latitudinal limits. Woodward (1987, 1995) attributed this more to increasingly non-competitive carbon balance rather than to direct damage from low temperatures. Although the tree species present within the various forest biomes should not normally be experiencing climatic stress, climate remains of major importance to forestry and to the conservation of forest ecosystems for three main reasons. First, within a climatic region foresters are able to select species that grow better (faster) or provide other desirable benefits for particular sites, as influenced by local conditions of topography and soils. Second, the effect of climate on tree species limits the use of introduced or exotic species, which otherwise has proved a very successful strategy for improving yields in commercial forestry worldwide. Third, and perhaps of greatest interest, is that climate changes, not only naturally on a geological time-scale but also rapidly as a result of human activity, i.e. global climate change.

Landsberg and Gower (1997) followed Melillo *et al.* (1993) in adopting a classification of forest biomes based on major climatic region (tropical, temperate and boreal) and tree physiognomy (broadleaved, evergreen, broadleaved deciduous and needle-leaved evergreen conifer). Seven forest biomes were identified as follows: boreal, temperate deciduous, temperate coniferous, temperate mixed, temperate broadleaved evergreen forest, tropical evergreen and tropical deciduous. Of these, the boreal and tropical evergreens cover the greatest area. Forest biomes are covered in detail in Chapter 2.

The increase of global surface land and ocean temperatures that has been observed during the last 100 years (see Chapter 10) is faster than that observed during the geological past. Predictions of future climate indicate that some tree species, and some parts of the forest biomes, will experience climates outside those of their natural range. Over significant areas climatic stress will occur and dieback is likely. A number of authors have used existing relationships (correlations) between climatic factors and biome distribution to predict the future distribution of terrestrial biomes from the climate predictions provided by general circu-

lation models (e.g. Woodward 1987; Landsberg & Gower 1997). Changes in biome distribution might be expected to occur by migration, although migration at rates of 5 km annually will be required and palaeoecological records indicate postglacial migration rates of tree species of only 40–150 m annually. Thus migration rates are likely to be insufficient, resulting in significant areas of stress and dieback, or rather, that stress and dieback will be part of the process of migration. A recent example of such predictions is that of Friend *et al.* (1997) who mapped the global distribution of eight biomes and of desert for existing (today's) climate and for climate predicted for the year 2080. This model, called 'HYBRID', predicts vegetation type at any particular location on the basis of the outcome of competition for light, nitrogen and water and of the plants' ability to survive extreme conditions. Unlike earlier, less sophisticated work of this type, HYBRID predicts that tropical forests will decline in area, with dieback beginning in the 2050s as a result of decreased rainfall and increased temperatures. Temperate and mixed forests are likely to move northwards and increase their total area. Boreal forest area is predicted to change very little provided that these forests can establish in more northern areas with shallow poorly developed soils (Plate 1, facing p. 298). The palaeoecological records show that the species composition of forest ecosystems varied during their postglacial advance so we could also predict some changes of species composition as a result of the stress caused by climate change.

These predictions suggest that forest managers will need to plan for, and respond to, climate change if extensive areas of forest dieback are to be avoided; some such problems are likely to occur in any event. Indeed, there is firm evidence that climatic conditions have been a major, or perhaps *the* major, factor responsible for forest dieback in Europe and the north-eastern USA (Johnson *et al.* 1994). Changes of species choice will be an obvious course of action in plantation forestry. The successful use of 'exotics' in commercial forestry, for example Sitka spruce (*Picea sitchensis*) in western Europe, *Pinus radiata* in New Zealand and Chile, and *Eucalyptus* in Portugal and South America, indicates that forest managers can, in any event, achieve increased yields by planting species outside their current natural distribution. Species choice and the use of the 'right tree for the site' has long been recognized as a very effective mechanism of avoiding stress to plantations and of optimizing yield (Anderson 1950). Various systems of biogeoclimatic ecosystem classification (BEC) have recently been adopted in different parts of the world (Pojar *et al.* 1987; Pyatt & Suárez 1997). Essentially such systems provide site classification based on our understanding of the influence of environmental factors on forest ecosystems and thus allow sound management of plantation and seminatural stands for timber production, conservation or other benefits. The basis of BEC, and the UK equivalent, ecological site classification (ESC), is consistent with our understanding of the climatic control of biome distribution and the influence of other factors, particularly nutrient availability (but also topography and windiness), on species composition within the major biomes. Sites are classed on the basis of three principal factors considered to control site quality: climate, soil moisture regime (on which climate exerts a significant influence) and soil nutrient regime. Choice of species is arguably the most important management and silvicultural decision. ESC provides a list of species suited to the climatic and soil qualities and capable of producing good-quality timber. Tree species are listed as *optimal* if they will grow at the upper rate given by models of general yield class (GYC) without undue risk of disease and pest attack. Species are rated as *suitable* if they will grow on the lower part of GYC curves. Where ESC predicts increased risk of injury from pests, diseases or drought species are considered *unsuitable*. Predictions of GYC can be made for a particular species.

Growth rates and productivity are influenced by climatic stress, although current indications are of increased growth prior to dieback, perhaps even prior to catastrophic decline. The important study by Spiecker *et al* (1996) indicated that the growth rates and timber yields of European forests are currently increasing relative to those achieved in earlier rotations. This is probably the result of better silviculture and planting stock, the use of fertilizers and suitable thinning regimes.

However, the effects of warmer weather, nitrogen depositions and elevated CO_2 have almost certainly also contributed to this effect. Perhaps hard to believe is the prediction of the HYBRID model that the potential biomass between latitudes 30° and 60° N will increase by 70% by the year 2080 *cet. par.*

Biogeoclimate classifications can be developed further to provide advice on other aspects of management, in particular risk from windthrow, ground preparation, regeneration techniques and management of open space. Currently, historical climatic data (weather records) are used to 'paramaterize' these systems of site classification; however, consideration of rotation length relative to the predicted rates of climate change indicates that the use of predicted climate of the site might provide more accurate advice on species choice and future management. Over a number of years, the evidence from average global surface temperatures has become increasingly clear and 1997 was the hottest year on record to date. Average global temperatures have risen by 0.6°C in the last 130 years and general circulation models are now able to predict the impact of increasing greenhouse gas concentrations on global temperatures. It is now accepted that anthropogenic climate change is occurring, with the UN Intergovernmental Panel on Climate Change (IPCC) concluding in its 1995 report that 'the balance of evidence suggests a discernible human influence on global climate' (Houghton *et al.* 1995). To understand the potential impacts of climate change on forests it is essential to have accurate predictions of future climate at the regional scale. Internationally, predictions or scenarios have been provided by the IPCC; within the UK, they have been provided by the Meteorological Office based at the Hadley Centre, Reading, at a spatial resolution of 10 min (i.e. 10 km grid squares). Mean winter temperatures and average precipitation (mm daily) are available for the 1961–90 baseline period and for the years 2025, 2050 and 2100. Relatively small changes in temperature (+1.0–1.8°C) and rainfall (−9% in summer to +10% in winter) will lead to much larger changes in potential evapotranspiration (40% increase predicted for southern England by 2050), while the annual distribution of rainfall and the occurrence of storm events may also change. Projected temperature increases should therefore not be viewed in isolation.

The changes predicted for the UK would have very important implications for forestry. The occurrence of late spring frost is already a problem in forestry and restricts the use of some tree species and provenances (Murray *et al.* 1995). Similarly, increased windiness and particularly storms could have devastating effects (e.g. the 1987 storm in southern England). Two important features in forestry are that a number of introduced species are used because they grow rapidly over short rotation lengths and, secondly, that native species have ranges which span considerable climate differences. Seed origins (provenance) are used to select the most appropriate genotype to maximize growth potential. Therefore, sourcing seeds from different origins is a possible proactive response to climate change: for example for Sitka spruce it is suggested that more Oregon or Washington provenances should be planted in place of those from the more northerly Queen Charlotte Islands. Similarly, species choice could be adjusted, with species such as *Eucalyptus* and *Nothofagus*, natives of regions in warmer climates, becoming suitable for Europe (for further discussion see Broadmeadow & Freer-Smith 1998).

In addition to temperature and extremes of climate (unusually late spring frosts, storms and drought), there are a number of other environmental factors that are changing as a result of human activities. These include UV-B radiation and air pollutant concentrations and depositions Such factors and their interacting effects on trees and forest insects and pathogens are likely to have significant effects on tree growth and forest productivity.

11.3 POLLUTION

Table 11.1 lists the eight types of pollution or specific pollutants of greatest concern globally. During the last 20 years it has become clear that vegetation, soils and the aquatic environment are more important sinks than the atmosphere for many pollutants. Pollutants are transferred from the atmosphere to the terrestrial environment by three processes.

Table 11.1 The eight types of pollution or specific pollutants of greatest concern globally and the World Health Organization health guidelines (where set) and the UK Expert Panel on Air Quality standard guidelines for hydrocarbons and particulates.

WHO health guidelines	
Sulphur dioxide (SO_2) (smoke)	10-min mean, 175 ppb 1-h mean, 122 ppb
Ozone (O_3)	1-h mean, 76–100 ppb 8-h mean, 50–60 ppb
Carbon monoxide (CO)	1-h mean, 25 ppm 8-h mean, 10 ppm
Nitrogen dioxide (NO_2)	1-h mean, 110 ppb Annual average, 20–30 ppb

UK Expert Panel on Air Quality standard guidelines	
Hydrocarbons, e.g. benzene, 1,3-butadiene, toluene and xylene	5 ppb as a rolling average
Particulates (PM_{10})	$50 \, \mu g \, m^{-3}$

1 Dry deposition: the movement of particles and gases by molecular diffusion and turbulent transfer to wet and dry surfaces.
2 Wet deposition: the movement of particles and dissolved gases in rainfall.
3 Occult deposition: the movement of particles and dissolved gases in cloud water, fog and mist.

For particles (dust, smoke and smaller material), impaction and gravitational deposition (sedimentation) are also important processes. Additionally, tree roots show preferential and active uptake of some pollutants. Trees and forests are more effective at removing pollutants from the atmosphere because of their high leaf areas and because their height increases turbulent mixing, enhancing the deposition of those pollutants for which surface resistance does not limit deposition (i.e. SO_2, O_3, NO and NO_2). Pollutant flux by wet deposition is largely unaffected by land use, while uptake by forests will be greater via both dry and occult deposition and for impaction of particles (for further details see Chapter 10). Our principal concern here is how pollutants affect trees and forest ecosystems and how foresters may be able to respond to adverse effects.

Method of exposure and exposure concentration of the various pollutants listed in Table 11.1 are important in determining their mode of action and phytotoxicity, hence the relevance of deposition mechanisms. It is helpful to distinguish direct effects, resulting from the entry of the pollutant itself into the tree, from indirect effects, usually considered to be principally the result of alterations to soil quality caused by the presence of pollution. It is logical to include pollutant effects on any other component of the ecosystem (e.g. damage to mycorrhizal fungi or insect pests and fungal pathogens) as indirect effects. Possible pathways of exposure and uptake, as well as physiological effects, metabolic impacts and the possible outcome of both direct and indirect effects, are presented in Figs 11.1–11.4.

11.3.1 Direct effects

Taylor and Constable (1994) have provided estimates of the proportions of various pollutants that enter vegetation through stomata or are deposited on the leaf surface. Pollutants can damage foliage directly when deposited on leaf surfaces, the best-established impacts being surface lesions caused by highly acidic rain and the erosion of epicuticular wax structure. However, most pollutants deposited here are likely to be washed down the stem to the soil where they contribute to soil-mediated effects or are lost in water flow (surface run-off or to ground water). Clearly the potential for pollutants to have detrimental effects on plant metabolism and physiology are strong once they have entered leaves via stomata. For a number of the major pollutants, pathways through which metabolism is damaged are now well established (Wellburn 1994). Essentially, the reactions of photosynthesis, respiration, energy distribution, storage and biosynthesis are inhibited by the presence of dissolved pollutants or their dissociation products. In a number of cases, the control of pH within cell compartments is disrupted or damage occurs to cell membranes resulting in leakage of metabolites and cell death.

11.3.2 Indirect effects and soil acidification

The indirect effects of wet deposited pollutants

are illustrated on the right-hand side of Figs 11.1, 11.2 and 11.4, and also in the root region of the diagrams. Effects on epicuticular wax, litter decomposition in the upper soil horizon and cation leaching from foliage are all potentially damaging. However, for both sulphur and nitrogen pollution soil-mediated effects are probably of overriding importance. The importance of soil processes and, in particular, the concept of the 'mobile anion effect' and associated 'aluminium' toxicity were developed by Ulrich and his coworkers in the early 1980s (Ulrich 1989). It is hard to underestimate the importance of this work: it formed the conceptual basis for soil critical load calculations now used to focus scientific understanding of pollutant impacts and to guide agreed international deposition and emission targets.

In brief, the mobile anion effect suggests that soil solution pH is usually buffered or controlled by the exchange of protons (H^+) with exchangeable base cations (Ca^{2+}, Mg^{2+}, Na^+ and K^+); these are normal cation exchange processes. If strong acids (H_2SO_4) enter the system, sulphate (SO_4^{2-}) and base cations will be present in soil solution and will be lost in soil drainage water. However, if the inputs of strong acid anions (sulphate and perhaps nitrates) persist, soil acidification continues until buffering of pH in soil solution is no longer via cation exchange mechanisms but rather by exchange with aluminium ions (Al^{3+}) in the $Al(OH)_3$ pool in soil solution (Al buffering range when pH < 4.2). Soils then acidify (H^+ concentrations increase), Al concentrations in soil solution increase and percentage base saturation decreases with loss of Mg^{2+}, Ca^{2+} and Na^+ along with SO_4^{2-} in soil leachate. In even more acid conditions there is a further Al/Fe buffer range (pH < 3.8). The resulting free Al^{3+} ions are toxic to roots and the loss of base cations and overall change is destabilizing to soil nutrient cycling. Full soil chemistry models are very complex, but an insight into the fundamental process is important since it informs understanding of soil status and of suitable management. There is now a very large body of data to support these ideas and they can be regarded as established science.

It is now clear that the atmospheric deposition of sulphur and nitrogen have a strong influence on

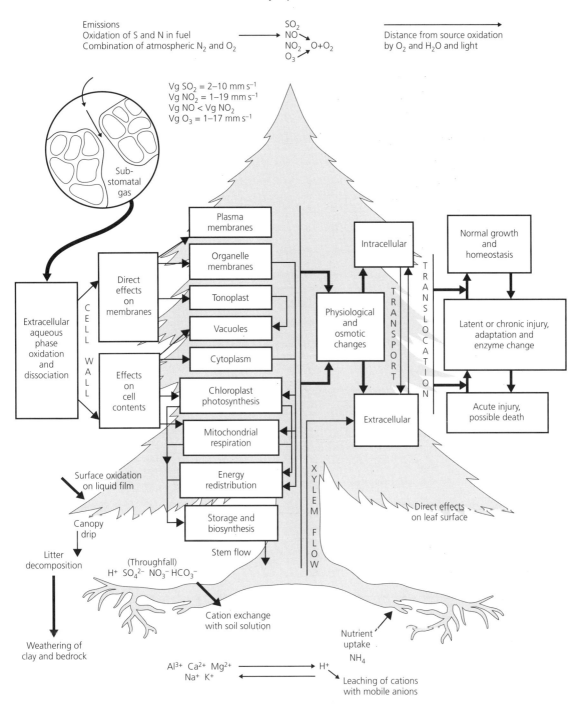

Fig. 11.1 Schematic representation of the exposure, uptake, physiological effects, metabolic impacts and possible outcome of exposure of trees to gaseous air pollutants. See text for further explanation.

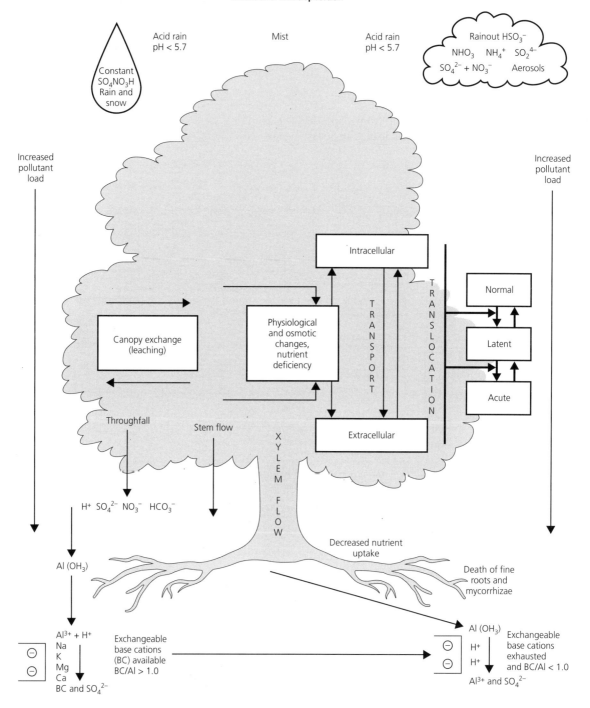

Fig. 11.2 Schematic representation of the exposure, uptake, physiological effects, metabolic impacts and possible outcome of exposure of trees to soil acidification. See text for further explanation.

the status of soils and on forest nutrition (Schulze *et al.* 1989; Zottl & Huttl 1991). Sulphur and nitrogen depositions and forest nutrition are strongly associated globally, although depositions and soil sensitivity to damage vary geographically. Recently, policy on emission abatements has been guided by the critical loads approach rather than, as previously, by an ill-defined concept of protecting ecosystems from damage. For forest ecosystems, soil critical loads can be set using a mass balance equation such that deposition of strong acid anions does not exceed the rate of release of base cations by weathering of minerals (after allowing for cation loss from soil in nutritional uptake of cations and nitrogen) and an allowable leaching of alkalinity (Sverdrup & De Vries 1994). In acidified soils low availability and uptake of base cations may limit tree growth, and the ratio between the divalent base cations (Ca and Mg) and Al in soil solution reaches a critical value (a molar ratio of 1.0). Below this value nutrient uptake is suboptimal, root damage can occur from free Al in soil solution, tree growth is reduced and visible forest damage may start to occur. The mass balance equation can be used to calculate S and N depositions (critical loads) below which (Ca+Mg)/Al is greater than 1.0 so that forest soils are protected from 'damage'. Information is now becoming available on S and N critical loads throughout Europe and initial work is being done on the soils of South-East Asia where major problems may occur over the next few years (Freer-Smith 1998). The above summary is sufficient to indicate the strong link between acidic deposition, forest nutritional status and the ability of forest soils to sustain successive rotations without loss of their capacity to supply base cations and nitrogen. Evaluation of the status of forest soils and the consideration of suitable management options (e.g. liming) becomes possible when these mechanisms are understood.

11.3.3 Atmospheric concentrations of carbon dioxide

Increases in atmospheric concentrations of CO_2 are probably the primary driving force of the anthropogenic greenhouse effect and thus of climate change; however, the atmospheric concentration of CO_2 also directly affects plant growth. Atmospheric CO_2 is the source of the plant's carbon and its active uptake in photosynthesis depends on diffusive flux across a concentration gradient. Trees grow better in elevated CO_2 compared with preindustrial ambient values (≈ 240 ppm) when other environmental factors, particularly nutrient and water availability, are not limiting. However, when these factors are below optimum, trees suffer from increased stress in elevated CO_2. Furthermore, CO_2 concentrations directly affect many plant processes other than photosynthesis, including respiration, stomatal conductance, ratio of root to shoot mass, rates of leaf emergence, leaf area index (ratio of leaf area to ground area), growth, senescence and tree phenology. The potential for the effects of elevated CO_2 concentrations to interact with those of other factors is considerable. Experimental evidence from controlled exposures is now showing that such interactions occur and can result in enhanced susceptibility to a number of abiotic stresses, particularly drought, nutrient deficiency, frost damage and damage from other air pollutants (Chappelka & Freer-Smith 1994).

Although the growth responses of trees to elevated CO_2 are more often positive than negative, they vary greatly with species (Ceulemans & Mousseau 1994) and the assumption that the overall effects of the current rise in CO_2 concentrations will be beneficial to forestry is an oversimplification that cannot be justified by detailed consideration of the evidence. The data reported by Broadmeadow and Freer-Smith (1999) are representative of similar interactions reported in the literature. Oak (*Quercus petraea*) showed a positive growth response to twice current ambient CO_2 concentrations (700 ppm); moreover, the detrimental effects of realistic peaks of ozone (O_3), i.e. a diurnal cycle peaking at 80 ppb, were lost when *Q. petraea* was grown in 700 ppm CO_2. Two other European species (ash, *Fraxinus excelsior*, and Scots pine, *Pinus sylvestris*) showed no detrimental effects of O_3 exposure and smaller stimulations of growth in 700 ppm CO_2. As has been reported for most trees exposed to realistic doses of these pollutants, the stimulation of growth in oak was the result of increased photo-

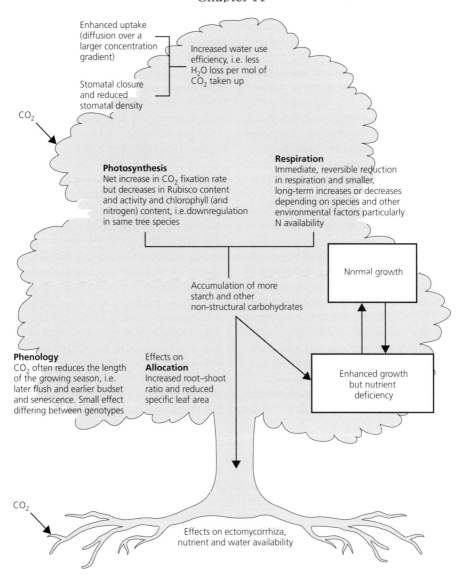

Enhanced uptake (diffusion over a larger concentration gradient)

Stomatal closure and reduced stomatal density

Increased water use efficiency, i.e. less H_2O loss per mol of CO_2 taken up

CO_2

Photosynthesis
Net increase in CO_2 fixation rate but decreases in Rubisco content and activity and chlorophyll (and nitrogen) content, i.e.downregulation in same tree species

Respiration
Immediate, reversible reduction in respiration and smaller, long-term increases or decreases depending on species and other environmental factors particularly N availability

Normal growth

Accumulation of more starch and other non-structural carbohydrates

Phenology
CO_2 often reduces the length of the growing season, i.e. later flush and earlier budset and senescence. Small effect differing between genotypes

Effects on **Allocation**
Increased root–shoot ratio and reduced specific leaf area

Enhanced growth but nutrient deficiency

CO_2

Effects on ectomycorrhiza, nutrient and water availability

Fig. 11.3 Schematic representation of the exposure, uptake, physiological effects, metabolic impacts and possible outcome of exposure of trees to elevated CO_2 concentrations. Rubisco, ribulose 1,5-bisphosphate carboxylase oxygenase. See text for further explanation.

synthesis, while the chronic growth reduction due to O_3 resulted from reduced photosynthetic capacity. A perhaps unique feature of this particular experiment was that soil-rooted (rather than pot-rooted) plants were also grown with two different soil water regimes in all combinations of CO_2 and O_3 (droughted and irrigated). The detrimental effect of O_3 on photosynthesis was reversed by elevated CO_2 but only for plants experiencing drought. Protection by CO_2 from O_3 probably results from the increased internal leaf CO_2 concentrations, closing stomata and limiting O_3 uptake (Stitt 1991); a classic stress avoidance mechanism in Levitt terminology. In well-watered trees, elevated CO_2 failed to decrease stomatal conductance and thus failed to provide protection from O_3. Unfortunately, tree responses to CO_2 and the interacting effects of different abiotic factors are even more complicated in forest stands.

Figure 11.3 illustrates the effects of elevated CO_2 concentrations on trees; as for the other figures, the illustration is greatly simplified. A comprehensive consideration of the biochemical and physiological effects of CO_2 is beyond the scope of this chapter. This overview is based on the recent compilation of results on elevated CO_2 effects on trees resulting from the investigation by the European Union (EU) ECOCRAFT project (Jarvis 1998). Within this programme, comprehensive studies of eight or more different tree species were conducted using a number of different techniques. The key processes affected were stomatal conductance (both stomatal opening and stomatal density), photosynthesis (ribulose 1,5-bisphosphate carboxylase oxygenase (Rubisco) content and activity, chlorophyll content and the light reactions), respiration (although CO_2 effects seem variable and small), phenology (particularly where nutrient limitations towards the end of the growing season give early bud set and senescence) and allocation. Compiled as a single volume, the studies conducted within ECOCRAFT present a coherent picture of increased growth in elevated CO_2 and of the very major way in which nutrient availability and water deficit can modify this response. The growth increases in elevated CO_2 in all the experiments conducted varied between 5% (Sitka spruce) and 58% (beech). In forest ecosystems water and nutrient availability are closely linked and experiments have also been conducted on the three-way interactions of elevated CO_2 concentrations, nutrient availability and drought on tree growth (Heath & Kerstiens 1997). These interactions are of major significance in determining stress response and forest condition.

11.3.4 Nitrogen depositions

Nitrogen is important as a plant nutrient because it occurs in many plant proteins (particularly Rubisco, the enzyme that catalyses the first carbon fixation reaction of photosynthesis) and chlorophyll. Although N_2 makes up about 78% of the atmosphere by volume, this is not available to plants. Plants absorb almost all N from the soil as nitrate (NO_3^-) or, in acid soils, as ammonium (NH_4^+) (see Fig. 11.4). First, atmospheric N_2 has to be fixed (reduced) to organic nitrogen by soil microorganisms or in the roots of N-fixing plants (e.g. alders, *Acacia*). Most of the N in soils is in the form of organic N, which is unavailable to tree roots. Organic N is then converted to NH_4^+ by ammonification, which in warm near-neutral soils is rapidly oxidized to NO_3^- (nitrification) by soil bacteria that derive energy from the conversion. In acid forest soils, nitrifying bacteria are less abundant and NH_4^+ becomes a more important source of N. Low temperatures in some forest soils will also lower NO_3^- availability. Ammonification consumes protons but nitrification generates protons. The net result of the conversion of organic material to NH_4^+ and then NO_3^- is acidification. As a cation, NH_4^+ can be held on cation exchange sites, while NO_3^- is more likely to be leached from soils. Nitrogen is also lost from soils by denitrification, the production of N_2, NO, N_2O and NO_2 from NO_3^- by anaerobic bacteria, often deep in soils or in waterlogged or compact soils.

Small amounts of N are also passed from the atmosphere to soil as NO_3^- and NH_4^+ in rain (up to $30\,kg\,ha^{-1}$ annually) and these ions can then be absorbed by plants. Natural sources of atmospheric NO_3^- and NH_4^+ are forest fires, volcanic activity and the oceans and NO_3^- production in lightning; however, industrial processes and the agricultural use of nitrogenous fertilizers and intensive animal husbandry are major anthro-

Nitrogen

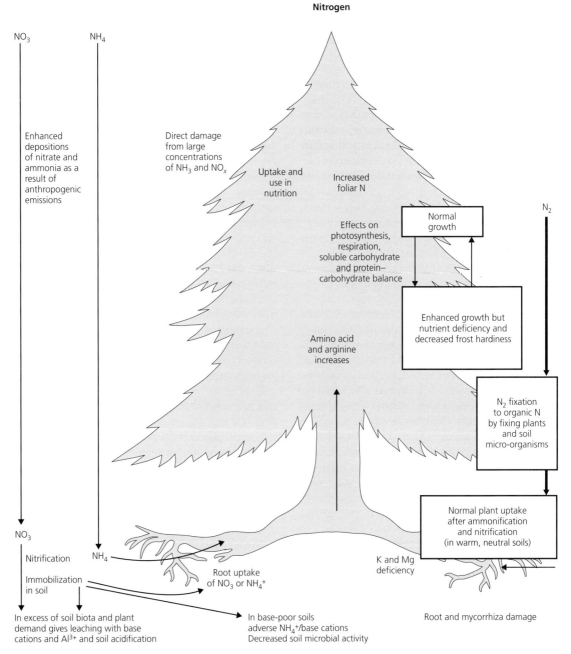

Fig. 11.4 Schematic representation of the exposure, uptake, physiological effects, metabolic impacts and possible outcome of exposure of trees to nitrogen depositions. See text for further explanation.

pogenic sources of both ions. In Europe and North America a number of studies have now linked high N deposition with declining tree condition (Cowling 1988; Heij & Erisman 1997). The acidity and chemistry of forest soils, including the balance between NH_4^+ and NO_3^-, are thus affected by depositions of N, S and other pollutants, by the preferred form of N uptake and by other soil/site factors (temperature, moisture content, etc.). NH_4^+ uptake enhances soil acidification because the cation is exchanged for protons. However, since so many forest soils are N deficient the most common initial effect of increased N inputs is likely to be increased growth.

The overall effect of NH_4^+, NO_3^- and ammonia (NH_3) on soils depends on the form of N deposition, the utilization of N by plants and microbes, and whether or not NO_3^- is leached. Soils have a limited capacity to immobilize and store NO_3^-; in contrast, NH_4^+ is not normally mobile. The eventual fate of deposited N determines the degree of soil acidification that occurs. Deposition of NH_4^+ followed by plant uptake will not affect the acidity of the system. However NH_4^+ deposited and converted to NO_3^- which is then leached from the system is acidifying. The balance between NO_3^- and NH_4^+ inputs and outputs thus determines the degree of acidification and this balance is affected by a large number of canopy and soil processes. Reuss and Johnson (1986) have suggested that soil acidification is only caused by N deposition if NO_3^- is leached from soil. Thus N depositions will only lead to acidification once NO_3^- is present in excess of the demand of the forest stand for N in biomass. The key considerations are thus whether any deposited NH_4^+ is nitrified and whether any mineral N (NO_3^-) is immobilized in the soil or organic matter. If deposition is mainly NO_3^- and inputs exceed the soil biota plus plant demand, then the anion will move through the soil leaching base cations and Al, which can lead to root damage. In contrast, if deposition is mainly NH_4^+, initially it will be immobilized, mainly on the soil's ion exchange sites or in mineral lattices. If inputs are great enough and uptake is slow, significant amounts may be nitrified to NO_3^-. In base-poor soils adverse NH_4^+:base cation ratios may

result from large NH_4^+ inputs. This will inhibit base cation uptake resulting in K or Mg deficiency. Soil acidification from NO_3^- or NH_4^+ inputs will also reduce the decomposition of organic matter so that it can accumulate in soils; litter accumulation.

There is a limit to the amount of N that a forest ecosystem can use and this limit is likely to be set by the availability of other nutrients, water, light or other environmental factors. A forest is said to be N saturated when it cannot retain any additional N and leaching occurs. The concept of N saturation has become central to discussion of forest soil critical loads.

11.4 WATER DEFICIT

Water must inevitably be lost from leaves in transpiration if stomata are to allow CO_2 to be fixed. The amount of water lost for each molecule of CO_2 assimilated into biomass is known as the water-use efficiency. Although water-use efficiency varies from plant to plant and indeed within species, larger plants with larger leaf areas require more water. Water availability (soil water potential), which for a particular soil is determined by water budget (precipitation minus evapotranspiration, run-off and leakage below the root zone), will thus limit the type of vegetation that can grow on a site. For example, Spurr and Barnes (1980) showed that an annual rainfall of 380 mm or more is required for open woodland in central areas of the USA, while over 640 mm was required for closed forest to occur. The global distributions of forest ecotypes described at the beginning of this chapter (see Plate 1) show how, with increasing length of the dry season, rain forests give way to semi-evergreen and then deciduous types. In the strongly seasonal dry tropics, forests of simple structure and fewer woody species occur and, ultimately, closed forest is replaced by open woodland and then savannah. In many areas of the world, sclerophyllous forests give way to scrub and then desert (Mediterranean Basin, North and South Africa, parts of South America) in a similar way. Desert, areas of sparse vegetation, have less than 200–400 mm of rain per year depending on temperature and potential evaporation. The most extensive deserts occur 20–30° north and south of the equator. However, in much of the temperate zone, climate patterns can be uncertain so that areas which normally have plenty of rain to support forest biomes may experience droughts for weeks, months or even several years. Examples are the widespread summer droughts of 1988 in the USA and of 1983 in Europe. In these areas, forest composition is controlled by these extreme events rather than the average condition.

The different physiology and anatomy of tree genera and species confer different degrees of drought tolerance, and as is the case for climate overall smaller-scale differences in water availability caused by topography and soil will influence species composition, growth rate and survival of trees within the major forest biomes. A major anatomical difference that influences drought tolerance is the presence of tracheids in conifers. Tracheids have border pits in the walls that close during drought. In contrast, the vessels of hardwood species have no tracheids so that water columns cannot be re-established after the withdrawal of water during drought. The buffering of leaf water potential by the storage of water in the stems of conifers has also been shown to minimize leaf loss during drought (Waring & Running 1978). In hardwoods, leaf abscission is a response to drought. A number of temperate zone species (e.g. *Ulmus americana*, *Tilia americana*, *Acer platanoides*, *Populus deltoides* and *Fagus sylvatica*) show whole or partial defoliation during summers of extreme drought (Woodward 1987). This response is not normally seen in conifers. A number of conifer leaves have thick waxy cuticles and additional chambers (antechambers) above the pair of guard cells that control gas exchange, including water loss, between the leaf and atmosphere. Water-use efficiency is often greater in conifers than broadleaved trees, although a net uptake of CO_2 may be maintained for larger periods of the year.

The cessation of leaf growth and stomatal closure are physiological responses to drought that normally precede leaf abscission. All three responses will limit growth but can be regarded as 'elastic' since they can be reversed. Cacti show many of the anatomical characteristics generally associated with drought tolerance (thick cuticles, presence of leaf hairs, high water storage) and many of them also show the principal physiologi-

cal adaptation to drought, crassulacean acid metabolism (CAM). This is a fascinating combination of anatomical and physiological adaptations that allows stomatal opening for CO_2 uptake to occur at night but the light reactions of photosynthesis to occur during the day while stomata are shut. As a result, transpiration losses are minimized. Only one tree has been reported as showing this adaptation, the tropical *Clusia rosea* which germinates from seed on the branches of other trees and thus experiences extreme drought between rain storms. *Clusia* has been shown to be a facultative CAM plant, switching between normal C3 photosynthesis and CAM in response to changes in water availability and light (Schmitt *et al.* 1988).

Adverse effects of water deficit (drought stress) occur when trees experience levels of water deficit to which they are not adapted. Drought has undoubtedly been a major factor in many of the forest declines that have occurred across the world but its role is particularly clear in the decline of oak in eastern and central USA during the last 20 years (Ciesla & Donaubauer 1994). In Europe the good records of forest condition have resulted from the monitoring work under the EU air pollution regulations and the Forests Programme (ICP) of the Convention on Long–range Transboundary Air Pollution. These data show that there was a general worsening of forest health in Europe between 1990 and 1995, attributed mainly to a series of dry years with drought and high summer temperatures (Freer-Smith 1998).

11.5 SOIL CONDITIONS, NUTRIENT DEFICIENCY/ENRICHMENT AND SOIL DEGRADATION

Plant nutrients are normally considered in two categories: macronutrients (N, P, K, Mg, Ca and S, to which C, H and O should be added for completeness) and trace elements or micronutrients (Mo, Ni, Cu, Zn, Mn, B, Fe and Cl). The availability of these elements varies with soil type, pH, cation exchange capacity and organic matter content, and with the exchange and other soil processes (e.g. weathering and mineralization, processes that are often rate-limiting) which determine availability from soil solution. Given the number of elements required for growth and the complexity of atmospheric and soil chemistry, it is perhaps inevitable that some degree of nutrient limitation and imbalance can result in stress to forests on a global scale. Indeed nutrient availability can result in natural successions in forest ecosystems, as seen in the *Metrosideros* rain forests of the Hawaiian islands and in *Eucalyptus* forests of eastern Australia. In the *Metrosideros* rain forests of the Hawaiian islands, low N availability on thin volcanic soils limits forest development and decreasing availability of base cations and P become a major causal factor in stand dieback. In eastern Australia, P availability on ancient soils limits the growth of indigenous *Eucalyptus*. Here fertilization can increase growth rates but causes increased susceptibility to insect pests which, along with other factors, cause dieback (Heatwole & Lowman 1986). There is also some evidence that the common age-related decline in forest growth may be related to increased nutrient constraints. *Eucalyptus* decline is a good example of a very common event, i.e. fertilizer application based on an identified mineral deficiency failing to alleviate decline or slow growth.

Clearly, it is important for forest managers to have a general understanding of the processes that determine nutrient availability and to be aware of the likely or actual nutrient limitations of regions and stands. Earlier sections of this chapter have already touched on a number of the key processes so that the links between water availability, climate, pollutant inputs and nutrient availability will already be clear. Pedogenesis (soil formation) occurs through the weathering of parent bedrock or deposited material, along with the decomposition and mineralization of organic matter (fallen leaves, etc.). Boreal soils are referred to as young, being postglacial and having been least weathered; temperate forest soils are moderately weathered, while warm temperate and subtropical soils have been strongly weathered. A number of the nutrients listed above are commonly available in soils and taken up by trees as positively charged cations. These are the base cations (Ca^{2+}, K^+, Mg^{2+}, Na^+) and NH_4^+, and also H^+ and Al^{3+}, which are not plant nutrients. The availability of these elements to tree roots in soil solution is determined by their release from the surface of soil mineral and organic material. Both

organic material and minerals normally present negatively charged surfaces on which the cations are held and released into soil solution in exchange for H^+. Minerals are mainly made up of aluminium octahedral and silica tetrahedral layers, making Al and Si the most common elements in most mineral soils. The minerals of most tropical soils are predominantly oxides of Al that have been strongly weathered so that the small cation exchange capacity which such soils have is provided by their organic matter. Of the macronutrients listed above, availability of N, P and S is not via cation exchange reactions (except for N uptake as NH_4^+). We have already considered the dependence of N supply on fixation, ammonification, nitrification and denitrification, processes largely determined by soil microbial activity. Phosphorus does occur in soil minerals (e.g. apatite) but it has low solubility and dissolved P is normally at very low concentrations. Plant uptake can rapidly deplete any dissolved P. The movement of P into solution is commonly the rate-limiting (perhaps growth-limiting) process. Significant amounts of P enter soils in leaf and branch litter and, like N, availability is then strongly dependent on soil microbial activity. Phosphorus deficiency can thus be a significant stress in a number of forest systems. Like N and P, S also occurs primarily as an anion in soil solution and can be immobilized in organic matter in soils, so that the rate of microbial mineralization may determine availability for plant nutrition. Sulphur is present in soil minerals as metal sulphides (e.g. FeS_2 and ZnS), particularly in igneous and sedimentary rocks. Sulphur differs from P in that sulphate, SO_4^{2-} its common form, can be rapidly leached from soils. Even in soils not experiencing anthropogenic S inputs, SO_4^{2-} inputs in rain are commonly greater than availability from mineralization of organic S.

Nutrient cycling is known to be 'tight' in forest ecosystems (Miller 1989), meaning that availability is commonly limiting or close to it and reuse following litterfall, decomposition and mineralization is rapid and efficient. Leaching of nutrients (loss in soil water movement) is uncommon in natural ecosystems and leaching is commonly an indication of stress and disturbance. The quantification of elemental sinks and fluxes, particularly atmospheric inputs and losses from forest ecosystems, is thus of major importance. Understanding the importance of atmospheric inputs has occurred only relatively recently (see for example Ulrich 1989). In the last 20 years there have been a number of major forest catchment-scale studies that provide valuable data on forest nutrient budgets, on elemental losses from forest ecosystems and thus on the biological sustainability of the systems. The development of this 'ecosystem approach' can be directed at understanding and indeed treating specific nutritional problems (see Zöttle & Hüttl 1991). The mass balance and critical load approach to acidification has developed from the understanding of nutrient cycling and leaching and has drawn on the datasets provided.

We have already seen that deposition of anthropogenic S and N can result in soil acidification, raised Al concentrations in soil solution and loss of base cations through leaching. Magnesium deficiency has commonly been seen as the first nutritional imbalance to occur in this sequence. Nitrogen, like P and K, can limit productivity in forests. As with S, human activities have resulted in substantial increases in the deposition of inorganic N from the atmosphere. Smith (1999) estimated that the global fixation of atmospheric N_2 for use in fertilization in agriculture rose from less than 10 Tg annually (1 Tg = 1 Mt) to 80 Tg annually in 1980, and this value is likely to rise to 135 Tg annually by 2030 with an estimated additional 20 Tg annually being added through the deposition of pollutant N from the burning of fossil fuels. Thus, human activity adds as much N to the terrestrial ecosystems as natural sources. It is not surprising then that some forests are showing clear symptoms of excess N and, furthermore, that anthropogenic N inputs have been proposed, along with rising CO_2 concentrations and warming temperatures, as a major cause of increased forest growth throughout Europe.

11.6 CONCLUSIONS

In many ways the last two sentences of the preceding section would have made a suitable closing statement: the deposition of elements derived from anthropogenic emissions (S and N) are linked with changes of atmospheric chemistry (CO_2) and climate to produce an impact on forests (decreased or increased growth rate, declines, etc.). However this statement places too much emphasis on nutrient enrichment. In many parts of the world forests persist on poor soils where nutrient availability is a common limitation and anthropogenic inputs are likely to be of limited relevance. In such systems, sensitive management is essential if productivity is to be sustained over successive rotations (Evans 1999). Furthermore, consideration of the four abiotic factors in this chapter has also illustrated a number of more fundamental and overarching points.

Firstly, when abiotic stress is sufficient to cause system shifts (increased growth, decline, biome migration, etc.) it is often, perhaps always, associated with changes in environmental conditions. These may be natural, resulting from extremes of climate (ice ages or droughts), or anthropogenic, caused by the introduction of exotic species, repeated rotations of fast-growing plantations, pollutant deposition or greenhouse gas emissions. Secondly, forest ecosystems and the effects of the interactions of abiotic factors on them are complex. Interactions in effects and causal chains are the norm not the exception. Faced with this, the objective of providing an understanding of ecological processes on which forest managers can make decisions that achieve the desired objectives proves difficult but not impossible. In many cases our understanding can be aided by the construction of conceptual and mathematical models and a number of these have been mentioned, for example Levitt's conceptual model of stress, general circulation models, models linking climate to vegetation (e.g. HYBRID, BEC, ESC), pollutant deposition and soil acidification models, critical load models, plant growth models and stand nutrient models. Today, modelling is often an effective way forward and we see the development of increasingly sophisticated forest growth models and decision support systems. We are fortunate to be at a stage in the development of forest science where such approaches are possible. Such models provide mechanisms by which we can integrate and draw on our complex knowledge of processes and of key interactions when making policy and management decisions.

ACKNOWLEDGEMENTS

I would like to thank Andrew Friend and Andrew White for permission to use Plate 1 and John Williams (Forest Research) for drawing Figs 11.1–11.4.

REFERENCES

Anderson, M.L. (1950) *The Selection of Tree Species.* Oliver & Boyd, Edinburgh.

Broadmeadow, M. & Freer-Smith, P.H. (1998) Climate change: the evidence so far and predictions of tree growth. *Irish Forestry* **55**, 122–32.

Ceulemans, R. & Mousseau, M. (1994) Tansley Review no. 71. Effects of elevated atmospheric CO_2 on woody plants. *New Phytologist* **127**, 425–46.

Chappelka, A.H. & Freer-Smith, P.H. (1994) Predisposition of trees by air pollutants to low temperatures and moisture stress. *Environmental Pollution* **87**, 105–17.

Ciesla, W.M. & Donaubauer, E. (1994) *Decline and Dieback of Trees and Forests.* Forestry Paper 120, Food and Agriculture Organisation, Rome.

Cowling, E.B. (1988) Ecosystems and their response to airborne chemicals: the current situation in North America and Europe. In: Mathy, P., ed. *Air Pollution and Ecosystems: Proceedings of an International Symposium, Grenoble, France, 18–22 May 1987,* pp. 18–38. D. Reidel Publishing Company, The Netherlands.

Daubenmire, R. (1978) *Plant Geography.* Academic Press, New York.

Evans, J. (1999) *Sustainability of Forest Plantations, the Evidence: a Review of Evidence Concerning the Narrow-sense Sustainability of Planted Forests.* Department for International Development, Issues Paper, DFID, London.

Freer-Smith, P.H. (1998) Do pollutant-related forest declines threaten the sustainability of forests? *Ambio* **27**, 123–31.

Friend, A.D., Stevens, A.K., Knox, R.G. & Cannell, N.G.R. (1997) A process based, terrestrial biosphere model of ecosystem dynamics (HYBRID v3.0). *Ecological Modelling* **95**, 249–87.

Heath, J. & Kerstiens, G. (1997) Effects of elevated CO_2 on leaf gas exchange in beech and oak at two levels of nutrient supply: consequences for sensitivity to drought in beech. *Plant, Cell and Environment* **20**, 57–67.

Heatwole, H. & Lowman, M. (1986) *Dieback: Death of an Australian Landscape.* Reed Books, Sydney.

Heij, G.J. & Erisman, J.W. (eds) (1997) Acid rain research: do we have enough answers? *Studies in Environmental Science* Vol. 64. Elsevier, Amsterdam.

Holdridge, L.R. (1947) Determination of world plant formations from simple climatic data. *Science* **105**, 367–8.

Houghton, J. (1997) *Global Warming. The Complete Briefing.* Cambridge University Press, Cambridge.

Houghton, J.T., Meira Filho, L.G., Bruce, J. *et al.* (eds) (1995) *Climate Change 1994: Radiative Forcing of Climate Change and an Evaluation of the IPCC IS92 Emission Scenarios.* Cambridge University Press, Cambridge.

Huettl, R. & Mueller-Dombois, D. (1992) *Forest Decline in the Atlantic and Pacific Region.* Springer-Verlag, Berlin.

Jarvis, P.G. (1998) *European Forests and Global Change. The Likely Impacts of Rising CO_2 and Temperature.* Cambridge University Press, Cambridge.

Johnson, A.H., Friedland, A.J., Miller, E.K. & Siccama, T.G. (1994) Acid rain and soils of the Adirondacks. *Canadian Journal of Forest Research* **24**, 663–9.

Landsberg, J.J. & Gower, S.T. (1997) *Applications of Physiological Ecology to Forest Management.* Academic Press, London.

Larcher, W. & Bauer, H. (1981) Ecological significance of resistance to low temperature. In: Lange, O.L., Nobel, P.S., Osmond, C.B. & Ziegler, H.,eds. *Encyclopedia of Plant Physiology,* Vol. 12A, pp. 403–37. Springer-Verlag, Berlin.

Levitt, J. (1972) *Responses of Plants to Environment Stresses,* 2nd edn. Academic Press, New York.

Levitt, J. (1980) *Responses of Plants to Environmental Stress,* 2nd edn. Academic Press, New York.

Melillo, J.M., McGuire, A.D., Kicklighter, D.W., Moore, B.I. II, Vorosmarty, C.J. & Schloss, A.L. (1993) Global climate change and terrestrial net primary production. *Nature* **363**, 234–40.

Miller, H.G. (1989) Internal and external cycling of nutrients in forest stands. In Landsberg, J.S.P. & Landsberg, J.J., eds. *Biomass Production by Fast-Growing Trees,* pp. 73–80. Academic Press, Dordrecht.

Murray, M.B., Smith, R.I., Leith, I.D. *et al.* (1995) Effects of elevated CO_2, nutrition and climatic warming on bud phenology in Sitka spruce and their impact on the risk of frost damage. *Tree Physiology* **14**, 691–707.

Pojar, J., Klinka, K. & Meidinger, D. (1987) Biogeoclimatic ecosystem classification in British Columbia. *Forest Ecology and Management* **22**, 119–54.

Pyatt, D.G. & Suarez, J.C. (1997) *An Ecological Site Classification for Forestry in Great Britain.* Technical Paper 20, Forestry Commission, Edinburgh.

Reuss, J.O. & Johnson, D.W. (1986) *Acid Deposition and the Acidification of Soils and Waters.* Springer-Verlag, New York.

Richards, J.H. & Bliss, L. (1986) Winter water relations of a deciduous timberline conifer *Larix lyallii* Parl. *Oecologia* **69**, 16–24.

Schmitt, A.K., Lee, H.S.J. & Luttge, U. (1988) The response of the C_3–CAM tree, *Clusia rosea,* to light and water stress. *Journal of Experimental Botany* **39**, 1581–90.

Schulze, E.D., Lange, O.L. & Oren, R. (1989) *Forest Decline and Air Pollution: A Study of Spruce* (Picea abies) *on Acid Soils*. Springer-Verlag, Berlin.

Schulze, E.D., Fuchs, M. & Fuchs, M.I. (1997) Spatial distribution of photosynthetic capacity and performance in a mountain spruce forest of Northern Germany. III. The significance of the evergreen habit. *Oecologia* **30**, 239–48.

Smith, V.H., Tilman, G.D. & Nekola, J.C. (1999) Eutrophication: impacts of excess nutrient inputs on freshwater, marine and terrestrial ecosystems. *Environmental Pollution* **100**, 179–96.

Spiecker, H., Mielikainen, K., Kohl, M. & Skovsgaard, J. (eds) (1996) *Growth Trends in European Forests*. Springer-Verlag, Berlin.

Spurr, S.H. & Barnes, B.V. (1980) *Forest Ecology*, 3rd edn. John Wiley & Sons, New York.

Stitt, M. (1991) Rising CO_2 levels and their potential significance for carbon flow in photosynthetic cells. *Plant, Cell and Environment* **14**, 741–62.

Sverdrup, H. & de Vries, W. (1994) Calculating critical loads for acidity with the simple mass balance method. *Water Air Soil Pollution* **72**, 143–62.

Taylor, G.E. & Constable, J.V.H. (1994) Modelling pollutant deposition to vegetation: scaling down from the canopy to the biochemical level. In: Percy, K., Cape Jagels, N. & Simpson, eds. *Air Pollutants and Leaf Cuticles*, pp. 15–37. Springer-Verlag, Heidelberg.

Ulrich, B. (1989) Effects of acidic precipitation on forest ecosystems in Europe. In: Adriano, D.C. & Johnson, A.H., eds. *Acidic Precipitation, Vol. 2 Biological and Ecological Effects*, pp. 189–269. Springer-Verlag, Berlin.

Waring, R.H. & Running, S.W. (1978) Sapwood water storage: its contribution to transpiration and effect upon water conductance through the stems of old-growth Douglas-fir. *Plant, Cell and Environment* **1**, 131–40.

Wellburn, A. (1994) *Air Pollution and Climate Change. The Biological Impact*. Longman Scientific & Technical, Harlow, Essex.

Woodward, F.I. (1987) *Climate and Plant Distribution*. Cambridge University Press. Cambridge.

Woodward, F.I. (1995) Ecophysiological controls of conifer distributions. In: Smith, W.K. & Hinckley, T.M., eds. *Ecophysiology of Coniferous Forests*, pp. 79–94. Academic Press, San Diego.

Zöttl, H.W. & Hüttl, R.F. (eds) (1991) *Management of Nutrition in Forests Under Stress*. Kluwer Academic, The Netherlands.

12: Forest Hydrology

L. A. (SAMPURNO) BRUIJNZEEL

12.1 INTRODUCTION

Scientists and the public alike have long appreciated that a forest cover influences climatic and soil conditions and therefore the amount of water flowing from forested areas. In one of the earliest reviews of forest influences, Zon (1927) stated that 'forests increase both the abundance and frequency of local precipitation over the areas they occupy'. Furthermore it was not only claimed that 'forests increase underground storage of water' but also that 'forests tend to equalize the flow throughout the year by making the low stages higher'. Even after several catchment experiments completed during the next few decades had shown that both annual and summer flows increased after forest clearing (Hoyt & Troxell 1932; Hoover 1944), the classical text on forest influences by Kittredge (1948) still concluded that 'forests prolong increased flow in low-water periods'. Moreover, the belief that forest removal automatically results in diminished dry-season flows and increased flooding is still regularly encountered today, particularly in the tropics (cf. Hamilton & King 1983; Bruijnzeel 1990). Whilst the current wealth of experimental evidence on the hydrological role of forests may have revealed some serious shortcomings in these earlier beliefs, it basically confirms the contention of Zon (1927) that 'of all the direct influences of the forest the influence upon the supply of water in streams and upon the regularity of their flow is the most important in human economy'. With populations rising explosively in some parts of the world, and per-capita demands increasing in others as living standards continue to improve, optimization of water resources (both streamflow and groundwater reserves) is becoming increasingly important, particularly during the dry season. Similarly, the rising demands for industrial wood (pulpwood, saw and veneer logs), fuelwood and charcoal (mostly in the tropics) require the establishment of large areas of fast-growing plantation forests, often on land that is currently not forested (Brown et al. 1997). Coupled with the continued indiscriminate clearing of the world's forests (Jepma 1995; Nepstad et al. 1999), which in many areas serve as the traditional supplier of high-quality water, the associated deterioration of soil and water quality due to erosion and pollution (cf. Oldeman 1994) and the possibility of gradually less dependable precipitation inputs due to global change (Wasser & Harger 1992), a sound understanding of the hydrological functioning of forests is arguably even more important now than ever before.

A classic example of the conflicting interests of wood producers and water supply institutions is provided by the controversy following the finding of Law (1956) that a spruce plantation in the British Pennines consumed almost 300 mm more water per year than an adjacent grass sward. He concluded, on the basis of a lysimeter experiment, that if the whole catchment had been afforested with spruce this might have led to a reduction in water yield to a downstream reservoir by as much as 42%. The debate that followed the publication of these results was so heated that it not only began to threaten upland afforestation plans in the UK but also stimulated a major forest hydrological and micrometeorological research programme (see summaries by Rutter 1968; Shuttleworth 1989; Calder 1990). Results from these programmes both confirmed Law's original findings and greatly increased our understanding of the underlying hydrological processes.

The first part of this chapter reviews the current state of knowledge with respect to the major hydrological processes that occur in forests and how these processes affect amounts and timing of streamflow. The second part of the chapter discusses the effects of (i) various forest management options (thinning, selective logging, clearfelling) and (ii) replacement of forest by grass or crops, and vice versa. The principal focus is on

the more humid parts of the world (both temperate and tropical).

12.2 FOREST HYDROLOGICAL CYCLE

The principal features of the forest hydrological cycle are illustrated in Fig. 12.1. Rain (P) is the main precipitation input to most forests, supplemented by snow at higher altitudes and latitudes and by 'occult' precipitation (fog) in coastal or montane fog belts. A small part of the precipitation reaches the forest floor directly without

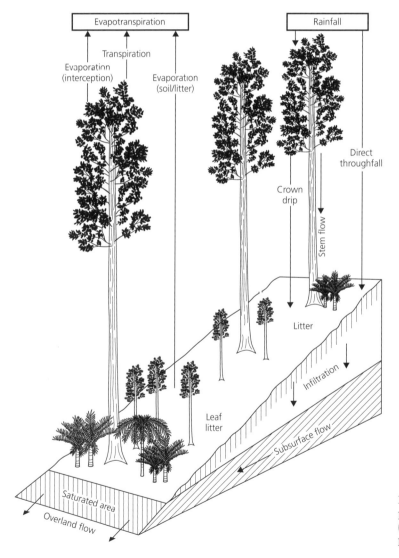

Fig. 12.1 Key hydrological processes on a forested hillslope. (After Vertessy *et al.* 1998 with permission.)

touching the canopy, the so-called 'free' or 'direct' throughfall. Another (usually small) part travels along the branches and trunks as stemflow (Sf). A substantial portion of the precipitation intercepted by the canopy is evaporated back to the atmosphere during and shortly after the storm (called interception loss, E_i), whereas the remainder reaches the soil surface as crown drip after the storage capacity of the canopy has been filled. Because direct throughfall and crown drip cannot be determined separately in the field, the two are usually taken together and referred to as throughfall (Tf). However, the distinction between the two is useful in modelling the interception process (see section 12.4.3). The sum of throughfall and stemflow is commonly called net precipitation and is usually substantially smaller than amounts of incident precipitation unless there are significant (unmeasured) contributions by fog precipitation (cf. section 12.3):

$$P = Tf + Sf + E_i \qquad [12.1]$$

and

$$E_i = P - (Tf + Sf) \qquad [12.2]$$

If the intensity of the total throughfall and stemflow reaching the forest floor exceeds the infiltration capacity of the soil, the unabsorbed excess runs off as 'Hortonian' or 'infiltration-excess' overland flow (HOF). Due to the generally very high infiltration capacity of the organic-rich topsoil in most forests, this type of flow is rarely observed in undisturbed forest unless there is an unusually dense clayey substrate (Elsenbeer *et al.* 1992) or an excessive concentration of stemflow (Herwitz 1986). Not all of the water infiltrating into the soil emerges as streamflow. A large part is taken up by the vegetation and returns to the atmosphere via the process of transpiration (E_t). The term 'evapotranspiration' (ET) is used to denote the sum of transpiration (evaporation from a dry canopy), interception loss (evaporation from the exposed surfaces of a wet canopy) and evaporation from the litter and soil surface (E_s). The latter term is often small, especially in dense forests where little radiation penetrates to the forest floor, humidity is high and the air virtually stagnant. It is essential to distinguish between E_t and E_i because the former is strongly governed by

stomatal control and the latter mostly by the aerodynamic properties of the vegetation (Jarvis & Stewart 1979). Thus:

$$ET = E_i + E_t + E_s \qquad [12.3]$$

The remaining soil moisture drains towards the nearest stream via 'throughflow', the result of downward-moving water meeting an impermeable layer of subsoil or bedrock and being deflected laterally (see Fig. 12.1). Such water drains slowly and steadily, thus accounting for the 'baseflow' of streams. In humid temperate climates, baseflow generally reaches a minimum in the summer period, whereas under seasonal tropical conditions this occurs during the dry season. These low flows are usually referred to as 'minimum flow', 'dry-season flow' or simply 'low flow'. During rainfall, infiltrated water may take one of several routes to the stream channel, depending on soil hydraulic conductivity, slope morphology and the spatial distribution of soil moisture (Dunne 1978; Fig. 12.2). 'Saturation' overland flow (SOF) is caused by rain falling on an already saturated soil. This situation typically occurs in hillside hollows or on concave footslopes near the stream where throughflow tends to converge and maintain near-saturated conditions. Occasionally, widespread hillside SOF (i.e. outside concavities and depressions) has been observed during and after intense rainfall in the tropics in places where an impeding layer is found close to the surface (Bonell & Gilmour 1978; Elsenbeer & Cassel 1990). Rapid throughflow during storms (subsurface stormflow, SSSF) usually consists of a mixture of 'old' (i.e. already present before the start of the rain) and 'new' water travelling through 'macropores' and 'pipes' (Bonell 1993; Brammer & McDonnell 1996). As a result of contributions by SOF, SSSF and, in extreme cases, HOF, streamflow usually increases rapidly during rainfall. This increase above baseflow levels is called 'stormflow' or 'quickflow'. The highest discharge observed in association with a precipitation event is referred to as 'peak flow'. Peak discharges may be reached during the rainfall event itself or as late as a few days afterwards, depending on catchment characteristics, soil wetness and the duration, intensity and quantity of the rainfall (Dunne 1978; Pearce *et al.* 1982;

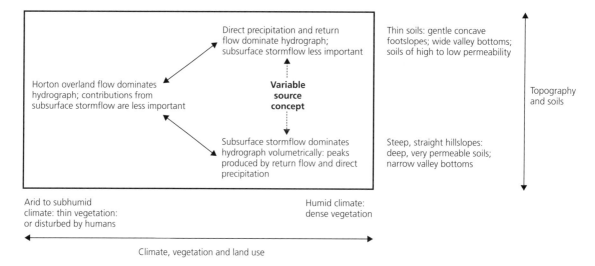

Fig. 12.2 Schematic representation of the occurrence of various streamflow generating processes in relation to their major controls. Note that 'direct precipitation and return flow' are equivalent to saturation overland flow. (After Dunne 1978 with permission.)

Brammer & McDonnell 1996). The total volume of water produced as streamflow from a catchment over a given period of time (usually a month, season or year) is called 'water yield'.

The interrelated character of the chief components of the hydrological cycle is summarized by the catchment or site water budget equation:

$$P = E_i + E_t + E_s + Q + \Delta S + \Delta G \qquad [12.4]$$

where Q is streamflow or drainage to deeper layers, ΔS change in soil water storage and ΔG change in groundwater storage, with the remaining terms as defined previously. All values are expressed in millimetres of water per unit of time (hour, day, week, month or year). Note that ΔS and ΔG may assume positive (gain) or negative (loss) values. In view of the seasonal cycle of soil water and groundwater storages in many areas, the values of ΔS and ΔG tend to approach zero on an annual basis (Lee 1970). Therefore, the annual water balance is often simplified to:

$$P = ET + Q \qquad [12.5]$$

12.3 FORESTS AND 'OCCULT' PRECIPITATION

Forest–climate interactions in general have been dealt with in Chapter 10. Therefore the present discussion does not focus on the extent to which a forest cover generates its own rainfall (and over what scale) but rather on the importance of fog and other forms of cloud moisture ('occult' precipitation) as an additional input of water. Wherever fog impacts a forested area, particularly where the fog occurs in the form of an orographic cloud belt or advective coastal fog, additional moisture is intercepted by plant surfaces (or indeed any other obstacle) and precipitation may occur in the form of 'occult' precipitation or 'fog drip', even if no rainfall is recorded on adjacent open ground (Zadroga 1981). An extreme example has been described by Aravena *et al.* (1989) whose isotope studies have demonstrated that the frequent occurrence of advective sea fogs along the arid coast of northern Chile has given rise to a patchy forest that, in the almost complete absence of ordinary rainfall, thrives primarily on fog. Several studies of fog drip beneath trees along the US Pacific Coast have shown that the amount is directly related to the area and density of the tree profile and to the degree of exposure of the trees to windblown fog. During times of fog, fog drip at the edge of a forest may be several times the rainfall in the open. However, amounts often decrease sharply towards the forest interior.

Therefore, reports of increased net precipitation beneath vegetation need to be considered carefully in terms of gauge positions with respect to the edge of the forest (Grünow 1955; Weathers *et al.* 1995). At favourably exposed locations the extra inputs stripped from the fog by trees (conifers especially) may be considerable. Examples include 425 mm in 46 rainless summer days below 18-m tall Douglas fir in northern California (Azevedo & Morgan 1974), 760 mm annually measured in Hawaii below *Araucaria* by Ekern (1964) and 880 mm annually for tall old-growth Douglas fir in Oregon by Harr (1982). Bruijnzeel and Proctor (1995) and Bruijnzeel (2000) recently reviewed the hydrological characteristics of 'cloud forests' found in orographic cloud belts on wet tropical mountains and reported extremely variable net precipitation fractions for these forests (55–130% of incident rainfall). As such, although tropical montane cloud forests are supposedly exposed to frequent fog (Zadroga 1981), the associated amounts intercepted by the trees are not always sufficient to raise net precipitation totals significantly above the 70–80% commonly recorded for montane rain forests on cloud-free locations. Such variations may be interpreted in terms of exposure to and persistence of fog, as well as to contrasts in canopy epiphyte biomass (Bruijnzeel & Proctor 1995).

Quantification of the amount of fog intercepted by a forest is difficult, particularly if the fog occurs together with rain. The usual approach is to compare amounts of net precipitation (often throughfall only) beneath the forest with the rainfall measured in the open (e.g. Harr 1982) or to subtract the latter from the catch obtained with some kind of fog gauge placed next to an ordinary rain gauge. A number of different fog gauge designs have been proposed, including wire harps (Goodman 1985), cylindrical screens (Russell 1984), 1 m^2 polypropylene screens (Schemenauer & Cereceda 1994) and louvered metal gauges (Juvik & Nullet 1995). The throughfall method essentially gives an estimate of 'net' fog drip since it includes an unmeasured amount of water lost to evaporation from the wet vegetation (i.e. E_i, see also next section). In addition, the result is site specific. The use of a fog gauge (regardless of the type adopted) also requires a

site-specific conversion factor to relate the catch of the gauge to that of the actual forest canopy. Some progress has been made with physical fog deposition models (Joslin *et al.* 1990; Mueller 1991; Walmsley *et al.* 1996, 1999) but the data requirements of these models are such that their use at remote forest locations must remain limited to a few well-researched sites. Therefore there may be some merit in alternative approaches to the quantification of fog contributions, such as the sodium/chloride budget method (Hafkenscheid *et al.* 1998) or the contrasting stable-isotope signatures of rain and fog water (Dawson 1998; Ingraham 1998). For example, by comparing the stable isotope compositions of rain, fog, soil moisture and groundwater with those of xylem water in the dominant species of a Californian redwood forest, Dawson (1998) was able to demonstrate the importance of fog to the different plant groups. During a year with average climatic conditions, 25–50% of the water uptake by the tall redwood trees was shown to consist of fog whereas these figures roughly doubled during a drought year. Understorey herbs were even more dependent on fog water: 19–50% in an average year and 28–100% in a drought year.

As discussed more fully in section 12.4.3 (Eqn 12.8), transpiration (E_t) ceases whenever the canopy is fully wetted. Because of this and the reduced radiation and high relative humidity normally associated with such conditions, E_t of trees exposed to frequent fog is low (Hutley *et al.* 1997). Together with the additional inputs via occult precipitation, this makes for an especially favourable water balance, i.e. amounts of streamflow or groundwater recharge will be comparatively high. For these reasons Zadroga (1981) expressed the fear that clearing tropical montane cloud forest could well result in substantially reduced water yield, especially during the dry season. Some support for this contention could come from the contrasting dry-season flows emanating from two pairs of cloud-forested and cleared catchments in Guatemala and Honduras reported by Brown *et al.* (1996). However, both catchment pairs were too different in size and elevational range (and therefore exposure to fog and rainfall) to be sure that the approximately 50%

reduction in flow was entirely due to the replacement of the forest by vegetable cropping. A more
convincing case was provided by Ingwersen (1985)
who observed a (small) decline in summer flows
after a 25% patch clearcut operation in the same
catchment in the Pacific North-west region of the
USA for which Harr (1982) had inferred an annual
contribution by fog of about 880 mm. The effect
disappeared after 5–6 years. Because forest cutting
in the Pacific North-west is normally associated
with strong increases in water yield (Harr 1983),
this anomalous result was attributed to an initial
loss of fog stripping upon timber harvesting, followed by a gradual recovery during regrowth.
Interestingly, the effect was less pronounced in
an adjacent (but more sheltered) catchment and
it could not be excluded that some of the condensation not realized in the more exposed catchment was 'passed on' to the other catchment
(Ingwersen 1985). The observation of Fallas (1996)
of enhanced throughfall in patches of forest surrounded by pasture in the montane cloud forest
belt of northern Costa Rica is probably pertinent
in this respect. Further (process-based) work is
needed to decide on the importance of occult precipitation for water yield from fog-affected areas
(cf. Bruijnzeel 2000).

12.4 THROUGHFALL, STEMFLOW AND INTERCEPTION LOSS

12.4.1 Amounts and measurement of rainfall interception

Arguably, no subject in (temperate zone) forest
hydrology has received as much attention as the
measurement of throughfall (Tf), stemflow (Sf)
and derived estimates of interception loss (E_i).
Yet, until the pioneering work of Rutter and
colleagues in the 1960s (Rutter 1967; Rutter
et al. 1971, 1975) the increase in understanding
of the underlying processes was not commensurate. Before then, interception was mainly
studied in terms of tree species and age, or the
amount, intensity and duration of the rain. For
example, in a review covering more than 50 interception studies in the USA spanning half a
century, Zinke (1967) arrived at essentially the
same generalizations as Kittredge (1948) and

Horton (1919) had before him. These can be summarized as follows:
1 E_i is a function of incident rainfall (P) and typically declines more or less hyperbolically when
expressed as a percentage of P;
2 both E_i and the canopy storage capacity (S) are
generally larger for coniferous forests (E_i =
15–40%; S = 0.5–3.0 mm; average ≈ 2 mm) than for
deciduous forests (growing season, E_i = 20–25%;
S = 0.5–2.0 mm; average ≈ 1.0 mm);
3 winter values of E_i and S for deciduous forests
are roughly half to two-thirds of summer values;
4 Sf generally constitutes only a modest fraction
of P (typically < 3%), with the exception of
smooth-barked species such as beech or maple
(5–12%) or young trees;
5 E_i increases with stand density (cf. section
12.7);
6 snowfall interception storage by coniferous
canopies exceeds that for rainfall.

The extensive reviews of early European work
by Delfs (1955) and Molchanov (1963) and of
British studies by Hall and Roberts (1990) broadly
support the above generalizations obtained for
forests in the USA.

The main contribution by Rutter and
coworkers to our current, improved, understanding of the rainfall interception process has been to
move away from the empirical site-specific
approach of the past to a more explicitly physical
description of the process. Central to the Rutter
approach is a running canopy (and trunk) water
balance in which hourly inputs (rainfall) and
outputs (drainage, evaporation) determine the
amount of water stored on the canopy (and the
tree trunks). The latter governs both the rate of
drainage from the wet canopy and the rate of evaporation (for storage less than S) (Rutter et al. 1971,
1975). We come back to this in section 12.4.3.

Usually, the results of interception studies are
expressed in relation to gross rainfall, either as a
ratio, percentage or through various types of
regression equations. Comparisons between different studies, even for the same species and age
class, are rendered difficult because of the more-
or-less unique character of each forest stand in
terms of density, presence or absence of undergrowth, exposure to prevailing air streams, and
rainfall regimes. In addition, there is a method-

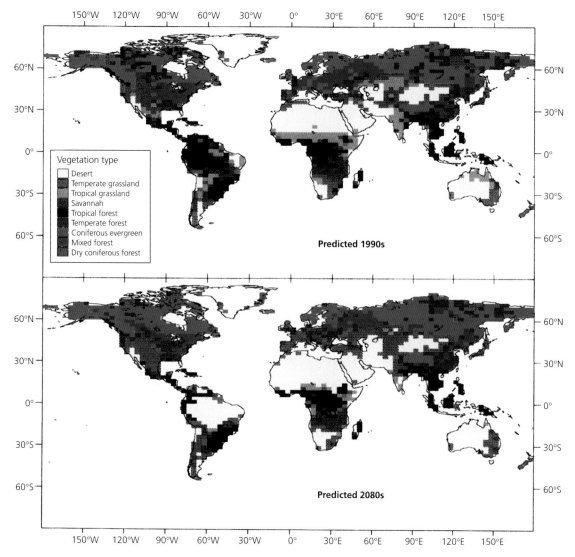

Plate 1 Global distribution of vegetation types predicted by the ITE ecosystem model HYBRID for the present day and the 2080s. Many regions that are currently tropical forests are predicted to change to savannah grassland or even desert. The model predicts vast areas of dieback in northern Brazil, beginning in the 2040s as a result of decreases of rainfall and increased temperatures. Temperate grasslands will expand into regions of Europe and North America currently dominated by temperate or coniferous forest. There is a northwards expansion of coniferous and temperate forest in both North America and Asia, and a large area where these forest types remain dominant. (With permission from Andrew Friend and Andrew White.)

Plate 2 Application of dolomitic limestone to forests dominated by *Picea abies* in order to correct magnesium deficiency caused by soil acidification and diagnosed through yellowing and elemental analysis of needles. Photograph taken in the Black Forest, Germany in 1989.

ological problem in that the notoriously uneven distribution of *Tf* (both in space and time) requires a rigorous sampling strategy that not all studies have achieved. In general, the spatial variability of *Tf* increases with forest density, i.e. it is greater in summer than in winter for deciduous stands (Helvey & Patric 1965) and much larger in tropical rain forest than in temperate plantations (Lloyd & Marques 1988; Fig. 12.3). The vegetation in most natural forests is made up of a mosaic representing different stages of growth, ranging from young rapidly growing trees in gaps to old-growth emergents on the decline. Burghouts *et al.* (1998) reported an inverse relationship between amounts of *Tf* and litterfall in a spatially heterogeneous lowland rain forest in Borneo. Delfs (1955) observed a gradual decrease in *Tf* over an age sequence from young regenerating spruce in the sapling and pole stages to mature stands. At the microscale, Eschner (1967) reviewed a range of studies in deciduous and coniferous forests

showing that beyond the immediate sphere around the trunks (where *Sf* and drip from the points of juncture of branches and bole tend to concentrate), *Tf* often increases away from the trunks to reach a maximum just within the perimeter of the crown. Hutchinson and Roberts (1981) reported that 98% of all *Sf* generated by a 9-m tall Douglas fir originated in the upper half of the tree, whereas the sheltered and more depressed branches of the lower half (representing 31% of total branch interception area) contributed little to stemflow production.

Czarnowski and Olszewski (1970) demonstrated that using up to 30 fixed standard gauges gave different average *Tf* values for a natural deciduous forest in central Europe, whereas the estimate stabilized after using 30 gauges or more. Helvey and Patric (1965) suggested that six gauges during the dormant season and 15 during the growing season would be sufficient for a proper estimation of *Tf* in mixed hardwood stands in the eastern USA, provided the gauges were relocated regularly. Other investigators relied on increasing the sampling surface area, be it through the use of large metal plates (Crockford & Richardson 1990), elongated gutters (Rowe 1983) or plastic sheets (Calder & Rosier 1976). However, the use of a large collecting surface *per se* is not a guarantee of a good estimate of E_i. For example, Bruijnzeel and Wiersum (1987) used 10–12 trough-type *Tf* gauges (with a total surface area equivalent to 140–168 standard gauges) in an *Acacia auriculiformis* plantation in Indonesia. Measurements were made during two consecutive rainy seasons with fixed trough arrangements that differed between the two periods because the gauges were removed during the dry season. Although both estimates of seasonal *Tf* had a standard error of less than 5%, the mean values were very different. Because the difference was only partially explicable in terms of increases in canopy development and differences in rainfall patterns, it had to be concluded that the two fixed trough configurations sampled the spatial variation in *Tf* differently (cf. Fig. 12.3). Such findings indicate the need for a sufficiently large sampling plot commensurate with the heterogeneity of the forest under study. A 10 × 10 m sampling plot could be sufficient for a uniform tree plantation of high stem density but

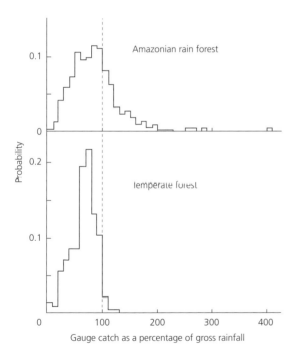

Fig. 12.3 Probability distribution of throughfall gauge catch in a random grid expressed as a percentage of coincident gross rainfall for an Amazonian rain forest and a pine forest in the UK. (After Lloyd & Marques 1988 with permission.)

entirely inappropriate for a heterogeneous species-rich lowland rain forest (Roberts 1999). Lloyd and Marques (1988), using a 100×4 m sampling transect in a central Amazonian rain forest, showed that about 40 fixed standard gauges would be needed to sample Tf with a standard error of 10% but that the same result could be achieved with only 10 'roving' gauges (Fig. 12.4). Interestingly, Lloyd and Marques (1988) obtained an E_i of only 9% using 36 roving bottle gauges where Leopoldo *et al.* (1982) had previously reported a value of 19% for the same forest on the basis of 20 fixed collectors. The large difference between these two estimates may well be caused by a more representative sampling of so-called 'drip points' (i.e. $Tf > P$, Fig. 12.3) by the roving-gauge arrangement. Not many studies have used a roving-gauge approach and published estimates of E_i must therefore be viewed with caution.

Figure 12.5 summarizes results for a number of British interception studies (expressed as a function of annual rainfall). Despite the large variation encountered in the deciduous group, E_i in deciduous stands is invariably lower than in coniferous forest for the same amount of P (Hall & Roberts

1990; Roberts 1999; cf. Zinke 1967). A similar contrast was noted for coniferous ($E_i > 20$–25%) and (semi)deciduous plantations such as teak and mahogany in the tropics ($E_i \approx 20\%$; Bruijnzeel 1997). Although evergreen, the relatively light crowns and smooth stems of (young) eucalypt plantations intercept only modest quantities of rainfall (typically about 10–15%), both in the tropics (Bruijnzeel 1997) and in their native Australia (Dunin *et al.* 1988; Crockford & Richardson 1990). However, recent work in south-eastern Australia has shown that E_i in an age series of natural eucalypt forest may rise to a peak of 25% at age 30 years after which it declines to about 15% at age 235 years (Haydon *et al.* 1996). Although some of the variation in E_i between different deciduous forests in Fig. 12.5 may be related to the presence or absence of a well-developed understorey, little is known of the intercepting capacity of the latter. Noirfalise

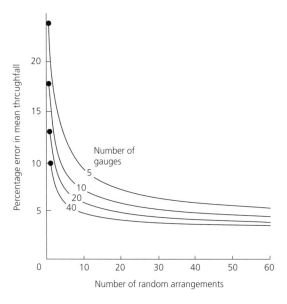

Fig. 12.4 Probable standard errors of the mean of throughfall in an Amazonian rain forest for increasing sample size and numbers of gauge relocations. (After Lloyd & Marques 1988 with permission.)

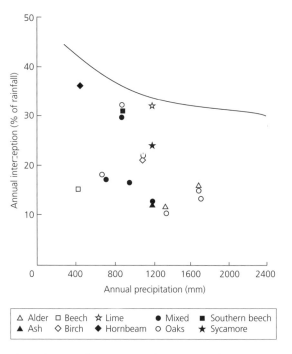

Fig. 12.5 Annual interception loss expressed as a percentage of annual rainfall vs. annual rainfall for European broadleaved trees. The solid line represents the annual interception percentage for coniferous forests in the UK. (After Hall & Roberts 1990 with permission.)

(1959) obtained values of 10.4 and 12.6% for E_i below bracken fern (*Pteridium aquilinum*) between late May and mid September and between mid September and mid November, respectively. Using rather more indirect means, Gash and Stewart (1977) derived a value of 12.3% of P for bracken fern below Scots pine in the eastern UK between mid May and early October, representing 12.6% of the annual E_i of the forest.

Bruijnzeel (1990) derived an average Tf of 85% (range 77–93%) from 13 investigations conducted in tropical lowland rain forest that used either a roving-gauge arrangement or more than 20 fixed gauges. Stemflow in these forests was usually less than 2%. Corresponding average values for tall montane forests experiencing little to no cloud incidence are 75% (range 67–81%) and less than 1%, respectively (Bruijnzeel 2000). As discussed in section 12.3, net precipitation in mossy montane cloud forests may be much higher due to the effect of 'occult' precipitation. The most stunted forms of ridge-top montane cloud forest, which frequently have crooked stems and trees with multiple stems, may exhibit exceptionally large amounts of Sf (up to 18%; Bruijnzeel 2000).

It should be noted that no distinction was made by Bruijnzeel (1990) between forests at mid-continental vs. continental edge and island locations. Shuttleworth (1989) advanced the idea that forest evaporation (notably E_i) may be much greater in the latter situations compared with the former. He based his contention on a comparison of the micrometeorology of a lowland forest in central Amazonia (Shuttleworth 1988) and a spruce plantation in Wales, UK (Shuttleworth & Calder 1979), as well as on the reported contrast in E_i between the Amazonian forest (Lloyd & Marques 1988) and a secondary lowland rain forest on the island of Java, Indonesia (Calder *et al.* 1986). The argument was rendered inconclusive at the time because of the use of markedly different methodologies for the estimation of E_i in Indonesia and Amazonia. However, since the late 1980s a number of rainfall interception studies in tropical forests located at continental edge and island locations with high rainfall (>3000 mm annually) have confirmed the possibility of very high E_i under such wet 'maritime' tropical condi-tions, both in the lowlands (Scatena 1990; Dykes 1997) and at higher elevations (Cavelier *et al.* 1997; Clark *et al.* 1998). We return to this impor-tant distinction in section 12.6.

12.4.2 Litter interception

Helvey and Patric (1965) argued that evaporation from the litter layer (E_s) may constitute a signifi-cant component of the overall interception loss and should therefore be determined separately. When measured *in situ* (as opposed to using litter trays which tend to give overestimates), E_s was shown to reach 2–5% of incident P in hardwood stands in the eastern USA ($\approx$50 mm annually). Interestingly, litter drying rates in winter were higher than in summer, presumably because the leafless condition of the forest permitted increased ventilation and irradiation of the forest floor at a time of maximum Tf (Helvey & Patric 1965). Later micrometeorological work by Black and Kelliher (1989) in Douglas fir forests con-firmed the importance of radiant energy and degree of turbulence to forest floor evaporation. These authors also showed that E_s and under-storey ET are largely complementary processes, with high values of E_s occurring in stands having little to no understorey vegetation and vice versa. Using weighing lysimetry Schaap (1996) showed E_s in a dense but well-ventilated Douglas fir forest in The Netherlands to be mainly dependent on forest floor moisture content, with (modelled) average annual values ranging from 0.21 mm daily on drier microsites to 0.38 mm daily on wetter sites. The corresponding stand-averaged annual total of 112 mm represented about 12% and 22% of incident P and Tf, respectively. Similar values (9.5–11% of P, or $\approx$ 155 mm yearly) were obtained for plantations of *Pinus caribaea* under warm non-equatorial tropical conditions in Fiji (Waterloo *et al.* 1999). Conversely, much lower annual totals (50–70 mm) have been reported for dense equatorial rain forest (Jordan & Heuveldop 1981; Roche 1982).

12.4.3 Interception modelling

Despite the numerous studies of rainfall intercep-tion conducted prior to the 1960s, little progress had been made in understanding the physics underlying the observed contrasts and inconsistencies in E_i, both between and within species and events. Explanations offered in the early literature were usually worded in terms of differences in canopy storage capacity (in turn related to canopy density, deciduousness, etc.) or the intensity and duration of the rain. However, by the mid 1960s profound changes were to occur. In the same year that Helvey and Patric (1965) still suggested that two simple linear *Tf–P* regression equations were sufficient to adequately quantify E_i for hardwoods in the eastern USA under a wide range of conditions, Monteith (1965) advanced a more physical description of the process of evaporation from a vegetation cover while Rutter (1967) presented a process-orientated approach to the study of E_i at the 1965 International Symposium on Forest Hydrology (Sopper & Lull 1967). The regression approach of Helvey and Patric (1965) was later criticized by Jackson (1971, 1975) for not incorporating rainfall intensity and duration or the length of intervals beween storms. However, Jackson himself was not very successful in raising the coefficients of determination (R^2), and therefore the predictive capacity, of his own more elaborate regression equations, thus illustrating the limitations of the empirical approach. Arguably therefore the two papers by Monteith (1965) and Rutter (1967), complemented by the interception modelling work of Rutter *et al.* (1971, 1975), mark a turn from the previous largely empirical experimental approach to a more physically orientated, process-based, approach (see below).

Monteith (1965) showed that the latent heat flux density (evaporation) from a vegetated surface could be described by:

$$\lambda E = \frac{\Delta A + \rho C_p \cdot VPD/r_a}{\Delta + \gamma(1 + r_s/r_a)} \quad [12.6]$$

where λE is latent heat flux flux density (W m^{-2}), A available energy (W m^{-2}), Δ slope of the saturation vapour pressure curve at air temperature T (Pa K^{-1}), ρ density of air (kg m^{-3}), C_p specific heat of air (J kg^{-1} K^{-1}) γ psychrometric constant (Pa K^{-1}), VPD vapour pressure deficit (Pa), r_a aerodynamic resistance (s m^{-1}) and r_s surface resistance (s m^{-1}). The surface resistance of the entire canopy (r_s) is represented by:

$$r_s = r_{sto}/LAI \quad [12.7]$$

where r_{sto} is the stomatal resistance to evaporation at the leaf level (s m^{-1}) and LAI is leaf area index (m^2 m^{-2}).

In the absence of advected heat, the energy available for evaporation (A) is considered equal to ($R_n - G$), i.e. the flux density of net radiation minus the heat flux density into the ground. For a wet canopy the surface resistance r_s becomes zero (Monteith 1965) and Eqn 12.6 reduces to:

$$\lambda E_{wet} = \frac{\Delta A + \rho C_p \cdot VPD/r_a}{\Delta + \gamma} \quad [12.8]$$

The aerodynamic resistance r_a, whose magnitude is critical to the outcome of Eqn 12.8, is ideally derived from the wind speed profile as measured above the canopy under (near)-neutral conditions of atmospheric stability (Thom 1975) according to:

$$r_a = \frac{\left(\ln\left[\dfrac{z-d}{z_0} \right] \right)^2}{k^2 \cdot u} \quad [12.9]$$

where z is observation height above the ground surface (m), d zero-plane displacement height (m), z_0 roughness length (m), k von Kàrmàn's constant (0.41) and u wind speed as measured at height z (m s^{-1}). In the absence of above-canopy wind data, d and z_0 are often assumed to be equal to 0.75 and 0.1 times the canopy height (h_v) (Brutsaert 1982). Alternatively, Eqn 12.8 may be solved inversely when λE_{wet} is known by other methods:

$$r_a = \frac{\rho C_p \cdot VPD}{\gamma \cdot \lambda E_{wet}} \quad [12.10]$$

although this method requires highly precise measurements of the atmospheric VPD under near-saturated conditions and is thus prone to errors (Gash *et al.* 1980).

With increasing vegetation height there is a corresponding increase in roughness length d and displacement length z_0 and thus, for a given wind speed, a decrease in r_a. Also, r_a decreases with wind speed; approximate values for different vegetation covers at a wind speed of 2.5 m s^{-1} measured at 10 m above the surface are 115, 50 and 10–15 s m^{-1} for short grass, field crops and forest, respectively (Pearce & Rowe 1979). Because values of the surface resistance to evaporation from a dry canopy (r_s) for grass and forest are much more similar (50–90 and 70–120 s m^{-1} respectively; Pearce & Rowe 1979), the magnitude of r_s/r_a in Eqn 12.6 is larger for a forest cover. This would lead to lower values of forest λE were it not for the compensating effect of the lower albedo and thus R_n of the forest. As a result, as long as soil water is not limiting and atmospheric humidity deficits remain low, transpiration rates for grass and forest do not differ much (Fig. 12.6). This is in great contrast to the finding that as a result of the much lower r_a of the forest, evaporation from a wet forest canopy will proceed much faster than from a wet grassland (Fig. 12.6).

Wet canopy evaporation rates of 0.2–0.5 mm h^{-1} have been observed under temperate conditions, even in cloudy winter weather when λE_{wet} must greatly exceed net radiant energy (Rutter 1967, 1968; Shuttleworth & Calder 1979). Under maritime tropical conditions, values as high as 0.7–1.2 mm h^{-1} have been inferred from measurements of net precipitation (Dykes 1997; Schellekens *et al.* 2000; Waterloo *et al.* 1999). As shown by Rutter (1967) and Stewart (1977), temperatures of wet forest canopies may be slightly cooler than the air passing overhead. This, together with the low aerodynamic resistance conducive to rapid evaporation, allows the development of a downward sensible heat flux capable of maintaining evaporation rates well in excess of available radiant energy. The degree to which the phenomenon influences annual E_i obviously depends on the frequency of wetting of the canopy (i.e. rainfall regime) and thus on the overall setting of the forest. For a pine plantation in the relatively dry eastern part of the UK, Stewart (1977) measured instantaneous rates of λE_{wet} of 127% of net radiant energy but on an annual scale E_i was modest at 214 mm or 38% of total forest ET (Gash & Stewart 1977). For a spruce plantation in the more maritime setting of wet central Wales with frequent passage of warm frontal rain, Shuttleworth and Calder (1979) reported instantaneous rates of λE_{wet} of about five times the available radiant energy. Annual ET exceeded R_n by 12%. Not surprisingly, E_i made up a large portion (61%) of total forest ET (Calder 1990). Recently, Schellekens *et al.* (2000) reported an even more extreme example from eastern Puerto Rico, where annual E_i from a colline rain forest varied between 1365 and 1790 mm (39–49% of P). Although no direct measurements were made of a downward sensible heat flux, there were strong indications that non-radiant energy dominated wet canopy evaporation. Not only were net precipitation fractions not statistically different for daytime and night-time events but also the average evaporation equivalent of net radiant energy (0.11 mm h^{-1}) was far smaller than the value of λE_{wet} as inferred from measurements of E_i (0.93–1.13 mm h^{-1}). In addition, annual ET as estimated from the water budget of a small watertight catchment exceeded R_n by 42%. The origin of the extra energy is still a matter of speculation. Advected warm air from the nearby Atlantic Ocean is a likely source (cf. Shuttleworth &

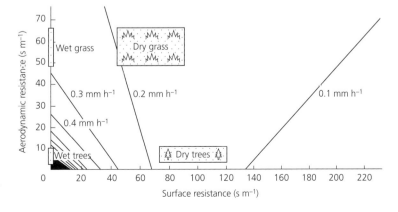

Fig. 12.6 Evaporation rates calculated from the Penman–Monteith equation as a function of the aerodynamic and surface resistances to evaporation for cool summer daytime conditions. (After Calder 1979 with permission.)

Calder 1979) although, unlike the situation in Wales, wind speeds in eastern Puerto Rico are generally low. Another possible energy source may be heat released upon condensation of water vapour in the air above the forest. Some support for this contention comes from the observation that Tf/P was independent of storm size, i.e. larger rainfall events were accompanied by higher interception totals (Schellekens *et al.* 2000). This would suggest a positive feedback of rainfall amount (and thereby condensation) on the magnitude of E_i. Further work is needed to test these speculations. However, even if it is accepted that for situations like those encountered in Wales and Puerto Rico, where the total energy available for evaporation during rain can be much larger than provided by radiant energy alone (Shuttleworth & Calder 1979; Schellekens *et al.* 2000), it is not clear how the aerodynamic resistance of the forest can be so low as to enable these high wet canopy evaporation rates. For the spruce plantation in Wales, Calder (1990) required an optimum value for r_a of 3.5 s m^{-1} (vs. a theoretical value of 5.4 s m^{-1} using Eqn 12.9 at an average wind speed of 3.75 m s^{-1}). Using Eqn 12.10, Schellekens *et al.* (2000) obtained values of 2.1–3.2 s m^{-1} (depending on VPD) vs. 18–22 s m^{-1} when using Eqn 12.9. Whilst in the Puerto Rican case r_a may be lowered somewhat by the irregular character of the canopy due to former hurricane damage, Calder (1990) and McNaughton and Laubach (1998) have pointed to the possibility of enhanced upward transport of evaporated moisture (i.e. low r_a) by wind gusts and eddies. Such gusts would assume extra importance under conditions of relatively low wind speeds such as in Puerto Rico. Determination of λE_{wet} using eddy correlation equipment in combination with an energy budget approach (Gash *et al.* 1999) recently confirmed the basic soundness of the momentum transfer theory underlying Eqn 12.9. More work is needed to resolve the matter.

Physically-based models may help to elucidate the relative importance of different factors in the interception process, which are difficult to assess from an actual interception record. Two well-known models are the dynamic model developed by Rutter *et al.* (1971, 1975) and its simpler derivative, the analytical model of Gash (1979).

The Rutter model calculates a running water balance for the canopy (and trunks). It requires hourly rainfall and meteorological data (for the computation of evaporation using Eqn 12.8) as inputs, as well as values of four parameters describing the forest (free throughfall and stem-flow coefficients, canopy and trunk saturation values). In addition, it uses an empirical expression to compute drainage from the canopy. The Rutter model has been used successfully for such contrasting environments as coniferous plantations in the UK (Gash & Morton 1978) and tropical rain forest in central Amazonia (Lloyd *et al.* 1988). However, it greatly underestimated E_i under wet maritime tropical conditions (Java, Puerto Rico) where unrealistically high canopy storage capacities or very low aerodynamic resistances were required to match predicted and measured E_i (Calder *et al.* 1986; Schellekens *et al.* 1999). Valente *et al.* (1997) used a reformulated version of the model to predict rainfall interception by sparse canopy forests in Portugal.

The effect of varying rainfall intensities and evaporation rates, canopy storage capacity and the distribution of rain (continuous vs. intermittent) on the magnitude of E_i using the Rutter model is illustrated in Fig. 12.7 (Rutter 1975). The limiting effect of low rainfall intensity at faster evaporation rates is clearly borne out by the simulations, as is the increase in E_i at higher rainfall intensities and for higher canopy storages, especially in the case of intermittent rain.

A major practical drawback of the Rutter model is its high data requirement. The analytical model of rainfall interception developed by Gash (1979) is conceptually similar to the Rutter model but much less data-demanding. It is usually applied on an event basis and uses average instead of hourly rates of rainfall on to, and evaporation from, a saturated canopy. Apart from evaporation from a saturated canopy (estimated using Eqns 12.8 & 12.9), the model computes the evaporation before the canopy is fully wetted and after rainfall has ceased, the effect of small storms insufficient to saturate the canopy, and evaporation from trunks. Total E_i is then estimated as the sum of the respective interception components. The analytical model and a derivative for sparse canopies developed by Gash *et al.* (1995) have been applied

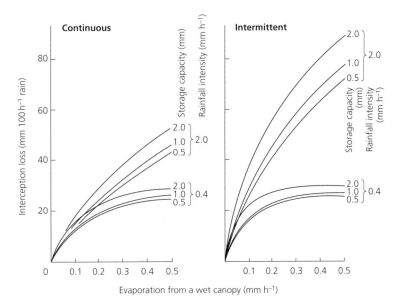

Fig. 12.7 Interaction of wet canopy evaporation rate, rainfall intensity, rainfall distribution and canopy storage capacity on interception loss from 100 h of rain. (After Rutter 1975 with permission.)

to a wide range of environments, ranging from temperate coniferous and deciduous forests (Gash *et al.* 1980; Carlyle-Moses & Price 1999) to sparse Mediterranean pine and eucalypt forests (Valente *et al.* 1997) and tropical rain forests (Lloyd *et al.* 1988; Hutjes *et al.* 1990). As in the case of the Rutter model, predicted amounts of E_i under more maritime conditions were underestimated by the model when evaporation from a saturated canopy was computed using Eqn 12.8. For example, Gash *et al.* (1980) had to roughly double the average value of λE_{wet} to match measured and predicted E_i in a spruce plantation in Wales whereas Schellekens *et al.* (1999) required a fourfold to fivefold increase for the Puerto Rican case discussed above (cf. Pearce *et al.* 1980a; Bruijnzeel & Wiersum 1987; Dykes 1997). A much better match of cumulative Tf was obtained in the latter case by optimizing for λE_{wet} than for canopy storage capacity S (cf. Calder *et al.* 1986).

In the Rutter and Gash models the canopy capacity is treated as a fixed parameter but in a more recent development in interception modelling (the so-called stochastic model proposed by Calder 1986) the size of the canopy capacity is allowed to vary. In the stochastic model the (repeated) wetting of the canopy is described in terms of the probability with which raindrops of different sizes strike the canopy. Larger drop sizes are considered less efficient at wetting the canopy (i.e. filling the storage capacity) than smaller drops. Therefore, for the same depth of rain, larger drops will not only take longer to (re)fill the canopy capacity but also the amount of water retained on the canopy will be smaller when drops are larger. In a recent modification of the model (Calder 1996), the wetting functions were adjusted to include the possibility of wetting of lower canopy layers by secondary drops falling from the upper layers. Although the two-layer version of the model was applied successfully in the simulation of E_i from a forest in Sri Lanka (Hall *et al.* 1996), the precise physical meaning of some of the model parameters remains somewhat vague. In this respect it is pertinent that Uijlenhoet and Stricker (1999) recently pointed out that the stochastic model employed an erroneous rainfall intensity vs. drop size relationship. In theory, however, the stochastic approach holds considerable promise for use in humid tropical situations where raindrop sizes tend to be larger than in the temperate conditions under which the Rutter and Gash models originated. It would be interesting therefore to examine the performance of the (improved) stochastic model under the contrasting climatic conditions prevailing in central

Amazonia (mid-continental equatorial conditions with relatively high rainfall intensities where the Rutter and Gash models gave satisfactory predictions; Lloyd & Marques 1988; Lloyd *et al.* 1988) and eastern Puerto Rico (maritime tropical site with frequent low-intensity rain where the Rutter and Gash models produced large underestimates; Schellekens *et al.* 1999, 2000).

Models describing rates of E_s (evaporation from the forest floor) range from site-specific exponential drying curves (Helvey 1964; Waterloo *et al.* 1999) to a more explicitly physical approach that is conceptually similar to the Rutter canopy interception model (Schaap 1996).

12.5 TRANSPIRATION

12.5.1 Amounts and measurement of transpiration

With few exceptions (e.g. Gash & Stewart 1977; Dunin & Greenwood 1986; Shuttleworth 1988), forest hydrological studies before the end of the 1980s generally evaluated forest transpiration (E_t) as a residual term in the water budget equation (Eqn 12.4) by subtracting streamflow (or drainage), soil water storage changes and rainfall interception from incident rainfall. Estimates of E_t based on the catchment or site water balance approach are prone to considerable uncertainty because of the possibility of error accumulation. Moreover, there is the danger of ungauged subterranean water transfers in the case of catchments (Lee 1970) and, for site water balances, extraction by deep roots beyond the range of the soil water measurements (Hodnett *et al.* 1996; Calder *et al.* 1997) as well as problems related to spatial variability of soil hydraulic properties and soil water contents (Cooper 1979; Rambal *et al.* 1984). The widespread application of micrometeorological approaches such as the Bowen ratio–energy balance method (Angus & Watts 1984) and eddy correlation techniques (Shuttleworth 1992) is limited not only by their requirement of homogeneous, smooth and even fetch conditions but also by the fact that the measurements need to be made above the forest canopy. Similarly, attempts to obtain *in situ* estimates of leaf transpiration using porometers to determine leaf stomatal

resistance to evaporation (r_{sto}), and the extrapolation of such results to entire canopies using Eqns 12.6 and 12.7 (Monteith 1965), have met with mixed success because of in-canopy variations in water use that require adequate characterization of climatic gradients and leaf age distribution within the canopy (cf. Schulze *et al.* 1985; Roberts *et al.* 1993; Dawson 1996). An additional problem relates to the often poorly quantified influence on whole-plant transpiration of the diffusive resistance of the boundary layers surrounding each leaf and the canopy. The sensitivity of leaf or canopy transpiration to a marginal change in r_{sto} has been described in terms of a dimensionless 'decoupling coefficient' (Ω) by Jarvis and McNaughton (1986). The value of Ω ranges between 0 (full stomatal control) and 1 (no stomatal control). Decoupling of transpiration from stomatal control increases sharply with increasing leaf size, both because larger leaves are associated with lower boundary layer conductance (i.e. higher r_a) and because r_{sto} tends to be smaller for large-leaved species than for fine-leaved species such as conifers (Jarvis & McNaughton 1986). Therefore, porometric measurements of transpiration in coniferous forests having low values of Ω (high stomatal control) may be interpreted as actual canopy-scale fluxes (Schulze *et al.* 1985), although reliance on porometric data alone for such large-leaved (tropical) species as *Tectona grandis* and *Gmelina arborea* (Ω values of 0.87–0.94; Jarvis & McNaughton 1986) could lead to misinterpretation of the significance of stomatal activity to transpiration (Wullschleger *et al.* 1998).

Partly in response to the limitations of conventional measuring techniques, efforts to determine forest transpiration via extrapolation to stand level of single-tree estimates of water use have been stepped up considerably, especially since the arrival of thermally based methods in the last two decades (see reviews by Swanson 1994; Smith & Allen 1996; Wullschleger *et al.* 1998; Grime & Sinclair 1999). Although estimates of water use derived from heat-balance, heat-dissipation and heat-pulse techniques suffer from uncertainties because of empirical calibrations, thermal gradients and variation in water flux with sapwood depth, these systems are relatively inexpensive, easy to use and readily linked up with data loggers for remote operation (Wullschleger *et al.* 1998). In addition, relatively good agreement has been reported between estimates of stand water use based on whole-tree measurements and eddy correlation methods in several cases (Köstner *et al.* 1992; Berbigier *et al.* 1996; Dawson 1996). Because hydrologists need stand water fluxes to be expressed on a ground area basis (L m^{-2} or mm), estimates of the water use of individual trees (usually expressed in kilograms or litres per unit of time) need to be scaled up to the stand level. A variety of scalars have been proposed to achieve this, ranging from tree domain (defined either as distance between stems or projected crown area), basal area or stem diameter at breast height, to leaf or sapwood area. The suitability of a particular scalar can be expected to depend on the situation (e.g. even-aged plantations vs. complex natural forests) but Hatton *et al.* (1995) demonstrated that four out of five different scalars performed satisfactorily for scaling up water use in a *Eucalyptus populnea* woodland in Australia. In contrast to the number of studies dealing with spatial aspects of scaling, there are comparatively few investigations of the temporal aspects of scaling (Wullschleger *et al.* 1998). Hatton and Wu

(1995) addressed this problem in various eucalypt woodlands and found that water use of individual trees was linearly related to leaf area during periods of abundant soil water, but that the relationship broke down during periods of water stress. Not only did large trees take up proportionately less water per unit leaf area than small trees under such conditions but also the shape of the relationship appeared to depend on soil water content and time of year, indicating that scalars like leaf area may exhibit strong short-term temporal dependence because of the possible occurrence of soil water stress. However, similar problems may also arise in areas where water stress is uncommon. Vertessy *et al.* (1995, 1997) derived rather contrasting relationships between tree water use and stem diameter for 15-year-old and 57-year-old stands of mountain ash (*Eucalyptus regnans*). Haydon *et al.* (1996) circumvented the long-term temporal dependence of stem diameter as a scalar by using tree sapwood area instead. Other long-term changes in forest water use may be related to the development or disappearance of understorey vegetation (Fig. 12.8; see also below).

In contrast to the wet canopy component of total evapotranspiration (rainfall interception, E_i),

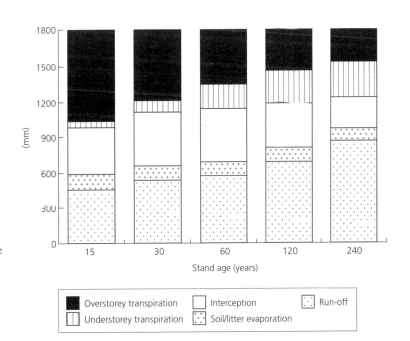

Fig. 12.8 Changes in water balance components for mountain ash forest stands of various ages in south-eastern Australia, assuming annual rainfall of 1800 mm. (After Vertessy *et al.* 1998 with permission.)

annual totals of evaporation from a dry canopy (transpiration, E_t) show comparatively little variation between species and sites within a given climatic zone. For example, Roberts (1999) lists results for 13 deciduous and seven coniferous stands in western Europe, together representing 29 years of observations. Leaving aside for the moment the high values obtained for three broadleaved and coniferous forests in southern England (possibly reflecting overestimates related to the methodologies used; cf. Jarvis & McNaughton 1986) and one excessively low value related to drought, but adding estimates for Austrian pine, Scots pine and Douglas fir in The Netherlands (Mulder 1983; Tiktak & Bouten 1984; Elbers *et al.* 1996), yields an average of 303 ± 37 (range 240–360) mm year^{-1} for the E_t of deciduous forests and 334 ± 37 (range 280–360) mm year^{-1} for conifers. Roberts (1999) ascribed the surprisingly similar E_t for deciduous and evergreen species of varying age and growing on different soils to three main factors: (i) the frequently observed positive correlation between atmospheric VPD and stomatal/surface resistance; (ii) the relative insensitivity of E_t to soil water availability; and (iii) the compensatory effects of the presence or absence of understorey vegetation. The respective factors are discussed briefly below.

A distinct response in stomatal/surface resistance to increases in VPD has been observed in vegetation types as diverse as northern conifers (Tan & Black 1976; Lindroth 1985) and tropical rain forests (Shuttleworth 1988; Meinzer *et al.* 1993). Coastal species seem to be more sensitive to changes in VPD than continental species, possibly because they are adapted to maritime conditions with relatively low vapour deficits (Whitehead & Jarvis 1981). Also, conifers have been reported to be more sensitive to changes in VPD than broadleaved species (at least in the Pacific North-west region of the USA), although this was to some extent compensated by a greater sensitivity of the latter to soil water deficits (Marshall & Waring 1984). Although the precise underlying mechanism is as yet unclear (Grantz 1990; Meinzer *et al.* 1993, 1997a,b), the strong feedback response between r_s and VPD results in a significant dampening of instantaneous transpiration rates to the extent that daily totals typically remain below 4 mm in most humid climates (both temperate and tropical), even though potential evaporative demands by the atmosphere may be much higher (Roberts 1983; Granier *et al.* 1996). On a related note, Federer (1977) showed that stomatal resistances in five broadleaved deciduous species from the north-eastern USA increased more rapidly with increasing VPD when initial values of r_{sto} were low and vice versa. Roberts (1999) suggested that, as a result, values of stomatal resistance and instantaneous leaf transpiration rates for the respective species could be much more similar under conditions of intermediate ('typical') VPDs than would be expected on the basis of the contrasts in minimum resistances alone. This in turn might constitute another reason for the similarity in E_t between species noted above. However, it is inconsistent with the considerable reductions in streamflow (up to 100 mm annually) accompanying a shift in forest composition from a dominance by (mature) maple and beech (having relatively high r_s) to pin cherry and birch (pioneer species with relatively low r_s) following clearcutting and regrowth in the same area (Hornbeck *et al.* 1993; cf. section 12.7.3).

Although most vegetation types exhibit similar increases in r_s in response to increases in VPD (Grantz 1990), there are notable exceptions. Poplars grown in a short-rotation coppice system in the UK have been reported to maintain low values of r_s and high transpiration despite increased humidity deficits as long as soil water was not limiting (Hall & Allen 1997; cf. Meinzer *et al.* 1997b). Similarly, Elbers *et al.* (1996) obtained an annual value of about 510 mm for the E_t of a poplar stand with a high groundwater table in The Netherlands, i.e. well above the 300 mm derived earlier for temperate broadleaved species in western Europe. Plantations of *Eucalyptus camaldulensis* and *E. tereticornis* in southern India exhibited similar behaviour (with transpiration values of up to 6 mm daily) when unrestricted by soil water deficits at the end of the monsoon. However, values fell off to 1 mm daily upon reaching low soil water contents during the subsequent long dry season (Robert & Rosier 1993). Because of this regulating mechanism, the annual water use of the eucalypt plantations on soils of intermediate depth (≈3 m) was not significantly different from that of indigenous dry deciduous forest (Calder *et al.* 1992). On deep soils (>8 m) the annual water use of the eucalypts exceeded annual rainfall considerably, suggesting 'mining' of soil water reserves that had accumulated previously in deeper layers during years of above-average rainfall. Moreover, the rate of root penetration was shown to be at least as rapid as 2.5 m annually and roughly equalled above-ground increases in height (Calder *et al.* 1997). Similar observations have been made below *Eucalyptus grandis* in South Africa (Dye 1996). It is pertinent that Viswanatham *et al.* (1982) observed strong decreases in streamflow after coppicing *E. camaldulensis* in northern India. Together, these findings confirm earlier contentions in the popular environmental literature that eucalypts can be 'voracious consumers of water' (Vandanashiva & Bandyopadhyay 1983). Planting of eucalypts, particularly in subhumid climates, should therefore be based on judicious planning (cf. section 12.7.6).

Although soil water stress has been shown to have a marked effect on r_s and E_t in seasonal climates (Harding & Stewart 1986; Roberts & Rosier 1993; Hatton & Wu 1995), the modest transpiration rates effected by the r_s–VPD feedback mecha-

nism referred to earlier normally imply that a substantial fraction of soil water reserves can be used before transpiration starts to level off due to increased stomatal resistance. As shown in Fig. 12.9 for a young eucalypt forest in southeastern Australia, the point at which transpiration becomes reduced tends to be reached earlier when rates of water uptake are high compared with during times of more modest uptake. Rutter (1968) presented a number of examples from the USA illustrating the same phenomenon. As such, critical soil water deficits affecting E_t will only be rarely attained in humid climates, and may therefore be expected to play only a minor role in generating differences between species and sites (Roberts 1983).

Finally, the contribution from the understorey (and to a lesser extent the litter layer, cf. section 12.4.2) to stand transpiration constitutes a third mechanism through which differences between sites with strongly contrasting forest structures tend to be evened out. Transpiration totals for a dense forest with little to no undergrowth can be similar to those of open forests with much more vigorous understoreys (Roberts *et al.* 1982; Roberts & Rosier 1994; cf. Vertessy *et al.* 1998). Contributions by understories can indeed be substantial: Tan and Black (1976) showed how a salal

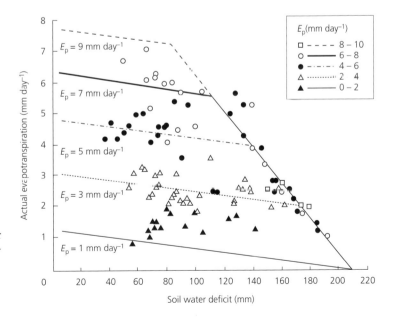

Fig. 12.9 Actual evaporation rates for a young eucalypt plantation in south-eastern Australia as a function of coincident potential evaporation rates and soil water deficits. (After Dunin *et al.* 1985 with permission.)

(*Gaultheria shallon*) understorey of a Douglas fir forest in western Canada could account for up to 70% of the transpiration. Similarly, Roberts *et al.* (1980) reported that bracken fern (*Pteridium aquilinum*) below Scots pine in East Anglia, UK contributed up to 60% of forest transpiration during dry spells, whereas under normal conditions transpiration from the bracken still made up 25% of the total. Further examples have been given by Black and Kelliher (1989).

A particularly powerful (and still underused) tool in the study of relative rates and sources of water uptake by the differently sized plants of the overstorey and understorey of a forest is stable isotope (δD, $\delta^{18}O$) analysis (Jackson *et al.* 1995; Dawson 1996). For example, using naturally occurring contrasts in stable hydrogen isotope abundances in soil and groundwater below a sugar maple forest in the north-eastern USA, Dawson (1996) was able to demonstrate that the tallest trees drew their water from groundwater found at more than 3 m below a fragipan whereas smaller trees (whose roots were unable to reach the groundwater table) used soil water exclusively, except during two periods of drought stress. Interestingly, the 'groundwater' taken up by the younger trees during these drought periods had previously been lifted to the upper layers of the soil by the larger trees via a process known as 'hydraulic lift', i.e. the nocturnal uptake of water by roots from the subsoil and later released from shallow roots into the upper layers of the soil (Dawson 1993).

In contrast to the relative abundance of transpiration estimates for humid temperate forests (Roberts 1999), information on tropical rain forests is scarce and mostly based on indirect (water-budget) computations. For nine studies in lowland rain forests, Bruijnzeel (1990) derived an average annual E_t of 1045 mm (range 885–1285 mm). For montane rain forests with limited to negligible cloud incidence, Bruijnzeel (1990, 2000) reported annual values of 510–895 mm ($n = 8$) compared with 255–315 mm for montane cloud forests. There are very few direct estimates of long-term water uptake by tropical forests, be it for natural forests (Shuttleworth 1988; Roberts *et al.* 1993; cf. Granier *et al.* 1996) or plantations (Waterloo *et al.* 1999; cf. Bruijnzeel 1997).

12.5.2 Transpiration modelling

With the incorporation of a physiological control (represented by the surface resistance term r_s) in the energy balance cum aerodynamic transport combination equation of Penman (1948), Monteith (1965) established an effective framework for the quantification of evaporation from any (vegetated) surface, as summarized in Eqn 12.6. Unlike evaporation from a wet vegetation, which under given weather conditions is largely controlled by the aerodynamic resistance of the vegetation (r_a), transpiration (evaporation from a dry canopy) is chiefly governed by the magnitude of r_s, which in turn is influenced by a range of environmental and plant variables, including solar radiation, leaf area, leaf temperature and leaf water potential, as well as atmospheric VPD and soil water deficit (Stewart 1988; Shuttleworth 1989). As shown in Fig. 12.10, canopy conductance g_c (i.e. the reciprocal of r_s) increases with increasing light intensity and temperature (up to a maximum value) and decreases with increasing atmospheric humidity deficit and (as mentioned in the previous section, after exceeding a threshold value) soil water deficit. Various, increasingly complex, submodels have been developed to describe the behaviour of r_s in response to ambient environmental conditions. However, field application of such models is often hampered by their need for highly detailed input data (Whitehead 1998).

Under humid temperate or tropical conditions the combined effect of the respective variables influencing r_s generally gives rise to distinct diurnal patterns that change comparatively little with time as long as soil water stress is absent (Dolman *et al.* 1991). The most common approach is to solve Eqn 12.6 inversely for r_s for all (half-hourly to hourly) periods for which λE is known through micrometeorological or other means (e.g. lysimetry, sap flow measurements) during short-term intensive measuring campaigns, using:

$$r_s = \frac{\rho C_p \cdot VPD}{\gamma \cdot \lambda E} + \left(\frac{\Delta A}{\gamma \cdot \lambda E} - \frac{\Delta}{\gamma} - 1 \right) \qquad [12.11]$$

Next, the resulting values of r_s are related to either the time of day (Gash & Stewart 1975;

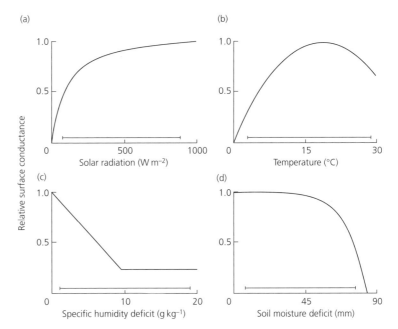

Fig. 12.10 General form of the dependence of the relative surface conductance on (a) solar radiation, (b) temperature, (c) specific humidity deficit and (d) soil moisture deficit. (After Stewart 1988 with permission.)

Dolman *et al.* 1991) or environmental variables (Waterloo *et al.* 1999) using simple or multiple-regression equations to enable the computation of r_s and thus λE during times when direct estimates of either are not available. Such an approach also avoids the need for quantitative estimates of the decoupling coefficient Ω (cf. section 12.5.1) when scaling up from leaf to canopy resistance (Jarvis & McNaughton 1986).

12.6 TOTAL EVAPOTRANSPIRATION

Although it is beyond the scope of the present process-orientated chapter to discuss amounts of total evapotranspiration (*ET*) associated with different forest types and climatic zones, it can be concluded from the previous sections that changes in the interception component of evaporation (E_i) will be much more pronounced than those of the transpiration component (E_t). Taking western Europe as an example, annual totals of E_t were shown to be quite similar for deciduous and coniferous forests over a wide range of soil and rainfall conditions (reported range 240–360 mm; see section 12.5.1). Conversely, absolute annual totals of E_i for the forests illustrated in Fig. 12.5 ranged from about 120 mm

(oak coppice under low rainfall conditions in the southern UK; Thompson 1972) to about 700 mm (spruce at a high rainfall location in upland Wales subjected to advected energy from the ocean; Calder 1990).

It is of interest to examine whether a similar contrast between variations in E_i and E_t also exists in the humid tropics where radiation loads, average temperatures and rainfall are all higher than in the temperate zone. In a survey of (mostly catchment) water budget studies in lowland tropical rain forests, Bruijnzeel (1990) encountered such large variations in annual *ET* that he was unable to detect any systematic differences between the main rain forest blocks of South-East Asia, Amazonia and West Africa. He suggested an average annual value of about 1400 mm for forests not subject to significant water stress, ascribing annual *ET* totals well above 1500 mm generally to catchment leakage rather than to specific climatic conditions favouring high evaporation (cf. Richardson 1982). More recently, Schellekens *et al.* (2000) combined amounts of *ET* and E_i as measured in 14 lowland tropical rain forests and derived approximate estimates of E_t by subtracting E_i from *ET* when direct measurements of E_t were lacking. As shown in Fig. 12.11(a), annual

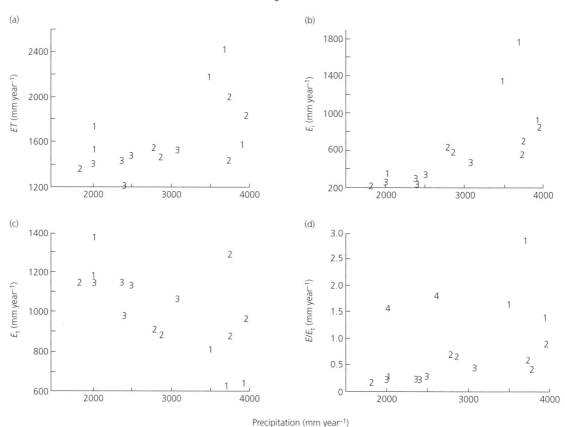

Fig. 12.11 Variations in (a) annual evapotranspiration (ET), (b) annual rainfall interception (E_i), (c) annual transpiration (E_t) and (d) interception–transpiration ($E_i : E_t$) ratio plotted against annual rainfall for selected water budget studies in the humid tropics. 1, type 1 data from coastal and island sites in the outer tropics; 2, type 2 data from continental edge sites at equatorial latitudes; 3, type 3 data from mid-continental equatorial sites; 4, type 4 data from temperate maritime sites. (After Schellekens *et al.* 2000, with permission.)

totals of ET increase with rainfall, although the scatter is considerable, especially for rainfall above 3500 mm. Interestingly, for a given rainfall, values of ET for forests located on islands in the outer tropics (i.e. at about 18° N or S, type 1 locations) tend to be higher than those associated with forests situated closer to the equator (0–10° N or S). The latter group can be subdivided into continental edge (type 2) and mid-continental (type 3) locations. The scatter in ET values is primarily due to the very large variation in intercepted rainfall (220–1790 mm year^{-1}; Fig. 12.11b), whereas the variability in the transpiration component is less pronounced (630–1375 mm year^{-1}; Fig. 12.11c). Although the values of E_t should be

treated with caution because of the crude manner in which they were derived, an interesting pattern emerges when the $E_i : E_t$ ratio is plotted against rainfall (Fig. 12.11d). At the low end of the rainfall spectrum ($P < 2000$ mm year^{-1}), rainfall interception typically makes up 20–25% of ET, regardless of site location in terms of proximity to the ocean or the equator (i.e. types 1, 2 or 3). However, E_i becomes gradually more important when rainfall exceeds a value of 2500–2700 mm year^{-1}, both in the absolute sense (Fig. 12.11b) and relative to E_t (Fig. 12.11d). Also, for similar amounts of rainfall above this 'threshold', both absolute and relative values of E_i tend to be somewhat higher for type 2 locations (equatorial continental edge sites) than for type 3 locations (mid-continental sites). In turn, values for type 1 locations (maritime outer tropical sites) exceed those associated with type 2 locations (Fig. 12.11b,d). Interestingly, two temperate maritime sites with high rainfall but lower absolute values of ET and E_t (Plynlimon, Wales, Calder 1990; and the north-western tip of South Island, New Zealand, Pearce *et al.* 1982) exhibited very similar values for $E_i : E_t$ (type 4 in Fig. 12.11d). Such findings confirm the importance of intercepted rainfall in both humid tropical and humid temperate maritime settings as postulated earlier by Shuttleworth (1989). However, the patterns emerging from Fig. 12.11 do not support the contention of Calder (1998) that 'climatic circulation patterns in the wet tropics do not favour large-scale advection of energy to support evaporation rates and so evaporation rates are likely to be closely constrained by the availability of solar radiation'. The very high ET values that were well in excess of radiant energy reported for colline rain forest in Puerto Rico and pine plantations in Fiji (Schellekens *et al.* 2000; Waterloo *et al.* 1999; cf. Richardson 1982) clearly indicate the importance of advected energy under wet maritime tropical conditions.

12.7 HYDROLOGICAL EFFECTS OF FOREST MANIPULATION

Within the limitations imposed by flow measuring techniques, the practical overall influence exerted by forests on hydrological processes is most clearly illustrated through comparison of streamflow amounts emanating from areas with contrasting proportions or types of forest. Because of the complications related to local climatic contrasts (rainfall, exposure to radiation and air streams) and ungauged subterranean transfers of water from one catchment to another, such direct comparison of results obtained for catchments with contrasting covers can be problematic (Lee 1970). The classical answer to these problems has been the paired catchment approach in which streamflows from two (often adjacent) catchments of comparable geology, exposition and vegetation are expressed in terms of each other (using regression analysis) during a 'calibration phase'. Once a good calibration relationship has

been established, one of the catchments is subjected to manipulation of its cover (e.g. strip cutting, clearfelling or afforestation) while the other remains undisturbed as a 'control'. Throughout this 'treatment' or 'experimental' phase, streamflow continues to be monitored and any effects of the treatment are evaluated by comparing actually measured flows from the manipulated catchment with those predicted by inserting the flows from the control catchment into the calibration relationship (Hewlett & Fortson 1983). Although a more rigorous comparison between catchments is obtained in this way, an underlying assumption is that differences in leakage between the two catchments remain unchanged with time, regardless of catchment cover status. Also, to avoid overstretching of the calibration statistics during the treatment phase to accommodate extremes in streamflow resulting from excessive rainfall or drought, it is important that the calibration period includes sufficient variation in rainfall (Hornbeck 1973; Lesch & Scott 1997). This, of course, renders the paired catchment method a time-consuming (>5–10 years) and expensive affair. In addition, the method is essentially a black-box technique requiring additional process research to reveal the relative importance of different causative factors to explain the obtained results (Lesch & Scott 1997). All this, plus the limited resolution afforded by the paired catchment approach (usually more than 20% cover change is required for effects on streamflow to be detectable), has led to a general decline in the number of such studies in the last two decades and a gradually greater emphasis on physically based modelling (cf. section 12.7.7).

In the following sections the hydrological effects of (i) forest manipulation (thinning, selective logging, undergrowth or riparian vegetation removal, clearcutting followed by regrowth or conversion to other land uses) and (ii) (re)forestation are reviewed from the experimental manipulation and modelling perspectives.

12.7.1 Effects of forest thinning on rainfall interception

It has long been recognized that amounts of throughfall tend to be inversely related to the stocking of a forest, be it in the context of thinning or logging operations (Wilm & Dunford 1948; Rogerson 1967; Teoh 1977; Aussenac *et al.* 1982; Teklehaimanot *et al.* 1991; Asdak *et al.* 1998) or forest age (Delfs 1955; Helvey 1967; cf. Haydon *et al.* 1996). However, the steady increase in rainfall interception with age observed for coniferous forests (Delfs 1955; Helvey 1967) is not paralleled in deciduous forest. Helvey and Patric (1965) reported that E_i in 11-year-old oak coppice in the south-eastern USA did not differ from that observed in 30-year-old and 50-year-old mixed poplar–hickory forests in the same area. Apparently, the oak coppice had already acquired leaf biomass and roughness characteristics similar to those of the much older forests. As indicated earlier (section 12.4.1), interception by deciduous forests during the dormant season is lower than during the growing season. Usually, an increase in throughfall of 5–10% occurs during winter (Helvey & Patric 1965). This increase is by no means proportional to the reduction in LAI, which can be up to sixfold (Monk & Day 1988). Similarly, the observed decrease in rainfall interception following forest thinning is often not commensurate with the degree of canopy opening. For example, a 50% reduction in basal area of a Douglas fir forest in France resulted in only a 13% drop in E_i (Aussenac *et al.* 1982). An even smaller effect was noticed in a regenerating eucalypt forest in south-eastern Australia, where E_i was reduced by as little as 4% following a uniform thinning operation that removed 50% of the overstorey biomass (Jayasurya *et al.* 1993). Likewise, a 13-fold reduction in basal area (corresponding to a change in planting interval from 2×2 m to 8×8 m) in a dense Sitka spruce plantation in Scotland was accompanied by only a 3.7-fold reduction in E_i (Teklehaimanot *et al.* 1991) who explained this phenomenon not so much in terms of decreases in canopy cover *per se* but rather as a function of a gradually decreasing turbulent exchange between the trees and the surrounding air after opening up the canopy. Evidence for this contention came from a strong negative correlation between rainfall interception and aerodynamic resistance (r_a), with r_a increasing linearly as the spacing between the trees increased up to 8×8 m

(Teklehaimanot *et al.* 1991; cf. Gash *et al.* 1995, 1999).

Research results from the humid tropics largely confirm the above findings. Veracion and Lopez (1976) and Florido and Saplaco (1981) found negligible increases in throughfall after thinning 30–50% of the biomass of 10–15-year-old and 30-year-old natural pine forests in the Philippines. Significant increases were obtained only after 70% of the trees had been removed. Ghosh *et al.* (1980) applied a 20% thinning treatment to a broadleaved (*Shorea robusta*) forest in northern India and found that throughfall increased from 72 to 81%. However, because of a concurrent reduction in stemflow of about 4% (and not necesssarily related to the thinning), the overall change in intercepted rainfall was a modest 5%. Recently, Asdak *et al.* (1998) compared rainfall interception from undisturbed and heavily logged lowland rain forest plots in central Kalimantan, Indonesia. Interception from the logged forest was 6% of gross rainfall compared with 11% in the unlogged forest. However, because the comparison did not pertain to the same forest before and after logging, it is not possible to assess to what extent the observed reduction in E_i reflects the change in canopy cover or pre-existing structural differences between the two stands (cf. Lloyd & Marques 1988).

12.7.2 Effects of thinning and selective logging on transpiration and water yield

Whilst the effect of thinning on E_i is rather limited, effects on soil water (and ultimately streamflow) are even smaller. Opening up of a stand not only enhances the penetration of radiation to the understorey vegetation and the forest floor, but also the remaining vegetation will start competing for the extra moisture supplied by the initially increased throughfall (Aussenac *et al.* 1982; Stogsdill *et al.* 1992). The magnitude and the duration of such effects will differ among locations, depending on the vigour of overstorey and understorey vegetation, climatic conditions (including site exposure) and the configuration of the cutting.

No changes were detected in the streamflow from a deciduous hardwood forest catchment of south-easterly exposure at Coweeta (south-eastern USA) after selective logging removed 27% of the basal area; only a 4.3% increase occurred after a 53% selective cut. Also, removing the entire understorey (representing 22% of forest basal area) from a 28-ha catchment of north-westerly exposure in the same area produced an equally modest change (Johnson & Kovner 1956). Similarly, in a Douglas fir forest on the west coast of Canada where substantial soil water deficits develop over the summer, Black *et al.* (1980) observed 'remarkably similar' transpiration rates for unthinned stands with little or no salal (*Gaultheria shallon*) undergrowth and thinned stands with a well-developed understorey (cf. section 12.5.1). In a related study, Kelliher *et al.* (1986) reported that following removal of the salal transpiration by the Douglas fir trees increased by 30–50%, with the greatest increases noted for the plots where the leaf biomass of the salal had been highest. The overall effect of the removal of the undergrowth on soil water content was therefore a mere 1–3% increase (cf. Adams *et al.* 1991). In the northern UK, Whitehead *et al.* (1984) compared the transpiration patterns of two 40-year-old Scots pine plantations of similar average height but with a more than fivefold difference in stocking. Transpiration in the widely spaced plantation averaged 67% of that of the denser stand, reflecting differences in canopy resistance and LAI at stand level. However, relative transpiration rates *per tree* were 3.3 times higher in the thinned plot and were intermediate in magnitude between the relative increases in average basal sapwood area per tree (2.9 times) and leaf area per tree (4.2 times), compared with the unthinned stand. Therefore, although the thinned plantation had not re-equilibrated completely to prethinning conditions at the time of the measurements, the clear increases in leaf and sapwood areas of the remaining trees could be seen as representing a tendency towards complete re-equilibration following a set of homeostatic relationships aimed at maximum site utilization (Whitehead *et al.* 1984). Lesch and Scott (1997) described an extreme case from South Africa where the rate of growth of young *Eucalyptus grandis* trees under seasonal rainfall conditions was such that any positive effects on streamflow of three rounds of

thinning (46, 34 and 50% at age 3, 5 and 8 years) were masked entirely by the steady reduction in flows resulting from the overall vigorous growth of the trees (see also section 12.7.6).

Working under wet tropical conditions, Gilmour (1977) was unable to detect any changes in streamflow after 'lightly' logging a rain forest in northern Queensland, Australia. More substantial timber extractions from hill rain forest in Peninsular Malaysia, equivalent to the removal of 33 and 40% of the commercial stocking, affected overall water yield positively, with increases of 40 and 70% respectively. However, during a 7-year postlogging observation period there was no sign of a decline in the streamflow gain, suggesting that the regrowth in the gaps created by logging remained well below that of the remainder of the forest (Abdul Rahim & Zulkifli Yusop 1994). This result is contrary to expectation in that soil water levels in gaps, although initially higher than in the surrounding undisturbed forest, tend to decline very rapidly as the regrowth is re-established (Fig. 12.12; cf. Stogsdill *et al.* 1992). Such findings illustrate the limitations of paired catchment studies if these are not complemented by detailed process studies (cf. Bruijnzeel 1996; Vertessy *et al.* 1998).

The influence of the configuration of the cutting on the magnitude and duration of any increases in flow has been investigated in some detail in the eastern USA (Fig. 12.13). The removal of 24% of the basal area from catchment LR2 at Leading Ridge (Pennsylvania) incurred a nearly twofold larger increase in flow than cutting 33% on catchment HB4 at Hubbard Brook (New Hampshire) or catchment FEF2 at Fernow Experimental Forest (West Virginia). The cutting at Leading Ridge consisted of a single block on the lowest portion of the catchment, whereas the cutting at Hubbard Brook took the form of a series of strips situated halfway up the catchment, and that at Fernow Experimental Forest involved harvesting trees from all over the catchment (Hornbeck *et al.* 1993). Therefore, increases in streamflow associated with strip cutting are smaller than for single blocks, which is in line with the idea of increased transpiration by surrounding trees upon opening up of the canopy (Stogsdill *et al.* 1992; cf. Fig. 12.12). No significant

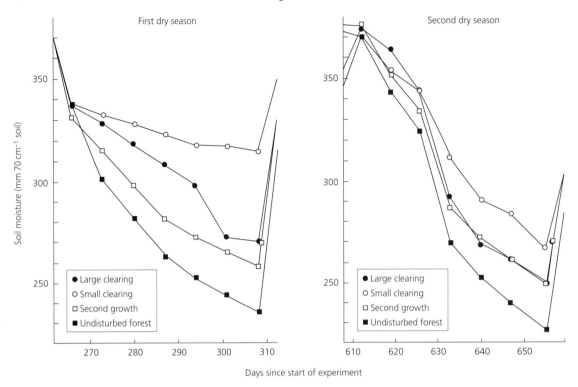

Fig. 12.12 Soil moisture contents in the top 70 cm of soil below undisturbed forest, 6-year-old secondary growth and in a narrow (10 × 50 m) and a large (50 × 50 m) clearing in lowland Costa Rica during two consecutive dry seasons. (After Parker 1985 with permission.)

differences were found between the cutting *en bloc* of the upper half of a catchment (such as Catchment 7 at Fernow, FEF7 in Fig. 12.13) or the lower half (catchment FEF6 in Fig. 12.13) (Hornbeck *et al.* 1993).

Of particular interest to water managers is the role of riparian vegetation which, in contrast to trees growing further upslope, generally has ready access to the groundwater table. Elimination of the riparian vegetation at Coweeta produced a decrease in diurnal fluctuations of baseflow, although the associated increase in water yield was not greater than that associated with the removal of an equal area of forest elsewhere in the catchment (Dunford & Fletcher 1947; cf. the similarity in results for catchments FEF 6 and 7 in Fig. 12.13). Apparently, the riparian effect is negligible in areas where soil moisture remains readily available in all parts of the catchment, such as at Coweeta with its well-distributed rainfall. A

similar result was obtained for a catchment with deep soils in the summer-rainfall zone of South Africa (Smith & Bosch 1989; Scott & Lesch 1996), but not for an area in the winter-rainfall zone where streamflow gains after cutting pine trees in the riparian zone exceeded those associated with the harvesting of pines away from the stream by 30% (Dye & Poulter 1995).

12.7.3 Effect of forest clearfelling on water yield

A basic summary of short-term results obtained from paired catchment experiments conducted before 1980 was given by Bosch and Hewlett (1982). Hornbeck *et al.* (1993) and Swank *et al.* (1988) discussed longer-term results, including the effects of multiple treatments, for catchments in the north-eastern and south-eastern USA respectively. Generally, increases in water yield during the first year after treatment are roughly proportional to percentage reductions in stand basal area, provided the latter exceed a threshold of 20–25% (cf. Fig. 12.14). Bosch and Hewlett (1982) made a distinction between coniferous and deciduous hardwood forests, suggesting that increases in flow associated with the cutting of

conifers were higher (40 mm per 10% change in cover) than for hardwoods (25 mm per 10% change in cover). However, the scatter around the two tentative regression lines was large, coefficients of determination modest (at 42% and 26% respectively) and no confidence limits were provided. In a later analysis of the same dataset plus an additional 50 studies (mostly from the tropics and Australia), Sahin and Hall (1996) used fuzzy linear regression analysis and obtained the following average changes in flow per 10% change in cover: conifers, 23 mm; mixed hardwoods–conifers, 22 mm; and hardwoods, 17–19 mm (depending whether annual precipitation exceeded 1500 mm or not). Based on this fuzzy regression approach, the overall differences between the respective vegetation groups were therefore smaller than suggested earlier by Bosch and Hewlett (1982). Stednick (1996) approached the problem of the heterogeneity in results for 95 paired catchment studies from the USA by deriving separate regression equations for each of eight geographically homogeneous regions. In doing so, a clearer picture was obtained of the regional variation in yield increases, which ranged from about 10 mm per 10% change in cover in the Rocky Mountains to over 60 mm in the Central Plains. Also, some of the regionally applicable regression equations had substantially higher coefficients of

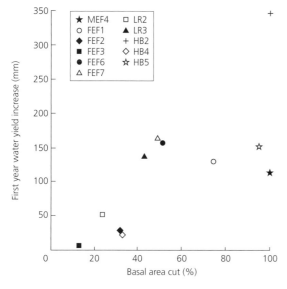

Fig. 12.13 First-year increases in water yield in response to forest cutting from 11 catchments in the north-eastern USA. (After Hornbeck *et al.* 1993 with permission.)

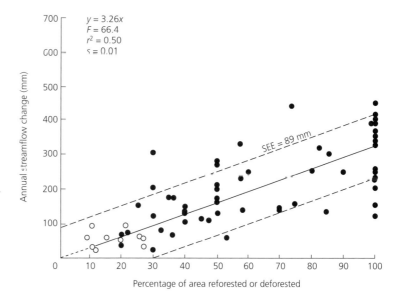

Fig. 12.14 Changes in annual water yield vs. percentage forest cover change (●, experimental data of Bosch & Hewlett 1982; ○, additional data of Timble *et al.* 1987). SEE, standard error of estimate. (After Trimble *et al.* 1987 with permission.)

determination (e.g. 0.65 for the Pacific North-west or Appalachia, which both had the largest number of experiments), although correlations for other regions were extremely low (e.g. 0.01 for the Rocky Mountains). As such, there is something to say for the approach of Trimble *et al.* (1987) who simply pooled the coniferous and mixed hardwood data of Bosch and Hewlett (1982) and added 10 data points of their own, representing reductions in annual flow from large non-experimental ('real world') catchment areas in the Piedmont of the south-eastern USA that had undergone 10–28% reforestation. As shown in Fig. 12.14, the overall coefficient of determination for the combined dataset attained the reasonable value of 0.50 by the inclusion of the Piedmont subset, with 8 of 10 data points from the latter falling within half the standard error of the estimate (compared with only 22 of 55 data points in the Bosch and Hewlett set). This finding is all the more remarkable because half the additional data involved a change in cover of less than 20%, a value below which Bosch and Hewlett (1982) had considered effects on streamflow to be more or less non-detectable (at least on small headwater streams). As could be expected after pooling data pertaining to different forest cover types, the average increase in flow of 33 mm per 10% change in forest cover predicted by the regression shown in Fig. 12.14 (Trimble *et al.* 1987) is intermediate between the 25–40 mm suggested for hardwoods and conifers by Bosch and Hewlett (1982).

Apart from contrasts in tree species (see section 12.7.4) and age of the vegetation (see below), much of the variability in the change in stream-flow observed for different sites and years after a change in cover relates to differences in climate, catchment exposure and soil depth. The climatic aspect is aptly summarized by the classic statement of Hewlett (1966) that 'it takes water to fetch water' (cf. Stednick 1996; Bruijnzeel 1990). The importance of catchment exposure is illustrated by the difference in first-year streamflow gain after cutting differently exposed deciduous hardwood forest catchments at Coweeta in the south-eastern USA. Flows from northerly exposed catchments increased by about 130 mm year^{-1} but increases from southerly

exposed catchments could be as high as 410 mm year^{-1} (Swank *et al.* 1988). As for the effect of soil depth, Trimble *et al.* (1963), when discussing differences in duration of the changes in streamflow following forest clearing in West Virginia, stated that 'the deeper the soil, the longer it takes roots of new growth or the expanding root systems of residual plants to occupy the soil mantle. When the soil mantle is reoccupied, transpiration reaches maximum levels again, and increases in streamflow disappear'. The role of soil depth can be illustrated further by examining the timing of the maximum and minimum increases in flow after forest clearance during the year (Fig. 12.15). At Coweeta, where precipitation is distributed evenly over the year and the loamy soils are deep (up to 6 m) and capable of storing large amounts of moisture, the soil does not become fully recharged until early spring (April–May). As a result, the presence or absence of a forest has little effect on the amounts of flow during this time of year (the soils being full of water anyway). However, at the onset of the growing season, increased flow due to reduced water use after clearcutting becomes evident and the contrast in streamflow from forested and cleared areas increases as the growing season advances. The effect continues well into the

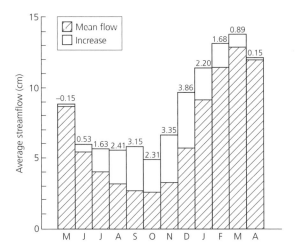

Fig. 12.15 Mean monthly streamflow from Watershed 17 at Coweeta, south-eastern USA before clearcutting and increases during a 7-year period of annual recutting. (After Swank *et al.* 1988 with permission.)

winter months (November–January), even though the trees have lost their leaves by then and evaporation is strongly reduced. This reflects the development of a much larger soil water deficit below the forest during the summer that takes an additional 2 months (February–March) to be refilled by the winter rains (Fig. 12.15). Conversely, on catchments with shallower or sandier soils (having less water storage and a lower water retention capacity), soil water recharge and depletion occurs much more rapidly and increases in flow after forest clearance follow the pattern of forest water use much more closely (Hornbeck *et al.* 1970).

The effect of the timing of the cutting and the importance of controlling regrowth are demonstrated by a comparison of the results obtained for catchments 2 and 5 at Hubbard Brook (HB 2 and HB 5 in Fig. 12.13; see section 12.7.2). On catchment HB 2, the forest was clearcut during the dormant season and regrowth eliminated by herbicide application the following spring. This produced a first-year increase in water yield of 347 mm. In contrast, on catchment HB 5, the removal of the trees took nearly a full year, while regrowth was not suppressed. The resulting first-year increase in flow thus amounted to only 152 mm, or 44% of that observed for catchment HB 2 (see Fig. 12.13). In addition, the gain in streamflow had largely disappeared after 3 years (Hornbeck *et al.* 1970, 1993). Such contrasting findings once more illustrate the effect of regenerating vegetation on soil water levels and streamflow (cf. Adams *et al.* 1991). However, it should be noted that regeneration was particularly vigorous in the example from Hubbard Brook as it took place mainly via sprouting. As demonstrated by comparative work in West Virginia, elevated streamflow levels lasted much longer (15–20 years) when regeneration had to originate from seeds because of previous herbicide applications than when sprouting was allowed (6–7 years) (Hornbeck *et al.* 1993). Similarly, when a regenerating hardwood forest at Coweeta was cut a second time after 23 years (when streamflow was still about 80 mm year^{-1} above that observed for mature mixed forest), the initial increase in water yield was nearly identical to that of the first cut (375 vs. 362 mm). However, the subsequent

decline in (extra) water yield was distinctly faster. The difference was attributable to a more rapid recovery of stand biomass during the second regeneration period in association with the greater sprouting potential of the even-aged forest, which contained numerous small stems prior to the second cut (Swank *et al.* 1988). As a result, the (projected) total duration of the second period of increased flows was estimated to be 18 years shorter than that associated with the first cutting (estimated at about 35 years). Such long durations reflect the time required by the roots to reoccupy the very deep soils at Coweeta compared with those of the north-eastern USA where effects are more likely to last 3–9 years (Swank *et al.* 1988; Hornbeck *et al.* 1993; cf. Trimble *et al.* 1963).

A rather different pattern of changes in forest water use and streamflow with time after clearing has been documented for native mountain ash (*Eucalyptus regnans*) forest in south-eastern Australia (Haydon *et al.* 1996; Vertessy *et al.* 1998). Following clearing and burning, there is an initial rise in streamflow lasting 4–6 years. This is followed by a rapid decrease in flows until the regrowth is about 27 years old, after which run-off totals slowly return to predisturbance levels. The latter may take as long as 150 years. Vertessy *et al.* (1998) provided a mechanistic explanation for this long-term pattern in terms of concurrent changes in overstorey and understorey leaf biomass and transpiration rates, as well as changes in rainfall interception and evaporation from the forest floor (cf. Fig. 12.8, see section 12.5.1).

Results from paired catchment experiments in the humid tropics also show a proportionality in flow increase with percentage of forest removed, as observed under humid temperate conditions. Reported first-year increases after clearfelling range from 125 mm in Nigeria to 820 mm in Peninsular Malaysia (Bruijnzeel 1990) and show little relation to amounts of precipitation. This led Bruijnzeel (1996) to suggest that perhaps the results were determined more by the degree of surface disturbance than by climatic or soil factors (cf. Malmer 1993; Van der Plas & Bruijnzeel 1993). However, in view of the large variability in total evapotranspiration among tropical forests apparent from Fig. 12.11 (section

12.6), it may be rewarding to reanalyse the data in the light of contrasts in rainfall interception between regions as well.

12.7.4 Effects of converting natural forest to other land cover types on water yield

Whilst streamflow totals are observed to eventually return to preclearing levels where regrowth is allowed, the conversion of native forest to other types of vegetation cover may produce permanent changes. For example, permanent increases in annual water yield are associated with the conversion of deciduous or evergreen native forest to agricultural cropping. Reported increases range from 60–125 mm year^{-1} under humid warm temperate conditions (Johnson & Kovner 1956) to 450 mm year^{-1} in the equatorial tropics (Edwards 1979). The diminished water use of annual crops compared with that of a full-grown forest (cf. Fig. 12.14) reflects the diminished capacity of low vegetation not only to intercept rainfall but also to extract water from deeper soil layers during periods of drought. The former relates primarily to the lesser aerodynamic roughness (cf. Eqns 12.8 & 12.9) of short annual crops (and possibly to their smaller leaf area and thus storage capacity as well), whereas the reduced water uptake of crops reflects their more limited rooting depth (Calder 1982). For the same reasons, conversion to pasture generally produces permanent increases in streamflow as well, although the magnitude of the increase depends on precipitation patterns and elevation (Holmes & Sinclair 1986). Interestingly, Hibbert (1969) noted that replacing deciduous hardwood forest in the south-eastern USA by vigorously growing *Festuca* grass did not lead to increased water yields, although increases of up to 125 mm year^{-1} were observed after the productivity of the grass declined. Such results can be explained in part by the fact that serious water shortages that could have reduced water uptake by shallow-rooted plants do not develop in the rainy climate at Coweeta, whereas in addition the contrast between forest and grass was lessened further by the deciduous character of the forest (cf. Swank *et al.* 1988). In contrast, the conversion of mixed deciduous hardwoods by eastern white pine (*Pinus strobus*) in the same area

brought about a reduction of 250 mm year^{-1} after 25 years. The decline in flow was ascribed both to a steady increase in intercepted rainfall as the pines grew older and to higher transpiration by the evergreen pine trees during the dormant season (Swank *et al.* 1988). Similarly, the higher total water use (*ET*) of mature *Pinus radiata* plantations in south-eastern Australia compared with that of native eucalypt forest has been explained in terms of the distinctly higher rainfall interception by the pines (Dunin & Mackay 1982; Pilgrim *et al.* 1982). However, where pine plantations replaced non-deciduous native forests elsewhere in the world, streamflows have been reported to return to preconversion levels within 8 years, such as in Kenya (Blackie 1979) and New Zealand (Rowe & Pearce 1994; Fahey & Jackson 1997).

The results presented thus far pertain mostly to small headwater catchment areas involving a unilateral change in cover. Although these experiments provide a clear and consistent picture of increased water yield after replacing tall vegetation by a shorter one, and *vice versa*, such effects are more difficult to discern in large river basins having a variety of land-use types and temporal changes therein (cf. Fig. 12.16). In addition, there are the complications associated with strong spatial and temporal variability in rainfall, especially in tropical areas with largely convective

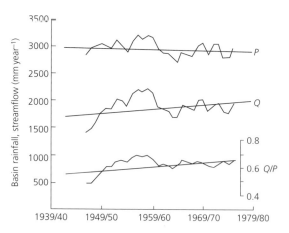

Fig. 12.16 Five-year moving averages of annual rainfall (*P*), streamflow (*Q*) and run-off ratios (*Q/P*) for the upper Mahaweli Basin above Peradeniya, Sri Lanka. (After Madduma Bandara and Kurupuarachchi 1988 with permission.)

rainfall, and the often large-scale withdrawals of water for municipal, agricultural and industrial purposes in populated areas. Qian (1983) was unable to detect any systematic changes in streamflow from catchments ranging in size from 7 to $727\,km^2$ on the island of Hainan, southern China despite a 30% reduction in tall forest cover over three decades. Dyhr-Nielsen (1986) arrived at the same conclusion for the $14\,500\,km^2$ Pasak River basin in northern Thailand, which lost 50% of its tall forest cover between 1955 and 1980. On the other hand, Madduma Bandara and Kurupuarachchi (1988) observed an increase in averaged annual flow totals for the $1100\,km^2$ upper Mahaweli catchment in Sri Lanka over the period 1940–80, despite a weak negative trend in rainfall over the same period. Although both trends were not statistically significant at the 95% level, the associated increase in annual run-off ratios was highly significant (Fig. 12.16). The increased hydrological response was ascribed to the widespread conversion of tea plantations (not forest) to annual cropping and home gardens without appropriate soil conservation measures (Madduma Bandara and Kurupuarachchi 1988). However, when analysing a longer time series (1940–97) for the $380\,km^2$ upper Nilwala catchment in Sri Lanka, which experienced a 35% reduction in forest cover during the observation period, a less consistent picture was obtained (Elkaduwa & Sakthivadivel 1999). Such contrasting findings illustrate the need for high-quality rainfall and streamflow data in historical hydrological data analyses, not only in the context of evaluating the effects of land-use transformations but also those associated with climatic change.

12.7.5 Effects of forest clearing on streamflow regimes

In areas with seasonal rainfall, the distribution of streamflow throughout the year is often of greater importance than the total annual amount *per se*. Reports of greatly diminished streamflows during the dry season after forest clearance abound in the literature, particularly in the tropics (see examples in Hamilton & King 1983; Bruijnzeel 1990). At first sight, this seems to contradict the evidence presented earlier, that forest removal leads to higher water yields (cf. Figs 12.13–12.15), even more so because the bulk of the increase in flow after *experimental* clearing is observed during baseflow or dry-season conditions (Bosch & Hewlett 1982; Bruijnzeel 1990; cf. Fig. 12.15). However, the controlled conditions imposed during catchment experiments differ from those encountered in many real-world situations. The continued exposure of bare soil after forest clearance to intense rainfall (Lal 1987), the compaction of topsoil by machinery (Malmer 1993; Van der Plas & Bruijnzeel 1993) or overgrazing (Gilmour *et al.* 1987), the gradual disappearance of soil faunal activity (Aina 1984) and increases in the area occupied by impervious surfaces such as roads and settlements (Ziegler & Giambelluca 1997) all contribute to gradually reduced rainfall infiltration opportunities in cleared areas. As a result, catchment response to rainfall becomes more pronounced and the increases in storm run-off during the rainy season may become so large as to seriously impair the recharging of soil water and groundwater reserves. When this critical stage is reached, diminished dry-season flow is the result (Fig. 12.17b), despite the fact that the removal of the forest has led to reduced evaporation and thus should have induced higher baseflows. On the other hand, if the surface characteristics after forest clearing are maintained sufficiently to allow the continued infiltration of (most of) the rainfall, then the effect of reduced *ET* associated with forest removal will show up as increased dry-season flow (Fig. 12.17a; Bruijnzeel 1989). This could be achieved through the establishment of a well-planned and maintained road system plus the careful extraction of timber in the case of logging operations (cf. Bruijnzeel 1992; Dykstra 1994), and by applying appropriate soil conservation measures (Young 1989; Hudson 1995) when clearing for agricultural purposes.

However, even with minimum soil disturbance, there will be local increases in stormflow volumes and peak flows after forest removal because the associated reduction in *ET* causes the soil to be wetter (cf. Fig. 12.12) and thus more responsive to rainfall. Relative increases in stormflow response associated with minimum

(a)

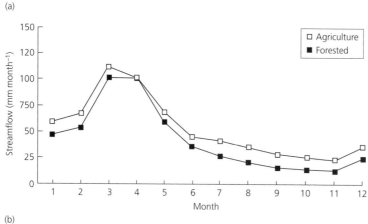

(b)

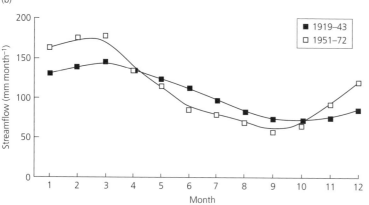

Fig. 12.17 Changes in seasonal distribution of streamflow following changes in land use: (a) after replacing montane rain forest by subsistence cropping in Tanzania (based on original data in Edwards 1979); (b) after replacing 33% of forest by rainfed cropping and settlement areas in East Java, Indonesia (after Bruijnzeel 1993 with permission.)

disturbance operations are largest for small events (up to 300%), declining to less than 10% for large events (Gilmour 1977; Pearce *et al.* 1980b; Hewlett & Doss 1984), although the effect will be more pronounced if soil disturbance is severe (Swindel *et al.* 1983; Malmer 1993). Also, whilst increases in stormflow response after (well-conducted) forestry operations tend to diminish or disappear altogether as the new vegetation cover becomes established (Swindel *et al.* 1983; Hsia 1987; Fritsch 1993; Fahey & Jackson 1997), they also may become structural. Examples include run-off contributions from roads or residential areas where there were none before (Ziegler & Giambelluca 1997) and permanently wetter soil conditions associated with reduced water use (e.g. grassland; Fritsch 1993). However, it is important to remember that caution is required when trying to extrapolate the undoubt-

edly adverse effects of indiscriminate forest clearing on local infiltration opportunities and storm run-off generation over large river basins (see Hewlett 1982 and Bruijnzeel 1990 for a fuller discussion).

12.7.6 Hydrological effects of (re)forestation

In response to the widely observed degradation of formerly forested land and the rising demands for industrial wood and fuelwood, the need for large-scale reforestation programmes has been expressed repeatedly (e.g. FAO 1986; Postel & Heise 1988). It is of interest therefore to examine to what extent such plantations are capable of restoring the original hydrological conditions, i.e. ameliorate peak flows and, above all, enhance low flows.

Whilst there is little doubt that annual water

yields from forested areas are reduced compared with those for non-forested areas (see sections 12.7.3 & 12.7.4), it must be granted that almost no catchment forestation experiments have been carried out on seriously degraded land. As such, it could be argued that the clear reductions in total and dry-season flows observed after afforesting natural grasslands and scrublands or pasture in South Africa (Dye 1996; Scott & Smith 1997), New Zealand (Dons 1986; Fahey & Jackson 1997), south-eastern Australia (Holmes & Sinclair 1986) and Fiji (Waterloo *et al.* 1999) only serve to demonstrate the difference in water use between the two vegetation types. In other words, the potentially beneficial effects on streamflow afforded by improved infiltration and soil water retention capacities during forest development could not become manifest in these experiments. The key question is therefore whether the reductions in storm run-off generating overland flow incurred by such soil physical improvements can be sufficiently large to compensate the extra water use by the new forest, and so (theoretically) boost low flows (Bruijnzeel 1989).

There is no easy answer to this question for several reasons. First, the effect of an increase in topsoil infiltrability on the frequency of occurrence of HOF (cf. section 12.2) depends equally on prevailing rainfall intensities. For instance, rainfall intensities in the Middle Hills of Nepal were generally so low that the 140 mm h^{-1} increase in infiltration capacity 12 years after reforesting a degraded pasture site with *Pinus roxburghii* made no difference in the generation of HOF (Gilmour *et al.* 1987). Similarly, whilst topsoil infiltrabilities had roughly doubled over 12 years since reforesting *Imperata* grassland with teak in Sri Lanka, at 30 mm h^{-1} the hydrological impact of this increase is probably limited (Mapa 1995). A much longer time span may be required therefore to restore infiltration capacities to the high values normally found in undisturbed forests. Secondly, the reductions in catchment response to rainfall observed after forestation (Tennessee Valley Authority 1961; Fahey & Jackson 1997) may well reflect the drier soil conditions prevailing under actively growing forest rather than a reduction in peak-generating HOF (cf. Hsia 1987). However, none of the studies documenting decreases or

increases in stormflows after land-use change has attempted to quantify the associated changes in relative contributions by subsurface and overland flow types (Bruijnzeel 1990, 1996). Third, there is also the effect of soil depth, which determines both the maximum amount of water that can be stored in a catchment under optimum infiltration conditions and the possibilities for water uptake by the developing root network of the new trees (Trimble *et al.* 1963). And last but not least, there is the confounding effect of rainfall patterns (seasonal vs. well distributed) and general evaporative demand, which both exert a strong influence on tree water use. Needless to say, the soil and climatic factors, plus the absence of detailed information on the prevailing stormflow mechanisms before and after the conversion, all render comparisons between different sites more complicated.

The only documented real-world case in which the kind of compensation mechanism supposed by Bruijnzeel (1989) seems to have occurred may be the White Hollow catchment in Tennessee, USA (Tennessee Valley Authority 1961). Prior to improvement of its vegetation cover, two-thirds of this catchment consisted of mixed secondary forest in poor condition (due to fire, heavy logging and grazing), with another 26% under poor scrub. About 40% of the catchment was estimated to be subject to more or less severe erosion. Following extensive physical and vegetative restoration works, peak discharges decreased considerably within 2 years, especially in summer. However, neither annual water yield nor low flows changed significantly over the next 22 years of forest recovery and reforestation. It was concluded that the extra water needed by the recovering and additionally planted trees was balanced by improved infiltration (Tennessee Valley Authority 1961). However, it is quite possible that the absence of major changes in total and low flows at White Hollow mainly reflects the lack of contrast between tall and short vegetation, as would have been the case when reforesting a truly deforested and degraded catchment. Support for this contention comes from the observation of Trimble *et al.* (1987) that reductions in flow from several large river basins in the south-eastern USA, which had suffered considerable erosion before being partially reforested (cf. Fig. 12.14), were largest during dry years. In addition, the effect became more pronounced as the trees grew older. A similar case was recently described for a once seriously degraded catchment in Slovenia (former Yugoslavia), where the spontaneous return of forest produced a steady reduction in annual water yield over a period of 30 years, with the strongest reductions again occurring during the dry summer months (Globevnik & Sovinc 1998). Therefore, in both these real-world examples the water-use aspect overrides the infiltration aspect, despite the rather modest contrasts in water use between forest and grassland/cropland in the south-eastern USA (see section 12.7.4).

Such findings offer little prospect for the possibility of significantly raising dry-season flows by forestation in the humid tropics. Observed maximum differences between annual water use of fast-growing pines or eucalypts and short vegetation (grass, crops) under well-watered (sub)tropical conditions attain values of 500–700 mm at the catchment scale (cf. Fig. 12.18) and even higher values (>1100 mm) on individual plots with particularly vigorous tree growth (Dye 1996; Waterloo *et al.* 1999). The hydrological benefits incurred by the limited increases in topsoil infiltration capacities observed after forestation of degraded land in Nepal and Sri Lanka cited earlier are dwarfed by such high water requirements. The conclusion that already diminished dry-season flows in degraded tropical areas will decrease even further upon reforestation seems inescapable. In view of the extent of the low-flow problem, the testing of alternative ways of increasing water retention in tropical catchments without the excessive water use associated with exotic tree plantations should receive high priority (cf. Van Noordwijk *et al.* 1998).

12.7.7 Modelling the hydrological impacts of forest manipulation and land-use change

As shown in the preceding sections, different types of forest manipulations affect the water (and material) flows through catchments differently. Traditionally, foresters have relied on time-consuming paired catchment experiments to evaluate such effects. Whilst this approach

enabled the construction of simple nomograms from which reductions in annual streamflow after afforestation could be read as a function of stand age and annual rainfall (Nänni 1970), or first-year increases in flow as a function of percentage basal area reduction and catchment aspect (Swank *et al.* 1988), the results were often so variable as to render their applicability for more detailed water resources planning rather limited (Bosch 1982). For example, Fig. 12.18 illustrates the considerable variation in the change in water yield after planting scrub and grassland with pines in South Africa, both between locations (curves 3–7) and years (curves 6 and 7). Afforestation with eucalypts produced similarly variable responses (curves 1 and 2) that also differed markedly from those for pine. The black-box nature of the paired catchment technique is unable to evaluate the relative importance of the various factors underlying such differences and this severely limits the possibilities for extrapolation of results to other areas or periods.

Process-based hydrological models represent an alternative way of predicting how catchments might respond to different forms of management. Because many practical forest management questions have a spatial dimension to them, and because landscapes usually comprise a complex

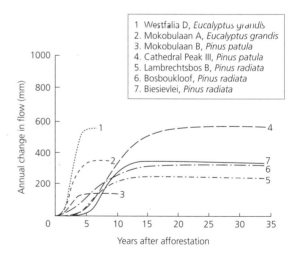

Fig. 12.18 Changes in catchment water yield after afforestation for seven paired catchment experiments in different forestry regions of South Africa (based on Bosch 1982; Dye 1996).

mosaic of different land uses, such models should preferably be of a 'distributed' nature, i.e. capable of taking into account spatial variations in topography, soils, vegetation and climate (Vertessy *et al.*, 1993). During the last 10–15 years, considerable progress has been made in the modelling of forest hydrological and ecological processes over a range of spatial and temporal scales (e.g. Running & Coughlan 1988; Shuttleworth 1988; Hatton *et al.* 1992; Sivapalan *et al.* 1996; Vertessy *et al.* 1996; Watson *et al.* 1999).

Arguably, one of the most comprehensive process-based distributed models is the topography-driven TOPOG model (Vertessy *et al.* 1993). TOPOG integrates the water, carbon, solute and sediment balances at the small catchment scale (typically < 10 km^2) and is thus particularly useful for exploring complex feedback mechanisms between system properties. Examples of within-catchment applications of TOPOG relevant to forestry include the prediction of steady-state soil moisture distributions and saturated zones in the landscape during wet and dry periods (O'Loughlin 1986; Moore *et al.* 1988), the spatial distribution of surface erosion, gully initiation and landslide hazards (Vertessy *et al.* 1990; Dietrich *et al.* 1992; Constantini *et al.* 1993), and the evaluation at the hillslope scale of the water balance and growth performance of different tree planting configurations (block planting vs. strip planting) under subhumid conditions (Silberstein *et al.* 1999). Off-site applications of TOPOG include the successful simulation of changes in tree growth and water yield for a small catchment after clear-felling during 20 years of regeneration (Vertessy *et al.* 1996). To address forestry-related questions of catchment water management at larger scales (e.g. 100–1000 km^2) as well, a simpler derivative of TOPOG (Macaque) has been developed recently. Macaque proved capable of effectively simulating the changes in water yield associated with the occurrence of a major wildfire (in 1939) and subsequent forest regeneration in a 163-km^2 catchment over an 82-year period (1911–92). It also enabled the prediction of the hydrological effect of the 1939 fires in a spatially explicit manner (Vertessy *et al.* 1998; Watson *et al.* 1999).

Models like TOPOG and Macaque constitute powerful tools whose predictions can help rational land-use decisions. However, whilst distributed models represent the only class of simulation models capable of capturing the complex feedbacks that occur upon disturbing hydrological systems, they are also data-demanding. Model outputs are especially sensitive to variations in LAI, soil hydraulic conductivity, the rainfall interception coefficient and canopy conductance, whereas at larger scales there is the problem of adequate spatial representation of rainfall inputs (Vertessy *et al.* 1998). However, there is reason for optimism as various remote sensing technologies that are currently still in their infancy can be expected to become widely available within the next decade. This will greatly facilitate data acquisition and scaling-up of results.

12.8 OUTLOOK

As evidenced by the material collated in this chapter, considerable progress has been made since the mid 1960s in the conceptual understanding and mathematical description of such key forest hydrological processes as evaporation, from a wet canopy (rainfall interception) and from a dry canopy (transpiration) and including from sparse canopies. Yet, the precise turbulent transfer mechanisms underlying the very high rates of wet canopy evaporation that have been observed under conditions of advected energy in maritime climates are not well understood. Neither do we know very much about the magnitude and hydrological significance of fog interception by the forests of montane cloud belts, particularly in the tropics. Furthermore, there are indications that low flows may be reduced considerably upon clearing such cloud forests (presumably due to the loss of the extra inputs of water afforded by fog interception), although experimental evidence based on process research is lacking. Whilst the database on transpiration by temperate forests is expanding steadily, such information is still very limited for tropical forests, especially for fast-growing plantation species and montane (cloud) forests. Given the adverse effect of reforesting degraded tropical areas with fast-growing evergreen trees on low flows, the identification and evaluation of alternative ways of improving dry-season flows should also receive high priority. The quantification of the water budgets of agroforestry systems assumes particular importance in this respect. Finally, very little information exists on the long-term changes in soil hydraulic properties associated with land-use change and how these affect storm run-off generation patterns.

The last decades have seen the waning of the empirical age of catchment treatment experimentation and ever more rapid developments in our ability to model complex natural systems in a spatially explicit way. The arrival of process-based distributed hydrological models and ongoing improvements in equipment, remote sensing technology and computational facilities guarantee that there are exciting times ahead for forest hydrologists.

ACKNOWLEDGEMENTS

I am grateful to Professor Julian Evans for his invitation to compile this chapter and for his considerable patience. I also thank Drs John Gash, John Roberts and Rob Vertessy for their comments on the manuscript and/or inspiration over the years. The manuscript also benefited from editorial suggestions by Dr Lawrence Morris. Special thanks are due to Mieke de Vries for numerous articles via interlibrary loan; Ron Williams for literature on early work carried out under the auspices of the Tennessee Valley Authority; Dr Stephen Swabey for generously giving me his personal copy of the classic book by Sopper and Lull (1967); and above all to Irene Sieverding for her continued support during the writing of the manuscript.

REFERENCES

Abdul Rahim & Zulkifli Yusop (1994) Hydrological response to selective logging in Peninsular Malaysia and its implications on watershed management. In: Ohta, T., Fukushma, Y. & Suzuki, M., eds. *Proceedings of the International Symposium on Forest Hydrology 1994*, pp. 263–74. IUFRO, Tokyo.

Adams, P.W., Flint, A.L. & Fredriksen, R.L. (1991) Long-term patterns in soil moisture and revegetation after a clearcut of a Douglas-fir forest in Oregon. *Forest Ecology and Management* **41**, 249–63.

Aina, P.O. (1984) Contribution of earthworms to porosity and water infiltration in a tropical soil under forest and long-term cultivation. *Pedobiologia* **26**, 131–6.

Angus, D.E. & Watts, P.J. (1984) Evapotranspiration: how good is the Bowen ratio method? *Agricultural Water Management* **8**, 133–50.

Aravena, R., Suzuki, O. & Pollastri, A. (1989) Coastal fog and its relation to groundwater in the IV region of northern Chile. *Chemical Geology* **79**, 83–91.

Asdak, C., Jarvis, P.G., Van Gardingen, P. & Fraser, A. (1998) Rainfall interception loss in unlogged and logged forest areas of Central Kalimantan, Indonesia. *Journal of Hydrology* **206**, 237–44.

Aussenac, G., Granier, A. & Naud, R. (1982) Influence of thinning on growth and water balance. *Canadian Journal of Forest Research* **12**, 222–31.

Azevedo, J. & Morgan, D.L. (1974) Fog precipitation in coastal California forests. *Ecology* **55**, 1135–41.

Berbigier, P., Bonnefond, J.M., Lousteau, D., Ferreira, M.I., David, J.S. & Pereira, J.S. (1996) Transpiration of a 64-year-old maritime pine stand in Portugal. 2. Evapotranspiration and canopy stomatal conductance measured by an eddy covariance techmique. *Oecologia* **107**, 43–52.

Black, T.A. & Kelliher, F.H. (1989) Processes controlling understorey evapotranspiration. *Philosophical Transactions of the Royal Society of London B* **324**, 207–31.

Black, T.A., Tan, C.S. & Nnyamah, J.U. (1980) Transpiration rate of Douglas-fir trees in thinned and unthinned stands. *Canadian Journal of Soil Science* **60**, 625–31.

Blackie, J.R. (1979) The water balance of the Kimakia catchments. *East African Agricultural and Forestry Journal* **43**, 155–74.

Bonell, M. (1993) Progres in the understanding of runoff generation dynamics in forests. *Journal of Hydrology* **150**, 217–75.

Bonell, M. & Gilmour, D.A. (1978) The development of overland flow in a tropical rain forest catchment. *Journal of Hydrology* **39**, 365–82.

Bosch, J.M. (1982) Streamflow response to catchment management in South Africa. In: *Proceedings of the Symposium on Hydrological Research Basins*, Sonderheft Landes-hydrologie, Bern **1982**, 279–89.

Bosch, J. & Hewlett, J.D. (1982) A review of catchment experiments to determine the effects of vegetation changes on water yield and evapotranspiration. *Journal of Hydrology* **55**, 3–23.

Brammer, D.D. & McDonnell, J.J. (1996) An evolving perceptual model of hillslope flow at the Maimai catchment. In: Anderson, M.G. & Brooks, S.M., eds. *Advances in Hillslope Processes*, Vol. 1, pp. 35–60. John Wiley & Sons, Chichester.

Brown, A.G., Nambiar, E.K.S. & Cossalter, C. (1997) Plantations for the tropics: their role, extent and nature. In: Nambiar, E.K.S. & Brown, A.G., eds. *Management of Soil, Nutrients and Water in Tropical Plantation Forests, pp. 1–23. ACIAR/CSIRO/CIFOR, Canberra/Bogor.

Brown, M.B., De la Roca, I., Vallejo, A. *et al.* (1996) *A Valuation Analysis of the Role of Cloud Forests in Watershed Protection, Sierra de las Minas Reserve, Guatemala and Cusuco National Park, Honduras.* RARE Center for Tropical Conservation, Washington, DC.

Bruijnzeel, L.A. (1989) (De)forestation and dry season flow in the tropics: a closer look. *Journal of Tropical Forest Science* **1**, 229–43.

Bruijnzeel, L.A. (1990) *Hydrology of Moist Tropical Forests and Effects of Conversion: a State of Knowledge Review.* UNESCO, Paris.

Bruijnzeel, L.A. (1992) Managing tropical forest watersheds for production: where contradictory theory and practice co-exist. In: Miller, F.R. & Adam, K.L., eds. *Wise Management of Tropical Forests*, pp. 37–75. Oxford Forestry Institute, Oxford.

Bruijnzeel, L.A. (1993) Land use and hydrology in warm humid regions: where do we stand? *International Association of Hydrological Sciences Publication* **216**, 3–34.

Bruijnzeel, L.A. (1996) Predicting the hydrological effects of land cover transformation in the humid tropics: the need for integrated research. In: Gash, J.H.C., Nobre, C.A., Roberts, J.M. & Victoria, R.L., eds. *Amazonian Deforestation and Climate*, pp. 15–55. John Wiley & Sons, Chichester.

Bruijnzeel, L.A. (1997) Hydrology of forest plantations in the tropics. In: Nambiar, E.K.S. & Brown, A.G., eds. *Management of Soil, Nutrients and Water in Tropical Forests*, pp. 125–67. ACIAR/CSIRO/CIFOR, Canberra/Bogor.

Bruijnzeel, L.A. (2000) Hydrology of tropical montane cloud forests: a re-assessment. In: Gladwell, J.S., ed. *Proceedings of the Second International Colloquium on Hydrology of the Humid Tropics.* CATHALAC, Panama City, Panama, in press.

Bruijnzeel, L.A. & Proctor, J. (1995) Hydrology and biogeochemistry of tropical montane cloud forests: what do we really know? In: Hamilton, L.S., Juvik, J.O. & Scatena, F.N., eds. *Tropical Montane Cloud Forests*, pp. 38–78. Springer-Verlag, Heidelberg.

Bruijnzeel, L.A. & Wiersum, K.F. (1987) Rainfall interception by a young *Acacia auriculiformis* A. Cunn. plantation forest in West Java, Indonesia: application of Gash's analytical model. *Hydrological Processes* **1**, 309–19.

Brutsaert, W.H. (1982) *Evaporation into the Atmosphere.* D. Reidel, Dordrecht, The Netherlands.

Burghouts, T.B.A., Van Straalen, N.M. & Bruijnzeel, L.A. (1998) Spatial heterogeneity of element and litter turnover in a Bornean rain forest. *Journal of Tropical Ecology* **14**, 477–506.

Calder, I.R. (1979) Do trees use more water than grass? *Water Services* **83**, 11–14.

Calder, I.R. (1982) Forest evaporation. In: *Hydrological Processes of Forested Areas*, pp. 173–93. National Research Council of Canada, Ottawa.

Calder, I.R. (1986) A stochastic model of rainfall interception. *Journal of Hydrology* **89**, 65–71.

Calder, I.R. (1990) *Evaporation in the Uplands.* John Wiley & Sons, Chichester.

Calder, I.R. (1996) Dependence of rainfall interception on drop size. 1. Development of the two-layer stochastic model. *Journal of Hydrology* **185**, 363–78.

Calder, I.R. (1998) Water use by forests: limits and controls. *Tree Physiology* **18**, 625–31.

Calder, I.R. & Rosier, P.T.W. (1976) The design of large plastic sheet net rainfall gauges. *Journal of Hydrology* **30**, 403–5.

Calder, I.R., Wright, I.R. & Murdiyarso, D. (1986) A study of evaporation from tropical rainforest: West Java. *Journal of Hydrology* **89**, 13–31.

Calder, I.R., Hall, R.L. & Adlard, P.G. (eds) (1992) *Growth and Water Use of Eucalyptus Plantations.* John Wiley & Sons, Chichester.

Calder, I.R., Rosier, P.T.W., Prasanna, K.T. & Parameswarappa, S. (1997) Eucalyptus water use greater than rainfall input: a possible explanation from southern India. *Hydrology and Earth System Sciences* **1**, 249–56.

Carlyle-Moses, D.E. & Price, A.G. (1999) An evaluation of the Gash interception model in a northern hardwood stand. *Journal of Hydrology* **214**, 103–10.

Cavelier, J., Jaramillo, M., De Solis, D. & Leon, D. (1997) Water balance and nutrient inputs in bulk precipitation in tropical montane cloud forest in Panama. *Journal of Hydrology* **193**, 83–96.

Clark, K.L., Nadkarni, N.M., Schaeffer, D. & Gholz, H.L. (1998) Atmospheric deposition and net retention of ions by the canopy in a tropical montane forest, Monteverde, Costa Rica. *Journal of Tropical Ecology* **14**, 27–45.

Constantini, A., Dawes, W., O'Loughlin, E. & Vertessy, R.A. (1993) Hoop pine plantation management in Queensland. I. Gully erosion hazard prediction and watercourse classification. *Australian Journal of Soil and Water Conservation* **6**, 35–9.

Cooper, J.D. (1979) Water use of a tea estate from soil moisture measurements. *East African Agricultural and Forestry Journal* **43**, 102–21.

Crockford, R.H. & Richardson, D.P. (1990) Partitioning of rainfall in a eucalypt forest and pine plantation in Southeastern Australia. I. Throughfall measurement in a eucalypt forest; effect of method and species composition. *Hydrological Processes* **4**, 131–44.

Czarnowski, M.S. & Olszewski, J.L. (1970) Number and spacing of rainfall gauges in a deciduous forest stand. *Oikos* **21**, 48–51.

Dawson, T.E. (1993) Hydraulic lift and water use by plants: implications for water balance, performance and plant–plant interactions. *Oecologia* **95**, 565–74.

Dawson, T.E. (1996) Determining water use by trees and forests from isotopic, energy balance and transpiration analyses: the roles of tree sizes and hydraulic lift. *Tree Physiology* **16**, 263–72.

Dawson, T.E. (1998) Fog in the California redwood forest: ecosystem inputs and use by plants. *Oecologia* **117**, 476–85.

Delfs, J. (1955) Die Niederschlagszurückhaltung im Walde. *Mitteilungen Des Arbeitskreises 'Wald und Wasser'* **2**, 1–54.

Dietrich, W.E., Wilson, C.J., Montgomery, D.R., McKean, J. & Bauer, R. (1992) Erosion thresholds and land surface morphology. *Geology* **20**, 675–9.

Dolman, A.J., Gash, J.H.C., Roberts, J.M. & Shuttleworth, W.J. (1991) Stomatal and surface conductance of tropical rainforest. *Agricultural and Forest Meteorology* **54**, 303–18.

Dons, A. (1986) The effect of large-scale afforestation on Tarawera river flows. *Journal of Hydrology (NZ)* **25**, 61–73.

Dunford, E.G. & Fletcher, P.W. (1947) The effect of removal of streambank vegetation upon water yield. *Transactions of the American Geophysical Union* **28**, 105–10.

Dunin, F.X. & Greenwood, E.A.N. (1986) Evaluation of the ventilated chamber for measuring evaporation from a forest. *Hydrological Processes* **1**, 47–62.

Dunin, F.X. & Mackay, S.M. (1982) Evaporation by eucalypt and coniferous forest communities. In: O'Loughlin, E.M. & Bren, L.J., eds. *The First National Symposium on Forest Hydrology*, pp. 18–25. Institution of Engineers, Barton ACT, Australia.

Dunin, F.X., O'Loughlin, E.M. & Reyenga, W. (1985) A lysimeter characterization of evaporation by eucalypt forest and its representativeness for the local environment. In: Hutchinson, B.A. & Hicks, B.B., eds. *The Forest–Atmosphere Interaction*, pp. 271–91. D. Reidel, Dordrecht, The Netherlands.

Dunin, F.X., O'Loughlin, E.M. & Reyenga, W. (1988) Interception loss from eucalypt forest: lysimeter determination of hourly rates for long term evaluation. *Hydrological Processes* **2**, 315–29.

Dunne, T. (1978) Field studies of hillslope flow processes. In: Kirkby, M.J., ed. *Hillslope Hydrology*, pp. 227–93. John Wiley & Sons, Chichester.

Dye, P.J. (1996) Climate, forest and streamflow relationships in South African afforested catchments. *Commonwealth Forestry Review* **75**, 31–8.

Dye, P.J. & Poulter, A.G. (1995) A field demonstration of the effect on streamflow of clearing invasive pine and wattle trees from a riparian zone. *South African Forestry Journal* **173**, 27–30.

Dyhr-Nielsen, M. (1986) Hydrological effect of deforestation in the Chao Phraya basin in Thailand. In: *International Symposium on Tropical Forest Hydrology and Application, June 1986, Chiangmai, Thailand*. 12 pp.

Dykes, A.P. (1997) Rainfall interception from a lowland tropical rain forest in Brunei. *Journal of Hydrology* **200**, 260–97.

Dykstra, D.P. (1994) *FAO Model Code of Forest Harvesting Practice*. Food and Agriculture Organization, Rome.

Edwards, K.A. (1979) The water balance of the Mbeya experimental catchments. *East African Agricultural and Forestry Journal* **43**, 231–47.

Ekern, P.C. (1964) Direct interception of cloud water at Lanaihale, Hawaii. *Proceedings of the Soil Science Society of America* **28**, 419–21.

Elbers, J.A., Dolman, A.J., Moors, E.J. & Snijders, W. (1996) *Hydrology and Water Economy of Forested Areas in the Netherlands. Phase 2: Measuring Design and First Results*. Staring Centre, Agricultural Research Service, Wageningen, The Netherlands (in Dutch).

Elkaduwa, W.K.B. & Sakthivadivel, R. (1999) *Use of Historical Data as a Decision Support Tool in Watershed Management: a Case Study of the Upper Nilgawa Basin in Sri Lanka*. International Water Management Institute, Colombo, Sri Lanka.

Elsenbeer, H. & Cassel, D.K. (1990) Surficial processes in the rainforest of western Amazonia. *International Association of Hydrological Sciences Publication* **192**, 289–97.

Elsenbeer, H., Cassel, D.K. & Castro, J. (1992) Spatial analysis of soil hydraulic conductivity in a tropical rain forest catchment. *Water Resources Research* **28**, 3201–14.

Eschner, A.R. (1967) Interception and soil moisture distribution. In: Sopper, W.E. & Lull, H.W., eds. *International Symposium on Forest Hydrology*, pp. 191–9. Pergamon Press, Oxford.

Fahey, B.D. & Jackson, R.J. (1997) Hydrological impacts of converting native forests and grasslands to pine plantations, South Island, New Zealand. *Agricultural and Forest Meteorology* **84**, 69–82.

Fallas, J. (1996) *Cuantificacíon de la Intercepcíon en un Bosque Ruboso, Monte de los Olivos, Cuenca del Río Chiquito, Guanacaste, Costa Rica*. CREED Notas Técnicas no. 6, Tropical Science Center, San Jose, Costa Rica.

FAO (1986) *Tropical Forestry Action Plan*. Committee on Forest Development in the Tropics, Food and Agriculture Organization, Rome.

Federer, C.A. (1977) Leaf resistance and xylem potential differ among broadleaf species. *Forest Science* **23**, 411–19.

Florido, L.V. & Saplaco, S.R. (1981) Rainfall interception

in a thinned 10–15-year-old natural Benguet pine (*Pinus kesiya* Roy. ex Gordon) stand. *Sylvatrop Philippines Forest Research Journal* **6**, 195–201.

Fritsch, J.M. (1993) The hydrological effects of clearing tropical rain forest and of the implementation of alternative land uses. *International Association of Hydrological Sciences Publication* **216**, 53–66.

Gash, J.H.C. (1979) An analytical model of rainfall interception by forests. *Quarterly Journal of the Royal Meteorological Society* **105**, 43–55.

Gash, J.H.C. & Morton, A.J. (1978) An application of the Rutter model to the estimation of the interception loss from Thetford forest. *Journal of Hydrology* **38**, 49–58.

Gash, J.H.C. & Stewart, J.B. (1975) The average surface resistance of a pine forest derived from Bowen ratio measurements. *Boundary-Layer Meteorology* **8**, 453–64.

Gash, J.H.C. & Stewart, J.B. (1977) The evaporation from Thetford forest during 1975. *Journal of Hydrology* **35**, 385–96.

Gash, J.H.C., Wright, I.R. & Lloyd, C.R. (1980) Comparative estimates of interception loss from three coniferous forests in Great Britain. *Journal of Hydrology* **48**, 89–105.

Gash, J.H.C., Lloyd, C.R. & Lachaud, G. (1995) Estimating sparse forest rainfall interception with an analytical model. *Journal of Hydrology* **170**, 79–86.

Gash, J.H.C., Valente, F. & David, J.S. (1999) Estimates and measurements of evaporation from wet, sparse pine forest in Portugal. *Agricultural and Forest Meteorology* **94**, 149–58.

Ghosh, R.C., Subba Rao, B.K. & Ramola, B.C. (1980) Interception studies in Sal (*Shorea robusta*) coppice forest. *Indian Forester* **106**, 513–25.

Gilmour, D.A. (1977) Effects of rainforest logging and clearing on water yield and quality in a high rainfall zone of north-east Queensland. In: *Proceedings of the Brisbane Hydrology Symposium 1977*, pp. 156–60. Institution of Engineers Australia, Canberra, Australia.

Gilmour, D.A., Bonell, M. & Cassells, D.S. (1987) The effects of forestation on soil hydraulic properties in the Middle Hills of Nepal: a preliminary assessment. *Mountain Research and Development* **7**, 239–49.

Globevnik, L. & Sovinc, A. (1998) Impacts of catchment land use change on river flows: the Dragonja River, Slovenia. In: *Hydrology in a Changing Environment*, pp. 525–33. British Hydrological Society.

Goodman, J. (1985) The collection of fog drip. *Water Resources Research* **21**, 392–4.

Granier, A., Huc, R. & Barigah, S.T. (1996) Transpiration of natural rain forest and its dependence on climatic factors. *Agricultural and Forest Meteorology* **78**, 19–29.

Grantz, D.A. (1990) Plant response to atmospheric humidity. *Plant, Cell and Environment* **13**, 667–79.

Grime, V.L. & Sinclair, F.L. (1999) Sources of error in stem heat balance sap flow measurements. *Agricultural and Forest Meteorology* **94**, 103–21.

Grünow, J. (1955) Precipitation in mountain forests. Interception and fog drip. *Forstwissenschaftliches Zentralblatt* **74**, 21–36.

Hafkenscheid, R.L.L.J., Bruijnzeel, L.A. & De Jeu, R.A.M. (1998) Estimates of fog interception by montane rain forest in the Blue Mountains of Jamaica. In: Schemenauer, R.S. & Bridgman, H.A., eds. *First International Conference on Fog and Fog Collection*, pp. 33–6. International Development and Research Council (IDRC), Ottawa.

Hall, R.L. & Allen, S.J. (1997) Water use by poplar clones grown as short-rotation coppice at two sites in the United Kingdom. *Aspects of Applied Biology* **49**, 163–72.

Hall, R.L. & Roberts, J.M. (1990) Hydrological aspects of new broad-leaf plantations. *SEESOIL* **6**, 2–38.

Hall, R.L., Calder, I.R., Gunawardena, E.R.N. & Rosier, P.T.W. (1996) Dependence of rainfall interception on drop size. 3. Implementation and comparative performance of the stochastic model using data from a tropical site in Sri Lanka. *Journal of Hydrology* **185**, 389–407.

Hamilton, L.S. & King, P.N. (1983) *Tropical Forested Watersheds: Hydrologic and Soils Response to Major Uses or Conversions*. Westview Press, Boulder, Colorado.

Harding, R.J. & Stewart, J.B. (1986) Transpiration from coniferous forest in a Mediterranean environment. In: *Proceedings of the Third Hellenic–British Climatological Congress*, pp. 213–22. Hellenic Climatic Society, Athens.

Harr, R.D. (1982) Fog drip in the Bull Run municipal watershed, Oregon. *Water Resources Bulletin* **18**, 785–9.

Harr, R.D. (1983) Potential for augmenting water yield through forest practices in western Washington and western Oregon. *Water Resources Bulletin* **19**, 383–93.

Hatton, T.J. & Wu, H.I. (1995) Scaling theory to extrapolate individual tree water use to stand water use. *Hydrological Processes* **9**, 527–40.

Hatton, T.J., Walker, J., Dawes, W. & Dunin, F.X. (1992) Simulations of hydro-ecological responses to elevated CO_2 at the catchment scale. *Australian Journal of Botany* **40**, 679–96.

Hatton, T.J., Moore, S.J. & Reece, P.H. (1995) Estimating stand transpiration in a *Eucalyptus populnea* woodland with the heat pulse method: measurement errors and sampling strategies. *Tree Physiology* **15**, 219–27.

Haydon, S.R., Benyon, R.G. & Lewis, R. (1996) Variation in sapwood area and throughfall with forest age in mountain ash (*Eucalyptus regnans* F. Muell.). *Journal of Hydrology* **187**, 351–66.

Helvey, J.D. (1964) *Rainfall Interception by Hardwood Forest Litter in the Southern Appalachians.* US Forest Service Research Paper SE-8, US Forest Service Southeastern Forest Experiment Station, Asheville, North Carolina.

Helvey, J.D. (1967) Interception by eastern white pine. *Water Resources Research* **3**, 723–9.

Helvey, J.D. & Patric, J.H. (1965) Canopy and litter interception of rainfall by hardwoods of eastern United States. *Water Resources Research* **1**, 193–206.

Herwitz, S.R. (1986) Infiltration-excess caused by stemflow in a cyclone-prone tropical rainforest. *Earth Surface Processes and Landforms* **11**, 401–12.

Hewlett, J.D. (1966) Will water demand dominate forest management in the East? *Proceedings Society of American Foresters* Seattle, 154–9.

Hewlett, J.D. (1982) Forests and floods in the light of recent investigations. In: *Hydrological Processes of Forested Areas*, pp. 543–60. National Research Council of Canada, Ottawa.

Hewlett, J.D. & Doss, R. (1984) Forests, floods, erosion: a watershed experiment in the southeastern Piedmont. *Forest Science* **30**, 424–34.

Hewlett, J.D. & Fortson, J.C. (1983) The paired catchment experiment. In: Hewlett, J.D., ed. *Forest Water Quality*, pp. 11–14. School of Forest Resources, University of Georgia, Athens, Georgia.

Hibbert, A.R. (1969) Water yield changes after converting a forested catchment to grass. *Water Resources Research* **5**, 634–40.

Hodnett, M.G., Tomasella, J., Marques Filho, A.O. & Oyama, M.D. (1996) Deep soil water uptake by forest and pasture in central Amazonia: predictions from long-term daily rainfall data using a simple water balance model. In: Gash, J.H.C., Nobre, C.A., Roberts, J.M. & Victoria, R.L., eds. *Amazonian Deforestation and Climate*, pp. 79–99. John Wiley and Sons, Chichester.

Holmes, J.W. & Sinclair, J.A. (1986) Streamflow from some afforested catchments in Victoria. In: *Hydrology and Water Resources Symposium, Brisbane*, pp. 214–18. Institution of Engineers, Barton ACT, Australia.

Hoover, M.D. (1944) Effect of removal of forest vegetation upon water yields. *Transactions of the American Geophysical Union* **6**, 969–75.

Hornbeck, J.W. (1973) *The Problem of Extreme Events in Paired-watershed Studies.* US Forest Service Research Note NE-175, USDA, Washington DC.

Hornbeck, J.W., Pierce, R.S. & Federer, C.A. (1970) Streamflow changes after forest clearing in New England. *Water Resources Research* **6**, 1124–32.

Hornbeck, J.W., Adams, M.B., Corbett, E.S., Verry, E.S. & Lynch, J.A. (1993) Long-term impacts of forest treatments on water yield: a summary for northeastern USA. *Journal of Hydrology* **150**, 323–44.

Horton, R.E. (1919) Rainfall interception. *Monthly Weather Review* **47**, 603–23.

Hoyt, W.G. & Troxell, H.C. (1932) Forests and streamflow. *Papers of the American Society of Civil Engineers* **58**, 1037–66.

Hsia, Y.J. (1987) Changes in storm hydrographs after clearcutting a small hardwood forested catchment in central Taiwan. *Forest Ecology and Management* **20**, 117–34.

Hudson, N.W. (1995) *Soil Conservation.* Batsford, London.

Hutchinson, I. & Roberts, M.C. (1981) Vertical variation in stemflow generation. *Journal of Applied Ecology* **18**, 521–7.

Hutjes, R.W.A., Wierda, A. & Veen, A.W.L. (1990) Rainfall interception in the Taï forest, Ivory Coast: application of two simulation models to a humid tropical system. *Journal of Hydrology* **114**, 259–75.

Hutley, L.B., Doley, D., Yates, D.J. & Boonsaner, A. (1997) Water balance of an Australian subtropical rainforest at altitude: the ecological and physiological significance of intercepted cloud and fog. *Australian Journal of Botany* **45**, 311–29.

Ingraham, N.L. (1998) Isotopic variations in precipitation. In: Kendall, C. & McDonnell, J.J., eds. *Isotope Tracers in Catchment Hydrology*, pp. 87–118. Elsevier, Amsterdam.

Ingwersen, J.B. (1985) Fog drip, water yield, and timber harvesting in the Bull Run municipal watershed, Oregon. *Water Resources Bulletin* **21**, 469–73.

Jackson, I.J. (1971) Problems of throughfall and interception assessment under tropical forest. *Journal of Hydrology* **12**, 234–54.

Jackson, I.J. (1975) Relationships between rainfall parameters and interception by tropical forest. *Journal of Hydrology* **24**, 215–38.

Jackson, P.C., Cavelier, J., Goldstein, G.C., Meinzer, F.C. & Holbrook, N.M. (1995) Partitioning of water resources among plants of a lowland tropical forest. *Oecologia* **101**, 197–203.

Jarvis, P.G. & McNaughton, K.G. (1986) Stomatal control of transpiration: scaling up from leaf to region. *Advances in Ecological Research* **15**, 1–49.

Jarvis, P.G. & Stewart, J.B. (1979) Evaporation of water from plantation forest. In: Ford, E.D., Malcolm, D.C. & Atteson, J., eds. *The Ecology of Even-Aged Forest Plantations*, pp. 327–49. Institute of Terrestrial Ecology, Penicuik, Scotland.

Jayasurya, M.D.A., Dunn, G., Benyon, R. & O'Shaughnessy, P.J. (1993) Some factors affecting water yield from mountain ash (*Eucalyptus regnans*) dominated forests in south-east Australia. *Journal of Hydrology* **150**, 345–67.

Jepma, C.J. (1995) *Tropical Deforestation: a Socioeconomic Approach.* Earthscan Publications, London.

Johnson, E.A. & Kovner, J.L. (1956) Effect on streamflow

of cutting a forest understory. *Forest Science* **2**, 82–91.

Jordan, C.F. & Heuveldop, J. (1981) The water budget of an Amazonian rainforest. *Acta Amazonica* **11**, 87–92.

Joslin, J.D., Mueller, S.F. & Wolfe, M.H. (1990) Tests of models of cloudwater deposition to forest canopies using artificial and living collectors. *Atmospheric Environment* **24A**, 3007–19.

Juvik, J.O. & Nullet, D.A. (1995) Comments on: 'A proposed standard fog collector for use in high-elevation regions'. *Journal of Applied Meteorology* **34**, 2108–10.

Kelliher, F.M., Black, T.A. & Price, D.T. (1986) Estimating the effects of understory removal from a Douglas fir forest using a two-layer canopy evapotranspiration model. *Water Resources Research* **22**, 1891–9.

Kittredge, J. (1948) *Forest Influences*. McGraw-Hill, New York.

Köstner, B.M.M., Schulze, E.D., Kelliher, F.M. *et al.* (1992) Transpiration and canopy conductance in a pristine broad-leaved forest of *Nothofagus*: an analysis of xylem sap flow and eddy correlation measurements. *Oecologia* **91**, 350–9.

Lal, R. (1987) *Tropical Ecology and Physical Edaphology*. John Wiley & Sons, New York.

Law, F. (1956) The effect of afforestation upon the yield of water catchment areas. *Journal of the British Waterworks Association* **38**, 344–54.

Lee, R. (1970) Theoretical estimates versus forest water yield. *Water Resources Research* **6**, 327–34.

Leopoldo, P.R., Franken, W., Matsui, E. & Salati, E. (1982) Estimation of evapotranspiration of 'terra firme' Amazonian forest. *Acta Amazonica* **12**, 333–7 (in Portuguese, with English summary).

Lesch, W. & Scott, D.F. (1997) The responses in water yield to the thinning of *Pinus radiata*, *Pinus patula* and *Eucalyptus grandis* plantations. *Forest Ecology and Management* **99**, 295–307.

Lindroth, A. (1985) Canopy conductance of coniferous forests related to climate. *Water Resources Research* **21**, 297–304.

Lloyd, C.R. & Marques, A. de O. (1988) Spatial variability of throughfall and stemflow measurements in Amazonian rain forest. *Agricultural and Forest Meteorology* **42**, 63–73.

Lloyd, C.R., Gash, J.H.C., Shuttleworth, W.J., de Marques, A. & O. (1988) The measurement and modelling of rainfall interception by Amazonian rain forest. *Agricultural and Forest Meteorology* **43**, 277–94.

McNaughton, K.G. & Laubach, J. (1998) Unsteadiness as a cause of non-equality of eddy diffusivities for heat and vapour at the base of an advective inversion. *Boundary-Layer Meteorology* **88**, 479–504.

Madduma Bandara, C. & Kurupuarachchi, T.A. (1998) Land-use change and hydrological trends in the upper Mahaweli Basin. Paper presented at the Workshop on Hydrology of Natural and Man-made Forests in the Hill Country of Sri Lanka, Kandy, October 1988.

Malmer, A. (1993) *Dynamics of hydrology and nutrient losses as rsponse to establishment of forest plantation. A case study on tropical rainforest land in Sabah, Malaysia*. PhD thesis, Swedish University of Agricultural Sciences, Umea, Sweden.

Mapa, R.B. (1995) Effect of reforestation using *Tectona grandis* on infiltration and soil water retention. *Forest Ecology and Management* **77**, 119–25.

Marshall, J.D. & Waring, R.H. (1984) Conifers and broadleaf species: stomatal sensitivity differs in western Oregon. *Canadian Journal of Forest Research* **14**, 905–8.

Meinzer, F.C., Goldstein, G., Holbrook, N.M., Jackson, P. & Cavelier, J. (1993) Stomatal and environmental control of transpiration in a lowland tropical forest tree. *Plant, Cell and Environment* **16**, 429–36.

Meinzer, F.C., Andrade, J.L., Goldstein, G., Holbrook, N.M., Cavelier, J. & Jackson, P. (1997a) Control of transpiration from the upper canopy of a tropical forest: the role of stomatal, boundary layer and hydraulic architecture components. *Plant, Cell and Environment* **20**, 1242–53.

Meinzer, F.C., Hinckley, T.M. & Ceulemans, R. (1997b) Apparent responses of stomata to transpiration and humidity in a hybrid poplar canopy. *Plant, Cell and Environment* **20**, 1301–9.

Molchanov, A.A. (1963) *The Hydrological Role of Forests*. Israel Programme for Scientific Translations, Jerusalem, Israel.

Monk, C.T. & Day, F.P. (1988) Biomass, primary production, and selected nutrient budgets for an undisturbed hardwood watershed. *Ecological Studies* **66**, 151–9.

Monteith, J.L. (1965) Evaporation and environment. *Symposium of the Society for Experimental Biology* **19**, 205–34.

Moore, I.D., O'Loughlin, E. & Burgh, G.J. (1988) A contour-based topographic model for hydrological and ecological applications. *Earth Surface Processes and Landforms* **13**, 305–20.

Mueller, S.F. (1991) Estimating cloud water deposition to subalpine spruce-fir forests. I. Modifications to an existing model. *Atmospheric Environment* **25A**, 1093–104.

Mulder, J.P.M. (1983) *A simulation of rainfall interception in a pine forest*. PhD thesis, University of Groningen, Groningen, The Netherlands.

Nänni, U.W. (1970) Trees, water and perspective. *South African Forestry Journal* **75**, 9–17.

Nepstad, D., Verissimo, A., Alencar, A. *et al.* (1999) Large-scale impoverishment of Amazonian forests by logging and fire. *Nature* **398**, 505–8.

Noirfalise, A. (1959) Sur l'interception de la pluie par le couvert dans quelques forêts belges. *Bulletin de la Société Royale Forestiere de la Belqique* **10**, 433–9.

Oldeman, L.R. (1994) The global extent of soil degradation. In: Greenland, D.J. & Szabolcs, I., eds. *Soil Resilience and Sustainable Land Use*, pp. 99–118. CAB International, Wallingford, UK.

O'Loughlin, E. (1986) Prediction of surface saturation zones in natural catchments by topographic analysis. *Water Resources Research* **22**, 794–804.

Parker, G.G. (1985) *The effect of disturbance on water and solute budgets of hillslope tropical rainforest in Northeastern Costa Rica*. PhD thesis, University of Georgia, Athens, Georgia.

Pearce, A.J. & Rowe, L.K. (1979) Forest management effects on interception, evaporation and water yield. *Journal of Hydrology (NZ)* **18**, 73–87.

Pearce, A.J., Rowe, L.K. & Stewart, J.B. (1980a) Night-time, wet canopy evaporation rates and the water balance of an evergreen mixed forest. *Water Resources Research* **16**, 955–9.

Pearce, A.J., Rowe, L.K. & O'Loughlin, C.L. (1980b) Effects of clearfelling and slashburning on water yields and storm hydrographs in evergreen mixed forests, western New Zealand. *International Association of Hydrological Sciences Publication* **130**, 119–27.

Pearce, A.J., Rowe, L.K. & O'Loughlin, C.L. (1982) Hydrologic regime of undisturbed mixed evergreen forests, South Nelson, New Zealand. *Journal of Hydrology (NZ)* **21**, 98–116.

Penman, H.L. (1948) Natural evaporation from open water, bare soil and grass. *Proceedings of the Royal Society of London A* **193**, 120–45.

Pilgrim, D.H., Doran, D.G., Rowbottom, I.A., Mackay, S.M. & Tjendana, J. (1982) Water balance and runoff characteristics of mature and cleared pine and eucalypt catchments at Lidsdale, New South Wales. In: O'Loughlin, E.M. & Bren, L.J., eds. *The First National Symposium on Forest Hydrology*, pp. 103–10. Institution of Engineers, Barton ACT, Australia.

Postel, S. & Heise, L. (1988) *Reforesting the Earth*. Worldwatch Institute, Washington, DC.

Qian, W.C. (1983) Effects of deforestation on flood characteristics, with particular reference to Hainan Island, China. *International Association of Hydrological Sciences Publication* **140**, 249–58.

Rambal, S., Ibrahim, M. & Rapp, M. (1984) Variabilité spatiale des variations du stock d'eau sous forêt. *Catena* **11**, 177–86.

Richardson, J.H. (1982) Some implications of tropical forest replacement in Jamaica. *Zeitschrift für Geomorphologie Neue Folge Supplement* **44**, 107–18.

Roberts, J.M. (1983) Forest transpiration: a conservative hydrological process? *Journal of Hydrology* **66**, 133–41.

Roberts, J.M. (1999) Plants and water in forests and woodlands. In: Baird, A. & Wilby, R., eds. *Eco-Hydrology: Plants and Water in Terrestrial and Aquatic Ecosystems*, pp. 181–236. Routledge, London.

Roberts, J.M. & Rosier, P.T.W. (1993) Physiological studies in young Eucalyptus stands in southern India and derived estimates of forest transpiration. *Agricultural Water Management* **24**, 103–18.

Roberts, J.M. & Rosier, P.T.W. (1994) Comparative estimates of transpiration of ash and beech forest at a chalk site in southern Britain. *Journal of Hydrology* **162**, 229–45.

Roberts, J.M., Pymar, C.F., Wallace, J.S. & Pitman, R.M. (1980) Seasonal changes in leaf area, stomatal and canopy conductances and transpiration from bracken below a forest canopy. *Journal of Applied Ecology* **17**, 409–22.

Roberts, J.M., Pitman, R.M. & Wallace, J.S. (1982) A comparison of evaporation from stands of Scots pine and Corsican pine in Thetford Chase, East Anglia. *Journal of Applied Ecology* **19**, 859–72.

Roberts, J.M., Cabral, O.M.R., Fisch, G., Molion, L.C.B., Moore, C.J. & Shuttleworth, W.J. (1993) Transpiration from an Amazonian rainforest calculated from stomatal conductance measurements. *Agricultural and Forest Meteorology* **65**, 175–96.

Roche, M.A. (1982) Evapotranspiration réelle de la forêt amazonienne en Guyane. *Cahiers ORSTOM, Série Hydrologie* **19**, 37–44.

Rogerson, T.L. (1967) Throughfall in pole-sized loblolly pine as affected by stand density. In: Sopper, W.E. & Lull, H.W., eds. *International Symposium on Forest Hydrology*, pp. 187–90. Pergamon Press, Oxford.

Rowe, L.K. (1983) Rainfall interception by an evergreen beech forest, Nelson, New Zealand. *Journal of Hydrology* **66**, 143–58.

Rowe, L.K. & Pearce, A.J. (1994) Hydrology and related changes after harvesting native forest catchments and establishing *Pinus radiata* plantations. Part 2. The native forest water balance and changes in streamflow after harvesting. *Hydrological Processes* **8**, 281–97.

Running, S.W. & Coughlan, J.C. (1988) A general model of forest ecosystem processes for regional applications. I. Hydrologic balance, canopy gas exchange and primary production processes. *Ecological Modelling* **42**, 125–54.

Russell, S. (1984) Techniques notebook: measurement of mist precipitation. *The Bryophytic Times* **25**, 4.

Rutter, A.J. (1967) An analysis of evaporation from a stand of Scots pine. In: Sopper, W.E. & Lull, H.W., eds. *International Symposium on Forest Hydrology*, pp. 403–16. Pergamon Press, Oxford.

Rutter, A.J. (1968) Water consumption by forests. In: Kozlowski, T.T., ed. *Water Deficits and Plant Growth*, Vol. II, pp. 23–84. Academic Press, London.

Rutter, A.J. (1975) The hydrological cycle in vegetation. In: Monteith, J.L., ed. *Vegetation and the Atmosphere*, Vol. 1, pp. 111–54. Academic Press, London.

Rutter, A.J., Kershaw, K.A., Robins, P.C. & Morton, A.J.

(1971) A predictive model of rainfall interception in forests. I. Derivation of the model from observations in a plantation of Corsican pine. *Agricultural Meteorology* **9**, 367–83.

Rutter, A.J., Morton, A.J. & Robins, P.C. (1975) A predictive model of rainfall interception in forests. II. Generalization of the model and comparison with observations in some coniferous and hardwood stands. *Journal of Applied Ecology* **12**, 367–80.

Sahin, V. & Hall, M.J. (1996) The effects of afforestation and deforestation on water yields. *Journal of Hydrology* **178**, 293–309.

Scatena, F.N. (1990) Watershed scale rainfall interception on two forested watersheds in the Luquillo Mountains of Puerto Rico. *Journal of Hydrology* **113**, 89–102.

Schaap, M.G. (1996) *The role of soil organic matter in the hydrology of forests on dry sandy soils*. PhD thesis, University of Amsterdam, Amsterdam, The Netherlands.

Schellekens, J., Scatena, F.N., Bruijnzeel, L.A. & Wickel, A.J. (1999) Modelling rainfall interception by a lowland tropical rain forest in northeastern Puerto Rico. *Journal of Hydrology* **225**, 168–84.

Schellekens, J., Bruijnzeel, L.A., Scatena, F.N., Bink, N.J. & Holwerda, F. (2000) Evaporation from a tropical rain forest, Luquillo Experimental Forest, eastern Puerto Rico. *Water Resources Research* **36** (in press).

Schemenauer, R.S. & Cereceda, P. (1994) A proposed Standard Fog Collector for use in high-elevation regions. *Journal of Applied Meteorology* **33**, 1313–22.

Schulze, E.D., Čermák, J., Matyssek, R. *et al.* (1985) Canopy transpiration and water fluxes in the xylem of the trunk of *Larix* and *Picea* trees: a comparison of xylem flow, porometer and cuvette measurements. *Oecologia* **6**, 475–83.

Scott, D.F. & Lesch, W. (1996) The effects of riparian clearing and clearfelling of an indigenous forest on streamflow, stormflow and water quality. *South African Forestry Journal* **175**, 1–14.

Scott, D.F. & Smith, R.E. (1997) Preliminary empirical models to predict reductions in total and low flows resulting from afforestation. *Water South Africa* **23**, 135–40.

Shuttleworth, W.J. (1988) Evaporation from Amazonian rain forest. *Philosophical Transactions of the Royal Society of London B* **323**, 321–46.

Shuttleworth, W.J. (1989) Micrometeorology of temperate and tropical forest. *Philosophical Transactions of the Royal Society of London B* **324**, 299–334.

Shuttleworth, W.J. (1992) Evaporation. In: Maidment, D.R., ed. *Handbook of Hydrology*, pp. 4.1–5.53. McGraw-Hill, New York.

Shuttleworth, W.J. & Calder, I.R. (1979) Has the Priestley–Taylor equation any relevance to forest evaporation? *Journal of Applied Meteorology* **18**, 639–46.

Silberstein, R.P., McJannet, D.L. & Vertessy, R.A. (1999) Trees on hills. Better growth = less waterlogging. *Journal of the Australian Water and Wastewater Association* May/June, 13–15.

Sivapalan, M., Viney, N.R. & Jeerjav, C.G. (1996) Water and salt balance moelling to predict the effects of land use changes in forested catchments. 3. The large catchment model. *Hydrological Processes* **10**, 429–46.

Smith, D.M. & Allen, S.J. (1996) Measurement of sap flow in plant stems. *Journal of Experimental Botany* **47**, 1833–44.

Smith, R.E. & Bosch, J.M. (1989) A description of the Westfalia catchment experiment to determine the effect on water yield of clearing the riparian zone and converting an indigenous forest to a eucalypt plantation. *South African Forestry Journal* **151**, 26–31.

Sopper, W.E. & Lull, H.W. (eds) (1967) *International Symposium on Forest Hydrology*. Pergamon Press, Oxford.

Stednick, J.D. (1996) Monitoring the effects of timber harvest on annual water yield. *Journal of Hydrology* **176**, 79–95.

Stewart, J.B. (1977) Evaporation from the wet canopy of a pine forest. *Water Resources Research* **13**, 915–21.

Stewart, J.B. (1988) Modelling surface conductance of pine forest. *Agricultural and Forest Meteorology* **43**, 19–35.

Stogsdill, W.R., Wittwer, R.F., Hennessey, T.C. & Dougherty, P.M. (1992) Water use in thinned loblolly pine plantations. *Forest Ecology and Management* **50**, 233–45.

Swank, W.T., Swift, L.W. Jr & Douglas, J.E. (1988) Streamflow changes associated with forest cutting, species conversions, and natural disturbances. *Ecological Studies* **66**, 297–312.

Swanson, R.H. (1994) Significant historical developments in thermal methods for measuring sap flow in trees. *Agricultural and Forest Meteorology* **72**, 113–32.

Swindel, B.F., Lassiter, C.J. & Riekerk, H. (1983) Effects of clearcutting and site preparation on stormflow volumes of streams in *Pinus elliottii* flatwood forests. *Forest Ecology and Management* **5**, 245–53.

Tan, C.S. & Black, T.A. (1976) Factors affecting the canopy resistance of a Douglas-Fir forest. *Boundary-Layer Meteorology* **10**, 475–88.

Teklehaimanot, Z., Jarvis, P.G. & Ledger, D.C. (1991) Rainfall interception and boundary layer conductance in relation to tree spacing. *Journal of Hydrology* **123**, 261–78.

Tennessee Valley Authority (1961) *Forest Cover Improvement Influences Upon Hydrologic Characteristics of White Hollow Watershed, 1935–1958*. Tennessee Valley Authority, Knoxville, Tennessee.

Teoh, T.S. (1977) Throughfall, stemflow and interception studies on *Hevea* stands in Peninsular Malaysia. *Malayan Nature Journal* **31**, 141–5.

Thom, A.S. (1975) Momentum, mass and heat exchange of plant communities. In: Monteith, J.L., ed. *Vegetation and the Atmosphere*, Vol. 1, pp. 57–109. Academic Press, London.

Thompson, F.B. (1972) Rainfall interception by oak coppice (*Quercus robur* L.). In: Taylor, J.A., ed. *Research Papers in Forest Meteorology*, Aberystwyth Symposium, University of Wales, Aberystwyth, Wales.

Tiktak, A. & Bouten, W.J. (1994) Soil water dynamics and long-term water balances of a Douglas fir stand in The Netherlands. *Journal of Hydrology* **156**, 265–83.

Trimble, G.R., Reinhart, K.G. & Webster, H.H. (1963) Cutting the forest to increase water yields. *Journal of Forestry* **61**, 635–40.

Trimble, S.W., Weirich, F.H. & Hoag, B.L. (1987) Reforestation and the reduction of water yield on the southern Piedmont since circa 1940. *Water Resources Research* **23**, 425–37.

Uijlenhoet, R. & Stricker, J.N.M. (1999) Dependence of rainfall interception on drop size: a comment. *Journal of Hydrology* **217**, 157–63.

Valente, F., David, J.S. & Gash, J.H.C. (1997) Modelling interception loss for two sparse eucalypt and pine forests in central Portugal using reformulated Rutter and Gash analytical models. *Journal of Hydrology* **190**, 141–62.

Vandanashiva, & Bandhyopadhyay, J. (1983) Eucalyptus: a disastrous tree for India. *Ecologist* **13**, 184–7.

Van der Plas, M.C. & Bruijnzeel, L.A. (1993) Impact of mechanized selective logging of rainforest on topsoil infiltrability in the Upper Segama area, Sabah, Malaysia. *International Association of Hydrological Sciences Publication* **216**, 203–11.

Van Noordwijk, M., Van Roode, M., McCallie, E.L. & Lusiana, B. (1998) Erosion and sedimentation as multiscale, fractal processes: implications for models, experiments and the real world. In: Penning de Vries, F.W.T., Agus, F. & Kerr, J., eds *Soil Erosion at Multiple Scales*, pp. 223–53. CAB International, Wallingford UK.

Veracion, V.P. & Lopez, A.C.B. (1976) Rainfall interception in a thinned Benguet pine (*Pinus kesiya*) forest stand. *Sylvatrop Philippines Forest Research Journal* **1**, 128–34.

Vertessy, R.A., Wilson, C.J., Silburn, D.M.O., Çonnolly, R.D. & Ciesiolka, C.A. (1990) Predicting erosion hazard areas using digital terrain analysis. *International Association of Hydrological Sciences Publication* **192**, 298–308.

Vertessy, R.A., Hatton, T.J., O'Shaughnessy, P.J. & Jayasuriya, M.D.A. (1993) Predicting water yield from a mountain ash forest catchment using a terrain analysis-based catchment model. *Journal of Hydrology* **150**, 665–700.

Vertessy, R.A., Benyon, R.G., O'Sullivan, S.K. & Gribben, P.R. (1995) Relationships between stem diameter, sapwood area, leaf area and transpiration in a young mountain ash forest. *Tree Physiology* **15**, 559–67.

Vertessy, R.A., Hatton, T.J., Benyon, R.G. & Dawes, W.R. (1996) Long-term growth and water balance predictions for a mountain ash (*Eucalyptus regnans*) forest catchment subject to clearfelling and regeneration. *Tree Physiology* **16**, 221–32.

Vertessy, R.A., Hatton, T.J., Reece, P., O'Sullivan, S.K. & Benyon, R.G. (1997) Estimating stand water use of large mountain ash trees and validation of the sap flow measurement technique. *Tree Physiology* **17**, 747–56.

Vertessy, R.A., Watson, F.G.R., O'Sullivan, S.K. et al. (1998) *Predicting Water Yield from Mountain Ash Forest Catchments*. Cooperative Research Centre for Catchment Hydrology, Clayton, Victoria, Australia.

Viswanatham, N.K., Joshie, P. & Ram Babu (1982) Influence of forest on soil erosion control: Dehradun. In: *Annual Report 1982*, pp. 40–3. Central Soil and Water Conservation Reearch and Training Institute, Dehradun, India.

Walmsley, J.L., Schemenauer, R.S. & Bridgman, H.A. (1996) A method for estimating the hydrologic input from fog in mountainous terrain. *Journal of Applied Meteorology* **35**, 2237–49.

Walmsley, J., Burrows, W.R. & Schemenauer, R.S. (1999) The use of routine observations to calculate liquid water content in summertime high-elevation fog. *Journal of Applied Meteorology* **38**, 369–84.

Wasser, H.J. & Harger, J.R.E. (1992) *Several Environmental Factors Affecting the Rainfall in Indonesia*. ROSTSEA/UNESCO, Jakarta, Indonesia.

Waterloo, M.J., Bruijnzeel, L.A., Vugts, H.F. & Rawaqa, T.T. (1999) Evaporation from *Pinus caribaea* plantations on former grassland soils under maritime tropical conditions. *Water Resources Research* **35**, 2133–44.

Watson, F.G.R., Vertessy, R.A. & Grayson, R.B. (1999) Large-scale modelling of forest hydro-ecological processes and their long term effect on water yield. *Hydrological Processes* **13**, 689–700.

Weathers, K.C., Lovet, G.M. & Likens, G.E. (1995) Cloud deposition to a spruce forest edge. *Atmospheric Environment* **29**, 665–72.

Whitehead, D. (1998) Regulation of stomatal conductance and transpiration in forest canopies. *Tree Physiology* **18**, 633–44.

Whitehead, D. & Jarvis, P.G. (1981) Coniferous forest and plantations. In: Kozlowski, T.T., ed. *Water Deficits and Plant Growth*, Vol. VI, pp. 49–152. Academic Press, London.

Whitehead, D., Jarvis, P.G. & Waring, R.H. (1984) Stomatal conductance, transpiration, and resistance to water uptake in a *Pinus sylvestris* spacing experiment. *Canadian Journal of Forest Research* **14**, 692–700.

Wilm, H.G. & Dunford, E.G. (1948) *Effect of Timber Cutting on Water Available for Streamflow from a Lodgepole Pine forest.* US Department of Agriculture Technical Bulletin 968, Washington, DC.

Wullschleger, S.D., Meinzer, F.C. & Vertessy, R.A. (1998) A review of whole-plant water studies in trees. *Tree Physiology* **18**, 499–512.

Young, A. (1989) *Agroforestry for Soil Conservation.* CAB International, Wallingford, UK.

Zadroga, F. (1981) The hydrological importance of a montane cloud forest area of Costa Rica. In: Lal, R. & Russell, E.W., eds. *Tropical Agricultural Hydrology*, pp. 59–73. John Wiley and Sons, Chichester.

Ziegler, A.D. & Giambelluca, T.W. (1997) Importance of rural roads as source areas for runoff in mountainous areas of northern Thailand. *Journal of Hydrology* **196**, 204–29.

Zincke, P.J. (1967) Forest interception studies in the United States. In: Sopper, W.E. & Lull, H.W., eds. *International Symposium on Forest Hydrology*, pp. 137–60. Pergamon Press, Oxford.

Zon, R. (1927) *Forests and Water in the Light of Scientific Investigation.* US Government Printing Office, Washington, DC.

Part 4
Social and Human Interface

No account of the world's trees and forests can neglect consideration of their interaction with people. Although treatment here is too brief, beyond introducing the subject, the three chapters present something of the breadth of the interface. Kjell Nilsson and colleagues (Chapter 13) recognize the particular and highly artificial environment which affects trees in our towns and cities. As an amenity to enjoy they directly impact the lives of many, many people. The fact that forest and woodlands are important in everyone's livelihoods and their social values needs full acknowledgement as central to any consideration of broad sense sustainability. This Stephen Bass does in outline in Chapter 14. However, for a great many people it is products other than timber which are of primary importance (Chapter 15). In this final chapter of Volume 1, Will Cavendish provides one of the very few economic analyses of the subject. These and other themes are touched upon and help provide background for understanding the central roles trees and forests play.

13: Trees in the Urban Environment

KJELL NILSSON, THOMAS B. RANDRUP AND BARBARA M. WANDALL

13.1 INTRODUCTION

Trees in the urban environment are often referred to as the urban forest, comprising trees in civic woodlands, parks and the street. Earlier, urban trees were mainly regarded as aesthetic elements, whereas today they are recognized as having a positive impact on the environment as well as providing economic and social benefits. Monetary evaluations reflect the various benefits arising from the urban forest, covering such aspects as reduction of pollution and energy use, environmental amelioration, savings in public health care and increase in economic investment. Hence, the value of the urban forest is being increasingly recognized as a vital component in the maintenance of a sustainable urban environment in cities around the world. Meanwhile, the population living in urban areas has increased rapidly since the 1950s and the lack of space makes it tempting to use green areas for infrastructure and buildings.

Collins (1997) emphasizes that urban forestry, the planning, management and maintenance of the urban forest, is more closely aligned to traditional forestry than might be immediately apparent. As implicit in the term, the principle of sustained yield has been adapted to the urban environment, applying rural land-use forestry to the management of the urban forest. Hence, the overall objective in urban forestry is not that of timber production but, through a balanced structure of age and species, a sustained production of environmental, social and economic benefits. These social benefits also accrue if local communities are encouraged to contribute to their own environment by promoting projects and activities involving local residents. These kinds of projects often prove effective in promoting social interaction and lead to increased local involvement in other aspects as well.

The planning and management of a healthy urban forest requires that coordinated strategies are agreed between the professions more or less directly involved, such as planners, landscape architects, arboriculturists, foresters, engineers, legislators, developers and utility managers. Management of the urban forest resource is moving towards the achievement of a healthy, well-distributed urban forest, which is increasingly becoming an essential and integral part of the urban infrastructure. In this chapter, we outline the science that underpins such forests.

13.2 DEFINITIONS

Trees in the urban environment have been defined in several ways, although 'urban forestry and arboriculture' is probably the most used term in relation to trees in or near the urban environment. Many different research disciplines are involved in the fields of arboriculture and urban forestry. Harris (1992) defines arboriculture as being 'primarily concerned with the planting and care of trees and more peripherally concerned with shrubs and woody vines and groundcover plants'. However, many people consider urban forestry to consist primarily of two types of planting: urban forests and urban trees. Also, definitions of urban green areas and urban forestry vary significantly throughout the world. There seem to be different opinions as to what urban forestry covers, depending on whether professionals have a background within or outside forestry or have

Table 13.1 The principal results of the different growing conditions associated with woodland trees, park trees and street trees.

	Growing conditions	Stress level	Average lifespan
Urban woodlands	Good	Low	Medium to high
Park trees	Good/fair	Medium	High*
Street trees	Fair/poor/very poor	High	Low (10–15 years)†

* In urban woodlands, fellings are performed in connection with timber production. In parks, trees are very seldom removed, unless they are potentially dangerous or hazardous.
† Moll (1989).

experience of working in the USA or other parts of the world such as Europe.

In many cases foresters have argued that urban forestry concerns 'forestry in urban areas'. Volk (1986) stated that:

> Green areas such as tree lined streets, cemeteries, and parks, which are not predominated by trees and as such do not come under forestry influence should be excluded from consideration [of being considered urban forestry]. On the other hand, park forests which come under forestry supervision, should be included in our considerations.

The fear of involvement of foresters in urban plantings has been described by Chambers (1987): 'Misguided resistance to the concept [of urban forestry] is still often mistakenly assumed to imply the take over of public open space and existing amenity trees for timber production.' Americans tends to look at urban forestry as the 'management of trees in urban areas on larger than an individual basis' (Harris 1992). In Europe this broad concept now seems to be accepted on a wider basis. Urban forest stands are now often considered as resources where an economic yield is not required, and traditional forestry practices are combined with both aesthetic and recreational considerations.

Costello (1993) suggested that urban forestry be defined as 'the management of trees in urban areas', including single trees. In this definition, 'management' is described as the planning, planting and care of trees; 'trees' as individuals, small groups, larger stands (e.g. green belts) and remnant forests; and 'urban areas' as those areas where people live and work. The location of urban forests can be anywhere, from urban settings to the countryside, as long as there are human structures on the site; these structures can be related to ecological functions, protection, merchandise or tourism and recreation.

Using this concept, we categorize trees in the urban environment as both urban forests and urban trees. Urban forests can be defined by their placement in or near urban areas and by their multifunctional aspects, giving shade, amenity values, etc. Therefore, we define urban forestry as the establishment, management, planning and design of trees and forest stands with amenity values, situated in or near urban areas (Nilsson & Randrup 1996).

Trees in the urban environment may be divided into three different types: trees in urban woodlands, park trees and street trees. These types of plantings differ distinctively in several ways. Most importantly, they have significantly different growing conditions, which means that they have different needs with regard to planning, management and maintenance (Table 13.1). The methods used and the research problems related to urban green areas are common throughout the world due to the fact that urban 'greening' concerns not only natural spheres (such as trees, growing conditions, etc.) but also the environment in close proximity to human populations.

13.3 URBAN GROWING CONDITIONS

Urban growing conditions differ significantly from those in the rural landscape, and produce difficulties as a result of both above- and below-ground influences.

13.3.1 Stress factors

The harsh soil and air conditions that exist in urban planting are problems that do not play the same role in landscape planting. Growing conditions may also be difficult due to shading effects, recreational users, etc. (Harris 1992; Bradshaw *et al.* 1995). The modified urban mesoclimate affects the quantity of contaminants in urban areas, which is raised by a factor of around 25 (Flint 1985; Harris 1992). In general, the average lifespan of a newly planted street tree may be as low as 10–15 years (Moll 1989).

During the last 30–40 years, the vitality of street trees has fallen drastically (Bradshaw *et al.* 1995). Heavier traffic patterns have increased demands for road construction, which consequently has changed the growing conditions of many roadside trees. Also, pollution from traffic has a highly detrimental impact on street trees (Pedersen 1990). The fact that 50% of the trees planted in an urban environment die within the first year emphasizes this point (Gilbertson & Bradshaw 1985). Nowak *et al.* (1990) found that 34% of 480 trees died within 2 years of planting, while Miller and Miller (1991) found that the mortality rate was 25–50% for a number of species planted in Wisconsin, USA.

Temperature extremes can occur, especially where trees are widely spaced and where heat is reflected from hard surfaces (Bradshaw *et al.* 1995). Harris (1992) described that, occasionally, tree limbs up to 0.6 m and trunks up to 1.2 m in diameter break and fall during hot calm summer and autumn afternoons and subsequent evenings. Roots are more sensitive to temperature extremes than the tops of plants (Harris 1992).

Wind speed will vary according to the shape and height of buildings. Areas with tall buildings will usually be relatively cool in summer due to shading effects, and warmer in winter due to wind-protection effects. On the other hand, winds are more variable and more extreme at exposed corners of tall isolated buildings. Buildings deflect strong winds downwards and concentrate their force at the base and corners of buildings, forming 'wind tunnels'. Trees planted in these exposed gaps may suffer scorched leaves and shoots, which lead to a stunted canopy, especially on the windward side. Newly planted trees will transpire more rapidly in windy situations, which can lead to the death of a tree already severely stressed by drought (Harris 1992; Bradshaw *et al.* 1995). The wind stability of trees is determined by tree species, stand structure, spacing, thinning regimes, soil classes, breeding and tree age at the time of anchorage.

The presence of airborne pollutants in the atmosphere has been a characteristic feature of the urban environment since the beginning of the Industrial Revolution. Air pollution can occur in a variety of forms but the principal ones are dust, SO_2 and NO_x (Bradshaw *et al.* 1995). Leaves are the plant parts most likely to show symptoms of air pollution injury. On broadleaved plants, leaves may develop interveinal necrotic areas, marginal or tip necrosis, stippling of the upper surface, or silvering of the lower surface (Harris 1992). However, trees in the urban environment also play a role in the quest for cleaner air in the cities. Scott *et al.* (1998) showed that daily uptake of NO_2 and particulate matter represented 1–2% of anthropogenic emissions for the county of Sacramento, California.

In areas with winter temperatures below 0°C, the use of de-icing salt is a well-known problem. De-icing salt is applied to the surrounding environment by surface run-off, wet spraying and airborne drifting. The initial and most common symptom of de-icing salt damage on trees and shrubs is reduced growth. This is often difficult to recognize or may be confused with other stress factors. Reduced growth is usually followed by early autumn colours and premature leaf fall. De-icing salt is usually accumulated on the windward side of trees. The damage is easily recognized because it faces the road and is normally regarded as the best indication of de-icing spray damage. The majority of trees and shrubs subjected to either soil salt or salt spray typically show necroses at the edges of the leaves or needles. Wounds, often related to pruning, are a common place for spray salt to infect the plants. The damage may cause lack of sprouting and eventually dieback. Conifers are very susceptible to de-icing salt spray damage because they are green all through the winter maintenance season. Trees and shrubs damaged by de-icing salt and showing dieback are difficult to cure. (Dobson 1991; Brod 1993; Gibbs & Palmer 1994; Pedersen & Fostad 1996; Randrup & Pedersen 1996).

13.3.2 Characteristics and restriction of rooting in the built environment

Urban soils as a growing medium are poorly understood and often misunderstood. Therefore plantings are carried out with little appreciation or attention to the character and quality of the material that lies beneath the surface (Craul 1992). One major problem in relation to planting in the urban situation is soil compaction, which may occur in small as well as large urban sites. Soil compaction can be divided into two types: (i) intentional soil compaction, which occurs when soil is deliberately compacted for site stabilization under roads, houses, etc. and (ii) unintentional soil compaction, which occurs when traffic uses areas intended for planting (Randrup 1997). In the urban situation, unintentional soil compaction is primarily found along roadsides and on construction sites.

When soil is compacted, its bulk density increases and its porosity decreases. These effects inhibit plant growth because the soil becomes impenetrable to root growth and, furthermore, restricts the water and oxygen available to the roots. For example, root growth of most plants is impeded once soil bulk density rises above 1.6. One consequence of compacted soil is waterlogging, which can kill roots around existing trees. Soil loosening has proved to be effective in alleviating compacted soil (Håkansson & Reeder 1994; Rolf 1994).

However, there is no doubt that the best treatment for compacted soil is to protect the soil from being compacted in the first place. Florgård (1987) suggested protecting trees growing on construction sites by dividing the site into zones in which different types of construction traffic are permitted. The principle of construction site zoning was adapted by Randrup and Dralle (1997) to protect the soil from being compacted. They suggested that the entire construction site be divided into a building zone, a working zone and a protection zone. In the protection zone no traffic is allowed, and special attention must be paid in the building and working zones because the soil is certainly going to be compacted during the construction of buildings.

13.4 URBAN CHARACTERISTICS/ENVIRONMENT ANALYSED

The city is characterized by paving and buildings, which variously results in decreased wind speed, but an increase in windthrow (Harris 1992; Bradshaw *et al.* 1995), raised temperatures (Nowak & McPherson 1993; Miller 1997) and precipitation, lowered humidity and shading in street canyons (Bradshaw *et al.* 1995; Nowak 1995). The extent of the influences of these factors depends on the structure and amount of vegetation, as well as on the size of the city. Green areas in the city accessible to the public include areas at schools, public libraries and social institutions, as well as parks and churchyards. These are areas that are typically relatively small and geographically widespread, which is why they often suffer great recreational pressure, most often due to the overall proportion of green areas in the city.

Throughout Europe the proportion of urban green areas varies greatly, from over 60% of the area of Bratislava, the capital of Slovakia, to about 5% in Madrid, the capital of Spain (Stanners & Bourdeau 1995). In comparison, the figure for Mexico City is only 2.2% (Benavides Meza 1992); in relation to the number of inhabitants, this only provides $1.94\,m^2$ per inhabitant, which is far below the $9\,m^2$ recommended by the World Health Organization. A suggested measure of urban environmental quality is the location of green areas within a walking distance of 15 min or less from all housing areas. This criterion is met for all citizens in Brussels, Copenhagen, Glasgow, Gothenburg, Madrid, Milan and Paris and, in general, for more than 50% of the population in most European cities (Stanners & Bourdeau 1995).

The quality of urban green areas are increasingly recognized as being important to the overall quality of human life in the cities. Beyond this, the urban forests and trees are important as ecosystems in relation to the conservation of biological diversity (Kuchelmeister & Braatz 1993; Collins 1997). Though the urban population benefits from the urban green areas, the increase in population places great pressure on the existing green areas as a result of urban and infrastructural development. The importance of the urban green space, of which urban forestry is an integral part, increases as population increases. The growing urban population needs the environmental and social benefits associated with urban forests, as shown by Ulrich (1984) and Grahn (1989). In developing countries, urbanization has had a dramatic influence on creating environments practically without any amenities, as is the case in Mexico City, where the growth in population has not been matched by an increase in green space (Caballero Deloya 1993). It is an accepted reality that the growth of the cities cannot be stopped. Instead, the challenge is to control urban growth so that it results in economic growth and a satisfactory environment.

As more than two-thirds of the population of Europe live in urban areas, the quality of the urban environment, including green areas, is becoming increasingly recognized as one key to the economic reconstruction of European cities (Commission of the European Community 1990). Urban areas constitute the everyday environment of the greater part of the population and, in recent years, this topic has received considerable attention in the European Union, United Nations and the Organization for Economic Cooperation and Development, which will hopefully lead to an improvement in our knowledge of urban environments.

13.5 AMENITY VALUES/BENEFITS OF URBAN FORESTS AND TREES

Trees in the urban environment contribute significantly to the aesthetic appeal of cities, helping to maintain the psychological health of the inhabitants (Miller 1997). Besides the aesthetic and environmental aspects, urban forestry is also of importance in helping populations poor in resources to meet basic needs (Kuchelmeister & Braatz 1993). Urban plantings of fruit trees to provide firewood is one way of utilizing obvious benefits from the available natural resources, as done in Kampala, the capital of Uganda (Haque 1987).

Research has shown that urban trees benefit communities economically (Dwyer 1992), socially (Schroeder & Cannon 1987; Michael & Hull 1994) and environmentally (Broderick & Miller 1989; Matthews 1991; Huang *et al.* 1992). The size, structure and condition of the urban forest directly affects the amount of benefits provided by the urban vegetation (Nowak & McPherson 1993). Nevertheless, the social and environmental benefits provided by the urban forest are often largely ignored in land-use planning (Dwyer *et al.* 1991; Tyrväinen 1997).

13.5.1 Economic impact

Though trees are not usually thought of as economic resources, the presence of trees has an impact on different economic factors, including the level of real-estate prices, economic investments and employment. Results of recent Finnish studies show that the benefits provided by urban forests are reflected in property prices, and that environmental variables, such as proximity to wooded recreation areas and water courses as well as an increasing proportion of total forested area in the housing district, had a positive influence on apartment prices (Tyrväinen 1997). The presence of trees may influence property values positively by as much as 20%, with an average increase of 5–10% (Ebenreck 1989; Kielbaso 1989).

Trees also provide different external environmental benefits. Costs for heating and cooling can be reduced by appropriate use of vegetation. Energy reductions for individual buildings have been shown to range from 5 to 15% for heating and 10–50% for cooling. Especially in areas with high summer temperatures, trees are important providers of shade. In areas with cool winters, it is of importance to use deciduous trees, as the loss of leaves in winter allows the sun to heat the house. Proper arrangements of vegetation around buildings can also reduce wind velocity and thereby reduce the heat loss from buildings (McPherson & Rowntree 1993).

13.5.2 Recreational use of green areas

Several studies have shown the importance of urban forests for participation in outdoor recreation activities (e.g. Gåsdal 1993; Sievänen 1993; Lindhagen 1996). In a Danish national survey it was found that around two-thirds of all forest visits take place in the forest situated nearest the housing area (Jensen & Koch 1997). Danish and Swedish studies have also showed that the use of green areas by people is associated with three highly esteemed values: (i) a high degree of natural elements, (ii) fresh air and (iii) the possibility of solitude (Grahn 1991; Holm 1998).

In general, people use green areas less than they would like. According to Swedish studies, distance and the fear of assault are the most prevalent reasons for people not visiting green areas. When the distance to the park exceeds 300 m, one person in four postpones a daily visit. As many as 56% refrain from regular walks in the park when the distance increases to 500 m (Grahn 1991). American studies indicate that the removal of vegetation at strategic points can reduce the risk of assault, although this should not be the reason for removing all the vegetation (Michael & Hull 1994).

13.5.3 Psychological aspects

City life is stressful but research shows that urban green areas have a beneficial influence on the health and well-being of the urban population. Studies indicate that visits to green areas can counteract stress, renew vital energy and speed healing processes.

In Sweden, Grahn (1989) conducted extensive studies on the significance of parks to different groups of the population. For instance, he persuaded 40 schools, hospitals, sports associations, cultural associations and day-care centres to keep diaries of their outdoor activities for 1 year. For example, the diaries show that periods spent outside had an actual medicinal value for patients and residents of hospitals, old people's homes and homes for the sick. People became happier, slept better, needed less medication, were less restless and far more talkative.

Ulrich (1984) showed that hospitalized patients recovered faster when they had a view through a window, allowing them to see trees. Ulrich *et al.* (1991) showed a gory film on industrial accidents to 120 people. Half the people were then shown a nature film, whereas the other half were shown a film on the city, with sequences of buildings and traffic. The subjects' heart beat, muscular tension and blood pressure were monitored throughout. All subjects exhibited strong signs of stress during the first film on industrial accidents. The stress levels of the subjects that then watched the nature film returned to a normal level after 4–6 min, whereas the half that watched the film on buildings and traffic continued to exhibit high stress levels.

Kaplan and Kaplan (1989) have formulated a theory on the interaction between human attention and the surroundings. This theory distinguishes between spontaneous attention and conscious attention. Spontaneous attention demands no effort and occurs without premeditation. Conscious attention demands energy and leads to psychological exhaustion in the long term. Rapid movements, strong colours, sudden noises and strong odours are typical stimuli that demand conscious attention. These signals have a powerful effect on our attention, consciously and unconsciously, because they are perceived as potential dangers to which we should react. This means that urban living, with fast vehicles, flashing neon signs and strong colours, causes constant stress. Kaplan and Kaplan's research indicates that vegetation and nature reinforce our spontaneous attention, allow our sensory apparatus to relax and infuse us with fresh energy. Visits to green areas bring relaxation and sharpen our concentration, since we only need to use our spontaneous attention. At the same time, we get fresh air and sunlight, which have significance for our diurnal and annual rhythms.

13.5.4 Environmental education

The change and the continuity in nature provides not only a sense of time but also a sense of security and confidence, through the predictable change of seasons and the repetition of natural processes. The urban trees and forests play a role in educating young people and children in understanding the basic processes in nature and the complexities of the environment, either informally during play or as part of the school curriculum. Playing in nature or natural settings also helps children to develop their motoric senses (Holm 1998) and to regain and experience the human connection to nature through their imagination. Experiences with environmental education in the USA show that access to even a small area of the urban forest is of importance when learning about natural processes (Ebenreck 1989).

In England in 1985 a research project supported by the Department of Education and Science, the Countryside Commission and a consortium of local authorities resulted in the organization Learning through Landscapes (LTL). LTL was founded on the recognition of the impact that the physical surroundings has on children. LTL helps schools to design schoolyards that improve the quality of the environment and helps to create additional resources for the formal curriculum, such as the promotion of more effective teaching and learning in the outdoor classroom (Adams 1989).

The English Community Forests are examples of integrated programmes in which environmental education is integrated in social programmes by well-coordinated planning. The forests are charged with running and coordinating social programmes, involving information, consultation, participation, art, education, sport, recreation and cultural activities (Davies & Vaughan 1998).

13.5.5 Community involvement

Other benefits of urban forestry are those provided by involving people in planning, planting and caring for the trees and urban forests in their own locality. By promoting social interaction and strengthening local pride and identity, local public engagement benefits management greatly, encouraging people to act protectively towards their local forests and providing a potentially huge volunteer workforce (Collins 1996).

The possibilities for creating public awareness in any town are endless. In the USA there is a tradition of involving the public in all processes, from the policy-making to the care and maintenance of the urban forest. Through various types of tree programmes, public awareness has proved to be a valuable tool in the justification of tree activities because it develops a public interest group that helps to lobby for support. The projects

that involve citizens have various aspects in common, such as planting, preserving, pruning, maintenance, public education, etc. (Dawe 1989; Sievert 1989). Examples of initiatives that enhance public awareness and engagement in a municipal tree programme are the Tree Boards, municipal non-profit groups with participants from a wide range of professions. The close cooperation between professionals and non-professionals provides direct communication of new ideas and research and helps to solve associated problems as experienced in a practical situation. Another way of involving the public is to encourage houseowners to adopt a tree, by taking part in the care and maintenance of a tree in their neighbourhood.

13.6 ENVIRONMENTAL ASPECTS

Trees influence the urban environment in various ways, whether individual trees or the entire urban forest. Urban trees can mitigate the environmental impacts of urban development by moderating climate, improving air quality, providing habitat for wildlife and reducing noise levels (Nowak *et al.* 1994). The environmental benefits associated with urban trees and woodlands are widely recognized. In large cities such as Chicago (Nowak & McPherson 1993) and Mexico City (Caballero Deloya 1993), the environmental aspects are integrated in the city greening programmes. At the UN Conference on the Environment and Development in Rio de Janeiro in 1992, all participating countries adopted Agenda 21, an action plan that obliges them to work towards sustainable development, an obligation that in turn devolves to the administrators of urban green areas (United Nations 1992).

13.6.1 Local-scale climate

The replacement of natural surfaces with buildings and roads, as typified by cities, has altered the thermal and moisture properties of the area, which modifies the local atmosphere and generates the 'urban climate', with poorer air quality and increased air temperatures (Nowak *et al.* 1994). By transpiring water and shading surfaces, trees lower local air temperatures (Nowak 1995).

In areas where the summer is hot, trees are important because of the shade they provide. Trees can even reduce energy use in buildings by lowering summertime temperatures, shading buildings during the summer and blocking winter winds and consequently reduce the emission of pollutants from power-generating facilities (McPherson & Rowntree 1993; Nowak 1995).

Trees and shrubs can be used for manipulating air movement in order to control the impact of winds, by obstruction, guidance, deflection and filtration. The resulting effects depend on plant size, shape, foliage density and retention, as well as the placement of the plants. Combinations of trees and shrubs provide the most effective barriers, which may even be created by making use of already existing landforms. Windbreaks provide protection for considerable distances, up to 20 times the average windbreak height (Jensen 1955; Grey & Deneke 1978). A medium-porous windbreak (40–60% reduction) reduces wind speed nearly as much as an impenetrable barrier. Meanwhile the shelter from this type of windbreak extends over a longer distance downwind, with much less turbulence present. However, the type of optimum windbreak depends on the situation. Hence, in some settings an impenetrable windbreak will provide the most appropriate protection, whereas in others the highly porous windbreak will be the optimal solution. Thus, the structure of windbreaks in terms of suitable provenances and clones of species, number of rows and density can be designed to provide the type of shelter needed (Forman 1997).

13.6.2 Air quality

Trees intercept particulate matter and absorb such gaseous pollutants as O_3, SO_2 and NO_2, thus removing them from the atmosphere. Trees also emit various volatile organic compounds, such as isoprene and monoterpenes, that can contribute to O_3 formation in cities. Studies conducted in Frankfurt am Main have shown that streets along which trees are planted contained 3000 polluting particles per litre of air, whereas streets that lacked trees had as many as 10 000–12 000 particles per litre of air (Bernatzky 1978). Protective

plantations along heavily trafficked roads and around industrial areas are therefore an effective means of reducing air pollution. However, this should obviously not be taken as an excuse for neglecting to combat pollution at its source. In The Netherlands, some cities operate a tree-planting plan, the objective of which is to plant as many trees as needed in order to absorb the quantity of CO_2 emitted in providing the city with heating. To attain this goal, support is also given to reforestation programmes in eastern Europe and Africa, as well as to the extensive tree-planting carried out in the cities themselves.

Mexico City's Urban Forest Programme is another example of increasing awareness of the needs of urban greening as a means of altering the overall quality of life in the city. The fundamental goals of the programme are to mitigate the effects of the severe air pollution in the metropolitan area, to establish more green areas within the city and to give the city a more welcoming appearance (Caballero Deloya 1993).

Because trees can store carbon for long periods, they are an important carbon sink. Even though plants absorb CO_2 and produce O_2, it is important not to assign excessive significance to their role in the urban environment. Harris (1992) reminds us that plants really have only a minor effect on the CO_2 and O_2 content of urban air. Photosynthesis in the oceans accounts for 70–90% of the world's total O_2 production, for which reason it is absolutely vital that they be protected against pollution. However, even a minor reduction in the O_2 content of the air will cause a large percentage increase in its CO_2 content, which would reinforce the greenhouse effect, thus leading to a rise in the global temperature. Harris (1992) also stresses that urban vegetation has an especially beneficial effect on air pollution through its ability to reduce the quantity of airborne particles. Studies of 9000 trees in Chicago have shown that the trees reduce air pollution by 12 t of CO_2 and 10.8 t of O_3 per day (Nowak *et al.* 1994).

13.6.3 Biodiversity

Green areas play a vital role in urban biodiversity because these are the main habitats of urban plants and animals. The size and location of the urban green areas are of importance for their qualities as biotopes. Vacant urban wasteland in inner cities often possess a high diversity of species adapted to urban environmental conditions and pressures. For instance, older well-established installations attract birds and mammals whose natural habitat is the forest. The urban forest structures are also of great importance, providing wildlife corridors and stepping stones that connect the habitats in the urban green areas. Swedish examples emphasize that only if the urban infrastructure is integrated with the green infrastructure is it possible to achieve biotope protection within cities (Stanners & Bourdeau 1995).

Since a major part of Europe's population lives in urban areas and receives its daily perception of nature therein, nature in urban areas is important for environmental awareness and an understanding of nature. Urban areas contain more of nature than is immediately apparent. Older gardens and parks, not to mention churchyards, often have noticeably rich biodiversity. Nature created by humans is often considered to be inferior to nature that evolved without human intervention. In support of this, it is often asserted that, for example, the number of species is often greater in untouched nature. This is mainly related to the fact that many urban green areas do not boast particularly rich biodiversity. Most of them were established with large paved areas, gravelled areas, well-mown greens and isolated individual trees. However, Owen (1992) showed that in an urban private garden with variegated flower beds and a well-proportioned mixture of cultivated and uncultivated plants it was possible to attract a large proportion of the indigenous species of butterflies (34%), moth (30%) and hover flies (36%).

The presence of birds in the immediate environment is a vital element in the context of recreation. Their presence in urban areas depends on the character of the local vegetation. Variation in the structure of plantations also forms a basis for a rich assemblage of birds. As it takes many years for new plantations to mature, it is vital to preserve as much of the existing vegetation as possible when establishing new areas (Dwyer 1995).

13.6.4 Sustainable urban forests

Sustainable urban forests are defined by Clark *et al.* (1997) as 'the naturally occurring and planted trees in cities which are managed to provide the inhabitants with a continuing level of economic, social, environmental and ecological benefits today and into the future'. They will provide long-term net environmental, ecological, social and economic benefits (Clark *et al.* 1997; Miller 1997). To the components of a sustainable forest, McPherson (1998) includes features of adequate species and age diversity, a large percentage of healthy trees that are well adapted to local growing conditions, and a climate-appropriate treecover with native forests stands as one component of overall canopy cover.

13.7 THREATS TO GREEN AREAS

Besides the harsh growing conditions that plants experience in the urban environment, there are other, more immediate threats to urban trees and woodlands. These are the pressures of urbanization, which compete with the green areas in demand for land; hence the high recreational pressure on the areas left for urban green spaces.

13.7.1 Urbanization pressure

There is great pressure on urban space resources. Even where this is a question of public buildings, such as museums, where green area remains accessible to the public, it would in many cases detract from the overall recreational quality of the area. In Mexico City, the proportion of green areas within the city is falling by about 3.7% annually. These are often replaced with buildings, especially in the poorest quarters of the city (Chacalo *et al.* 1994). Traffic installations and noise are other threats to green areas. Roads can isolate green areas from each other, which reduces their recreational value and their value as corridors for the propagation of flora and fauna. The annoyance caused by noise is more indirect. Dutch studies indicate that road noise annoys about 20% of the population, whereas about 11% is annoyed by air traffic (Stanners & Bourdeau 1995). These figures apply to indoor annoyance. Outdoors, people are exposed to even more noise, especially since part of the urban green areas consist of 'residual areas' along traffic constructions.

13.7.2 Social factors

Another important urban stress factor is vandalism. This is usually perceived to be a serious factor, although surveys shows that it is not: it is usually the result of careless play and affects only the most exposed trees (Bradshaw *et al.* 1995). Vandalism is predominantly a social problem. Successful community landscaping and gardening in densely populated inner-city neighbourhoods have shown that one deterrent to vandalism is the development of a spirit of proprietorship in residents (Flint 1985). Nowak *et al.* (1990) noted the highest tree mortality in areas of lower socioeconomic status. Percentage tree mortality was most strongly correlated with percentage unemployment.

13.7.3 Economic cuts

In addition, economic cuts are an obtrusive threat to green areas. Park administrations responsible for a large part of city green areas have been hit hard in recent years by cuts in appropriations and personnel. As an example it is typical of the trend in Denmark over the past 20 years that economic resources have dropped by 10–20% while green areas have increased by 20–40%. Similar trends have been observed in other countries, for instance Sweden, Germany and the UK. The sector responsible for the planning, establishment and operation of green areas in Denmark (including the administration of natural and recreational areas) has an annual turnover, public and private, of $US1.35 billion and employs the equivalent of 30 000 full-time employees. The maintenance costs of urban green areas amount to approximately $US16 per inhabitant (Juul 1995). In this respect, if Denmark can be regarded as an average country in European terms, this gives a total European urban greening maintenance cost of approximately $US3.84 billion. In 1991, it was estimated that at least $US300–350 million was spent annually to enhance amenity tree resources in the UK (Ball *et al.* 1999).

13.8 PLANNING AND MANAGEMENT OF URBAN GREEN AREAS

The vigorous growth of urban areas has not only increased the need for intensified urban planting but also highlighted major problems related to species selection, establishment techniques, care and maintenance, and planning. Proper species selection, tree care and maintenance will ultimately result in a greater understanding of urban tree management. Consequently, this will ensure that planted trees remain healthier and live longer, thus reducing labour and replacement costs, while still providing an overall ameliorating effect. Money saved is therefore equivalent to money earned, since it becomes available for other purposes.

In Denmark, the major part of the public green area is administered centrally by a municipal park administration. The park administration is typically located in a local council's technical department and has its own operating organization that is responsible for the practical care of green areas. However, throughout Europe there is a trend towards an ever-more privatized operation of green areas, demonstrated by increasing use of private contractors, first and foremost when establishing new installations but also in the maintenance of green areas. Nevertheless, it is characteristic of issues concerning urban forests or green areas that there is a high level of public involvement. Also, self-administration is gradually gaining a foothold, for example in daycare centres, where the management and parents' committees are increasingly able to decide for themselves how their green areas should be operated.

In addition, there is a growing tendency towards reducing costs related to tree establishment and care. Urban trees and uneconomic forest stands are particularly exposed to such cuts. Therefore, better planning and management systems for urban green areas are often currently lacking (Konijnendijk 1999). Small individual economic cuts may not be very significant in themselves, but considered over an extended period the decrease in budgets can have detrimental consequences not apparent in the isolated individual case. One approach for solving this problem could be the introduction of overall green infrastructure planning.

13.8.1 Green infrastructure planning

The EU Green Book ascertains that there are many threats to green areas, for example from vehicles and advertising (Commission of the European Community 1990). It also stresses the need for an element of nature in the urban environment. Even if urban nature perhaps cannot compare to 'natural' nature, it is still of great importance. The nature that exists in the landscape is the source we should use for disseminating wild plants and animals into urban areas. This is why green corridors leading from the countryside into urban areas are vital, as is the transitional zone between urban and rural areas. In many places, the green structure is not cohesive and thus cannot provide the necessary transport routes. Just a minor interruption of a green corridor can prevent it from functioning. Although binding the green network together in urban areas demands a great effort, it is absolutely essential if we are to succeed in developing urban biotopes and creating the conditions needed for a richer flora and fauna. The EU Green Book recommends that public green space should be increased as much as possible. Hence, green infrastructure planning includes a holistic assessment of the green infrastructure, current conditions and plans for development that could be achieved, for instance, by making it possible for local councils to give a green plan formal status as a part of council planning.

Examples

Germany is one of the countries at the forefront of green infrastructure planning. The German nature and environmental protection legislation also regulates parks and urban green areas. Their green plans are included on several different levels in landscape planning (which is linked to general plans) and in green structure planning (which is linked to local plans). Although these plans are not legally binding they still play a significant role. Examples from three German cities illustrate the divergent ways in which these plans originate and how they are used (Nilsson *et al.* 1998).

In Hannover, the overall green structure is included in two planning phases. The park administration has the option of conducting a landscape analysis before work is started, in all forms of area utilization. In this way, planning takes landscape potential into account from the very beginning. Plans for the green structure are combined with the plans for building development in the next phase. Both phases are laid before the politicians, who thus gain the opportunity of seeing how consideration is given in the final proposal to the original, overall intentions. In cases where it is impossible to avoid damaging nature, the authorities can demand compensation. Advantage of this is often taken by the city head gardener to demand improvements elsewhere, for instance a builder can be obliged to undertake the improvement of a neighbouring park. In this way, nature – the green area – becomes a commodity that can be traded, for better or worse.

Stuttgart approaches superior green planning another way. This city has a long tradition of securing the expansion and improvement of its green structure through exhibitions held every 10 years, most recently in 1993. Stuttgart also offers a classical example of the fact that it is not only the lack of green areas that can present a problem; the problem can also be related to access to these areas and their lack of mutual cohesion. The major problem in Stuttgart was the cohesion of the old royal gardens in the centre of the city. For this reason 10 bridges, which linked the parks together across various arterial roads, were built in conjunction with the Bundesgartenschau of 1977. This was a phase in the long-term 'Green U' plan, which was completed in connection with the 1993 exhibition. The Green U now makes it possible to pass through green areas, from the central royal gardens to the forests at the edge of the city.

In Munich, a plan was adopted in 1992 for 14 green strips, totalling 584 ha, which are to link the city's green areas to the surrounding landscape. The green strips are to be established over a 25-year period and will cost a total of $US430 million. The reason for this major effort is the fact that the development of the city's green areas has not increased in proportion to building development. At the same time, increased awareness of the significance of green areas to ecology and recreational activities has meant that greater attention is being paid to the importance of the cohesion of green areas. A large part of the green areas are on the edge of the city and are owned by it. Only the innermost 10-m belt of this resembles a park and consists of small tree plantations, a green and perhaps a stream. It also contains a network of footpaths and cycle paths. The remainder of these areas is rented to farmers, who operate them with only limited use of pesticides.

In Denmark, a successful example of green infrastructure planning is the 'Finger Plan' for the area outside Copenhagen. The name relates to the fact that the overall structure is shaped like a hand, with the five 'fingers' being planned for urban development. The Finger Plan, which was born as long ago as 1947, had the goal of stopping the layered growth of the city and ensuring that urban growth would thereafter be concentrated in narrow urbanized areas ('fingers') along traffic corridors. The wedges between the fingers were to be used for agricultural and recreational purposes. This idea was very clear and was beneficial to the green areas. Subsequent developments have largely adhered to the guidelines of the plan.

In England the planning of the new towns was founded, even as early as 1900, on the concept of using trees and woodland as a setting for urban development. The aims of the new towns were to relieve the urbanization pressure on large cities such as London, Birmingham and Glasgow, and to create self-contained and balanced communities for living in, for employment and for recreation. The new towns were designed and developed by development corporations. The success of the new towns is to be found in the fact that they were produced as integrated structures with good communications, high levels of amenities and higher than average physical environment, based to a large degree on social commitment and the belief that good physical environment was good for people as well as for business (Simson 1998).

13.8.2 Conflicts in management

In the policy-making, planning and development of urban forestry, some general trends and challenges can be recognized. These comprise a growing complexity, an increasing role of public participation and conflict management, and the transition towards closer-to-nature forests. In a study of urban forest planning and policy-making in 16 European cities, the main conflicts were discovered to be conflicts in relation to nature conservation, forestry vs. urban development, conflicts between different types of recreation and recreation vs. conservation management (Konijnendijk 1998).

13.9 CONCLUSIONS

Urban green areas are of great importance to the people living in cities. Nevertheless, urban green areas are also increasingly exposed to threats related to the development and progress of three main factors: the densification of cities, the construction of infrastructure and decreasing resources. Even though individual actions can seem insignificant by themselves, they can lead to detrimental consequences if an overall view of the green structure and its different functions is lacking. Only by emphasizing the environmental, social and cultural importance of urban green areas is it possible to create a base for the needed priorities, in order to protect and develop the urban green structure.

The importance of an overall green structure lies in its various functions: environmental, social and cultural. The ecological functions are conditioned by the whole system of plants, animals, soil, water, climate and human impact. Hence, the possibility of developing and utilizing the green structure in the long term depends on creating a complete picture of this system in order to utilize and consider the needs of the cities. The green structures of the cities are of great importance when it comes to quality of life. The social functions have been emphasized in studies from different cities, showing that the majority of people spend most of their spare time within the city or in nearby areas. The cultural and aesthetic characteristics related to urban green areas and structures are also of great importance for the quality of life in cities. The urban green areas, the gardens, parks and cemeteries, all add to and reflect the cultural history and identity of the city.

In general, an increasing interest in urban environmental problems can be traced. Over the coming decades this will probably cause a considerable increase in the development of parks and green areas, most of which will occur in the larger cities of Asia, Africa and Latin America. Similar expansion of the green infrastructure was carried out in North America and Europe during the years after the Second World War and up to the 1970s. It was done with a strong belief in high technology, heavy machines and the liberal application of chemical aids. Over the coming years, the main challenge will be to ensure that the development of green infrastructure in the world's urban areas is carried out within the framework of sustainable development, without the use of technology destructive to humans and nature and with as few chemical aids as possible. Humans are a vital resource in this context, not only because manual work is an alternative to many of the methods harmful to the environment but also because knowledge can in many instances replace the use of artificial methods.

ACKNOWLEDGEMENTS

The following experts arc gratefully acknowledged for reviewing and commenting on the manuscript or for contributing to the preparation of this chapter in other ways: Mr Rune Bengtsson, Swedish Agricultural Univeristy, Sweden; Mrs Alicia Chacalo, Universidad Autónoma Metropolitana–Azcapotzalco, Mexico; Kevin D. Collins, Tree Council of Ireland, Ireland; Dr Larry Costello, University of California, USA; Dir. Clive Davies, Cleveland Community Forest, England; Dr Dirk Dujesiefken, Institut für Baumpflege, Germany; Mr Cecil C. Konijnendijk, European Forest Institute, Finland; Dr Palle Kristoffersen, Danish Forest and Landscape Research Institute, Denmark; Dr Frank S. Santamour, US National Arboretum, USA; Mr Allan Simson, Leeds Metropolitan University, England; Dr Jozef van Slycken, Institut voor Bosbouw en Wildbeheer, Belgium; Dr Frank Søndergaard Jensen, Danish Forest and Landscape Research Institute, Denmark; Dr Liisa Tyrväinen, University of Joensuu, Finland; Dr Gary Watson, Morton Arboretum, USA.

REFERENCES

Adams, E. (1989) Learning through Landscapes. *Landscape Design* June no. **181**, 16–19.

Ball, R., Bussey, S.C., Patch, D., Simson, A. & West, S. (1999) United Kingdom. In Forrest, M., Konijnendijk, C.C. & Randrup, T.B., eds. COST Action E12. *Research and Development in Urban Forestry in Europe*, pp. 325–40. European Commission, Luxembourg.

Benavides Meza, H.M. (1992) Current situation of the urban forests in Mexico City. *Journal of Arboriculture* **18**, 33–6.

Bernatzky (1978) *Tree Ecology and Preservation*. Elsevier, Amsterdam.

Bradshaw, A., Hunt, B. & Walmsley, T. (1995) *Trees in the Urban Landscape. Principles and Practice*. E. & F.N. Spon, London.

Brod, H. (1993) *Langzeitwirkung von Streusalz auf die umwelt. [Long-term environmental effect from de-icing salt.]* Berichte der Bundesanstalt für Strassenwesen. Verkehrstechnik, Vol. 2. [In German]

Broderick, S.H. & Miller, D.M. (1989) *Trees, cities and global warming*. Fact Sheet no. 2, Cooperative Ext. System, Urban and Community Forest File, University of Connecticut, Connecticut.

Caballero Deloya, M. (1993) Urban Forestry in Mexico City. *Unasylva* **44** (173), 28–32.

Chacalo, A., Aldama, A. & Grabinsky, J. (1994) Street tree inventory in Mexico City. *Journal of Arboriculture* **20**, 222–6.

Chambers, K. (1987) Urban forestry in the London Borough of Tower Hamlets: an account of the European Urban Forestry Project sponsored by the EEC. *Arboricultural Journal* **11**, 1–14.

Clark, J.R., Matheny, N.P., Cross, G. & Wake, V. (1997) A model of urban forest sustainability. *Journal of Arboriculture*. **23**, 17–30.

Collins, K. (1996) Creating urban woodlands: a practical perspective. *Irish Timber and Forestry* **5**, 10–12.

Collins, K. (1997) Editorial. *Irish Forestry* **54**, 1.

Commission of the European Community (1990) *Green Paper on the Urban Environment*. Commission of the European Communities, Brussels.

Costello, L.R. (1993) Urban forestry: a new perspective. *Arborist News* April, Vol. 2, 33–6.

Craul, P.J. (1992) *Urban Soil in Landscape Design*. John Wiley & Sons, New York.

Davies, C. & Vaughan, J. (1998) England's community forests. A case study: community forests in Northeast England. Paper presented at the 1st European Forum on Urban Forestry, IUFRO, Wuppertal, Germany, May 1998.

Dawe, N.A. (1989) Citizens with a vision. In: Moll, G. & Ebenreck, S., eds. *Shading Our Cities, A resource guide for urban and community forests*, pp. 229–35. Island Press, Washington, DC.

Dobson, M.C. (1991) *De-icing Salt Damage to Trees and Shrubs*. Forestry Commission Bulletin No. 101, Forestry Commission, Edinburgh.

Dwyer, J.F. (1992) Economic benefits and costs of urban forests. In: Rodbell, P.D., ed. *Proceedings of the Fifth National Urban Forest Conference, Los Angeles, November 1991*, pp. 55–8. American Forestry Association, Washington, DC.

Dwyer, J.F. (1995) The significance of trees and their management in built environments. In: Watson, G. & Neely, D., eds. *Trees and Building Sites. Proceedings of an International Workshop on Trees and Buildings*, pp. 3–12. International Society of Arboriculture, Savoy, Illinois.

Dwyer, J.F., Schroeder, H.W. & Gobster, P.H. (1991) The significance of urban trees and forests: towards a deeper understanding of values. *Journal of Arboriculture* **17**, 276–84.

Ebenreck, S. (1989) The values of trees. In: Moll, G. & Ebenreck, S., eds. *Shading Our Cities*, pp. 49–58. Island Press, Washington, DC.

Flint, H.L. (1985) Plants showing tolerance of urban stress. *Journal of Environmental Horticulture* **3**, 85–9.

Florgård, C. (1987) *Så gick det med naturmarken! Naturmarken i 1960-talets och 1970-talets planering. [What Happened to the Non-cultivated Soils! The Non-cultivated Soils in the 1960–1970s Planning.]* Swedish Agricultural University, Stad & Land No. 54. [In Swedish]

Forman, R.T.T. (1997) *Land Mosaics: the Ecology of Landscapes and Regions*. Cambridge University Press, Cambridge.

Gåsdal, O. (1993) Uteliv i byen. [Outdoor recreation in the city.] In: Kaltenborn, B.P. & Vorkinn, M., eds. *Vårt Friluftsliv. [Our Outdoor Recreation.]*, pp. 33–41. Norwegian Institute for Nature Research, NINA Thematic Issue No. 3. [In Norwegian], Oslo.

Gibbs, J.N. & Palmer, C.A. (1994) A survey of damage to roadside trees in London caused by the application of de-icing salt during the 1990/91 winter. *Arboricultural Journal* **18**, 321–43.

Gilbertson, P. & Bradshaw, A.D. (1985) Tree survival in cities: the extent and nature of the problem. *Arboricultural Journal* **9**, 131–42.

Grahn, P. (1989) *Att uppleva parken. Parkens betydelse för äldre, sluka och handikappade skildrede genom dagböcker, intervjuer, teckninger och fotografier. [Experiencing the Park. The Significance of the Park for the Elderly, Weak and Handicapped: a Description Based on Diaries, Interviews, Drawings and Photos].* Swedish Agricultural University, Alnarp. [In Swedish]

Grahn, P. (1991) *Om parkers betydelse [The Significance of Parks].* Swedish Agricultural University, Alnarp, Stad & Land No. 93. [In Swedish with English summary]

Grey, G.W. & Deneke, F.J. (1978) *Urban Forestry*. John Wiley & Sons, New York.

Håkansson, I. & Reeder, R.C. (1994) Subsoil compaction by vehicles with high axle load: extent, persistence and crop response. *Soil and Tillage Research* **29**, 277–304.

Haque, F. (1987) Urban forestry: 13 city profiles. Kampala, Uganda: fuelwood and ornamentals. *Unasylva* **39**, 22–3.

Harris, R.W. (1992) *Arboriculture: Integrated Management of Landscape Trees, Shrubs and Vines*, 2nd edn. Prentice Hall, Englewood Cliffs, New Jersey.

Holm, S. (1998) *Use and importance of urban parks*. PhD thesis, Royal Veterinary and Agricultural University, Copenhagen.

Huang, J., Richard, R., Sampson, N., Taha, H. (1992). The benefits of urban trees. In: Akbari, H. *et al.*, eds. *Cooling Our Communities: a Guidebook on Tree Planting and Light-Coloured Surfacing*, pp. 27–42. US Environmental Protection Agency, Office of Policy Analysis, Climate Change Division, Pittsburgh, Pennsylvania.

Jensen, F. & Koch, N.E. (1997) *Friluftsliv i skoven 1976/77–1993/94. [Outdoor Recreation in the Forest 1976/77–1993/94.]* Forskningsserien no. 20, Danish Forest and Landscape Research Institute. [In Danish]

Jensen, M. (1955) *Lævirkning: læets aerodynamik. [Shelter Effect: the Aerodynamic of the Shelter and its Impact on Climate and Crops].* Det danske Hedeselskab. [In Danish]

Juul, J.O. (1995) *Nøgletal: Kommunal forvaltning af grønne områder. [Key-numbers: Municipal Management of Green Areas].* Park og Landskabsserien no. 9, Danish Forest and Landscape Research Institute. [In Danish]

Kaplan, R. & Kaplan, S. (1989) *The Experience of Nature*. Cambridge University Press. Cambridge.

Kielbaso, J.J. (1989) City tree care programs: a status report. In: Moll, G. & Ebenreck, S., eds. *Shading Our Cities*, pp. 35–46. Island Press, Washington, DC.

Konijnendijk, C.C. (1998) Urban forestry policy-making: a comparative study of selected cities in Europe. *Arboricultural Journal* **23**(1), 1–15.

Kuchelmeister, G. & Braatz, S. (1993) Urban forestry revisited. *Unasylva* **44** (173), 3–12.

Lindhagen, A. (1996) *Forest Recreation in Sweden. Four Case Studies Using Quantitative and Qualitative Methods*. Swedish Agricultural University, Uppsala, Report no. 64.

McPherson, E.G. (1998) Structure and sustainability of Sacramento's urban forest. *Journal of Arboriculture* **24**, 174–90.

McPherson, E.G. & Rowntree, R.A. (1993) Energy conservation potential of urban tree planting. *Journal of Arboriculture* **19**, 321–31.

Matthews, J.R. (1991) Benefits of amenity trees. In: Hodge, S.J., ed. *Research for Practical Arboriculture: Proceedings of a Seminar, University of York, April 1990*, pp. 74–80. Forestry Commission Bulletin no. 97, Forestry Commission, HMSO, London.

Michael, S.E. & Hull, R.B. (1994) *Effects of Vegetation on Crime in Urban Parks*. Department of Forestry, College of Forestry and Wildlife Resources, Virginia Polytechnic Institute and State University, Blacksburg, Virginia.

Miller, R.H. & Miller, R.W. (1991) Planting survival of selected street tree taxa. *Journal of Arboriculture* 17, 185–91.

Miller, R.W. (1997) *Urban Forestsry. Planning and Managing Urban Greenspaces*, 2nd edn. Prentice Hall, Upper Saddle River, New Jersey.

Moll, G. (1989) The state of our urban forest. *American Forests* November/December, 61–4.

Nilsson, K. & Randrup, T.B. (1996) Urban forestry: definitions, European research initiatives and organisational matters. In: Randrup, T.B. & Nilsson, K., eds. *Urban Forestry in the Nordic Countries. Proceedings of a Nordic Workshop on Urban Forestry, Reykjavik, Iceland, September 21–24, 1996*. Danish Forest and Landscape Research Institute, pp. 10–17. Hoersholm Denmark.

Nowak, D.J. (1995) Urban trees and air quality. In: Korpilhati, E., Salonen, T. & Oja, S., eds. *Caring for the Forest: Research in a Changing World*, p. 476. Abstracts of invited papers, IUFRO XX World Congress, 6–12 August 1995, Tampera, Finland.

Nowak, D.J. & McPherson, E.G. (1993) Quantifying the impact of trees: the Chicago Urban Forest Climate Project. *Unasylva* 44 (173), 39–44.

Nowak, D.J., McBride, J.R. & Beatty, R.A. (1990) Newly planted street tree growth and mortality. *Journal of Arboriculture* 16, 124–9.

Nowak, D.J., McPherson, E.G. & Rowntree, R. (1994) Chicago's urban forest ecosystem: results of the Chicago Urban Forest Climate Project. In: McPherson, E.G., Nowak, D.J. & Rowntree, R.A., eds. *Chicago's Urban Forest Ecosystem: Results of the Chicago Urban Forest Climate Project*, pp. iii–vi. US Department of Agriculture, Forest Servcice, North Eastern Forest Experiment Station, Radnor, Pennsylvania.

Owen, J. (1992) *The Ecology of a Garden*. Cambridge University Press, Cambridge.

Pedersen, P.A. (1990) *Roadside pollution and vegetation*. DSc thesis, Agricultural University of Norway. [In Norwegian with English abstract.]

Pedersen, P.A. & Fostad, O. (1996) *Effekter av veisaltning på jord, vann og vegetasjon. Hovedrapport del I. Undersøkelser av jord og vegetasjon. [Effects of De-icing Salts on Soil, Water and Vegetation. Main Report Part I. Examination of Soil and Vegetation.]* Centre for Research, Ås, Norway. [In Norwegian with English summary], Oslo, Norway.

Randrup, T.B. (1997) Soil compaction on construction sites. *Journal of Arboriculture* 23, 207–10.

Randrup, T.B. & Dralle, K. (1997) Influence of planning and design on soil compaction in construction sites. *Landscape and Urban Planning* 38, 87–92.

Randrup, T.B. & Pedersen, L.B. (1996) *Vejsalt, træer og buske. En litteraturundersøgelse om NaCl's effekter på vedplanter langs veje. [De-icing Salts, Trees and Shrubs. A Literature Study of NaCl's Effects on Woody Plants along Roads.]* Directorate for Roads, Department of Roads, Report no. 64. [In Danish with English summary]

Rolf, K. (1994) *Recultivation of Compacted Soils in Urban Areas*. Report no. D6:1994, Swedish Council for Building Research/Department of Agricultural Engineering, University of Agricultural Sciences, Alnarp, Sweden.

Schroeder, H.W. & Cannon, W.N. (1987) Visual quality of residential streets: both street and yard trees make a difference. *Journal of Arboriculture* 13, 236–9.

Scott, K.I., McPherson, E.G. & Simson, J.R. (1998) Air pollutant uptake by Sacramento's urban forest. *Journal of Arboriculture* 24, 224–34.

Sievänen, T. (1993) *Outdoor Recreation Household Survey in the City of Hämeenlinna*. Folia Forestalia 824. [In Finnish with English summary]

Sievert, R.C. Jr (1989) Public awareness and urban forestry in Ohio. In: Moll, G. & Ebenreck, S., eds. *Shading Our Cities*, pp. 247–252. Island Press, Washington DC.

Simson, A. (1998) Urban forestry in the UK new towns. In: Randrup, T.B., Korijnendijk, C.C. & Nilsson, K., eds. *Proceedings of COST Action E12: Urban Forests and Trees*. Damish Centre for Forest, Landscape and Planning, Horsholm, Denmark (in press)

Stanners, D. & Bourdeau, P. (eds) (1995) *Europe's Environment. The Dobris Assessment*. European Environment Agency, Copenhagen.

Tyrväinen, L. (1997) The amenity value of the urban forest: an application of the hedonic pricing method. *Landscape and Urban Planning.* 37, 211–22.

Ulrich, R.S. (1984) View through a window may influence recovery from surgery. *Science* 224, 420–1.

Ulrich, R.S., Simons, R.F., Losito, B.D., Fiorito, E., Miles, M.A. & Zelson, M. (1991) Stress recovery during exposure to natural and urban environments. *Journal of Environmental Psychology* 11, 201–30.

United Nations (1992) *Agenda 21: the United Nations Programme of Action from Rio*. United Nations, New York.

Volk, H. (1986) Task and goals of the IUFRO Project group 'Urban Forestry'. Note made by the Urban Forestry Project Group P1.05.00, Ljubljana, September 6, 1986.

14: The Importance of Social Values

STEPHEN BASS

14.1 SUSTAINABLE FOREST MANAGEMENT INVOLVES POLITICAL AND SOCIAL PROCESSES

The past decade may come to be known by foresters as the period when they, and just about everybody else, sought to define or to prescribe sustainable forest management (SFM). Many of these efforts have painted very detailed pictures of what a well-managed forest should look like. Others have established general principles. Whatever their form, the key ingredients are sustaining multiple values through forest management, and the rights of multiple actors.

The earliest initiatives to define SFM were unilateral, undertaken by industry associations or by environmental non-governmental organizations (NGOs), and consequently were mistrusted by other stakeholders. Most of the currently accepted SFM initiatives result from multistakeholder processes, and consequently reflect a range of these stakeholders' values. They generally comprise sets of principles, criteria and indicators that have to be interpreted in detail at the local level, offering further scope for incorporating local values. They include global/intergovernmental processes (notably the UN Forest Principles 1992), regional processes (e.g. the Ministerial Conference on the Protection of Forests in Europe 1993) and national standards (e.g. the UK Forest Standards 1998). Others are led by civil society (e.g. the Forest Stewardship Council's Principles and Criteria).

The International Institute for Environment and Development (IIED 1996) analysed 17 such initiatives and found that all had the following in common:

- sustaining yields of all socially valued goods and services from forests;
- ensuring positive impacts of forest use on different social groups;
- maintaining the state of the forest to enable continued production for future generations.

These initiatives to define SFM all state or imply two premises.

1 The purpose of forestry is to provide the highly varied needs of society; this requires that demands are signalled effectively in policy and markets, and necessitates means for making trade-offs.

2 The basic principles of SFM need to be interpreted at the most local level at which specific goods and services are needed, e.g. the community level for recreation and firewood, the national level for major watersheds, and the global level for climate regulation. 'Sustainability' can be likened to liberty or justice, a goal we all understand and aspire to, but which has to be negotiated and defined locally to understand and achieve it in practice (Holmberg *et al.* 1991).

This all points to an emphasis on determining local values, on participation among interest groups, on using local decision-making processes, and on ensuring that forest management systems, markets and policies reflect social values.

Only recently, however, have foresters been expected to focus on social issues. There is little prior experience of foresters taking on roles of social development, although there are historical examples: the famous *taungya* system developed in Burma in the 1850s was a response to local people's needs for farmland. Yet foresters are being held increasingly accountable for social

Table 14.1 Forest values depend upon who you are (developed from WCFSD 1997).

Forest interest groups	Approximate numbers (million)	Important forest values
Urban people	2500	Wood, NTFPs Water, climate moderation
Rural poor/landless	1000	Food/fibre/health components of livelihood Support to farm systems
Shifting cultivators	250	Much of livelihood Spiritual and cultural values
Forest communities	60	Sole means of livelihood Spiritual and cultural values
Agribusiness	10	Land, water Support to farm systems
Oil/mining business	7	Minerals
Logging business	5	Timber
Retailers of forest products	3	Timber, NTFPs, environment
Ecotourism business	3	Landscape, recreation Biodiversity, culture
Environmental groups, scientists	1	Carbon storage, climate Biodiversity, culture

NTFPs: non-timber forest products.

conditions (in various new regulations, voluntary codes and certification). There are many unresolved arguments. To what extent can forestry be an instrument of social development? For which communities should foresters be held accountable for social conditions, those close to the forest or also those further away? Clearly, forest managers have at least to understand which groups are most dependent upon the forest and who could contribute most to the success or failure of forest management.

14.2 FORESTS PROVIDE MULTIPLE SOCIAL VALUES

The social values of forests differ widely, depending on the culture and social group in question and the roles which forests play in their livelihoods and quality of life (Table 14.1). These values are not static, but change over time. Key change factors appear to be the following.

• Shortages of, and opportunities to develop, the five main forms of capital (natural, physical, financial, human/individual and social/community capital). Where natural capital such as forest is abundant, it tends to be valued as a resource for conversion to generate other forms of less-abundant capital, and vice versa.

• In a similar vein, the absolute and relative scarcity of particular forest resources, such as timber or biodiversity.

• The availability and price of substitutes for forest goods and services.

• Scientific discovery, education and technology changes.

• Access rights, and resources such as labour that alter capacities to exploit forests.

• An individual's allegiance to specific groups and their shared values.

• Factors that change individual allegiances: campaigning and the political influence of certain groups, tastes and associated communications and media. The prevalence of individual consumerism has changed values away from those of the community.

• Changes in political culture.

- Changes in national development conditions.
- Changes in individual income.

Table 14.2 attempts to categorize the varied social values that can be obtained from forests. Whilst the table begins with basic livelihood values such as food, and finishes with recreational and aesthetic values, it would be a mistake to assume that there is a universal, rigid hierarchy of needs. For example, for some groups, spiritual values are fundamental, while for others these are much less significant. It is therefore important to identify the relevant values from the point of view of the group in question. Indeed, certain groups define themselves (in part) by the values they hold in common. This is a source of many recent clashes between forest 'stakeholders'. These clashes can be exacerbated when some groups promote groundswells of broader opinion in favour of the values they hold most strongly, as we have witnessed with many environmental campaigning groups or forest peoples' groups. In other words, group values can be influenced by those of others. The outcome is, in large part, a function of power relations, something the forest manager may be able to do little about but must be aware of.

It is also crucial to be aware of the broader trends that affect social values. Policy, institutional and market conditions frequently make the realization of some values difficult. Furthermore, the role of forests in providing such values can be substituted by other sources (e.g. imported foods, non-forest employment and entertainment). The possibilities of substitution tend to be much greater for richer communities than for poorer, forest-based groups. Hence the need for an emphasis on working with the poor.

In developing countries, subsistence domestic products such as fuelwood, poles, fencing and fodder are a priority for forestry, the other priority area usually being food security from forest management or conversion: 'Over 1 billion people (about 20% of the world's population) depend either wholly or to a significant extent on forests, woodland or farm trees for their subsistence needs and/or livelihood' (WCFSD 1997). As Filer and Sekhran (1998) point out, many people who are perceived as rather passive 'indigenous forest dwellers' think of themselves more as forest gardeners and forest developers, with food a number one concern. Forest managers ignore these pressing social needs at their peril.

Some demands for forest values are now far stronger than they were even a few years ago. For example, 'biodiversity' is a word that was coined only in the late 1980s, and the idea of carbon sequestration as a prime value is only beginning to be accepted. Yet now there are all sorts of international environmental agreements, especially the Biodiversity Convention and Climate Change Convention, with significant financial transfer mechanisms emerging to match the political obligations to produce biodiversity and carbon storage values. At present, those who are pressing for such 'global' forest values tend to be the politically more powerful countries and groups such as international environmental NGOs; much remains to be done to ensure that the production of these values also benefits local groups, rather than squeezing out their (more parochial) concerns.

It is quite a challenge for the forest manager to make decisions about the relative weights of the values held by different groups, from local to global, and how to integrate them or how to make choices between them where integration proves impossible. Three trends are helping here. Firstly, as noted in Table 14.2, there is the trend for certain markets to recognize some social/environmental values of forest management that were previously unrewarded, for example the market for certified forest products and the emerging market for carbon storage. Secondly, there is development of more inclusive systems of forest policy and management decision-making, such as Round Tables, forest fora, and stakeholder liaison groups, where consultation with different groups about their values is on the increase (even if there is as yet insignificant participation in decision-making and management). Thirdly, there is the work by economists to categorize values and to develop methodologies to assess them on a single (financial) scale. While these methodologies remain contentious (it is not so much the financial magnitude that counts as who bears the cost and who gains the benefits) and indeed have so far made little difference to major policy decisions, especially those regarding private forests, the economic categorization of multiple values is useful to information science if

Table 14.2 The spectrum of social values associated with forests.

Social values	Forest management provides	Typical trends and conditions
Livelihood basics		
Staple food	Carbohydrates and protein for forest-dwelling communities Fuelwood for cooking	• Dwindling in most areas due to penetration of rural markets by new products, taste changes, reduced supply, high labour costs, loss of traditional knowledge • Still key for remote developing country communities
Supplementary food	Variety/palatability to diet through meat, fish, fruit Seasonal buffers/famine foods	• Important in many developing countries experiencing economic/climate uncertainty and food insecurity, especially for marginalized groups including women
Health	Water supplies Climate moderation Medicine Vitamins and minerals	• Water and climate roles increasingly critical in most countries; global markets for these services developing • Medicinal value important in many forest communities, associated with culture
Shelter	Poles, thatch	• Important in practically all forest communities, strongly associated with culture
Economic security		
Main income	Forest products for sale Forest services, e.g. tourism business	• Where incomes are rising generally, most people move away from forestry; a few (richer) local groups specialize in forestry • Where incomes declining, and where many are landless, still high dependence on forests • Community systems/rules for management of common pool resources tend to be in decline • Globalization, taste changes, and market price fluctuations constrain investment in sustainable forest management
Supplementary income	Forest products for sale	• Very common for rural communities, especially where pressure on farmland makes farm income inadequate; here, access to natural forests/fallow is key • Farmers with adequate farmland but labour constraints often plant trees • Government/corporate control of land limits access by poorer groups
Savings/social security	Timber stocks Land value	• Traditionally key for periodic expenditure, e.g. dowry, feasts, house building • Forest stocks more important for their timing, not the size of income
Risk reduction	Biodiversity Multiple products Soil conservation Water conservation	• Increasing understanding of the value of diversification, but as yet inadequately developed markets • Risk reduction should be seen in context of whole-livelihood/farm system

Table 14.2 *continued*

Social values	Forest management provides	Typical trends and conditions
Cultural/social identity		
Cultural/historical/spiritual associations	Forest landscapes Forests as sacred groves Individual species and their products	• Many developing country cultures are forest-based; suffering tensions when faced with Western cultures based on forest removal and standardization of products
Social identity and status	Forest as source of power from ownership/cultivation/clearance Ability to pass forest on to future generations	• Outside involvement in local forests can draw on, and exacerbate, local power inequities
Quality of life		
Education/science	Biodiversity conservation Means of access to forest	• Where livelihood and economic security are obtained largely by non-forest means, forests are valued largely for their 'quality of life' attributes, as in most developed countries
Recreation	Biodiversity conservation/control Forest-based facilities	• Where livelihood/economic security are closely dependent on forests, there can be big clashes with outsiders' 'quality of life' demands
Aesthetic values	Landscape design/management Biodiversity conservation	• Seen as a key area for government intervention, but markets are also developing • Local/government partnerships are key

Source: Arnold (1997); Mayers (1997); Bass (1997); Segura *et al.* (1996).

not yet to decision science. Broadly speaking, it includes the following.

• Direct-use values: where the value is derived directly from the forest, either in a consumptive manner (timber, nuts, fodder, game, fish, etc.) or a non-consumptive manner (tourism, recreation, etc.).

• Indirect-use values: where the value derives from environmental services such as watershed and social protection, carbon sequestration or biodiversity protection, rather than from the forest directly.

• Passive-use values: where value is accorded by the mere fact that the forest exists (existence value), or for the future possibilities that the forest represents (option value) or for the ability to pass the forest on to future generations (bequest values) (Gregersen *et al.* 1997).

14.3 THE CHALLENGE OF ACHIEVING SECURITY OF FOREST VALUES

Despite the vast range of values that people seek in forests, few values have been evident in policies, markets and institutions until recently. This is largely because of the prevalence of governments and corporations in power structures and their relatively narrow requirements from forests (notably timber and foreign exchange).

Over time, the procedures of forest authorities that serve these interests have often become increasingly anomalous in the face of needs for multiple values. Their procedures have tended to become objectives in themselves, officers' rewards being based on observing hierarchical norms of behaviour and not on innovation or results, with an emphasis on narrow technical issues leading to a legitimation of increasing isolation from other forest interest groups and their values (Hobley 1996). The continued allocation and management of forests for timber alone, bolstered by such 'fortress forestry' authorities, hinders stakeholders from using forests to produce other goods and services. There are usually several overlapping customary and legal interests attached to a piece of forest territory. Rarely is more than one set of interests, usually timber or mining, taken into consideration in forest policy, planning and land allocation. This is often exacerbated by local stakeholder rights having been removed through legislation. Moreover, because many governments tend to be weak, powerful (private sector) operators can appropriate these same rights. The net result is inequity in bearing the costs of forest use, and in enjoying the benefits. Means for dialogue, negotiation and partnerships between stakeholders that might improve the situation are weak; thus either the *status quo* is perpetuated or clashes over forest values escalate.

Consequently, many groups are experiencing insecurities in supplies of forest goods and services (Bass 1997). These insecurities will worsen as population grows, or more particularly as the consumption of richer groups increases, whilst forest resources continue to diminish. By the middle of this century, if present trends continue, there will be twice as many people relying on a much smaller area of forest. It is imperative that demands for wood production in particular are consistent with supplying the other forest values that people need, especially those that currently have no market or can never operate within a market, such as local livelihood needs (Table 14.1).

In agriculture, the concept of 'food security' at household, national and global levels helped to galvanize action against food shortages and famine. The concept of 'security of forest goods and services' may help in making more precise decisions about forests, i.e. what goods and services forests should produce and who should share the costs and benefits. The value of the recent initiatives towards SFM principles, criteria and indicators is that they force different

groups to be clear about what they seek, in terms of goods and services and the state of forest values.

The challenge, then, must be to achieve security of those forest goods and services that are socially valued at specific levels. There are significant policy and fiscal challenges regarding security of the (indirect-use) values that are required at national level, notably the production of watershed, biodiversity and other public benefits from private lands or management companies. In addition, divisions between countries of the North and South have hindered attempts to secure global forest values in the face of the assertions of many countries that forests are sovereign territory.

It is suggested that the local level, especially the community and household in forest-dependent areas, should be accorded priority, for forests are key for livelihood security and poverty reduction; furthermore, local forest producers are often best placed to generate the values that are required at national and global levels (although they will require incentives). Forestry that improves social values and reduces poverty will be of a type that empowers poorer groups as forest stewards, improving their security of access to forest goods and services necessary for livelihoods. It will also mitigate or restrain the power of those in whose hands forests are currently over-concentrated and who are wasting or asset-stripping them (Mayers 1997).

14.4 PEOPLE'S MEANS TO ACHIEVE SECURITY OF FOREST VALUES

People have always planted and managed trees *where* they want, for the goods and services they want, *if* they have the means at their disposal, i.e. the technology and resources, and recognized rights, responsibilities, rewards and relationships with other groups.

14.4.1 Traditional knowledge

There is much discussion about the need to reconcile traditional indigenous knowledge with 'scientific' forestry in order to improve local action to secure forest values. It is ironic that, at a time when we are seeking means to sustain multiple values, many traditional management systems that do precisely this are on the decline. For example, Lacandones forest farmers in Mexico cultivate 75 species in a single hectare, and a farmer will need to clear no more than 10 ha of rain forest in his entire farming career (Clay 1988). In Sweden, until the nineteenth century, rotational shifting cultivation sustained grain, root crops and livestock as well as timber (Hamilton 1997), and the government saw it as a legitimate form of forest management. As Ghillean Prance has said, traditional forest peoples 'who depend upon so many of the species in the forest will hardly want to destroy more than the absolute minimum necessary for their agriculture. The settler from outside the region, on the other hand, is perplexed by such diversity and tends to destroy rather than use forest [diversity]' (quoted in Bass 1997)

Traditional forest management systems comprise both knowledge and rules, some of which may be transferable to new SFM approaches. The challenge is partly a democratic one: to admit local groups into policy and management decisions and to ensure that their rights can be exercised. But it is also a challenge to the prevailing systems of knowledge, which by and large are scientific and commercial, to admit such knowledge into the pantheon of accepted practices that guides our forest management.

14.4.2 Rights

Different groups have particular rights associated with obtaining forest values. Some rights directly concern the forest and its use:
• territorial rights;
• ownership of trees and other resources, such as minerals, grazing and wildlife;
• rights of access, use and control (for specific purposes and times);
• intellectual property rights.
Several problems are commonly associated with rights. One is that different rights overlap on one territory. Furthermore, many rights may not be established in legal documents, especially customary rights.

Other relevant stakeholder rights concern communities and their cohesion:
• rights to protection of cultural heritage, landscapes and folklore;
• collective rights, e.g. a community's rights to represent itself through its own institutions;
• rights to religious freedom;
• rights to self-determination and to development;
• rights to privacy.
National laws tend to cover specific social issues linked to forests, notably rights to public consultation (planning laws), rights of access (public rights of way) and forest worker rights, etc.

Internationally, there is a growing body of relevant legally binding agreements, to which countries may be party, and which eventually find their way into national legislation. These include:
• UN Covenant on Economic, Social and Cultural Rights (1966);
• UN Covenant on Civil and Political Rights (1966);
• UN Declaration on the Human Right to Development (1986);
• International Labour Organization Convention 169 concerning indigenous and tribal peoples in independent countries (1989);
• UN Convention on Biological Diversity (1992).
Many new laws are based on recent legal principles deriving from international consensus on sustainable development (such as the 'intergenerational equity', 'polluter pays' and 'precautionary' principles). Laws on land, environment and forestry are being reviewed in many countries, especially regarding the links or conflicts between formal and customary rights.

Some groups are forceful in exerting their rights and/or have skills and resources to do so. Others are less well equipped. Irrespective of the stakeholders' powers and resources, forest managers are likely to experience problems if they ignore or violate these rights, and indeed claims. An informed approach is better. This would include anticipating future legal requirements by making provision for best practice in dealing with local groups, and continuous improvement of practice (see Vol. 2, Chapter 10).

14.5 REASONS FOR NURTURING SOCIAL VALUES IN COMMERCIAL FOREST MANAGEMENT

Apart from the policy, legal and ethical reasons noted above, there are good commercial reasons to consider social values (Higman *et al.* 1999). Forest managers are increasingly subject to pressures from other interest groups, frequently concerning social values.

• Demands from forest peoples' groups may include greater respect for local populations' rights, carrying out or desisting from specific management practices, and support for their own environmental/social projects.

• Demands from local groups to contribute to social and economic development may include support for local enterprise, employment opportunities, excision of specific areas from management, and use of company infrastructure.

• Demands from trades unions and their initiatives may include greater respect for workers' rights to organize and negotiate, fair wages and benefits, health and safety at work, and the right to skills development throughout the organization.

• Demands to recognize other forest actors' rights to monitor, control and negotiate may include the development of agreements, and procedures to manage conflicts and make compensation.

Unless an active, organized approach is taken to respond to such pressures, forest managers may find themselves facing:

• slow-downs, strikes, blockades, boycotts, legal battles;

• damage to forest stock;

• arson, sabotage or vandalism to equipment and infrastructure;

• reappropriation of forest lands for cultivation, or migration;

• development of 'cultures of resistance' among forest-dependent people and their supporters (such as some consumers).

Considerable management skill and time may need to be invested in dealing with disputes, legal challenges and compensation claims, stalled negotiations, 'bad press' and political backlashes and general hostility towards forest managers and companies.

In contrast, participatory approaches that develop people's potential contributions can benefit forest managers, by:

• improving the reputation of forest managers and companies;

• uncovering and sharing useful local information;

• broadening the base of ideas, skills and inputs applied to forestry;

• efficient apportioning of responsibility, e.g. local groups may be better suited to managing recreation and harvesting of non-timber forest products;

• better understanding of broader social, and thereby market, trends;

• improving transparency, accountability and therefore trust between all parties;

• longer-run cost savings and risk reduction.

By taking an active approach to people as well as to trees, forest managers greatly increase the potential social benefits from forest management and their chances of support from others. Such experience should also increase their capacity to anticipate future developments regarding social values, for example in legislation or the market, and thus to gain 'first-mover' advantage from the situation.

If some social values are indicators of market trends (and they frequently are), then it behoves forestry organizations to keep close track of them. Many companies have made good business ventures by diversifying into recreation provision in particular. And it is increasingly clear that there is considerable financial value attached to brand names of certain leading companies which are known to produce social and environmental benefits alongside fibre. For example, the Greenpeace name has been estimated to be worth hundreds of millions of dollars (Anon. 1998).

14.6 CODES OF PRACTICE AND CERTIFICATION STANDARDS ON SOCIAL ISSUES

As noted in section 14.1, all initiatives to define SFM principles and criteria cover social values. Some of these are becoming enshrined in legisla-

Table 14.3 Comparison of social issues addressed by the Helsinki Process, Montreal Process, FSC Principles and Criteria and ITTO guidelines.

Social issues	Helsinki Process	Montreal Process	FSC	ITTO
Impact on indigenous people				
Traditional land rights protected		D	•	•
Customary rights maintained		⊐	•	•
Compensation for traditional knowledge			•	
Impact on local communities				
Community consultation and involvement		⊐	•	•
Compensation for grievances			•	⊐
Recreational use opportunities	D	D		D
Subsistence use opportunities		D	•	•
Sites of special significance respected (historical, cultural, spiritual)		D	•	•
Aesthetic values of forest maintained	D	D	•	D
Employment provided	D	D	•	D
Impact on employees				
Health and safety protected		D	•	•
Wages fair		D	•	•
Right to organize respected			•	
Economic and financial impacts				
Stable generation of revenue		D	•	•
Contribution to gross national product	D	D		⊐
Investment in forestry operations		D	•	•
Investment in local economy			•	•

D, Measurement criteria; •, performance requirement; ⊐, policy criteria.
FSC, Forest Stewardship Council; ITTO, International Tropical Timber Organization. *Source:* Nussbaum *et al.* (1996).

tion, while others face forest enterprises through market relations.

Current systems of forest certification specify various social requirements at the level of the forest management unit. Even so, social issues remain perhaps the most contentious of certification standards, and there are differences between standards, particularly regarding the required degree of involvement of different interest groups in forestry and treatment of peoples' rights. There are also differences in interpretation of the standards by certifiers/assessors. Whilst some confine themselves to assessing the local social outcomes of forest management, others assess the roles of local stakeholders in the management of the forest and indeed the enterprise, and call for changes that appear to reflect their own biases. Whilst it is increasingly clear that forestry should

not be a principal means of social engineering, and less still should forest certification, it is generally agreed that social standards for forestry will become more stringent in future. Social standards included in four key initiatives are summarized in Table 14.3 (Nussbaum *et al.* 1996). Note that some require precise performance thresholds to be met, while others require only that the issue shall be measured. Still others consider that the social issue is a policy concern and require it to be covered in policies only.

14.7 CONCLUSIONS

There is a strong political and legislative tendency towards 'subsidiarity', i.e. making decisions at, and for, the most local level appropriate to the goods and services in question, and being held accountable for this. The trend implies an important role for the forest manager in understanding local groups and their values and reconciling them with national and global values.

Flurries of activity in recent years have resulted in a kind of 'policy inflation' facing local forest managers. Indeed, many managers might now feel, with some legitimacy, that they can hardly get started in forest management without consulting with various groups, conducting multiple assessments and being held up to scrutiny by public and market bodies alike. Whilst it is clear that this 'policy inflation' needs rationalization, and here the principles and criteria initiatives are a promising means to find common denominators and allow for local interpretation, the 'capacity paralysis' that is so out of balance with the new policies also needs addressing. At a practical level, many forest managers have just not made any effort at all in relations with their neighbours. Others have invested heavily in major public relations but without substantially changing management to manage social risk and opportunity in the first place. There is much disagreement on the limits to social responsibility of forest managers.

Some practical approaches now need to be installed and monitored, so that they can reduce the 'policy inflation' by indicating what is working and desirable. Chapter 10 in Volume 2 gives some guidance on how forest managers can get started in this respect.

REFERENCES

Anon. (1998) The limits to growth? *The Economist* 1 August 1998.

Arnold, J.E.M. (1997) Social dimensions of forestry's contribution to sustainable development. Paper for World Forestry Congress, Antalya, Turkey, October 1997. FAO, Rome.

Bass, S. (1997) Not by wood alone: how can wood production be consistent with social and environmental demands? Paper presented at the 1997 Marcus Wallenburg Prize Symposium, Stockholm, 14 October 1997. Marcus Wallenburg Foundation, Stockholm.

Clay, J. (1988) *Indigenous Peoples and Tropical Forests: Successes Forgotten and Failures Misunderstood*. Cultural Survival, Cambridge, Massachusetts.

Filer, C. & Sekhran, N. (1998) *Loggers, Donors and Resource Owners*. Policy that works for forests and people Series no. 2, IIED, London.

Gregersen, H., Lundgren, A., Kengen, S. & Byron, N. (1997) Measuring and capturing forest values: issues for the decision-maker. Paper for World Forestry Congress, Antalya, Turkey, October 1997. FAO, Rome.

Hamilton, H. (1997) *Slash-and-burn in the History of Swedish Forests*. Rural Development Forestry Network Paper 21f, Overseas Development Institute, London.

Higman, S., Bass, S., Judd, N., Mayers, J. & Nussbaum, R. (1999) *The Sustainable Forestry Handbook*. Earthscan, London.

Hobley, M. (1996) *Institutional Change Within the Forest Sector: Centralised Decentralisation*. ODI Working Paper 92, London.

Holmberg, J., Bass, S. & Timberlake, L. (1991) *Defending the Future: a Guide to Sustainable Development*. Earthscan, London.

IIED (1996) *The Sustainable Paper Cycle*. IIED/WBCSD, London.

Mayers, J. (1997) Poverty and forests: a preliminary review of the issues, and how they can be tackled. Unpublished, IIED, London.

Nussbaum, R., Bass, S., Morrison, E. & Speechly, H. (1996) *Sustainable Forest Management: an Analysis of Principles, Criteria and Standards*. Towards a Sustainable Paper Cycle Sub-study Series No 4, IIED/WBCSD, London.

Segura, O. *et al.* (1996) *Politicas Forestales en Centro America: Restricciones para el Desarrollo del Sector*. CCAB-AP, San Jose, Costa Rica.

WCFSD (1997) Our forests: our future (unpublished draft final report, June 1997). World Commission on Forests and Sustainable Development, Geneva.

15: Non-timber Forest Products and Rural Poverty: an Economic Analysis[1]

WILLIAM CAVENDISH

15.1 WHY NON-TIMBER FOREST PRODUCTS AND RURAL POVERTY?

The term 'non-timber forest products' (NTFPs) refers to the wide range of species, both flora and fauna, that are produced by (or exist in) forests and woodlands and which are available to humans for uses other than commercial timber. In the past decade, there has been an enormous upsurge in research on NTFPs, reflected in, and promoted by, works such as Nepstad and Schwartzman (1992), Panayotou and Ashton (1992) and Godoy and Bawa (1993). This upsurge was caused by a number of factors, including a shift in forestry research away from the traditional concentration on commercial or single-use forests, the rise of social forestry, and an increasing interest in indigenous forests and woodlands. However, the most important reason for the rise in work on NTFPs has been the prominence they are given in discussions surrounding the global protection of biodiversity in partnership with local forest-based communities. Indeed, the greater commercial use of NTFPs has been widely promoted by researchers and international agencies as a means of simultaneously promoting the economic development of poor rural areas and of conserving critical tropical and temperate forests and woodlands. Exploring this argument forms the subject matter of this chapter.

If we are to understand the potential role that NTFPs might play in the socioeconomic development of rural areas, we need first to examine the role that these resources currently play in rural livelihoods, and the economic underpinnings of NTFP use in rural households. Here there is a disjunction between international and local perspectives. For the international community, NTFPs have two potential roles. The first is what one can term the 'development imperative', namely that NTFPs are seen as a set of resources that are potentially exploitable to promote the socioeconomic development of poor, resource-dependent communities. NTFPs are often thought of as 'hidden' resources, whose production technology and ecology are well understood by local users and producers but whose economic potential has yet to be fully realized. More broadly, forests generally are regarded as having a wide range of potential products, particularly medicinal, whose utility can be expressed presently only as option values. Assuming that these products can be commercialized and that the benefits of commercialization can be captured by local communities, NTFPs it is believed might form the basis of significant income-raising trades for a reasonably substantial class of the rural poor.

The second reason for the focus on NTFPs can be termed the 'environmental imperative'. Since NTFP uses are generally non-destructive to the resource stock (unlike timber), utilization of NTFPs adds value to the forest without undermining the resource stock. If rural agents face the portfolio choice of either clearing land for agricultural purposes or leaving land as forest, then a higher return to land under forest use will lead to more forest being preserved. This property of non-destructive use value is what has led some inter-

[1] This chapter was originally prepared as part of a Centre for International Forestry Research (CIFOR) programme on NTFPs and is reprinted with revisions by kind permission of CIFOR.

Table 15.1 A classification of forest products by economic function: Shindi, Zimbabwe.

Consumption goods	Inputs	Output (sold) goods	Durables and stocks
Wild fruits and nuts	Firewood (beer brewing)	Wild fruit and nut sales	Furniture
Wild vegetables	Firewood (brick burning)	Wild vegetable sales	Large utensils (wood)
Large wild animals	Leaf litter	Wild animal sales	Firewood store
Small wild animals	Termitaria	Wine sales	Construction wood
Wine	Livestock browse	Firewood sales	Fencing (wood)
Other wild foods	Thatching grass	Construction wood sales	
Firewood (cooking/heating)		Thatching grass sales	
Agricultural tools (wood)		Other wild good sales	
Small utensils (wood)		Carpentry sales	
Mats (reeds)		Woven goods sales	
Woven baskets		Pottery sales	
Pottery			
Wild medicines			

Source: Taken from Cavendish (1997), which also presents a more detailed list by type of forest product, including botanical names.

national agencies to suggest that the development of NTFPs can be used as part of a strategy for forest and woodland conservation. So, given their potential role in promoting economic development and environmental conservation, the recent international interest in NTFPs hardly needs explaining.

However, from the perspective of rural households, there is nothing special or unique about NTFPs. In fact, rural households use NTFPs for prosaic reasons. That is to say, NTFPs are goods and services that households utilize according to the economic nature of the NTFP in question.[2] Indeed this is true of all forest products, whether timber or non-timber. As an example, an economic classification of the type of forest products generally used in Zimbabwe is given in Table 15.1, based on research in the south-east of the country. As this table suggests, some forest products are used as consumption goods, some forest products are inputs into other economic production activities, some forest products can be sold and some forest products are transformed

to produce durable goods. Note that although the particular items categorized in this table reflect utilizations based on species endemic to miombo woodlands, similar lists could be drawn up for forest products derived from tropical forests, including of course the Amazonian rain forest.

Thus NTFPs are simply goods and services with particular physical, technical and economic characteristics that may or may not make households desire to use them. In deciding whether and how to use NTFPs, rural households will assess the economic characteristics of these NTFPs against other uses of their time, labour or income, and against other competing products. For example, take the decision of the household to collect and consume wild fruits. The decision to do this will depend, amongst other things, on the physical qualities of the wild fruit (taste, nutrients, etc.), the accessibility of the wild fruit in terms of the time taken to collect it, the availability of alternative wild or exotic fruits, the cost to the household of collecting the fruit in terms of other tasks forgone, and the cost to the household of consuming the wild fruit rather than selling it if a market is available. Similarly, the use of firewood by rural households will depend on the quality of firewood species available, their collection costs, and the quality, availability and cost of alternative technologies such as kerosene stoves, paraffin lamps,

[2] Owing to the focus on products, we exclude from consideration here the non-use benefits that forests bring, such as soil protection, watershed regulation and carbon sequestration. While these would constitute a separate class of economic good in Table 15.1, their inclusion would not invalidate the arguments of this chapter.

solar-powered batteries and so on. In short, rural households will use NTFPs if it is economically rational for them to do so.

Furthermore, the use and value of NTFPs is just one, possibly quite small, part of a whole set of economic activities within the household. Rural households in developing countries have the distinct characteristic of being at the same time both a consumption and a production unit, that is to say they have the characteristics simultaneously of both firms and households (the classic economic model of rural households is analysed in Singh *et al.* 1986). In consequence, rural household economic activities are multifarious, usually comprising agricultural production for own consumption and for sale; livestock rearing for own use, own consumption and sale; a variety of non-farm enterprises; migration and economic connections to formal labour markets; goods trading; involvement in networks of reciprocal exchange; investment in education, agricultural equipment and other productive capital; and the collection and use of freely provided environmental goods, some of which come from forests and woodlands. From the perspective of rural households, then, NTFPs form just one component of a set of livelihood activities.

So, if we are going to understand the role and value of NTFPs to rural households, we need to ask some questions. What are the general economic characteristics of these households? How do NTFP activities fit into the broader set of economic decisions or portfolio allocations that are being made by these households? Understanding this will allow us to relate the economic characteristics of rural households to the economic characteristics of NTFPs, and hence understand why households use these resources. This is the basis for sections 15.2 and 15.3. (Note that this chapter draws particularly on evidence from Zimbabwe, for which there is a comprehensive dataset linking NTFP use to the economy of rural households.) In section 15.4, this understanding of the economic underpinnings of NTFP use is employed to examine the potential of NTFPs to drive rural socioeconomic development. As I shall argue, there are many serious economic constraints to the use of NTFPs for this purpose, with the result that I believe that the

recent enthusiasm of the international community for NTFPs in development projects is considerably misplaced.

15.2 ECONOMIC CHARACTERISTICS OF RURAL HOUSEHOLDS

It may seem pointless to attempt to generalize about the characteristics of rural households, given the many different production activities rural households undertake and the many different agroclimatic zones in which such households live. For example, in sub-Saharan Africa alone the term 'rural household' encompasses settled agriculturalists, settled agropastoralists, nomadic pastoralists and extractivist forest-dwellers. These households can be found in ecological zones as varied as the arid Sahel, the Central African rain forests, and the semiarid plains and woodlands of southern Africa.

However, despite the huge variety of production types and climatic zones in which households are located, there are many ways in which rural households share quite similar economic characteristics. (For a comprehensive overview of the economic analysis of rural households, see Dasgupta 1993.) The first, and perhaps most important, is that rural households are generally either poor or have very low standards of living. Exact figures on rural household incomes and rural poverty are hard to find, not least thanks to a lack of rural household survey data, the incommensurate nature of many of the rural household surveys that are carried out, and genuine problems in comparing the welfare of rural households across countries and across time. However, what evidence there is proves fairly conclusive. For example, the United Nations Development Programme (UNDP) has estimated that 72% of rural households in the developing countries as a whole have standards of living below the poverty line (UNDP 1992, p. 161). In Africa, the same organization estimated average incomes to be $US475 per capita in 1992; this is low enough, and yet we know that rural incomes are quite a bit lower than these country-wide means. Zimbabwe itself has reasonably good data on rural standards of living. The latest reliable statistics come from the national Income, Consumption and Expenditure

Survey (ICES) of 1995/96, which shows mean rural per-capita consumption levels of $US217 per annum and median levels of only $US146 per annum (CSO 1998, p. 33). That is to say, half of the individuals living in Zimbabwe's rural areas exist on per-capita consumption levels of less than $US146 per annum. Not surprisingly, rural poverty rates are high. The same survey calculated that 86.4% of rural dwellers could be classified as suffering from poverty, while an astonishing 62.8% of rural dwellers suffered extreme poverty (CSO 1998, p. 33). (Extreme poverty is defined as an inability to reach the food poverty line, defined as the minimum food intake an individual needs to satisfy basic calorific requirements.) Note that by African standards, Zimbabwe is by no means a poor economy, suggesting that in many other countries on the continent the situation is still worse.

Alongside this poverty and these low standards of living is the second feature of rural households, namely their economic remoteness. By this is meant the fact that many rural households are a considerable economic distance from the formal markets in which serious income-raising trades can be made (Dasgupta 1993, pp. 234–45). This is often a result of physical distance: rural households are simply a long way away from urban markets. However, it can also be the result of poor infrastructure provision, essentially a lack of roads (Barwell 1996). For example, in Zimbabwe a government study found that rural households were on average 20 km away from a tarred road and 3 km away from a dirt road. The results of this are that the unit transactions costs of trading in formal markets are high: rural households must traverse substantial physical distances if they are to connect to formal markets. Simply getting produce to the market is expensive, and this is what is meant by the term 'economic remoteness'. The fact that many rural households face such high transactions costs severely restricts their economic opportunities and their exploitation of gains to trade (Platteau 1996; Risopoulos *et al.* 1999). (Conversely, rural households close to formal markets, especially urban areas, are often better off than most others.) Indeed, these high mobility/transactions costs lead to pervasive formal product market failures (de Janvry *et al.*

1991; Dasgupta 1993, p. 234), which impose significant welfare costs on rural households and constrain their responses to economic incentives. (Concomitantly, rural households are also distant from basic amenities such as banks, health services and grid electricity and this limits use of these facilities and services by rural households; Barwell 1996.)

The third feature of rural households is their high exposure to risk.[3] Almost all rural households are highly dependent on natural resources, whether for crop production, livestock fodder and browse, or the use of environmental resources. These in turn are strongly influenced by climatic conditions, particularly rainfall, which are highly variable. As well as this natural risk, due to their comparative remoteness rural households also face considerable idiosyncratic risks, such as exposure to health shocks (both human and animal) and crop pests that can have devastating consequences. However, the main problem is that in the face of this natural variability, rural households' exposure to risk is high because they do not have access to formal insurance or credit markets. Indeed, given the structure of rural economies, there are good economic reasons why formal markets in risk fail to exist. These reasons are well known in the economics literature and include the high covariance of rural households' agricultural and livestock outputs, a lack of collateral to underpin credit markets, and the high costs of information acquisition and contract monitoring in remote rural areas. The consequence of these factors is that formal trading in risk is rendered unprofitable.[4]

---

[3] There is huge literature on rural households and risk. For reviews that trace the impact of risk on the economic choices of rural households and on the existence or non-existence of important markets and institutions (of which the following paragraph is an extremely brief summary), see Hoff *et al.* (1999), Dasgupta (1993) and Collier and Gunning (1999).

[4] In the absence of formal risk markets, rural households are not powerless. There is a substantial literature on the various methods used by rural agents both to reduce risk (e.g. crop diversification, migration) and spread risk (e.g. inter-household contracts and the 'moral' economy). For a literature review, see Alderman and Paxson (1992).

The combination of missing formal product markets and missing markets in risk explains the fourth feature of rural households, namely the fact that their economic production is only partially monetized and the fact that local goods trading is often vibrant. The low share of cash transactions in rural households' total budgets is explained by the lack of formal product markets: rather than trade goods and incur the high transactions costs of trading, it is more economical for the household to consume the goods it has produced itself. Thus, for example, in Zimbabwe the 1995/96 ICES suggested that amongst poor rural households, fully 42.5% of food consumption was derived from the household's own production. These sorts of figures are typical of other African rural households. However, if a household does wish to trade in goods or factor services, the absence of formal markets means that it has no other choice than to trade in local markets. It is common therefore to find active local product markets and markets in labour. In almost all settled agricultural or agropastoral areas, there is localized trading in crops, livestock and in labour for agricultural tasks such as ploughing, weeding, harvesting and threshing. Likewise, local markets also tend to exist for other goods and services, in particular NTFP goods and the goods made from NTFPs. Thus, Cavendish (1999a) found that of the wide variety of NTFPs and NTFP-based goods used by households in his survey, many were traded locally and therefore had recognizable local prices. However, the fact that trading is restricted to the rural life space means that these goods and services face low demand levels, so prices also tend to be low.

A fifth feature of rural households is that they tend to have lower levels of human capital than their urban counterparts. There are two reasons for this. Development patterns in general in poorer countries have been urban-centred, so that public and private education has been much more available to urban households than to rural (Chambers 1983). However, there is also the impact of urban households' higher income levels, which allows them to invest much more heavily in their families' education than poorer, cash-constrained rural households. The lack of human capital again constrains the economic

welfare of rural households. In general, higher educational levels raise labour productivity in all activities (Psacharopoulos 1994). Even in agriculture, it has been shown that higher education levels lead to greater output (Jamison & Lau 1982). Perhaps more important though is that a low education level disbars rural individuals from competing for more highly paid formal sector jobs. Low-education individuals are only qualified to compete for relatively unskilled jobs, which pay low wages due to the excess of supply over demand in this segment of the labour market. Thus, low human capital levels often act to 'lock in' the lower economic status of rural households (Eswaran & Kotwal 1986).

In consequence, rural households tend to share a sixth feature, namely low levels of physical or productive capital. Both because of the low incomes of these households and also due to the failure of capital markets for rural agents, rural households find capital accumulation extremely hard.[5] So the main factor of production for rural households is relatively unskilled labour: it is this that they must allocate to the highest return activities in their economic reach.

15.3 VALUE OF NTFPS TO RURAL HOUSEHOLDS AND THE CAUSES OF NTFP USE

15.3.1 Evidence on rural households' use of NTFPs

There is much descriptive evidence that NTFPs are used widely by rural households. Two themes run through the case study literature on rural households' NTFP use: first, that the level of NTFP use by rural households could be substantial; and second that these NTFP utilizations are seldom quantified or valued in conventional household surveys, so that little is known about their economic contribution. Looking at Africa first, the breadth of the goods and services that

---

[5] Note that the assets that rural households are able to hold (cash and livestock) are also quite risky. Cash may easily be undermined by inflation; livestock can suffer from disease, theft, or death during drought. For this reason, neither are ideal vehicles for capital accumulation.

environmental resources potentially offer rural households is vividly apparent from the overarching literature review of Sale (1981). In this work, Sale presents a list of the various different utilizations of wild plants and animals, most of which are derived from forests or woodlands, in order to summarize the work of many other researchers (Table 15.2). It is clear from this table that wild resources provide rural households with a wide range of sources of economic value. Naturally the exact resources that households use, and the extent of their use, will differ across regions, if only due to differences in species distribution across Africa's various ecological zones. None the less, Table 15.2 charts a very substantial list of consumption, production input and asset formation uses for wild resources that can be drawn on by rural households. Indeed, from a review of the evidence on rural resource use, Sale concludes:

> . . . in many cases, wild animals and plants are regularly and permanently utilized in an

Table 15.2 Uses of wild plants and animals in Africa.

Wild plants	Wild animals
Food plants	Foods (mammals, reptiles, fish, birds, molluscs, arthropods, etc.)
Condiments, spices and flavourings	Beverages
Beverages	Fats, oils and waxes
Fodder plants for domestic stock	Fuel (dung)
Bee plants	Cosmetics
Plants for silkworms and caterpillars	Perfumes
Fats, oils and waxes	Dyes and stains
Medicinal and veterinary plants	Medicinal and veterinary uses
Poisons and antidotes	Poisons and antidotes
Saponin-producing plants	Weapons
Tannin-producing plants	Cultural uses (instruments, carvings, etc.)
Latex-producing plants	House construction
Gums, resins and waxes	Clothing
Dyes, stains and inks	Personal adornment
Cosmetics and pomades	Domestic uses (implements, whisks, mats, furniture, etc.)
Perfumes	
Mucilage plants	
Vegetable salts	
Fibres (for baskets, cloth, brushes, mats, pulp, etc.)	
Decorative plants	
Fencing and boundaries	
Hut poles, rafters, walls	
Timber	
Furniture and carving	
Domestic uses (tools, implements, utensils, instruments, etc.)	
Charcoal	
Firewood	
Tinder and torches	
Land reclamation	
Manure	
Weeds	
Weather signs	
Plants for smoking fish and meats	
Sacred plants	

Source: Sale (1981), Appendices I and II.

impressive variety of ways and to such a degree that loss of access would result in a complete collapse of the traditional economy.

The existence of widespread NTFP utilization by African rural households has been confirmed by a number of other studies. For example, the literature on the contribution of woodlands and woodland products has been exhaustively reviewed in Townson (1994). This study quotes many references as to the extent of rural households' use of forest foods, firewood use and sales, basketry and handicrafts, furniture and carpentry and other extractive products, all based on NTFPs. In similar fashion, Falconer (1990) has detailed the importance of a wide range of usually unnoticed forest products to the livelihoods of rural households in West Africa. Arnold *et al.* (1994) review data from surveys in six countries in eastern and southern Africa suggesting that forest products have an important role to play in the growth and functioning of rural households' small-scale enterprises.

More evidence on the importance of wild foods and wild products to African rural households is given in Eltringham (1984) and Scoones *et al.* (1992), the latter referring to the 'hidden harvest' because much of the resource utilizations that they review are not usually uncovered by conventional research programmes. In Zimbabwe, Campbell *et al.* (1991, 1997) have used various methods to estimate the contribution of a wide range of woodland products to rural households, suggesting these might generate around $US50 per household per annum, equivalent to roughly one-quarter of household cash incomes. More recently, Brigham *et al.* (1996) have reported a small number of traders in Zimbabwe and South Africa earning substantial cash incomes from wood and wood-based products, and Shackleton *et al.* (1998) found a small number of traders in South Africa earning high incomes from edible wild herbs. However, in none of these case studies has an accurate analysis been carried out of the role and value of NTFPs in the context of the broader household economy.

African households are not unique in using NTFPs so extensively. In a classic work, Jodha (1986) calculated that amongst poor Indian families lies the proportion of household income derived from common property resources, of which NTFPs would be a large subset, was between 9 and 26%. Working in one forest area of India, Hegde *et al.* (1996) estimated that the cash income from NTFPs potentially comprised a stunning 48–60% of household incomes. Falconer and Arnold (1991), in their review of rural households generally, argue that local forest products are central to household livelihoods, and draw out the particular importance of NTFPs to rural women and children. There is a substantial case study literature on households and NTFP use from Latin America, summarized in Godoy *et al.* (1993) and Lampietti and Dixon (1995). These suggest that rain forests are rich in products used by rural households; that these NTFP use values are seldom captured in conventional measures of household consumption or well-being; and that poorer households depend on these resources more heavily than richer households. The latter work reviews 16 studies of the economics of NTFP extraction in the Amazonian rain forest and suggests that the value of these minor forest products is approximately $US70 per hectare annually.[6]

However, the literature on rural households' NTFP use has not focused on the value of NTFPs in the context of the household's overall level of economic well-being, so rigorous quantifications of NTFP values *vis-à-vis* other sources of economic value are rare. The only source of detailed economic data on rural households and NTFP use comes from a panel household survey conducted in 1993/94 and 1996/97 by the author in Shindi Ward, Zimbabwe. These data are examined at some length, as it is the only study that systematically links NTFP use to the broader household economy; indeed, the whole purpose of this par-

---

[6] Lampietti and Dixon lament that there is very little reliable data on NTFP values at the household level and argue that this leaves significant research questions unanswered. This concern has been expressed by a number of authors, including Dasgupta (1993, p. 273), Godoy and Bawa (1993), Hegde *et al.* (1996) and, more recently, Wollenberg and Nawir (1997) in their review of selected best-practice studies of rural households and NTFP product incomes.

Table 15.3 Total household income by quintile and by income source: Shindi, Zimbabwe.

	Household quintile					
	Lowest 20%	20–40%	40–60%	60–80%	Top 20%	All households
Total non-environmental cash income	2418	6041	6789	12197	28580	56025
Total net gifts/transfers to and from the household	204	−758	1174	1549	2496	4665
Total household use of own-produced goods	6971	10025	10859	12541	20858	61254
Household use of non-NTFP environmental goods	2601	2931	4600	4828	7449	22409
Household use of NTFPs	3619	5377	6551	9264	10604	35415
Cash income from NTFP sales	689	831	1302	2740	1874	7436
Consumption of own-collected NTFP foods	314	298	352	533	454	1951
Consumption of own-collected firewood	1430	1991	2239	2498	3679	11837
Consumption of other own-collected NTFPs	132	178	190	190	214	904
Use of NTFPs as housing inputs	412	326	543	580	983	2844
Use of NTFPs for fertilizer	43	30	41	72	73	259
Livestock browse/graze in woodlands	599	1722	1884	2652	3328	10185
Total income	15813	23616	29974	40380	69986	179769
Summary data						
Total environmental income	6220	8308	11151	14092	18053	57824
Quintile share of total Shindi income	8.8	13.1	16.7	22.5	38.9	100.0

All data are in 1993/94$Z. Household incomes are adjusted for household composition, household size and economies of scale in household production. For explanation of sampling, data descriptions, valuations methods, etc. see Cavendish (1999a).

ticular data survey was to examine the use, value and management of environmental resources generally in the context of a rigorous and quantitative household survey.[7] It must be remembered of course that the results of this study represent only one set of households in a particular agroclimatic zone at particular points in time; none the less it is the best dataset available to analyse the issues covered in this chapter. Thus, Tables 15.3 and 15.4 contain summarized data from this study, presenting the aggregate value of NTFP use and other sources of household income across all households and by income quintiles, and the budget shares of NTFP use and other income sources expressed as shares of household total

income. (Total income refers not just to cash income but to cash income plus the household's use of its own production plus net gifts and transfers to the household plus the use of freely provided environmental resources collected by the household itself. It is therefore a very broad conception of income, and much closer to a conceptually accurate measure of household welfare than cash income alone.)

It is clear from these data that the households in question depend reasonably heavily on a broad set of environmental goods. On average, these goods comprise 35% of total household income, a share almost equivalent to that of subsistence consumption (37%) and substantially greater than non-environmental cash income (26%). Further, the majority of these environmental goods are NTFPs. The consumption and sale of NTFP goods alone comprises 21.5% of household total income on average. As we saw earlier, there is a very wide range of NTFPs being used by the households;

[7] In Tables 15.3 and 15.4, for brevity's sake data are presented for one year only, 1993/94. However, the 1996/97 data generate very similar conclusions to those here (see Cavendish 1999a for an analysis of both survey waves).

Table 15.4 Total household income shares by quintile and by income source: Shindi, Zimbabwe.

	Household quintile					
	Lowest 20%	20–40%	40–60%	60–80%	Top 20%	All households
Total non-environmental cash income	14.8	25.4	22.6	30.3	36.4	26.0
Total net gifts/transfers to and from the household	1.4	−3.0	3.9	3.6	2.5	1.7
Total household use of own-produced goods	44.1	42.4	36.4	31.0	32.1	37.2
Household use of non-NTFP environmental goods	16.1	12.5	15.4	12.2	12.5	13.7
Household use of NTFPs	23.6	24.8	21.7	22.9	16.5	21.5
Cash income from NTFP sales	4.9	3.5	4.3	6.8	3.1	4.5
Consumption of own-collected NTFP foods	2.5	1.3	1.2	1.4	0.8	1.3
Consumption of own-collected firewood	9.1	8.5	7.5	6.2	5.6	7.4
Consumption of other own-collected NTFPs	0.9	0.8	0.7	0.5	0.3	0.6
Use of NTFPs as housing inputs	2.7	1.3	1.8	1.5	1.5	1.7
Use of NTFPs for fertilizer	0.3	0.1	0.1	0.2	0.1	0.2
Livestock browse/graze in woodlands	3.8	7.3	6.2	6.4	5.2	5.8
Total income	100.0	100.0	100.0	100.0	100.0	100.0
Summary data						
Total environmental income share	39.7	37.3	37.1	35.0	29.0	35.2
All cash income share of total income	27.3	34.0	36.9	46.7	48.8	37.8

In this table we have calculated average income shares as the mean of the individual household's budget shares, rather than the simpler procedure of calculating the aggregate share of the income subcomponent in total income. This reduces the impact of extreme individual household values on the average budget share value.

thus this significant total share is made up for some households by a large number of small utilizations, the type of activities usually overlooked in conventional economic studies of rural households. None the less, there are certain NTFP uses that stand out for these households, namely the value of firewood use, the value of livestock browse and graze offered by woodlands, and the cash income arising from the sale of NTFP goods (e.g. thatching grass, carpentry products).

Another striking feature of these figures is the fact that in total value terms the use of NTFPs generally rises as overall household incomes rise. Take firewood for example. There is a systematic relationship between the total value of firewood used and the income quintile of the households. Indeed, top quintile households use roughly 2.5 times as much firewood per household as those in the bottom quintile. In similar fashion, top quintile households use roughly twice as much NTFPs for housing inputs as those in the bottom quintile,

and 5.5 times as much woodland browse and graze for their livestock. The value per household of cash income from NTFP sales appears to peak in the 4th quintile; none the less, top-quintile households still source over 2.5 times as much cash income from NTFPs as bottom-quintile households. So in absolute terms, NTFP resource demands appear to increase as households within the rural life space become more wealthy.

Conversely, dependence on NTFPs, and indeed environmental goods more broadly, appears to decline with income. That is to say, the income budget shares of NTFPs and total environmental goods basically decrease as the income quintile of households rises. The overall NTFP shares in Table 15.4 display this property less clearly than the overall environmental income shares. Thus, for the bottom four quintiles, the share of total income derived from NTFPs hovers between 22 and 24%. However, for the top quintile this falls fairly decisively to 16.5%. The decline in the share of total environmental income is more pronounced: from 40% for the bottom quintile it consistently decreases to reach 29% for the top quintile. Note that some of the individual NTFP subgroups also display this property of declining dependence as incomes rise. Income budget shares for the consumption of own-collected NTFP foods, own-collected firewood and other own-collected NTFPs all decline systematically with income. (Income budget shares for the use of NTFPs as housing inputs and fertilizers also basically decline as income rises but the relationship is not smooth.) What this means, of course, is that while it is true that richer households consume quantitatively greater amounts of NTFPs, it is poorer households that depend more heavily on NTFPs for their livelihoods.

15.3.2 Economic underpinnings of rural households' extensive use of NTFPs

We have seen that rural households, particularly poorer rural households, indeed use NTFPs extensively. What explains this pattern of use? We do not believe it is because rural households are innately more environmentally minded or naturally more orientated towards using environmental goods. Rather, it is because the economic characteristics of NTFPs match the economic characteristics and constraints of rural households that we listed earlier: NTFPs 'fit in' naturally to the current economic system of poor constrained rural households.

First, there is the fact that the great bulk of forests and woodlands in developing countries are held under communal tenure, with NTFPs (a largely non-destructive resource use) generally available under conditions of effective open access.[8] This implies that households do not have to purchase NTFP goods in order to consume them; rather they can be collected and consumed.

As we noted earlier, rural households' economic transactions are only partially monetized, thanks to the large number of missing product markets caused by the high unit transactions costs of trading. The economic remoteness of many rural households therefore results in them being cash constrained. When they do acquire cash, there are often economic priorities that can only be purchased using cash, such as basic foods (cooking oils, salt, sugar, etc.), clothing, school fees and the like. What freely provided NTFPs offer, then, is an alternative source of goods and services that economizes on the household's use of its scarce cash resources. Rather than purchase fuel, the household can collect firewood; rather than purchase agricultural implements, it can collect the wood and make them itself; rather than purchase fruits, it can collect them from the wild. Take for example the household data in Table 15.4. For the bottom quintile, only 27% of their total income is actually monetized; although this figure rises, even for the top quintile less than half of their total income is monetized. It is no coincidence that as the share of cash income in total income rises, the share of environmental resources and NTFPs in total income falls. Thus one reason for the widespread use of NTFPs is as a response by poor rural households to their cash-constrained nature, caused ultimately by the pervasive failure of formal markets and the absence of trading opportunities.

Second, and leading on from this point, NTFPs can overwhelmingly be collected using unskilled labour, and this matches the factor endowments of rural households as described earlier. For some NTFPs, collection involves simply the application of labour, such as picking up dead wood or plucking foods. This is one of the reasons why children and older people are often heavy users of NTFPs: not only do these groups tend to have less control over cash resources, but they can also

---

[8] Following the seminal contributions of Berkes (1989) and Ostrom (1990), the past decade has seen an enormous upsurge in research on effective management systems for collectively owned resources. However, although for many environmental resources communal management systems appear (under certain conditions) to be effective and sustainable resource management mechanisms, there are few examples for multiple-use forests of the type being discussed here. Recent research suggests that heterogeneity in forest user groups and differentiation in the economic characteristics of NTFPs together make communal management systems for forests extremely difficult to sustain (Baland & Platteau 1996; Ostrom 1999).

simply help themselves to resources using their own labour. For the majority of NTFPs, collection does require the use of some equipment, such as an axe for cutting wood, a machete for grasses and leaves, snares for hunting wild animals and so on. In general though this equipment is inexpensive and is fairly durable, and use of equipment does not need a high level of education or skills. Even where local rules exist about the harvesting or collection of NTFPs (e.g. about how to take bark from medicinal trees), these are simple to learn and do not stop individuals collecting the resource should they choose to do so.

Third, not just NTFP collection but NTFP processing also is a set of activities that requires the application of either unskilled labour alone or unskilled labour augmented by low-level capital equipment. As an example, a common set of NTFP-based products is baskets and mats made from leaves, creepers or rushes. Weaving these baskets and mats is labour-intensive but it is not highly skilled, most case studies reporting that these products are said by local respondents to be the type of things that anyone can make. Likewise the only capital equipment required is a weaving needle, which is a very cheap item to buy. Similarly, wood is often used to make a variety of household implements and agricultural tools, or is carved to make products aimed at the tourist market. Once again, interviews with resource makers stress that the skills used to make these objects can be acquired in a very short time and the capital equipment required is minimal. Finally, where NTFP foods are processed to make commercial products, such as fruit wines, palm wines, dried meats, dried fish and dried insects, again the skills and capital equipment required are low: it is essentially a matter of allocating sufficient household labour to complete the task. As discussed earlier, due to the difficulties that rural households face in capital accumulation, because formal credit markets generally fail in rural areas and because education levels are low, rural households are usually endowed with unskilled labour but poorly endowed with capital. So another reason for the extensive use of NTFPs by rural households is that the factor inputs required to collect and process NTFPs (unskilled labour, little capital equipment) match closely the factor endowments of the classic rural household. (These required factor inputs, combined with open access to the NTFP resource, means that NTFP activities have very low entry barriers. In such circumstances we would therefore expect the returns to NTFP activities to be correspondingly low.)

Fourth, households also use NTFPs in response to the general riskiness of rural economic activities. The use of NTFPs, particularly what have been classified as 'minor forest products', during times of household stress is an observation common to much of the NTFP case study literature. NTFPs display a certain degree of non-covariance with respect to agricultural output. When crops fail due to drought or disease, or when shocks hit the household such as unemployment, death or disease, some NTFPs will still be available for the household to either consume or use to generate cash income to purchase its essential needs. Thus it is economically rational for risk-averse rural households to hold a portfolio of production opportunities that exploits this non-covariance. In this sense, commonly held forests and woodlands can be regarded as providing a set of back-stop resources that insure households against the failure of other, higher return production activities. Thus, even if the returns per hectare of woodland were systematically below that of agricultural land, rural villages would still retain common lands for this purpose.

Finally, the argument comes full circle. One of the reasons that rural households use NTFPs so extensively is precisely because they are open-access resources. There are a number of reasons for the survival of communally held resources, for example the insurance element noted above; however, a major reason for the survival of such resources is the very low income levels of rural households. Resource privatization is an expensive business: households taking charge of privatized resources usually need to spend time and money on creating exclusion (e.g. by building fences), on enforcement (e.g. employing guards to monitor incursions) and on punishing infractions (by going through the courts). Particularly where a resource has previously been held in common, privatization is often strongly contested and consequently these various costs can be expected to be high. Where households are generally poor, the private costs of resource privatization are likely to be far higher than the potential private gains from control of the resource. Thus the low incomes of rural areas underpin the existence of commonly held resources, which in turn underpins the extensive use of NTFPs by those very same low-income households (on the relationship between efficient property rights regimes and transactions costs, see Coase 1960; Bromley 1989; de Meza & Gould 1992).

15.4 CAN RURAL DEVELOPMENT BE BASED AROUND NTFPS?

We have presented evidence that rural households' use of NTFPs is widespread and significant in terms of their overall livelihoods. We have also suggested that the cause of this extensive NTFP use by poor rural households is the match between the economic characteristics of NTFPs and the economic endowments and constraints of those same rural households. Can we therefore conclude from this that the commercialization of NTFPs is economically feasible and likely to bring substantial economic benefits to rural poor? Further, if commercialization were feasible, what would it be likely to do to the resource base that produces these NTFPs? In other words, what light does the analysis above shed on the possibility of using NTFPs to meet the twin goals of economic development and environmental conservation? I argue that there are substantial difficulties here, arising from the same economic characteristics of rural households and the causes of their NTFP use that were listed earlier. I look at a number of issues in turn and draw on the case study literature to illustrate where they have led to problems for various NTFPs.

15.4.1 The problem of preferences

We saw earlier that one of the reasons for rural households' NTFP use was their low incomes and the still-partial monetization of many of their economic transactions. However, it is a truism that as individuals and households become more affluent, their demands for goods and services

change. In particular, people choose to consume more highly preferred goods or goods of a higher quality. (There is abundant evidence for these demand shifts in the higher-income developed countries.) However, evidence from academic studies and from history suggests that NTFPs do not have high product quality or are not in general highly preferred goods. One way of showing this is to look at income elasticities of demand for NTFPs, as these elasticities reveal what the demand response of households will be to any increase in their incomes.

Unfortunately, demand elasticities for NTFPs are not easy to come by, due to the lack of attention paid to such goods by formal economic studies. Thus we have culled evidence from two sources. The first is the author's own study of households and environmental resources in Shindi, Zimbabwe and the second is from a recent literature review reported in Köhlin (1998) (Tables 15.5 & 15.6). Table 15.5 contains income elasticities for a range of NTFPs: wild fruits, wild vegetables, wild animals, firewood, NTFP-derived goods and thatching grass. In general, the income elasticities for these goods are low. In only one case, wild fish demands, is the recorded elasticity greater than 1 (i.e. a luxury good), although it is believed this has much to do with the supply conditions of wild fish in the study site. Otherwise most elasticities are closer to zero than to 1. For example, wild fruits collectively and mice have income elasticities of 0.3–0.4; woodland-derived wild vegetables have an elasticity of 0–0.2, firewood has an elasticity of 0.4; and NTFP-derived goods have income elasticities of 0.4–0.5. While these elasticities are positive, so that increases in income within the range of incomes spanned by the sample will increase overall NTFP demands, the budget shares of these goods will decline, to be replaced in importance by other more preferred goods. Table 15.5 gives one example of this, by comparing the income elasticities of woodland-derived and field-derived wild vegetables. The fact that the latter are much higher than the former reflects the higher preference households have for these foods, and also means that as their incomes rise rural households will switch to field-derived wild vegetables and away from their woodland-derived counterparts.

Table 15.5 Demand elasticities for non-timber forest products: Shindi, Zimbabwe.

Type of NTFP	Non-zero observations*	Mean budget share	Income elasticity	
			OLS†	Tobit
All wild fruits	206	0.00541	0.33	0.44
Diospyros mespiliformis (*suma*)	180	0.00094	0.49	0.64
Sclerocarya birrea nut (*shomwe*)	87	0.00166	0.28	0.28
Sclerocarya birrea wine (*mukumbi*)	68	0.00206	0.19	1.00
Berchemia discolor (*nyii*)	93	0.00040	0.41	1.00
Woodland-derived wild vegetables	93	0.00171	0.00	0.20
All wild animals	180	0.00793	1.00	1.00
Mice	120	0.00257	0.28	0.41
Game meats	74	0.00114	0.50	1.00
Wild fish	96	0.00394	1.53	1.31
All wild foods	213	0.06580	0.54	
Firewood	213	0.06110	0.42	
NTFP-derived goods	188	0.00480	0.36	0.46
Agricultural tools (wood)	99	0.00097	0.31	0.49
Small household utensils (wood)	84	0.00030	−0.06	0.50
Woven goods	125	0.00330	0.35	0.68
(of which woven mats	84	0.00250	0.25	1.00)
Pottery	141	0.00110	0.50	0.65
Thatching grass	91	0.00898	0.25	1.00
Addendum: field-derived wild vegetables				
Cucurbita pepa (*muboora*)	211	0.02088	0.75	0.76
Corcorus olitorius (*derere*)	168	0.00354	0.71	1.00
Gynandropsis gynandra (*rudhe*)	140	0.00435	1.00	1.00

To derive these income elasticities, a demographically augmented version of the Working-Leser form for demand functions was estimated using both OLS and Tobit estimation procedures. Note that Tobit estimation is unnecessary if there are no zero budget share observations. Diagnostic statistics are not reported here, but can be found in Cavendish (1999b).
* 213 total observations for all regressions.
† Ordinary least squares.

Further, the positive income elasticities in this table for NTFPs reflect the rather low income levels of Shindi households. Evidence from the regressions suggests that at higher income levels these elasticities would turn negative, so that at higher incomes these NTFPs would actually be inferior goods.

This suggestion is supported by the evidence of Table 15.6. While this table draws on the wider literature on NTFPs, most of the quoted elasticities refer to firewood, as this has been the main quantitative focus of the literature. Three studies of firewood demands conclude that income elas-

ticities turn negative as incomes rise (Amacher *et al.* 1993 for Nepal; Bagahawatte 1997 for Sri Lanka; Shyamsundar & Kramer 1996 for Madagascar). Further, the other elasticities cited are all either low and positive, or negative. This implies that as households get richer, the absolute quantities of NTFPs that they wish to consume will decrease until at some point NTFP demands will fall to zero. This evidence from formally estimated income elasticities is also backed up by classic observations of consumption shifts that occur as rural households become richer. For example, one immediate shift is from natural con-

Table 15.6 Demand elasticities for non-timber forest products: other studies. (Adapted from Köhlin 1998.)

Study	Place	Forest product	Income elasticity
Cooke (1998)	Nepal hills	Firewood	+ve
		Forage	−ve
		Fodder	+ve
Mekonnen (1998)	Ethiopia	Firewood	0.03–0.06
Amacher *et al.* (1993)	Nepal hills: low income	Firewood	−0.31 to −0.20
	Nepal hills: high income	Firewood	0.0005–0.002
Amacher *et al.* (1999)	Nepal: hills	Firewood	0.0005
	Nepal: tarai	Firewood	0.07
Shyamsundar & Kramer (1996)	Madagascar	Firewood	−0.01
		Palm leaves	−0.01
Bagahawatte (1997)	Sri Lanka: low income	Firewood	<1
		Medicinal plants	<1
		Mushrooms	<1
	Sri Lanka: high income	Firewood	<0
		Medicinal plants	<0

struction materials (earth bricks, poles, thatching grass) towards higher-quality purchased substitutes (breeze blocks, zinc roofing). Likewise, richer households choose to consume more purchased foods and beverages and less wild foods and locally produced alcohols. As the income elasticities suggested, richer households also prefer to switch away from firewood towards other sources of energy and lighting.

There are good economic reasons why NTFPs are regarded as inferior goods by richer households. Their production is variable, the quality of the goods cannot be controlled and other substitutes have been bred or produced with higher product quality in mind. Thus more affluent households shift to better substitutes when they can afford to do so. However, the key point is that the inferiority of NTFP goods creates a real problem for NTFP commercialization projects. Dependence on, and use of, NTFPs are linked to poverty and to market failure rather than to household choice: the current prevalence of NTFP use by rural households is a result of their low incomes rather than the attraction of NTFP products themselves. A dramatic illustration of this is given by Hegde *et al.* (1996), who argue that the very high rates of NTFP utilization of their sample households are caused by the fact that these households earn barely more than the Indian minimum wage, and present empirical evidence that as households get richer so their NTFP use declines. So where NTFPs are inferior, attempts to commercialize them will founder on a lack of long-term product demand.

15.4.2 High transactions costs of trading

In our discussion of the stylized features of rural areas, we stressed the role that economic remoteness played in underpinning low rural incomes and in causing the failure of formal product markets. This is also a major cause of the low monetization of rural households' activities, since trading in formal markets outside the rural area is so often uneconomic. These transactions costs will affect NTFP commercialization projects just as much as they would affect any other project aimed at commercializing rural produce. The high unit costs of transportation imply that NTFPs which have a low unit value relative to their weight will have low if not negative margins in the marketplace.

This may be the case for a number of NTFPs. On account of the open-access nature of the resource rules surrounding forests and woodlands in developing countries and the fact that many are inferior goods, NTFP prices tend to be fairly low. Some NTFPs will have high weight-to-price ratios, especially those made of wood. For example, it is noticeable that despite the reasonable prices being attained for carvings on the major roads of Zimbabwe, production activity is contained within a reasonably narrow area, close enough to the road to make transporting the wood inputs an economic proposition. High trading costs have also been implicated in the lack of NTFP commercialization in Latin America (Richards 1993). Of course, the converse of this point is the effect that road building has on reducing trading costs and thereby stimulating the marketization of rural produce generally. The impact of the Belem–Brasilia and the Transamazon highways on economic conditions in the Amazon basin has been widely noted. However, assuming that NTFP projects take the physical infrastructure as given, the impact of transactions costs on the profitability of NTFP commercialization will remain a serious constraint for remoter rural areas for a number of NTFPs.

15.4.3 Storage problems

Another feature of rural areas we noted was the generally low levels of physical capital. This is true both at the household level and at the community level. Rural areas not only lack household-based productive capital but also the area-wide capital stock, such as power supplies and communications, on which the productivity of household-based enterprises depends. One key problem that this raises is a lack of appropriate storage facilities, such as an absence of refrigeration and refrigerated transport. Indeed, storage of goods is often a major difficulty for rural households. Even for longer-lasting foods such as grains, losses are regularly incurred thanks to mould, rotting and crop pests despite the careful construction of granaries and the use of locally available goods to line and protect these granaries.

Problems with storage could be a major constraint to the commercialization of NTFPs where these goods are perishable. A substantial fraction of the NTFPs examined under the 'Hidden Harvest' programmes are exactly these types of goods: fruits, fruit-based beverages, wild meats

and even wild fish. All these would require reasonably capital-intensive storage if commercialization projects were to succeed. Appropriate storage provides two major economic functions, the absence of which would cause problems to emerge. The first function is that of product preservation. Given that the commercialization of NTFPs requires transportation of products to new markets, an absence of adequate storage would result in high losses of products due to spoilage (rotting, bruising in transit, excess fermentation of liquids, etc.). The other function that storage can play is that of interseasonal supply smoothing. Without it, products can only be supplied to the market during the season of their production, resulting in glutted markets, lower product prices and unsmoothed income streams. Thus seasonality and perishability lower the economic returns of NTFPs to rural communities, thereby making these activities less attractive.

15.4.4 Production risk

As discussed earlier, rural households face considerable levels of risk, and one reason for the preservation of woodlands for NTFPs is as a risk-spreading insurance mechanism against severe shocks to the other economic activities of the household. However, in the medium to long term risk-averse rural households will wish not just to spread risk but to reduce it. If NTFP activities themselves still involve considerable production risks, then such activities will be less attractive to rural households than other, more certain income flows. This will be a consideration where the supply of NTFPs is dependent on climatic conditions, such as some of the foods mentioned earlier, or where NTFPs are prone to disease.

15.4.5 Open access and the costs of privatization

A further concern derives from the fact that many forests and woodlands in developing countries are held under communal tenure, with usufruct rights to NTFPs very often being effectively open access. If NTFPs were to be commercialized, then one immediate concern would relate to the degree of exploitation of the resource. Raising the value of an NTFP will naturally increase supply pressures on the resource. Where NTFP use is non-destructive to the resource stock, this is not necessarily a problem. However, where NTFP use can diminish the resource stock, then the commercialization of the resource can have serious implications for objectives of environmental conservation. Indeed, there are a number of cases where increases in resource pressure consequent on increased demand for NTFP goods has had serious environmental consequences. One such case is the impact of basket-making on an indigenous tree in Botswana. The main source of leaves for baskets is *Hyphaene petersiana*, which stands up very well to intense harvesting pressure. However, the main source of dye for these baskets is the bark of *Berchemia discolor*, and harvesting the bark can harm the tree if done too frequently. The increase in demand for Botswana baskets has increased substantially the demand for dye inputs, and this in turn has led to the death of many *B. discolor* in basket-making areas (Cunningham & Terry 1995). Similarly, the high demand for *Warbergia salutaris* as a source of medicines in Zimbabwe has led to the almost complete extinction of this species within the country (Mukamuri & Kozanai 1999). There are numerous cases of this dynamic in the Amazonian rain forests, where palm heart production has resulted in the near elimination of *Euterpe edulis* and *Euterpe oleracea*; the cutting of palms for fruits in Peru has rapidly depleted the numbers of *Mauritia flexuosa*, *Mauritellia peruviana* and *Jessenia bataua*; and the felling of trees for rosewood oil production had led to the virtual extinction of *Aniba rosaeodora* and *Aniba ducke* in economically accessible parts of Brazil (Richards 1993). Commercialization of a destructive NTFP use when held under open access can thus have severe environmental consequences for the species in question (though not necessarily for the forest as a whole).

A longer-term concern relates to the question of who gains from NTFP commercialization. We noted earlier that the existence of communal tenure over woodlands is related to the high costs of exclusion, monitoring and enforcement relative to the benefits to be gained from private woodland control. In other words, the tenurial arrangements over woodlands are related to the ratio of costs and benefits faced by rural households and communities. If as a result of NTFP projects the per-hectare returns to woodland rise, then this ratio changes and economic incentives will arise to change tenurial arrangements. In particular, privatization may emerge as a cost-effective option. However, this is only likely to be cost-effective for households that are already more affluent than others, since only these will be able to bear the private costs of resource control. Two groups stand out here. The first is local élites. There is evidence that as open-access resources rise in value, these resources become co-opted by local élites. The classic study of this phenomenon is Ensminger's (1990) analysis of resource privatization in rural Kenya. In response to the lowering of trading costs via cheaper transportation costs, communal grazing lands were privatized with the consent of rural elders who themselves benefited from this process. However, this process of élite partitioning of commons resources in response to changing economic conditions has also been found in India (Jodha 1986) and Brazil (Hecht 1985). The second group is outsiders. Town-dwellers or traders/middlemen with superior technology and access to capital may gain control of either the NTFP resources or the NTFP goods' trade and thereby arrogate to themselves the economic returns to these goods. In either case, commercialization-induced changes in local tenure could not only stymie the expected socioeconomic gains from NTFP products to the generality of rural households, but also by triggering the privatization of woodlands and other common resources may leave the rural poor worse off by removing from them a vital consumption back-stop and a source of insurance goods.

15.4.6 Incentives for domestication and technical substitution

We have discussed the many reasons above why NTFPs are currently often low-value goods, such as missing markets, the prevalence of local trading, low local demand and the open-access nature of NTFP resources. One implication of these low values is that neither rural households nor outside agencies currently have an incentive to invest in any sort of resource management or product improvement, since the costs of such actions would far outweigh the benefits. However, this would change if NTFPs were to become higher-value goods.

There are two processes to mention here: domestication and the development of technical substitutes.[9] It is globally true that when natural resources become valuable, there is pressure to intensify production and to domesticate the resource. Take as an example a wild fruit. While prices for the fruit remain low, it is unprofitable for households to do anything other than collect the fruit from the wild. However, as fruit prices rise, at some threshold value it becomes economic for the household (or some other agency) to incur the time and resource costs of establishing the resource under their own control, through planting and tending the relevant species. Furthermore, domestication also allows resource managers to improve product quality through the development of a constant, uniform or standardized product, thereby maintaining or even increasing product prices. This dynamic is the general story of the development of agricultural products and there is no reason why NTFPs should be any different. Indeed, there are ample examples of cases where rising NTFP values have led to precisely this response. Perhaps the most famous is that of Brazilian rubber (*Hevea brasiliensis*) collection from natural forests. As rubber prices rose in the early part of the century, so it became economic to establish rubber plantations in South-East Asia; more recently, rubber plantations have been established in Brazil as well and these now supply 60% of Brazil's market (Browder 1992). Another classic NTFP affected in this way is brazil nuts. In response to rising

nut prices, several thousand hectares of *Bertholletia excelsa* plantations have now been established within Brazil itself. So neither rubber nor brazil nuts provide a solid foundation for NTFP commercialization in the Amazon in the long run. Indeed, it is difficult to point to a wild resource which, once valuable, has remained undomesticated (an extensive list of other Amazonian NTFPs that have been domesticated is given by Homma 1992). Beyond a certain point, then, the commercialization of NTFPs will almost certainly lead to the product being removed from the forest and being placed on the farm instead.

Additionally, if NTFPs become high-value goods, there is an economic incentive to invest in the search for cheaper synthetic substitutes, if this is technically feasible. Once such substitutes are developed, the market for NTFPs often collapses – the 'bust' following the 'boom'. Richards (1993) discusses many such examples of substitute-induced market collapses in commercialized Amazonian NTFPs. At various times these have included dyewood (*Caesalpinia echinata*), vegetable ivory (*Phytelephus macrocarpa*), balata (*Manilkara bidentada*), leche caspi (*Couma macrocarpa*), barbasco (*Lonchocarpus* spp.), babaçu oil (*Orbignya phalerata*), cumaru nuts (*Dipteryx odorata*), sorva latex (*Couma utilis* and *C. rigida*) and of course rubber. Such technical substitutions are problems for the entire goal of NTFP-based economic development, since they raise the spectre of entire markets disappearing just as rural households have come to depend on the NTFP resource for a high proportion of their livelihoods.

15.5 CONCLUSIONS

We should not therefore underestimate the problems associated with the use of NTFPs for the economic development of rural areas. Although rural households currently use NTFPs quite extensively, this is a consequence of the low incomes, capital scarcity, missing markets, riskiness in production and tenure conditions that characterize rural areas. It is these that make it economically rational for households currently to use NTFPs, as these are low value, depend on

---

[9] Homma (1992) provides the seminal analysis of the dynamics of the interaction between NTFP use, economic remoteness, domestication and substitution. Based on historical evidence from Brazil, he argues that NTFP utilizations will be locked into an inevitable cycle of expansion, stabilization and decline, followed by replacement with cultivated products or technical substitutes. The flavour of this work underpins our analysis here.

unskilled labour for their collection and processing, and offset to a degree the riskiness of other production activities. However, this is no guarantee of success when it comes to the commercialization of NTFPs: (i) there is little evidence that NTFP dependence in the long run is desired at all by rural households; (ii) there are economic problems associated with the costs of both trading and storing NTFPs; (iii) NTFPs still have levels of production risk that may make them unattractive; (iv) the commercialization of NTFPs is likely to lead to the elimination of some species; (v) NTFP development may also undermine the communal tenure system, thereby removing the wider benefits of NTFP use from poorer households; and (vi) NTFP projects aimed at raising product prices will have to face the twin pressures of domestication and technical substitution that have affected almost all other wild products. In the face of these problems, NTFP extraction-based rural economies are likely to be highly unstable, thereby undermining the goal of improved rural household welfare.

So what type of NTFPs might meet the twin desires of increased rural incomes and greater protection for forests? Based on this review, such products would need to have a number of characteristics:

1 a market amongst high-income consumers in order to surmount the low-income elasticities of rural households for these products;

2 prices high enough to overcome the transactions costs of collection and trading but which are not high enough to trigger investment in technical substitutes where these could feasibly be developed;

3 be a reasonably durable product, so that storage is not a binding constraint;

4 have a fairly low production-risk profile;

5 harvesting that is non-destructive to the resource stock;

6 returns per hectare that make it economically rational to conserve forests and woodlands but which are not high enough to trigger resource privatization;

7 a species ecology that rules out domestication of the resource.

Whether such products exist is a moot point. One goal for the next phase of forestry and NTFP research might be to identify and publicize such products. Only then are the initial aspirations for NTFP use to both protect the forest and alleviate rural poverty likely to be fulfilled.

REFERENCES

Alderman, H.A. & Paxson, C.H. (1992). *Do the Poor Insure: a Synthesis of the Literature on Risk and Consumption in Developing Countries*. Discussion Paper no. 164, Research Programme in Development Studies, Princeton University, Princeton, New Jersey.

Amacher, G.S., Hyde, W.F. & Joshee, B.R. (1993) Joint production and consumption in traditional households: fuelwood and crop residues in two districts in Nepal. *Journal of Development Studies* **30**, 206–25.

Amacher, G.S., Hyde, W.F. & Kanel, K.R. (1999) Nepali household fuelwood consumption and production: regional and household distinctions, substitution and successful intervention. *Journal of Development Studies* **35**, 138–63.

Arnold, J.E.M., Liedholm, C., Mead, D. & Townson, I.M. (1994) *Structure and Growth of Small Scale Enterprises in the Forest Sector in Southern and Eastern Africa*. OFI Occasional Paper no. 47, Oxford Forestry Institute, University of Oxford.

Bagahawatte, C. (1997) *Non-timber Forest Products and Rural Economy in Conservation of Wet Zone Forests in Sri Lanka*. Mimeo, Department of Agricultural Economics, University of Peradeniya, Sri Lanka.

Baland, J.-M. & Platteau, J.-P. (1996). *Halting Degradation of Natural Resources: Is There a Role for Rural Communities?* Clarendon Press, Oxford.

Barwell, I. (1996) *Transport and the Village: Findings from African Village-level Travel and Transport Surveys and Related Studies*. Discussion Paper no. 344, African Region Series, World Bank, Washington, DC.

Berkes, F. (ed.) (1989) *Common Property Resources: Ecology and Community-Based Sustainable Development*. Bellhaven Press, London.

Brigham, T., Chihongo, A. & Chidumayo, E. (1996) Trade in woodland products from the miombo region. In: Campbell, B., ed. *The Miombo in Transition: Woodlands and Welfare in Africa*, pp. 137–74. Centre for International Forestry Research, Bogor, Indonesia.

Bromley, D.W. (1989) Property relations and economic development: the other land reform. *World Development* **17**, 867–77.

Browder, J.O. (1992) The limits of extractivism: tropical forest strategies beyond extractive reserves. *Bioscience* **42**, 174–82.

Campbell, B.M., Vermeulen, S.J. & Lynam, T. (1991)

Value of Trees in the Small-scale Farming Sector of Zimbabwe. IDRC-MR302e, International Development Research Centre, Ottawa.

Campbell, B.M., Luckert, M. & Scoones, I. (1997) Local level valuation of savanna resources: a case study from Zimbabwe. *Economic Botany* **51**, 59–77.

Cavendish, W.P. (1997) *The economics of natural resource utilisation by Communal Area farmers of Zimbabwe.* DPhil thesis, University of Oxford.

Cavendish, W. (1999a) *Empirical Regularities in the Poverty–Environment Relationship of African Rural Households.* Working Paper Series WPS/99.21, Centre for the Study of African Economies, University of Oxford.

Cavendish, W. (1999b) *The Complexity of the Commons: Environmental Resource Demands in Rural Zimbabwe.* Working Paper Series WPS/99.8, Centre for the Study of African Economies, University of Oxford.

Chambers, R. (1983) *Rural Development: Putting the Last First.* Longman, New York.

Coase, R.H. (1960) The problem of social costs. *Journal of Law and Economics* **3**, 1–44.

Collier, P. & Gunning, J.W. (1999) Explaining African economic performance. *Journal of Economic Literature* **37**, 64–111.

Cooke, P.A. (1998) Intrahousehold labour allocation responses to environmental good scarcity: a case study from the hills of Nepal. *Economic Development and Cultural Change* **46**, 807–30.

CSO (1998) *Poverty in Zimbabwe.* Central Statistics Office, Harare, Zimbabwe.

Cunningham, A.B. & Terry, M.E. (1995) Basketry, people and resource management in southern Africa. In: Ganry, F. & Campbell, B.M., eds. *Sustainable Land Management in African Semi-Arid and Sub-Humid Regions.* Centre de coopération internationale en recherche agronomique pour le développement, CIRAD, Montpellier.

Dasgupta, P. (1993) *An Inquiry Into Well-being and Destitution.* Clarendon Press, Oxford.

de Janvry, A., Fafchamps, M. & Sadoulet, E. (1991) Peasant household behaviour with missing markets: some paradoxes explained. *Economic Journal* **101**, 1400–17.

de Meza, D. & Gould, J.R. (1992) The social efficiency of private decisions to enforce property rights. *Journal of Political Economy* **100**, 561–80.

Eltringham, S.K. (1984) *Wildlife Resources and Economic Development.* John Wiley & Sons, New York.

Ensminger, J. (1990) Co-opting the elders: the political economy of state incorporation in Africa. *American Anthropologist* **92**, 662–75.

Eswaran, M. & Kotwal, A. (1986) Access to capital and agrarian production organisation. *Economic Journal* **96**, 482–98.

Falconer, J. (1990) *The Major Significance of 'Minor' Forest Products: the Local Use and Value of Forests in the West African Humid Forest Zone.* Forests, Trees and People, Community Forestry note no. 6, FAO, Rome.

Falconer, J. & Arnold, J.E.M. (1991) *Household Food Security and Forestry: an Analysis of Socio-Economic Issues.* FAO, Rome.

Godoy, R. & Bawa, K.S. (1993) The economic value and sustainable harvest of plants from the tropical forest: assumptions, hypotheses and methods. *Economic Botany* **47**, 215–19.

Godoy, R., Lubowski, R. & Markandya, A. (1993) A method for the economic valuation of non-timber tropical forest products. *Economic Botany* **47**, 220–33.

Hecht, S. (1985) Environment, development and politics: capital accumulation and the livestock sector in Eastern Amazonia. *World Development* **13**, 663–84.

Hegde, R., Suryaprakash, S., Acoth, L. & Bawa, K.S. (1996) Extraction of non-timber forest products in the forests of Biligiri Rangan Hills, India. 1. Contribution to rural income. *Economic Botany* **50**, 243–51.

Hoff, K., Braverman, A. & Stiglitz, J.E. (1992) *The Economics of Rural Organizations.* Oxford University Press, Oxford.

Homma, A.K.O. (1992) The dynamics of extraction in Amazonia: a historical perspective. In: Nepstad, D.C. & Schwartzman, S., eds. *Non-timber Products from Tropical Forests: Evaluation of a Conservation and Development Strategy,* pp. 23–31. New York Botanical Garden, New York.

Jamison, D.T. & Lau, L.J. (1982) *Farmer Education and Farm Efficiency.* Johns Hopkins University Press, Baltimore.

Jodha, N.S. (1986) Common property resources and rural poor in dry regions of India. *Economic and Political Weekly* **21**, 1169–81.

Köhlin, G. (1998) *The Value of Social Forestry in Orissa, India.* Economic Studies 83, Göteborg University, Sweden.

Lampietti, J.A. & Dixon, J.A. (1995) *To See the Forest for the Trees: a Guide to Non-timber Forest Benefits.* Environment Department Paper no. 013, World Bank, Washington, DC.

Mekonnen, A. (1998) *Rural energy and afforestation: case studies from Ethiopia.* PhD thesis, Göteborg University, Sweden.

Mukamuri, B. & Kozanai, W. (1999) *Socioeconomic Issues Relating to* Warburgia salutaris: *a Powerful Medicinal Plant in Zimbabwe.* Mimeo, Institute for Environmental Sciences, University of Zimbabwe, Harare, Zimbabwe.

Nepstad, D.C. & Schwartzman, S. (eds) (1992) *Non-timber Products from Tropical Forests: Evaluation of a Conservation and Development Strategy*. New York Botanical Garden, New York.

Ostrom, E. (1990) *Governing the Commons: the Evolution of Institutions for Collective Action*. Cambridge University Press, Cambridge.

Ostrom, E. (1999) *Self-governance and Forest Resources*. Occasional Paper no. 20, Centre for International Forestry Research, Bogor, Indonesia.

Panayotou, T. & Ashton, P. (1992) *Not by Timber Alone: the Case for Multiple Use Management of Tropical Forests*. Island Press, Covelo, USA.

Platteau, J.-P. (1996) Physical infrastructure as a constraint on agriculture growth: the case of sub-Saharan Africa. *Oxford Development Studies* **24**, 189–219.

Psacharopoulos, G. (1994) Returns to investment in education: a global update. *World Development* **22**, 1325–43.

Richards, E.M. (1993) *Commercialisation of Non-timber Forest Products in Amazonia*. NRI Socio-economic Series 2, Natural Resources Institute, Chatham, UK.

Risopoulos, J., Al-Hassan, R., Clark, S., Dorward, A., Poulton, C. & Wilkon, K. (1999) *Problems of Market Access in Remote Areas*. Wye College, University of London.

Sale, J.B. (1981) *The Importance and Values of Wild Plants and Animals in Africa*. IUCN, Gland, Switzerland.

Scoones, I., Pretty, J.N. & Melnyk, M. (1992) *The Hidden Harvest: Wild Foods in Agricultural Systems. A Bibliography and Literature Review*. International Institute for Environment and Development, London.

Shackleton, S.E., Dzeferos, C.M., Shackleton, C.M. & Mathabela, F.R. (1998) Use and trading of edible wild herbs in the central lowveld savanna region, South Africa. *Economic Botany* **52**, 251–9.

Shyamsundar, P. & Kramer, R.A. (1996) Tropical forest protection: an empirical analysis of the costs borne by local people. *Journal of Environmental Economics and Management* **31**, 129–44.

Singh, I., Squire, L. & Strauss, J. (1986) *Agricultural Household Models: Extensions, Applications and Policy*. Johns Hopkins University Press, Baltimore.

Townson, I.M. (1994) *Forest Products and Household Incomes: a Review and Annotated Bibliography*. Oxford Forestry Institute, University of Oxford.

UNDP (1992) *Human Development Report 1992*. Oxford University Press, Oxford.

Wollenberg, E. & Nawir, A.S. (1997) *Estimating the Incomes of People Who Depend on Forests*. Mimeo, Centre for International Forestry Research, Bogor, Indonesia.

Synthesis and Conclusions

When we began *The Forests Handbook* project in 1997 we invited contributors 'to synthesize scientific principles and practical knowledge about the world's forests to demonstrate and provide foundations which underlie sustainable management. Put simply, how can an understanding of forests and forest processes lead to their better management and, ultimately, people's better stewardship of this immensely important resource? We went on to say that the message of Volume 1 should be 'this is what we know about forests and how we find out new things', and of Volume 2 'this is how we apply scientific knowledge to address issues and problems that keep cropping up in forestry.'

Now at the conclusion of the project it is fair to ask: what has been learnt from such a wide-ranging review of forest science and its application in practice collated in the two volumes that constitute the *The Forests Handbook*? How far have we travelled on the journey we began, while always recognizing that in the very nature of the exercise one rarely reaches a destination, but one does, to press the metaphor, pass important places on the way? It is in this sense that I attempt a synthesis of Volumes 1 and 2 to draw out some conclusions and pointers to help us in sustainable management of forests (SFM). In some cases I relate them to particular chapters (Volume:Chapter). I have tried to avoid anodyne or absurdly obvious remarks while being careful to restate well-known points where these have emerged with renewed force or greater clarity. However, the points are not a summary but rather emerging themes. This same synthesis occurs in both Volumes 1 and 2.

Of course, in one sense, in a book with 45 authors writing 33 chapters, agreed definitive conclusions cannot really be assembled. Each author presents cogently his or her analysis. But it is worthwhile drawing out the themes that emerge to inform sustainable forest management and to present them together to help forward our thinking and ultimately our striving towards this worthwhile goal. As Peter Attiwill and Jane Fewings (2:16) quote from the Australian state of Victoria's Regional Forest Assessment: '... sustainable forest management is a goal to be pursued vigorously, not an antique to be admired.'

THE RECORD OF HISTORY

1 Knowing what has happened to forests in the past and why (1:1, 2:1, 2:3 and 2:14) teaches crucial lessons, one of which is that a forest that is used, and the benefits of which are enjoyed, is a forest that survives (2:14).

2 The record of history is particularly important in understanding the evolution of policy towards forests and their management, and the way it has developed (2:2, 2:11 and 2:16). A fundamental underpinning of informed SFM is to know why things are done, as well as how.

3 SFM needs a policy framework—administrative, environmental, production, social—to function successfully (1:14, 2:1–2, 2:11 and 2:16). SFM won't just happen.

4 Recent history shows rapid migration to urban areas, particularly in developing countries, and that worldwide the majority of people will soon live in massive conurbations. While this is only touched on indirectly (1:13 and 2:12) the globalization of trade and fair sharing of resources increases the SFM imperative since most people will have no direct control and often very little say in what happens to forests.

ROLE OF FOREST SCIENCE

5 Knowledge about forests (1:2–3), their functioning as biological systems and their interaction with the environment (1:4–8, 1:10–12) is the only sound basis for informed and potentially sustainable management. It is obvious that scientific and related research is largely responsible for building this knowledge base.

6 Knowledge informs understanding of *impacts* and *consequences* of forest operations and processes, which is fundamental to SFM. However, general knowledge must be applied to the particular in terms of sites, forest and species types and circumstances (2:3 and Volume 2 case studies).

7 Great advance has been achieved in the last 30 years, not least in our understanding of forest ecological processes and interaction between forest and the wider environment (1:3–5, 1:7–8, 1:10–12, 2:5, 2:7). It has become clear that many processes need to be understood at the landscape scale (1:5–6, 1:8–10, 1:12), i.e. at a scale which has received relatively little attention compared with the molecular, cellular, tree or stand scale.

8 The capture of scientific knowledge with the long timespans of forestry may not be cheap, will often involve complex investigations of processes including large-scale field studies (1:7–12) and crucially requires some role for long-term experimentation in forest science (2:5, 2:13, 2:17–18).

9 Modelling is a powerful tool in forest science to help predict consequences of actions and to raise questions about different options, such as their sensitivity (1: 6–10, 2:7–9). It can assist but not substitute for 8 above. High data quality to calibrate models is crucial for informed decision taking – see examples in 1:10 and 2:7.

10 Decision support can only be developed on the basis of detailed relevant knowledge, e.g. 'understanding disease epidemiology is vital' for successful disease management (2:9). A crucial component is maintaining a critical mass of expertise as the whole of *The Forests Handbook* shows!

SILVICULTURE AND MANAGEMENT

11 Understanding of forest ecology informs silvicultural actions, for example (a) to allow improvements in biodiversity, whether in plantation restructuring to deliver multiple benefits (2:15), maintaining or rebuilding natural populations (2:16) or more generally (1:4–5, 2:4), or (b) to optimise the protective role of forests (2:3).

12 Holistic and integrated approaches are central to SFM, in pest management (2:8), in soil husbandry (2:7) and particularly of organic matter (1:7), and more generally in planning and executing forest operations. For example, divorce of logging from regeneration will almost always lead to unsustainable actions (1:7, 2:5, 2:6–7, 2:15 and 2:18), hence 13.

13 Sustained yield rests on two assumptions (2:5) which separate forest management from forest exploitation: (a) commitment to successful and productive regeneration; and (b) harvesting balance must not exceed growth increment.

14 In the vast majority of forests sustainability is largely about understanding how forests function and working with that knowledge, but in the massively altered and unnatural urban environment, it is more to do with understanding how the built environment impacts on urban forest, and especially street (shade) trees, and to modify management accordingly (1:13, 2:12).

PEOPLE AS STAKEHOLDERS

15 Involvement of stakeholders (people's participation) is today's received wisdom (1:14, 2:10), but clearly in the sense of stewardship or custodians 'communities of protection' in Saxena's terminology (2:11) it greatly aids adoption of SFM in many instances, since it confers both benefits and responsibilities.

16 Non-timber forest products (NTFPs) often oil the mechanism to develop successful SFM in less developed countries (1:15, 2:11, 2:17).

17 Stakeholder involvement is also essential for achieving SFM policies in developed countries (1:14, 2:2 and 2:16).

These very general points and principles are what, for me as a committed Christian, I see as going to the heart of stewardship in terms of humanity's responsibility towards the world's resources. We are to be careful and caring custodians and not exploiters of them or of people who benefit from them. Today we find that concern for such resources, and for forests and forest peoples especially, is becoming enshrined in new institutions and processes of which 'certification' is increasingly the guarantor of SFM, albeit perhaps not yet in a wholly satisfactory form.

The world's forests are more than of utilitarian value. SFM acknowledges and focuses attention on this fact. And we are not alone. Uniquely in the creation account (Genesis 2:9, NIV translation) is added precisely this extra dimension when the Lord God declares of trees (and of nothing else He made): '. . . trees that *are pleasing to the eye* and good for food.' We would humbly agree.

Julian Evans

Index

Note: This is primarily a subject and topic index. In the text there are numerous references to tree species, both by vernacular and scientific names as appropriate, and also reference to most countries of the world. Not all such references to species and countries have been indexed. The reader is referred to the relevant topic, e.g. boreal forest, savannahs, tropical forest, etc., or regions, e.g. Africa, North America, Oceania, etc., to help access information sought. Where appropriate, cross-references are made, and readers will often find reference to tree genera under a particular topic.

Page numbers in **bold** indicate tables; those in *italics*, illustrations